Volume I
Large Production and Priority Pollutants

Handbook of Environmental

FATE
and
EXPOSURE
DATA
For Organic Chemicals

Philip H. Howard

 LEWIS PUBLISHERS

Library of Congress Cataloging-in-Publication Data

Howard, Philip H.
   Handbook of environmental fate and exposure data for organic chemicals.

   Bibliography: v. 1, p.
   Includes index.
   Contents: v. 1. Large production and priority pollutants.
   1. Pollutants--Handbooks, manuals, etc.   I. Title.
TD176.4.H69   1989   363.7'38   89-2436
ISBN 0-87371-151-3   (v. 1)

Second Printing 1989

LEWIS PUBLISHERS, INC.
121 South Main Street, Chelsea, Michigan 48118

PRINTED IN THE UNITED STATES OF AMERICA

Associate Editors for Volume I

The following individuals from the Syracuse Research Corporation's Chemical Hazard Assessment Division either were authors of the individual chemical records prepared for the Hazardous Substances Data Bank or edited the expanded and updated chemical chapters in this volume. The order of names, which will vary in each volume, is by the number of chemicals for which the individual was responsible.

William F. Jarvis, Ph.D.

Gloria W. Sage, Ph.D.

Dipak K. Basu, Ph.D.

D. Anthony Gray, Ph.D.

William Meylan

Erin K. Crosbie

Preface

Many articles and books have been written on how to review the environmental fate and exposure of organic chemicals (e.g., [11] and [19]). Although these articles and books often give examples of the fate and exposure of several chemicals, rarely do they attempt to review large numbers of chemicals. These "how to" guides provide considerable insight into ways of estimating and using physical/chemical properties as well as mechanisms of environmental transport and transformation. However, when it comes to reviewing the fate and exposure of individual chemicals, there are discretionary factors that significantly affect the overall fate assessment. For example, is it reasonable to use regression equations for estimating soil or sediment adsorption for aromatic amine compounds? Is chemical oxidation likely to be important for phenols in surface waters? These discretionary factors are dependent upon the available data on the individual chemical or, when data are lacking, on chemicals of related structures.

This series of books outlines in detail how individual chemicals are released, transported, and degraded in the environment and how they are exposed to humans and environmental organisms. It is devoted to the review and evaluation of the available data on physical/chemical properties, commercial use and possible sources of environmental contamination, environmental fate, and monitoring data of individual chemicals. Each review of a chemical provides most of the data necessary for either a qualitative or quantitative exposure assessment.

Chemicals were selected from a large number of chemicals prepared by Syracuse Research Corporation (SRC) for inclusion in the National Library of Medicine's (NLM) Hazardous Substances Data Bank (HSDB). Chemicals selected for the first two volumes were picked from lists of high volume commercial chemicals, priority pollutants, and solvents. The chemicals in the first two volumes include most of the non-pesticidal priority pollutants and many of the chemicals on priority lists for a variety of environmental regulations (e.g., RCRA and CERCLA Reportable Quantities, Superfund, SARA). Pesticides, polycyclic aromatic hydrocarbons, and other groups of chemicals will be included in later volumes.

The chemicals are listed in strict alphabetical order by the name considered to be the most easily recognized. Prefixes commonly used in organic chemistry which are not normally considered part

v

of the name, such as ortho-, meta-, para-, alpha-, beta-, gamma-, n-, sec-, tert-, cis-, trans-, N-, as well as all numbers, have not been considered for alphabetical order. Other prefixes which normally are considered part of the name, such as iso-, di-, tri-, tetra-, and cyclo-, are used for alphabetical positioning. For example, 2,4-Dinitrotoluene is under D and tert-Butyl alcohol is under B. In addition, cumulative indices are provided at the end of each volume to allow the reader to find a given chemical by chemical name synonym, Chemical Abstracts Services (CAS) number, and chemical formula.

Acknowledgments

The following authors of the initial chemical records for the Hazardous Substance Data Bank were or are staff scientists with the Chemical Hazard Assessment Division of the Syracuse Research Corporation: Gloria W. Sage, William Meylan, William F. Jarvis, Erin K. Crosbie, Jeffery Jackson, Amy E. Hueber, and Jeffrey Robinson. We wish to thank several individuals at the National Library of Medicine (NLM) for their encouragement and support during the project. Special thanks go to our project officer, Vera Hudson, and to Bruno Vasta and Dalton Tidwell.

Explanation of Data

In the following outline, each field covered for the individual chemicals is reviewed with such information as the importance of the data, the type of data included in each field, how data are usually handled, and data sources. For each chemical, the physical properties as well as the environmental fate and monitoring data were identified by conducting searches of the Environmental Fate Data Bases of Syracuse Research Corporation (SRC) [12].

SUBSTANCE IDENTIFICATION

Synonyms: Only synonym names used fairly frequently were included.

Structure: Chemical structure.

CAS Registry Number: This number is assigned by the American Chemical Society's Chemical Abstracts Services as a unique identifier.

Molecular Formula: The formula is in Hill notation, which is given as the number of carbons followed by the number of hydrogens followed by any other elements in alphabetic order.

Wiswesser Line Notation: This is a chemical structure representation that can be used for substructure searching. It was designed back when computer notations had to fit into 80 characters and, therefore, is very abbreviated (e.g., Q is used for a benzene ring).

CHEMICAL AND PHYSICAL PROPERTIES

The Hazardous Substances Data Bank (HSDB) of the National Library of Medicine was used as a source of boiling points, melting points, and molecular weights. The dissociation constant, octanol/water partition coefficient, water solubility, vapor pressure, and Henry's Law constant were judiciously selected from the many values that were identified in SRC's DATALOG file. All values selected were referenced to the primary literature source when possible.

Boiling Point: The boiling point or boiling point range is given along with the pressure. When the pressure is not given it should be assumed that the value is at 760 mm Hg.

Melting Point: The melting point or melting point range is given.

Molecular Weight: The molecular weight to two decimal points is given.

Dissociation Constants: The acid dissociation constant as the negative log (pKa) is given for chemicals that are likely to dissociate at environmental pH's (between 5 and 9). Chemical classes where dissociation is important include, for example, phenols, carboxylic acids, and aliphatic and aromatic amines. Once the pKa is known, the percent in the dissociated and undissociated form can be determined. For example, for an acid with a pKa of 4.75, the following is true at different pH's:

1% dissociated at pH 2.75
10% dissociated at pH 3.75
50% dissociated at pH 4.75
90% dissociated at pH 5.75
99% dissociated at pH 6.75

The degree of dissociation affects such processes as photolysis (absorption spectra of chemicals that dissociate can be considerably affected by the pH), evaporation from water (ions do not evaporate), soil or sediment adsorption, and bioconcentration. Values from evaluated sources such as Perrin [21] and Serjeant and Dempsey [23] were used when available.

Log Octanol/Water Partition Coefficient: The octanol/water partition coefficient is the ratio of the chemical concentration in octanol divided by the concentration in water. The most reliable source of values is from the Medchem project at Pomona College [8]. When experimental values are unavailable, estimated values have been provided using a fragment constant estimation method, CLOGP3, from Medchem. Occasionally chemical octanol/water partition coefficients were not calculated because a necessary fragment constant for the chemical was not available. The octanol/water partition coefficient has been shown to correlate well with bioconcentration factors in aquatic organisms [26] and adsorption to soil or sediment [13], and recommended regression equations have been reviewed [15].

Water Solubility: The water solubility of a chemical provides considerable insight into the fate and transport of a chemical in the environment. High water soluble chemicals, which have a tendency to remain dissolved in the water column and not partition to soil or sediment or bioconcentrate in aquatic organisms, are less likely to volatilize from water (depending upon the vapor pressure - see Henry's Law constant) and are generally more likely to biodegrade. Low water soluble chemicals are just the opposite; they partition to soil or sediments and bioconcentrate in aquatic organisms, volatilize more readily from water, and are less likely to be biodegradable. Other fate processes that are, or can be, affected by water solubility include photolysis, hydrolysis, oxidation, and washout from the atmosphere by rain or fog. Water solubility values were taken from either the Arizona Data Base [27] or from SRC's DATALOG or CHEMFATE files. The values were reported in ppm at a temperature at or as close as possible to 25 °C. Occasionally when no values were available, the value was estimated from the octanol/water partition coefficient using recommended regression equations [15].

Vapor Pressure: The vapor pressure of a chemical provides considerable insight into the transport of a chemical in the environment. The volatility of the pure chemical is dependent upon the vapor pressure, and volatilization from water is dependent upon the vapor pressure and water solubility (see Henry's Law constant). The form in which a chemical will be found in the atmosphere is dependent upon the vapor pressure; chemicals with a vapor pressure less than 10^{-6} mm Hg will be mostly found associated with particulate matter [7]. When available, sources such as Boublik et al [3], Riddick et al [22], and Daubert and Danner [5] were used, since the data in these sources were evaluated and some of them provided recommended values. Vapor pressure was reported in mm Hg at or as close as possible to 25 °C. In many cases, the vapor pressure was calculated from a vapor pressure/temperature equation.

Henry's Law Constant: The Henry's Law constant, H, is really the air/water partition coefficient, and therefore a nondimensional H relates the chemical concentration in the gas phase to its concentration in the water phase. The dimensional H can be determined by dividing the vapor pressure in atm by the water solubility in mole/cu m to give H in atm-cu m/mole. H provides an indication of the partition between air and water at equilibrium and also is used to calculate the rate of evaporation from water (see discussion under Evaporation from Water/Soil). Henry's Law constants can be directly measured, calculated from the water solubility and vapor pressure, or estimated from structure using the method of Hine and Mookerjee [9], and this same order was used in selecting values. Some critical review data on Henry's Law constants are available (e.g., [16]).

ENVIRONMENTAL FATE/EXPOSURE POTENTIAL

Data for the following sections were identified with SRC's Environmental Fate Data Bases. Biodegradation data were selected from the DATALOG, BIOLOG, and BIODEG files. Abiotic degradation data were identified in the Hydrolysis, Photolysis, and Oxidation fields in DATALOG and CHEMFATE. Transport processes such as Bioconcentration, Soil Adsorption/Mobility, and Volatilization as well as the monitoring data were also identified in the DATALOG and CHEMFATE files.

Summary: This section is an abbreviated summary of all the data presented in the following sections and is not referenced; to find the citations the reader should refer to appropriate sections that follow. In general, this summary discusses how a chemical is used and released to the environment, how the chemical will behave in soil, water, and air, and how exposure to humans and environmental organisms is likely to occur.

Natural Sources: This section reviews any evidence that the chemical may have any natural sources of pollution, such as forest fires and volcanos, or may be a natural product that would lead to its detection in various media (e.g., methyl iodide is found in marine algae and is the major source of contamination in the ocean).

Artificial Sources: This section is a general review of any evidence that the chemical has anthropogenic sources of pollution. Quantitative data are reviewed in detail in Effluent Concentrations; this section provides a qualitative review of various sources based upon how the chemical is manufactured and used as well as the physical/chemical properties. For example, it is reasonable to assume that a highly volatile chemical which is used mostly as a solvent will be released to the atmosphere as well as the air of occupational settings even if no monitoring data are available. Information on production volume and uses was obtained from a variety of chemical marketing sources including the <u>Kirk-Othmer Encyclopedia of Chemical Technology</u>, SRI International's <u>Chemical Economics Handbook</u>, and the Chemical Profiles of the <u>Chemical Marketing Reporter</u>.

Terrestrial Fate: This section reviews how a chemical will behave if released to soil or groundwater. Field studies or terrestrial model ecosystems studies are used here when they provide insight into the overall behavior in soil. Studies which determine an individual process (e.g., biodegradation, hydrolysis, soil adsorption) in soil are reviewed in the appropriate sections that follow. Quite often, except with pesticides, field or terrestrial ecosystem studies either are not available or do not give enough data to make conclusions on the terrestrial fate of a chemical. In these cases, data from the sections on Biodegradation, Abiotic Degradation, Soil Adsorption/Mobility, Volatilization from Water/Soil, and any appropriate monitoring data will be used to synthesize how a chemical is likely to behave if released to soil.

Aquatic Fate: This section reviews how a chemical will behave if released to fresh, marine, or estuarine surface waters. Field studies or aquatic model ecosystems are used here when they provide insight into the overall behavior in water. Studies which determine an individual process (e.g., biodegradation, hydrolysis, photolysis, sediment adsorption, and bioconcentration in aquatic organisms) in water are reviewed in the appropriate sections that follow. When field or aquatic ecosystems studies are not available or do not give enough data to make conclusions on the aquatic fate of the chemical, data from the appropriate degradation, transport, or monitoring sections will be used to synthesize how a chemical is likely to behave if released to water.

Atmospheric Fate: This section reviews how a chemical will behave if released to the atmosphere. The vapor pressure will be used to determine if the chemical is likely to be in the vapor phase or adsorbed to particulate matter [7]. The water solubility will be used to assess the likelihood of washout with rain. Smog chamber studies or other studies where the mechanism of degradation is not determined will be reviewed in this section; studies of the rate of reaction with hydroxyl radical or ozone or direct photolysis will be reviewed in Abiotic Degradation and integrated into this section.

Biodegradation: The principles outlined by Howard and Banerjee [10] are used in this section to review the relevant biodegradation data pertinent to biodegradation in soil, water, or wastewater treatment. In general, the studies have been separated into screening studies (inoculum in defined nutrient media), biological treatment simulations, and grab samples (soil or water sample with chemical added and loss of concentration followed). Pure culture studies are only used to indicate potential metabolites, since the artificial nutrient conditions under which the pure cultures are isolated provide little assurance that these same organisms will be present in any quantity or that their enzymes will be functioning in various soil or water environments. Anaerobic biodegradation studies, which are pertinent to whether a chemical will biodegrade in biological treatment digestors, sediment, and some groundwaters, are discussed separately.

Abiotic Degradation: Non-biological degradation processes in air, water, or soil are reviewed in this section. For most chemicals in the vapor phase in the atmosphere, reaction with photochemically generated hydroxyl radicals is the most important degradation process. Occasionally reaction in the atmosphere with ozone (for olefins), nitrate radicals at night, and direct photolysis (direct sunlight absorption resulting in photochemical alteration) are significant for some chemicals [2]. For many chemicals, experimental reaction rate constants for hydroxyl radical are available (e.g., [1]) and are used to calculate an estimated half-life by assuming an average hydroxyl radical concentration of $5 \times 10^{+5}$ molecules/cu cm in non-smog conditions (e.g., [2]). If experimental rate constants are not available, they have been estimated using the fragment constant method of Atkinson [1] and then a half-life estimated using the assumed radical concentration. The reaction rate for ozone reaction with olefins may be experimentally available or can be estimated using the Fate of Atmospheric Pollutants (FAP) from the Graphic Exposure and Modelling System (GEMS) (available from the Exposure Evaluation Division, Office of Toxic Substances, U.S. Environmental Protection Agency). Using either the experimental or estimated rate constant and an assumed concentration of $6.0 \times 10^{+11}$ molecules/cu cm (FAP) or 7.2

x 10^{+11} molecules/cu m [2], an estimated half-life for reaction with ozone can be calculated. Nitrate radicals are significant only with certain classes of chemicals such as higher alkenes, dimethyl sulfide and lower thiols, furan and pyrrole, and hydroxy-substituted aromatics [2].

The possibility of direct photolysis in air or water can be partially assessed by examining the ultraviolet spectrum of the chemical. If the chemical does not absorb light at wavelengths provided by sunlight (>290 nm), the chemical cannot directly photolyze. If it does absorb sunlight, it may or may not photodegrade depending upon the efficiency (quantum yield) of the photochemical process, and unfortunately such data are rarely available. Indirect photolysis processes may be important for some chemicals in water [17]. For example, some chemicals can undergo sensitized photolysis by absorbing triplet state energy from the excited triplet state of chemicals commonly found in water, such as humic acids. Transient oxidants found in water, such as peroxy radicals, singlet oxygen, and hydroxyl radicals, may also contribute to abiotic degradation in water for some chemicals. For example, phenols and aromatic amines have half-lives of less than a day for reaction with peroxy radicals; substituted and unsubstituted olefins have half-lives of 7 to 8 days with singlet oxygen; and dialkyl sulfides have half-lives of 27 hours with singlet oxygen [17].

Chemical hydrolysis at pH's that are normally found in the environment (pH's 5 to 9) can be important for a variety of chemicals that have functional groups that are potentially hydrolyzable, such as alkyl halides, amides, carbamates, carboxylic acid esters, epoxides and lactones, phosphate esters, and sulfonic acid esters [18]. Half-lives at various pH's are usually reported in order to provide an indication of the influence of pH.

Bioconcentration: Certain chemicals, due to their hydrophobic nature, have a tendency to partition from the water column and bioconcentrate in aquatic organisms. This concentration of chemicals in aquatic organisms is of concern because it can lead to toxic concentrations being reached when the organism is consumed by higher organisms such as wildlife and humans. Such bioconcentrations are usually reported as the bioconcentration factor (BCF), which is the concentration of the chemical in the organism at equilibrium divided by the concentration of the chemical in water. This unitless BCF value can be determined experimentally by dosing water containing the organism and dividing the concentration in the organism by the concentration in the water once equilibrium is reached, or if equilibration is slow, the rate of uptake can be used to calculate the BCF at equilibrium. The BCF value can also be estimated by using recommended regression equations that have been shown to correlate well with physical properties such as the octanol/water partition coefficient and water solubility [15]; however, these estimation equations assume that little metabolism of the chemical occurs in the aquatic organism, which is not always correct. Therefore, when available, experimental values are preferred.

Soil Adsorption/Mobility: For many chemicals (especially pesticides), experimental soil or sediment partition coefficients are available. These values are measured by determining the concentration in both the solution (water) and solid (soil or sediment) phases after shaking for about 24 to 48 hours and using different initial concentrations. The data are then fit to a Freundlich equation to determine the adsorption coefficient, Kd. These Kd values for individual soils or sediments are normalized to the organic carbon content of the soil or sediment by dividing by the organic content (Koc), since of the numerous soil properties that affect sorption (organic carbon content, particle size, clay mineral composition, pH, cation-exchange capacity) [14], organic carbon is the most important for undissociated organic chemicals. Occasionally the experimental adsorption coefficients are reported on a soil-organic matter basis (Kom) and these are converted to Koc by multiplying by 1.724 [15]. When experimental values are unavailable, estimated Koc values are calculated using either the water solubility or octanol/water partition coefficient and some recommended regression equations [15]. The measured or estimated adsorption values are used to determine the likelihood of leaching through soil or adsorbing to sediments using the criteria of Swann et al [24]. Occasionally experimental soil thin-layer chromatography studies are also available and can be used to assess the potential for leaching.

The above discussion applies generally to undissociated chemicals, but there are some exceptions. For example, aromatic amines have been shown to covalently bond to humic material [20] and this slow but non-reversible process can lead to aromatic amines being tightly bound to the humic material in soils. Methods to estimate the soil or sediment adsorption coefficient for dissociated chemicals which form anions are not yet available, so it is particularly important to know the pKa value for chemicals that can dissociate so that a determination of the relative amounts of the dissociated and undissociated forms can be determined at various pH conditions. Chemicals that form cations at ambient pH conditions are generally thought to sorb strongly to clay material, similar to what occurs with paraquat and diquat (pyridine cations).

Volatilization from Water/Soil: For many chemicals, volatilization can be an extremely important removal process, with half-lives as low as several hours. The Henry's Law constant can give qualitative indications of the importance of volatilization; for chemicals with values less than 10^{-7} atm-cu m/mole, the chemical is less volatile than water and as water evaporates the concentration will increase; for chemicals around 10^{-3} atm-cu m/mole, volatilization will be rapid. The volatilization process is dependent upon physical properties of the chemical (Henry's Law constant, diffusivity coefficient), the presence of modifying materials (adsorbents, organic films, electrolytes, emulsions), and the physical and chemical properties of the environment (water depth, flow rate, the presence of waves, sediment content, soil moisture, and organic content) [15]. Since the overall volatilization rate cannot be estimated for all the various environments to which a chemical may be released, common models have been used in order to give an indication of the relative importance of volatilization. For most chemicals that have a Henry's Law constant greater than 10^{-7} atm-cu m/mole, the simple volatilization model outlined in Lyman et al [15] was used; this model assumes a 20 °C river 1 meter deep flowing at 1 m/sec with a wind velocity of 3 m/sec and requires only the Henry's Law constant and the molecular weight of the chemical for input. This model gives relatively rapid volatilization rates for this model river and values for ponds, lakes, or deeper rivers will be considerably slower. Occasionally a chemical's measured reaeration coefficient ratio relative to oxygen is available, and this can be used with typical oxygen reaeration rates in ponds, rivers, and streams to give volatilization rates for these types of bodies of water. For chemicals that have extremely high Koc values, the EXAMS-II model has been used to estimate volatilization both with and without sediment adsorption (extreme differences are noted for these high Koc chemicals). Soil volatilization models are less validated and only qualitative statements are given of the importance of volatilization from moist (about 2% or greater water content) or dry soil, based upon the Henry's Law constant or vapor pressure, respectively. This assumes that once the soil is saturated with a molecular layer of water, the volatilization rate will be mostly determined by the value of the Henry's Law constant, except for chemicals with high Koc values.

Water Concentrations: Ambient water concentrations of the chemical are reviewed in this section, with subcategories for surface water, drinking water, and groundwater when data are available. In general, the number of samples, the percent positive, the range of concentrations, and the average concentration are reported when the data are available.

Effluents Concentration: Air emissions and wastewater effluents are reviewed in this section. In general, the number of samples, the percent positive, the range of concentrations, and the average concentration are reported when the data are available.

Sediment/Soil Concentrations: Sediment and soil concentrations are reviewed in this section. In general, the number of samples, the percent positive, the range of concentrations, and the average concentration are reported when the data are available.

Atmospheric Concentrations: Ambient atmospheric concentrations are reviewed in this section, with subcategories for rural/remote and urban/suburban when data are available in such sources as Brodzinsky and Singh [4]. In general, the number of samples, the percent positive, the range of concentrations, and the average concentration are reported when the data are available.

Food Survey Values: Market basket survey data such as found in Duggan et al [6] and individual studies of analysis of the chemical in processed food are reported in this section. In general, the number of samples, the percent positive, the range of concentrations, and the average concentration are reported when the data are available.

Plant Concentrations: Concentrations of the chemical in plants are reviewed in this section. If the plant has been processed for food, it is reported in Food Survey Values.

Fish/Seafood Concentrations: Concentrations in fish, seafood, shellfish, etc. are reviewed in this section. If the fish or seafood have been processed for food, the data are reported in Food Survey Values.

Animal Concentrations: Concentrations in animals are reviewed in this section. If the animals have been processed for food, the data are reported in Food Survey Values.

Milk Concentrations: Since dairy milk constitutes a high percentage of the human diet, concentrations of the chemical found in dairy milk are reviewed in this section and not in Food Survey Values.

Other Environmental Concentrations: Concentrations of the chemical found in other environmental media that may contribute to an understanding of how a chemical may be released to the environment or exposed to humans (e.g., detection in gasoline or cigarette smoke) are reviewed in this section.

Probable Routes of Human Exposure: The monitoring data and physical properties are used to provide conclusions on the routes (oral, dermal, inhalation) of exposure.

Average Daily Intake: The average daily intake is a calculated value of the amount of the chemical that is typically taken in daily by human adults. The value is determined by multiplying typical concentrations in drinking water, air, and food by average intake factors such as 2 liters of water, 20 cu m of air, and 1600 grams of food [25].

Occupational Exposures: Monitoring data, usually air samples, from occupational sites are reviewed in this section. In addition, estimates of the number of workers exposed to the chemical from the two National Institute for Occupational Safety and Health (NIOSH) surveys are reviewed in this section. The National Occupational Hazard Survey (NOHS) conducted from 1972 to 1974 and the National Occupational Exposure Survey (NOES) conducted from 1981 to 1983 provided statistical estimates of worker exposures based upon limited walk-through industrial hygiene surveys.

Body Burdens: Any concentrations of the chemical found in human tissues or fluids are reviewed in this section. Included are blood, adipose tissue, urine, and human milk.

REFERENCES

1. Atkinson RA; Internat J Chem Kinet 19: 799-828 (1987)
2. Atkinson RA; Chem Rev 85: 60-201 (1985)
3. Boublik T et al; The Vapor Pressures of Pure Substances. Amsterdam: Elsevier (1984)
4. Brodzinsky R, Singh HB; Volatile organic chemicals in the atmosphere: an assessment of available data. SRI Inter EPA contract 68-02-3452 Menlo Park, CA (1982)
5. Daubert TE, Danner RP; Data Compilation Tables of Properties of Pure Compounds. Amer Inst Chem Engr pp 450 (1985)
6. Duggan RE et al; Pesticide Residue Levels in Foods in the U.S. from July 1, 1969 to June 30, 1976. Washington, DC: Food Drug Administ. 240 pp (1983)
7. Eisenreich SJ et al; Environ Sci Technol 15: 30-8 (1981)
8. Hansch C, Leo AJ; Medchem Project Issue No 26. Claremont CA: Pomona College (1985)
9. Hine J, Mookerjee PK; J Org Chem 40: 292-8 (1975)
10. Howard PH, Banerjee S; Environ Toxicol Chem 3: 551-562 (1984)
11. Howard PH et al; Environ Sci Technol 12: 398-407 (1978)
12. Howard PH et al; Environ Toxicol Chem 5: 977-88 (1986)
13. Karickhoff SW; Chemosphere 10: 833-46 (1981)
14. Karickhoff SW; Environ Expos from Chemicals Vol I. ed Neely WB, Blau GE, Boca Raton, FL: CRC Press p 49-64 (1985)
15. Lyman WJ et al; Handbook of Chemical Property Estimation Methods. McGraw-Hill, NY (1982)
16. Mackay D, Shiu WY; J Phys Chem Ref Data 10: 1175-99 (1981)
17. Mill T, Mabey W; In Environ Expos from Chemicals Vol I. ed Neely WB, Blau GE, Boca Raton, FL: CRC Press p 175-216 (1985)
18. Neely WB; In Environ Expos from Chemicals Vol I. ed Neely WB, Blau GE, Boca Raton, FL: CRC Press p 157-73 (1985)
19. Neely WR, Blau GE; Environ Expos from Chemicals Vol I. Boca Raton, FL: CRC Press (1985)
20. Parris GE; Environ Sci Technol 14: 1099-1105 (1980)
21. Perrin DD; Dissociation Constants of Organic Bases in Aqueous Solution. IUPAC Chemical Data Series, London: Buttersworth (1965)
22. Riddick JA et al; Organic Solvents: Physical Properties and Methods of Purification, 4th Edit. New York: J Wiley & Sons (1986)
23. Serjeant EP, Dempsey B; Ionisation Constants of Organic Acids in Aqueous Solution. IUPAC Chemical Data Series No 23, New York: Pergamon Press (1979)
24. Swann RL et al; Residue Reviews 85: 17-28 (1983)
25. U.S. EPA; Reference Values for Risk Assessment. Environ Criteria Assess Office, Off Health Environ Assess, Off Research Devel, ECAO-CIN-477, Cincinnati, OH: U.S. Environ Prot Agency (1986)
26. Veith GD et al; J Fish Res Board Can 36: 1-40-8 (1979)
27. Yalkowsky SH et al; ARIZONA dATABASE of Aqueous Solubility, U. Arizona, Tucson, AZ (1987)

Contents

Volume I
Large Production and Priority Pollutants

Handbook of Environmental

FATE
and
EXPOSURE
DATA

For Organic Chemicals

Acetyl Chloride

SUBSTANCE IDENTIFICATION

Synonyms:

Structure:

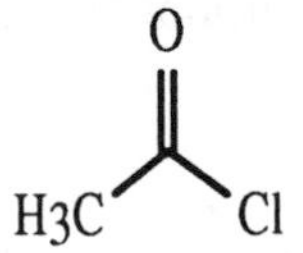

CAS Registry Number: 75-36-5

Molecular Formula: C_2H_3ClO

Wiswesser Line Notation: GV1

CHEMICAL AND PHYSICAL PROPERTIES

Boiling Point: 52 °C

Melting Point: -112 °C

Molecular Weight: 78.50

Dissociation Constants:

Log Octanol/Water Partition Coefficient:

Water Solubility:

Vapor Pressure: 352.5 mm Hg at 24 °C [1]

Henry's Law Constant:

ENVIRONMENTAL FATE/EXPOSURE POTENTIAL

Summary: Acetyl chloride may be released to the environment during its production and use as a catalyst in the chlorination of acetic acid and in chemical synthesis. It reacts violently with water

1

and other hydrogen active compounds such as amines, phenols, and alcohols and will not persist if released in water or on land. If released in the atmosphere, it will react with atmospheric moisture, and its half-life will depend on the humidity of the air. No half-lives for air with different water contents could be found; however, acetyl chloride is known to fume in moist air. Human exposure is primarily occupational via inhalation or dermal contact.

Natural Sources:

Artificial Sources: Acetyl chloride may be released to the environment during its production and use as a catalyst in the chlorination of acetic acid and in chemical synthesis (acetylating agent), particularly in the manufacture of pharmaceuticals and dyestuffs [3]. In use as a catalyst, it is usually produced on site [3].

Terrestrial Fate: No reports are available on the fate of acetyl chloride on land. In view of its violent decomposition in the presence of water and high reactivity towards molecules with active hydrogen groups such as amines, phenols, and alcohols that occur in soil [5], it is doubtful that acetyl chloride would persist for long in soil.

Aquatic Fate: Acetyl chloride violently decomposes with water and therefore will not persist in the aquatic environment.

Atmospheric Fate: Acetyl chloride violently decomposes in the presence of water and fumes in the presence of moist air, indicating rapid degradation. No information could be located on its lifetime in drier air.

Biodegradation:

Abiotic Degradation: Acetyl chloride violently decomposes with water and alcohol, fuming in moist air and being easily hydrolyzed to acetic acid and hydrochloric acid by body moisture (e.g., mucous membrane, skin) [3]. It will therefore not persist for any length of time in the environment where water is present. Reaction with photochemically produced hydroxyl radicals in the atmosphere should not be important, since the estimated half-life for this process is 3.3 mo [2].

Bioconcentration:

Soil Adsorption/Mobility:

Volatilization from Water/Soil:

Water Concentrations:

Effluent Concentrations:

Sediment/Soil Concentrations:

Atmospheric Concentrations:

Food Survey Values:

Plant Concentrations:

Fish/Seafood Concentrations:

Animal Concentrations:

Milk Concentrations:

Other Environmental Concentrations:

Probable Routes of Human Exposure: Exposure to acetyl chloride will be primarily occupational via inhalation and dermal contact.

Average Daily Intake:

Occupational Exposures: NIOSH has estimated that 2535 workers are exposed to acetyl chloride based on statistical estimates derived from the NIOSH survey conducted 1981-1983 in the U.S. [4].

Body Burdens:

REFERENCES

1. Boublik T et al; The Vapor Pressures of Pure Substances Vol 17 Amsterdam, Netherlands: Elsevier Science Publ (1984)
2. GEMS: Graphical Exposure Modeling System. FAP. Fate of Atmos Pollutants (1986)
3. Moretti TA; Kirk-Othmer Encycl Chem Technology 3rd ed 1:162-6 (1978)
4. NIOSH; National Occupational Exposure Survey (1983)
5. Thurman EM, Malcolm RL; Structural study of humic substances: New approaches and methods. In: Aquatic and Terrestrial Humic Materials Christman RF & Gjessing ET eds. pp. 1-23. Ann Arbor Science, Ann Arbor, MI (1983)

Acrolein

SUBSTANCE IDENTIFICATION

Synonyms:

Structure:

$$H_2C \diagup \diagdown \text{(CHO)}$$

O
‖
$H_2C\diagup\diagdown_H$

CAS Registry Number: 107-02-8

Molecular Formula: C_3H_4O

Wiswesser Line Notation: VH1U1

CHEMICAL AND PHYSICAL PROPERTIES

Boiling Point: 52.5 °C at 760 mm Hg

Melting Point: -88 °C

Molecular Weight: 56.06

Dissociation Constants:

Log Octanol/Water Partition Coefficient: -0.10 [14]

Water Solubility: 208,000 mg/L at 20 °C [27]

Vapor Pressure: 265 mm Hg at 25 °C [27]

Henry's Law Constant: 4.4 x 10^{-6} atm-m³/mole [31]

ENVIRONMENTAL FATE/EXPOSURE POTENTIAL

Summary: Acrolein is released to the environment: (a) in emissions and effluents from its manufacture plants and facilities which use this compound as an intermediate, (b) in exhaust gas

5

from combustion processes, (c) from direct application to water and wastewater during use as an aquatic herbicide and slimicide, and (d) as a photooxidation product of various hydrocarbon pollutants found in air including 1,3-butadiene. If released to moist soil, acrolein is expected to be susceptible to extensive leaching. Biodegradation under aerobic conditions may be an important fate process. Acrolein is predicted to volatilize rapidly from dry soil surfaces. If released to water, acrolein may biodegrade under aerobic conditions, volatilize (half-life 10 days from a model river), or undergo reversible hydration to beta-hydroxypropionaldehyde (half-life 21 days). The overall half-life of acrolein in water is reported to range between 2 to 6 days. Bioaccumulation in aquatic organisms, adsorption to suspended solids and sediments, reaction with singlet oxygen or alkylperoxy radicals, and photolysis are not expected to be important fate processes in water. If released to the atmosphere, the dominant removal mechanism is expected to be reaction of acrolein vapor with photochemically generated hydroxyl radicals (half-life 0.001 hr). Products of this reaction include carbon dioxide, formaldehyde, and glycolaldehyde, and in the presence of nitrogen oxides include peroxynitrate and nitric acid. Small amounts of this compound may be removed from the atmosphere by wet deposition. Reaction with ozone and direct photolysis are not expected to be important fate processes in atmosphere. The most probable routes of exposure to acrolein by the general population are inhalation of contaminated air and ingestion of food which contains this compound. Worker exposure may occur by dermal contact and/or inhalation.

Natural Sources: Aldehydes are reported to be common products of a variety of microbial and vegetative processes [12]. Acrolein has been identified as a volatile component of essential oil extracted from the wood of oak trees [18].

Artificial Sources: Acrolein is released to the environment (a) in emissions and effluents from its manufacturing and use facilities (this compound is an intermediate for glycerine, methionine, glutaraldehyde and other organic chemicals), (b) in exhaust gas from combustion processes (which includes tobacco smoke, emissions from forest fires, and auto exhaust), (c) from direct application to water and wastewater during use as an aquatic herbicide and slimicide, and (d) from formation in the atmosphere as a photooxidation product of various hydrocarbon pollutants including 1,3- butadiene [12,16,22,23,30].

Terrestrial Fate: If released to moist soil, acrolein is expected to be susceptible to extensive leaching. Biodegradation screening studies in aquatic media indicate that biodegradation in soil under aerobic conditions may be an important fate process. Acrolein is predicted to volatilize rapidly from dry soil surfaces.

Aquatic Fate: If released to water, acrolein may biodegrade under aerobic conditions, volatilize (half-life 10 days from a model river), or undergo reversible hydration to beta-hydroxypropionaldehyde (half-life 21 days). Bioaccumulation in aquatic organisms, adsorption to suspended solids and sediments, reaction with singlet oxygen or alkylperoxy radicals, and photolysis are not expected to be important fate processes. It is reported that acrolein applied to natural waters at rates suggested for herbicidal use will persist up to 6 days depending on water temperature [37]. Acrolein added to irrigation channels at initial concentrations of 6.1, 17.5 and 50.5 ppm underwent 100% loss in 12.5 days [37]. Removal rate constants ranging from 0.27 to 0.34 1/day were calculated by linear regression. These values correspond to half-lives of 2.0 to 2.5 days [4].

Atmospheric Fate: If released to the atmosphere, acrolein is expected to exist almost entirely in the vapor phase based upon its vapor pressure. The dominant removal mechanism is expected to be reaction with photochemically generated hydroxyl radicals (half-life 10-13 hr). Products of this reaction include carbon dioxide, formaldehyde, and glycolaldehyde. In the presence of nitrogen oxides, products include peroxynitrate and nitric acid. Detection of acrolein in rainwater samples suggests that small amounts of this compound may be removed from the atmosphere by wet deposition. Reaction with ozone and direct photolysis are not expected to be important fate processes.

Biodegradation: The half-life of acrolein in natural unsterilized water was 29 hours, compared with 43 hours in sterilized (thymol-treated) water [4]. These results suggest that biodegradation may be partially responsible for the degradation of acrolein in the environment. When 5 and 10 mg/L acrolein was statically incubation in the dark at 25 °C with sewage inoculum, 100% loss was observed [35]. Results of other biodegradation screening studies also indicate that acrolein would be readily degraded by mixed microbial populations [8,17,33]. In contrast, no BOD removal was observed during a 5-day BOD dilution test in which effluent from a biological waste treatment plant was used [6]. Acrolein, at an initial concentration of 50 mg/L as organic carbon, gave no evidence of degradation when incubated for 8 weeks in a 10% anaerobic sludge inoculum [29].

Abiotic Degradation: Acrolein contains no functional groups which would be susceptible to chemical hydrolysis under environmental conditions [8,21]. Acrolein will be susceptible to formation of beta-hydroxypropionaldehyde by hydration in water. Hydration is a reversible reaction. The half-life for hydration of acrolein has been calculated to be 21 days based on a pseudo-first order reaction rate constant of 0.032 day^{-1} [8]. Half-lives for acrolein reacting with singlet oxygen and alkylperoxy radicals in natural sunlit water have been estimated to be 8 and 23 years, respectively. These values are based on reaction rate constants of 1 x 10^{+7} and 3.4 x 10^{+3} L/mole-hr, respectively, a singlet oxygen concentration of 1 x 10^{-12} mole/L, and an alkylperoxy radical concentration of 1 x 10^{-9} mole/L [21]. Acrolein in hexane solvent show moderate absorption of UV light >290 nm [21], which indicated potential for photolytic transformation under environmental conditions. However, hydration of acrolein in water would destroy the chromophores which absorb light. As a result, the potential for direct photolysis would be slight [21]. The half-life for acrolein vapor reacting with photochemically generated hydroxyl radicals in the atmosphere has been estimated to be 10 to 13 hours based on experimentally determined reaction rate constants ranging between 1.90 x 10^{-11} and 2.53 x 10^{-11} cm^3/molecule-sec at 25-26 °C [2] and assuming an average ambient hydroxyl radical concentration of 8.0 x 10^{+5} molecules/cm^3 [11]. Products of the reaction of acrolein with hydroxyl radicals include carbon dioxide, formaldehyde, and glycolaldehyde. In the presence of nitrogen oxides, products include peroxynitrate and nitric acid [9]. The half-life for acrolein reacting with ozone in the atmosphere has been estimated to be 18 days based on an experimentally determined reaction rate constant of 7.4 x 10^{-19} cm^3/molecule-sec at room temperature [1] and assuming an average ambient ozone concentration of $6 \times 10^{+11}$ molecules/cm^3 [11]. The half-life for photodissociation of acrolein in the atmosphere has been estimated to be approx 3.5 days based on measured quantum yields [10].

Bioconcentration: A bioconcentration factor (BCF) of 344 has been measured for acrolein in bluegill sunfish [3]. However, this value may be an overestimate, since total ^{14}C was measured and may have included acrolein metabolites. A BCF of 0.6 can be estimated from the Kow [20]. These BCF values suggest that bioconcentration in aquatic organisms would not be significant.

Soil Adsorption/Mobility: A soil adsorption coefficient (Koc) of 24 was estimated for acrolein from the Kow [20]. This low Koc value and the relatively high water solubility of acrolein suggest that this compound would not adsorb significantly to suspended solids and sediments in water and would be highly mobile in soil [34].

Volatilization from Water/Soil: Loss of acrolein by volatilization from a large tank of water was approx 10% in 160 hours, initial concentration 100 ppm. However, the lack of turbulence in the tank was thought to have minimized volatility losses considerably [5]. Based on the Henry's Law constant, the volatilization half-life of acrolein from a model river 1 m deep flowing 1 m/sec, with a wind speed of 3 m/sec has been estimated to be approx 10 days [20]. The high vapor pressure suggests that acrolein should volatilize rapidly from dry soil surfaces.

Water Concentrations: SURFACE WATER: USEPA STORET database - 798 water samples, 0.25% pos., median concn <14 ug/L [32]. GROUND WATER: Detected in one out of five leachate samples from a Wisconsin municipal solid waste landfill [28].

Effluent Concentrations: Present in six out of 11 samples of municipal effluent from Dayton, OH, concn range 20-200 ug/L [18]. Detected in raw sewage in two sewage treatment plants in Chicago at concn ranging from 216-825 ug/L; although concn in final effluents were below 100 ug/L [19]. STORET database - 1265 effluent samples, 1.5% pos., median concn <10.0 ug/L [32]. Acrolein has been identified in emissions from plants manufacturing acrylic acid, not quantified; coffee roasting operations, ND-0.6 mg/m^3 (detection limit not reported); from a lithographic plate coater, <0.23-3.9 mg/m^3; and from an automobile spray booth, 1.1-1.6 mg/m^3 [18].

Sediment/Soil Concentrations: Detected in sediment/soil/water samples collected from Love Canal in Niagara Falls, NY during 1980 [15]. STORET database - 331 sediment samples, 0% pos. [32].

Atmospheric Concentrations: Air samples collected in Claremont, CA during Aug.-Sept. 1979 contained acrolein at a max concn of 34 mg/m^3 [36]. During June-July 1976, in air of Edison, NJ (near emission sources), 19 samples, mean concn 0.71 ng/m^3 [7]. During a 12-month period in 1968, acrolein was detected in air of the Los Angeles Basin at levels ranging from ND - 0.04 mg/m^3 (detection limit not reported), although most measurements were between 0.002 and 0.02 mg/m^3 [18]. Detected at a level of 0.14 mg/m^3 in an atmospheric grab sample obtained near an oil fire [26]. Levels of acetone and acrolein in rainwater samples obtained in CA ranged from ND to 0.05 mg/L (detection limit not reported) [13]. Levels of acetone-acrolein-propanal in cloud, mist and fog samples obtained in CA ranged from ND to 0.86 mg/L (detection limit not reported) [13].

Food Survey Values: Acrolein has been detected in sugarcane molasses, souring salted pork, the fish odor of cooked horse mackerel, the volatiles from white bread, the volatile components of chicken-breast muscle, the aroma volatiles of ripe arctic bramble berries, and the products from heating animal fats and vegetable oils [18]. This compound has also been detected in fresh lager beer at levels of 1.11-2.00 ug/L, mean concn 1.6 ug/L [18].

Plant Concentrations:

Fish/Seafood Concentrations: STORET database - 87 samples, 1% pos., median concn <1.0 ug/kg wet basis [32].

Animal Concentrations:

Milk Concentrations:

Other Environmental Concentrations: Found in tobacco smoke, 3-141 ug/cigarette; diesel engine exhaust gas, 0.06-19.6 mg/m^3; gasoline engine exhaust gas, 0.46-12.2 mg/m^3; rotary gasoline engine exhaust gas, 0.46 mg/m^3; combustion products of hydraulic fluid, no concentrations given; smoke from the combustion of wood, 115 mg/m^3; kerosene, <2.3 mg/m^3; and cotton, 138 mg/m^3; combustion products of cellophane used to seal meat packages; and decomposition products of overheated wax [18]. Acrolein emissions from a wood-burning fireplace ranged from 21 to 132 mg/kg of wood burned [18].

Probable Routes of Human Exposure: The most probable routes of exposure to acrolein by the general population are inhalation of contaminated air and ingestion of foods which contain this compound. Worker exposure may occur by dermal contact and/or inhalation.

Average Daily Intake:

Occupational Exposures: Acrolein was detected in the air of a truck maintenance shop at a mean concn of 4.6 ug/m^3 [18]. NIOSH has estimated that 1300 workers are potentially exposed to acrolein based on statistical estimates derived from a survey conducted 1981-83 in the U.S. [25]. NIOSH has estimated that 7301 workers are exposed to acrolein based on statistical estimates derived from a survey conducted during 1972-1974 in the U.S. [24].

Body Burdens:

REFERENCES

1. Atkinson R, Carter WP; Chem Rev 84: 437-70 (1984)
2. Atkinson R; Chem Rev 85: 69-201 (1985)
3. Barrows ME et al; pp. 279-92 in Dynamics, Exposure Hazard Assess Toxic Chem. Ann Arbor, IM: Ann Arbor Science (1980)
4. Bowmer KH, Higgins ML; Arch Environ Contam Toxicol 5: 87-96 (1976)
5. Bowmer KH et al; Weed Res 14: 325-28 (1974)
6. Bridie AL et al; Water Res 13: 627-30 (1979)

Acrolein

7. Brodzinsky R, Singh HB; Volatile Organic Chemicals in the Atmosphere: An Assessment of Available Data. Menlo Park, CA: SRI International (1982)
8. Callahan MA et al; Water-Related Environmental Fate of 129 Priority Pollutants pp. 20-1 to 20-11 USEPA-440/4-79-029A (1979)
9. Edney E et al; Atmospheric Chemistry of Several Toxic Compounds USEPA-600/53-82-092 (1983)
10. Gardner EP et al; Project Summary: The Primary Photochemical Processes of Acrolein pp. 1-3 USEPA-600/S3-86-005 (1986)
11. GEMS; Graphical Exposure Modeling System. FAP. Fate of Atmos Pollut (1986)
12. Graedel TE; pp.159-173 and 407 in Chemical Compounds in the Atmosphere NY: Academic Press (1978)
13. Grosjean D, Wright B; Atmos Environ 17: 2093-6 (1983)
14. Hansch C, Leo AJ; Medchem Project Issue No. 26 Claremont, CA: Pomona College (1985)
15. Hauser TR, Bomberger SM; Environ Monit Assess 2:249-72 (1982)
16. Hess LG et al; Kirk-Othmer Encycl Chem Tech 3rd ed NY: Wiley 1: 277-97 (1978)
17. Hultman B; Water Sci Tech 14: 79-86 (1982)
18. IARC; Acrolein, Inter Agency for Research on Cancer 36:133-61 (1985)
19. Lue-Hing C et al; AICHE Symp Ser 77: 144-50 (1981)
20. Lyman WJ et al; Handbook of Chemical Property Estimation Methods. NY: McGraw-Hill New York (1982)
21. Mabey WR et al; Aquatic Fate Process Data for Organic Priority Pollutants pp. 53-4 USEPA-440/4-81-014 (1981)
22. Maldotti A et al; Int J Chem Kinet 12: 905-13 (1980)
23. National Research Council; Formaldehyde and Other Aldehydes pp.4-9 USEPA 600/6-82-002 (1982)
24. NIOSH; National Occupational Hazard Survey (1974)
25. NIOSH; National Occupational Exposure Survey (1983)
26. Perry R; Mass Spectrometry in the Detection and Identification of Air Pollutants pp. 130-37 Int Symp Ident Meas Environ Pollut (1971)
27. Riddick JA et al; Organic Solvents: Physical Properties and Methods of Purification, 4th Edit. New York: J Wiley & Sons (1986)
28. Sabel GV, Clarke, TP; Waste Manag Res 2: 119-30 (1984)
29. Shelton DR, Tiedje JM; Development of Tests for Determining Anaerobic Biodegradation Potential USEPA 560/5-81-013 NTIS PB84-166495 (1981)
30. Shimada I et al; Kikai Gijutsu Kenyusho Shoho 32: 62-77 (1978)
31. Snider JR, Dawson GA; Environ Int 7: 237-58 (1982)
32. Staples CA et al; Environ Toxicol Chem 4: 131-42 (1985)
33. Stover EL, Kincannon DF; J Water Poll Control Fed 55: 97-109 (1983)
34. Swann RL et al; Res Rev 85: 17-28 (1984)
35. Tabak HH et al; J Water Pollut Control Fed 53: 1503-18 (1981)
36. Tuazon EC et al; Atmospheric Measurement of Trace Pollutants: Long Path Fourier Transform Infrared Spectroscopy USEPA 600/S3-81-026 (1981)
37. Weed Sci Soc of America; Herbicide Handbook 5th ed Champaign, IL Weed Sci Soc of America pp.8-12 (1983)

Acrylamide

SUBSTANCE IDENTIFICATION

Synonyms: 2-Propenamide

Structure:

CAS Registry Number: 79-06-1

Molecular Formula: C_3H_5NO

Wiswesser Line Notation: ZV1U1

CHEMICAL AND PHYSICAL PROPERTIES

Boiling Point: 87 °C at 2 mm Hg

Melting Point: 84.5 °C

Molecular Weight: 71.08

Dissociation Constants:

Log Octanol/Water Partition Coefficient: -0.67 [13]

Water Solubility: 2,151,000 mg/L at 30 °C [2]

Vapor Pressure: 7 x 10^{-3} mm Hg at 25 °C [2]

Henry's Law Constant: $3.2x10^{-10}$ atm-m³/mole (calculated from water solubility and vapor pressure)

ENVIRONMENTAL FATE/EXPOSURE POTENTIAL

Summary: Acrylamide may be released into the environment primarily in wastewater during its production and use in the manufacture of polyacrylamides and other polymers. Other releases may result from the disposal of the solid monomer on land or from leaching of residual monomer from use of polyacrylamides. If released on land, acrylamide would be expected to leach readily into the ground and biodegrade within a few weeks. If released into water, it should biodegrade in approx 8 to 12 days. Bioconcentration in fish and adsorption to sediment should not be significant due to its high water solubility. In the atmosphere, the vapor phase chemical should react with photochemically produced hydroxyl radicals (half-life 6.6 hr) and be washed out by rain. Human exposure will be primarily occupational via dermal contact and inhalation, although exposure to the general public has resulted from the leaching of the acrylamide monomer from polyacrylamide flocculants used in water treatment.

Natural Sources: All acrylamide in the environment is man-made [24].

Artificial Sources: Acrylamide may be released into wastewater during its production and use in the synthesis of dyes, manufacture of polymers, adhesives, paper, paperboard and textile additive, soil-conditioning agents, ore processing, oil recovery, and permanent press fabrics [14,19]. Acrylamide may also be released into water from the treatment of water with polyacrylamide as a flocculating agent [4]. Improvements in the polymerization process has reduced the monomer content of these polymers from 5% to 0.3% [4]. The largest end use is as a flocculent in facilitating liquid-solid separation for processing minerals in mining, waste treatment, and water treatment [19]. Other sources of release to water is from acrylamide-based sewer grouting and recycling of waste paper [4].

Terrestrial Fate: When released on soil, acrylamide would be expected to leach readily into the ground and biodegrade with a few weeks.

Acrylamide

Aquatic Fate: Since acrylamide has been found to degrade in distilled water over a period of 1-2 months and river water in 8-12 days, it is unlikely that acrylamide would have a long residence time in natural waters. However, acclimation of the microorganisms is important. In systems in which the residence time is relatively short such as in sewage works and water treatment facilities, acrylamide may not be completely degraded, and acrylamide has been detected in the effluent from a sewage treatment plant [4]. Adsorption to sediment and volatilization will not be appreciable.

Atmospheric Fate: If released into the atmosphere, acrylamide in the vapor phase should react with photochemically produced hydroxyl radicals (estimated half-life 6.6 hr). Due to its high solubility in water, it should be scavenged by rain and fog.

Biodegradation: Acrylamide degrades rapidly with acclimation in biodegradability screening tests [3,9,15]. In two 5-day screening tests using acclimated sewage seed, 69 and 75% of theoretical BOD was obtained [3,8]. In longer term screening tests, 72.8% of theoretical BOD was achieved in 2 weeks using the MITI test, the biodegradability test of the Japanese Ministry of International Trade and Industry [15], and 100% degradation was obtained in 16 days [9]. In two river die-away tests using aerated Thames River water, the lag time for biodegradation to begin was 220 and 50 hours [9]. In the latter case, 90% of the acrylamide disappeared in approx 150 hours. When the water was inoculated with cultures capable of degrading acrylamide, the lag period was reduced to 5 hours and degradation was 90% complete after 24 hours. In order to access the efficiency of sewage works in removing acrylamide, two sewage works were dosed for four times longer than the residence time [4]. Little loss of acrylamide occurred during initial or final settling. However, 50 to 70% was lost in the activated sludge plants. Further studies showed that high loss rates required high microbial activity or, in particular, contact with surfaces of high microbial activity [4]. Studies of the river into which the sewage works discharged its effluents suggest that microbial degradation is unlikely to affect the level of acrylamide in river water for several hours, and possibly days, even in a river into which acrylamide is continually discharged [4]. Degradation was, however, more marked in the summer [4]. Acrylamide disappeared from Hackensack River water within 12 days [8] and Thames River water within 9 days [9]. In another experiment in which 0.5 and 10.0 ppm of acrylamide was added to river and estuarine water with and without added sediment, all acrylamide had disappeared within 8 days [6], but only in the river water/sediment system was degradation observed within a day [6]. In seawater, 75% and 10% degradation occurred in 8 days with and without added sediment, respectively [6]. In five surface soils that were moistened to field capacity, 74-94% degradation occurred in 14 days in three soils and 79 to 80% degradation occurred in 6 days in the other two soils [1]. Similar results were obtained with air-dried soils. Under waterlogged (anaerobic) conditions, 64 to 89% degradation in 14 days were obtained [1]. Another investigator obtained half-lives of 18-45 hr in four Central New York soils moistened to 70% capacity [16]. Two of the soils were also incubated under anaerobic conditions for which 21 and 55% degradation were obtained in 14 days [16]. Acrylamide did not degrade in 56 days when incubated with digester sludge under anaerobic conditions [23].

Abiotic Degradation: Acrylamide was found to degrade over a period of 1-2 months at room temperature when stored in the dark in distilled water [5]. The product of the hydrolysis was acrylic acid and an unidentified compound [5]. Acrylamide does not absorb light >250 nm [7], so direct photolysis would not be expected to occur. Vapor phase acrylamide should react with photochemically produced hydroxyl radicals and ozone by addition to the double bond with an estimated atmospheric half-life of 6.6 hours [10].

Bioconcentration: When the uptake of radiolabelled acrylamide (0.338 mg/L) was studied in fingerling trout at 12 °C for 72 hr under static conditions, the BCF in the carcass and viscera was 0.86 and 1.12, respectively, indicating that no appreciable bioaccumulation had occurred [22]. The uptake was rapid in the first 24 hours and then leveled off to a plateau after 72 hours. When the fish were transferred to fresh water, levels of acrylamide declined to 75% after 96 hours [22].

Soil Adsorption/Mobility: No significant adsorption of acrylamide by natural sediments, industrial and sewage sludges, clays, (montmorillonite and kaolinite) was observed [6]. The retardation factor of acrylamide with respect to water in soil thin layer chromatography (TLC) experiments are: Hilton loam - 0.715, Williamston silt loam - 0.880, Crogham loamy fine sand - 0.805, and silt clay - 0.657 [16]. These results indicate that acrylamide is mobile in soil.

Volatilization from Water/Soil: Volatilization from water or moisture soils for chemicals with such low Henry's Law constants should not be appreciable [17].

Water Concentrations: DRINKING WATER: Cases of human poisoning from well water contaminated with acrylamide have been documented [4]. The source of the poisoning was sewer grouting [4]. No acrylamide was found in either the output of the Kansas City water treatment plant or tap water at Midwest Research Institute [12]. Tap water in Plymouth, England contained 0.75 ppb of acrylamide; however, no acrylamide was found in Devon tap water from areas not using polyacrylamides for water treatment [5]. SURFACE WATER: In a monitoring study of acrylamide around five industrial sites including two acrylamide producers, four polyacrylamide producers, and one polyacrylamide user, acrylamide was found in only one water sample, 1.5 ppm, downstream from the outfall of a polyacrylamide producer [12]. In an unspecified British river downstream from a clay pit, 0.03 ppb of acrylamide was found [9]. Acrylamide was not detected in the River Erme of Culm in Devon, England, but 3.4 ppb was found in the River Tavy, the source of which was unknown [5]. No acrylamide was detected in seawater from Plymouth Sound or estuarine water from a creek in Devon, England [5].

Effluent Concentrations: Acrylamide (17.4 ppb) was detected in sewage effluent in Devon, England [5].

Sediment/Soil Concentrations: In a monitoring study of acrylamide around five industrial sites including two acrylamide producers, four polyacrylamide producers, and one polyacrylamide user, no acrylamide was in any soil or sediment samples [12]. All sampling was performed beyond the plant site perimeter.

Atmospheric Concentrations: In a monitoring study of acrylamide around five industrial sites including two acrylamide producers, four polyacrylamide producers and on polyacrylamide user, no acrylamide was found in any air samples [12]. All sampling was performed beyond the plant site perimeter.

Food Survey Values:

Plant Concentrations:

Fish/Seafood Concentrations:

Animal Concentrations:

Milk Concentrations:

Other Environmental Concentrations: Residual acrylamide concentrations in 32 polyacrylamide samples approved for use as flocculants in water treatment plants ranged from 0.5 to 600 ppm [11].

Probable Routes of Human Exposure: Human exposure to acrylamide is primarily occupational from dermal contact with the solid monomer and inhalation of dust and vapor, especially when emptying bags and drums and in maintenance and repair [18,19]. Residual monomer is a concern in the polymer [19]. The general public may be exposed to acrylamide through drinking water that is contaminated with acrylamide from the manufacture or use of the monomer or polymers and polyelectrolytes based on acrylamide [4].

Average Daily Intake:

Occupational Exposures: According to the 1985 National Occupational Exposure Survey (NOES), 9776 workers are potentially exposed to acrylamide [21]. The 1972-1974 National Occupational Hazard Survey (NOHS) reported that 10,368 workers are exposed to acrylamide [20].

Body Burdens:

REFERENCES

1. Abdelmagid HM, Tabatabai MA; J Environ Qual 11: 701-4 (1982)
2. Bikales NM, Kolody ER; Kirk-Othmer Encycl Chem Tech 2nd ed 1: 274-84 (1963)
3. Bridie Al et al; Water Res 13: 627-30 (1979)
4. Brown L et al; Water Res 16: 579-91 (1982)
5. Brown L, Rhead M; Analyst 104: 391-9 (1979)
6. Brown L et al; Water Res 14: 779-81 (1980)
7. Carpenter EL, Davis HS; J Appl Chem 7: 671-6 (1957)
8. Cherry AB et al; Sewage Indust Wastes 28: 1137-46 (1956)
9. Croll BT et al; Water Res 8: 989-93 (1974)
10. GEMS; Graphical Exposure Modeling System. FAP. Fate of Atmos Pollut (1986)
11. Going JE; Environmental Monitoring Near Industrial Sites, EPA-560/6-78-001 Washington, DC (1978)
12. Going JE, Thomas K; Sampling and Analysis of Selected Toxic Substances. EPA-560/13-79-013 Kansas City, MO pp. 39 (1979)
13. Hansch C, Leo AJ; Medchem Project Issue No 26. Claremont CA: Pomona College (1985)
14. Hawley GG; Condensed Chem Dictionary 10th ed Van Nostrand Reinhold NY p. 16 (1981)
15. Kitano M; Biodeg Bioaccum Test on Chem Subs, OECD Tokyo Meeting Ref Book TSU-No. 3 (1978)
16. Lande SS et al; J Environ Qual 8: 133-7 (1979)
17. Lyman WJ et al; Handbook of Chem Property Estimation Methods Environ Behavior of Organic Compounds. McGraw-Hill NY (1983)
18. MacWilliams DC; Encycl Chem Tech 1: 298-311 (1978)
19. Morris JD, Penzenstadler RJ; Encycl Chem Tech 1: 312-330 (1978)
20. NIOSH; National Occupational Health Survey (1975)
21. NIOSH; National Occupational Exposure Survey (1985)
22. Petersen DW et al; Toxicol Appl Pharmacol 80: 58-65 (1985)
23. Shelton DR, Tiedje JM; Development of Tests for Determining Anaerobic Biodeg Potential EPA 560/5-81-013 E. Lansing, MI p. 92 (1981)
24. WHO; Environ Health Criteria: Acrylamide p.9 (1985)

Acrylic Acid

SUBSTANCE IDENTIFICATION

Synonyms: 2-Propenoic acid

Structure:

CAS Registry Number: 79-10-7

Molecular Formula: $C_3H_4O_2$

Wiswesser Line Notation: QV1U1

CHEMICAL AND PHYSICAL PROPERTIES

Boiling Point: 141.6 °C at 760 mm Hg

Melting Point: 14 °C

Molecular Weight: 72.06

Dissociation Constants: pKa = 4.247 [13]

Log Octanol/Water Partition Coefficient: 0.161 [5](estimated)

Water Solubility: Miscible [10]

Vapor Pressure: 7.76 mm Hg at 20 °C [10]

Henry's Law Constant: 3.2 x 10^{-7} atm-m^3/mole [15]

Acrylic Acid

ENVIRONMENTAL FATE/EXPOSURE POTENTIAL

Summary: Acrylic acid may be released into wastewater during its production and use in the manufacture of acrylic esters, water-soluble resins, dispersants, and flocculants. It is also produced naturally in marine algae and in the rumen fluid of sheep. If released on land, it would leach into the ground. It is amenable to aerobic and anaerobic biodegradation, but no rate data for soil systems are available. If released into water, acrylic acid will biodegrade, but rate information in environmental systems are lacking. Adsorption to sediment, volatilization, and bioconcentration in aquatic organisms should not be significant. In the atmosphere, acrylic acid will react with ozone and photochemically produced hydroxyl radicals (half-life 6.6 hr). Human exposure will be primarily occupational via inhalation and dermal contact.

Natural Sources: Acrylic acid has been reported to occur naturally in the following species of marine algae: 9 species of Chlorophyceae, 10 of Rhodophyceae, and 11 of Phaeophyceae [6]. It also has been found in the rumen fluid of sheep [6].

Artificial Sources: Acrylic acid is produced in large quantities (730 million lb in 1984 in U.S.) and may be released in waste water and as emissions during its production and use, 80% of which goes into the production of acrylate esters [1]. Most of this acrylic acid is used captively in the U.S. [6]. Other uses that may result in releases into wastewater or spills are in dispersants, flocculants in water treatment, and water-soluble resins [6].

Terrestrial Fate: If released on land, acrylic acid will leach into the ground and possibly biodegrade. No studies are available, however, concerning its biodegradation in soil samples.

Aquatic Fate: If released into water, acrylic acid should readily biodegrade, although no rate data are available. Adsorption to sediment or volatilization will not be significant.

Atmospheric Fate: If released into the atmosphere, acrylic acid will react with photochemically produced hydroxyl radicals and ozone resulting in an estimated half-life of 6.6 hr [5].

Biodegradation: In a 42-day screening study using a sewage seed inoculum, 71% of acrylic acid was mineralized [2]. After acclimation, 81% was degraded to CO_2 in 22 days [2]. Acrylic acid has been reported to be significantly degraded in the MITI test, a biodegradability screening test of the Japanese Ministry of International Trade and Industry [12]. When added to water, acrylic acid is rapidly oxidized and wastewater containing the compound can deplete reservoirs of oxygen [3]. Acrylic acid is amenable to anaerobic treatment [16] and in an anaerobic screening study utilizing 10% sludge from a secondary digester as an inoculum, acrylic acid was judged to be degradable with >75% of theoretical methane being produced in 8 weeks of incubation [14]. In another study, it was toxic to unacclimated anaerobic acetate-enriched cultures and was poorly utilized (21%) in a completely mixed anaerobic reactor with a 20-day hydraulic retention time after a 90-day acclimation period [2]. A possible resolution between the conflicting results for anaerobic degradation is the observation that acetate cultures have to exhaust the acetic acid as a carbon and energy source before utilizing a cross-fed compound [2].

Abiotic Degradation: The UV absorption band of acrylic acid extends to about 320 nm [11]. While direct sunlight photolysis is therefore possible, no experimental data on the photolysis of acrylic acid could be located. Acrylic acid reacts with photochemically produced hydroxyl radicals primarily by addition to the double bond and with atmospheric ozone resulting in an estimated overall half-life of 6.6 hr [4].

Bioconcentration: One can estimate a BCF of 0.78 using a recommended regression equation [7] with the octanol/water partition coefficient. This indicates that bioconcentration in aquatic organisms should be negligible.

Soil Adsorption/Mobility: Acrylic acid is miscible with water and therefore would not be expected to adsorb significantly to soil or sediment [7].

Volatilization from Water/Soil: Chemicals with such low Henry's Law constants are essentially nonvolatile [7]. The vapor pressure of acrylic acid would suggest that it should volatilize to some extent from surface and dry soil.

Acrylic Acid

Water Concentrations:

Effluent Concentrations: Acrylic acid occurs in wastewater effluents from its production by the oxidation of propylene at concn exceeding 0.5 ppm [17].

Sediment/Soil Concentrations:

Atmospheric Concentrations:

Food Survey Values:

Plant Concentrations:

Fish/Seafood Concentrations:

Animal Concentrations:

Milk Concentrations:

Other Environmental Concentrations: Acrylic acid has been reported in trace quantities in commercial-grade propionic acid [6].

Probable Routes of Human Exposure: Exposure to acrylic acid is primarily occupational via inhalation and dermal contact.

Average Daily Intake:

Occupational Exposures: NIOSH has estimated that 52,469 workers are potentially exposed to acrylic acid based on statistical estimates derived from a Survey conducted in 1972-1974 in the U.S. [8]. This survey sampled 5000 businesses in 67 metropolitan areas throughout the United States for the manufacture and use of chemicals, grade name products known to contain the compound and generic products suspected of containing the compound. NIOSH has estimated that 54,352 workers are potentially exposed to acrylic acid based on statistical estimates derived from a survey conducted in 1981-1983 in the U.S. [9].

Body Burdens:

REFERENCES

1. Chemical Profile: Acrylic Acid Chemical Marketing Reporter. October 15 (1984)
2. Chou WL et al; Biotech Bioeng Symp 8: 391-414 (1978)
3. Ekhina RS, Ampleeva GP; Combined effect of Six Acrylates on the Sanitary Status of a Reservoir Vodemov 157-64 (1977)
4. GEMS: Graphical Exposure Modeling System FAP Fate of Atmos Pollut (1986)
5. GEMS; Graphical Exposure Modeling System. CLOG3. (1986)
6. IARC Monograph on the Evaluation of the Carcinogenic Risk of Chemicals to Humans. Some Monomers, Plastics and Synthetic Elastomers, and Acrolein 19: 47-71 (1971)
7. Lyman WJ et al; Handbook of Chem Property Estimation Methods Environ Behavior of Organic Compounds McGraw-Hill NY (1982)
8. NIOSH; National Occupational Health Survey (1975)
9. NIOSH; National Occupational Exposure Survey (1985)
10. Riddick JA et al; Organic Solvents: Physical Properties and Methods of Purification, 4th Edit. New York: J Wiley & Sons (1986)
11. Sadtler NA; Sadtler Standard Spectra
12. Sasaki S; The Scientific Aspects of the Chemical Substance Control Law in Japan in Aquatic Pollutants Transformation and Biological effects. Hutzinger O et al; (eds) Oxford Pergamon Press pp. 283-98 (1978)
13. Serjeant EP, Dempsey B; Ionization Constants of Organic Acids in Aqueous Solution. IUPAC Chemical Data Series No 23 New York,NY: Pergamon Press pp989 (1979)
14. Shelton DR, Tiedje JM; Appl Environ Microbiol 47: 850-7 (1984)
15. Singh HB et al; Reactivity/Volatility Classification of Selected Organic Chemicals: Existing Data. EPA-600/3-84-082 Menlo Park, CA SRI INTER p. 190 (1984)
16. Speece RE; Environ Sci Technol 17: 416A-27A (1983)
17. Wise HE, Fahrenthold PD; Occurrence and Predictability of Priority Pollutants in Wastewater of the Organic Chemicals and Plastics/Synthetic Fibers Industrial Categories. Amer Chem Soc Div Indust (1981)

Acrylonitrile

SUBSTANCE IDENTIFICATION

Synonyms: 2-Propenenitrile

Structure:

$$H_2C = C \overset{CN}{\underset{H}{<}}$$

CAS Registry Number: 107-13-1

Molecular Formula: C_3H_3N

Wiswesser Line Notation: NC1U1

CHEMICAL AND PHYSICAL PROPERTIES

Boiling Point: 77.3 °C at 760 mm Hg

Melting Point: -82 °C

Molecular Weight: 53.06

Dissociation Constants:

Log Octanol/Water Partition Coefficient: 0.25 [15]

Water Solubility: 75,000 mg/L at 25 °C [36]

Vapor Pressure: 107.8 mm Hg at 25 °C [9]

Henry's Law Constant: 1.10 x 10^{-4} atm-m³/mole at 25 °C [2]

ENVIRONMENTAL FATE/EXPOSURE POTENTIAL

Summary: Acrylonitrile is an important industrial chemical used in the production of acrylic and modacrylic fibers and other important chemicals and resins and is released as fugitive emissions and in wastewater during its production and use. More general sources such as auto exhaust, cigarette smoke, and release from fibers and plastics have been identified, but their importance has not been evaluated. When released to the atmosphere, acrylonitrile will degrade primarily by reacting with photochemically produced hydroxyl radicals. The half-life for this process is estimated to be 3.5 sunlit days under relatively clean atmospheric conditions to somewhat over a day with smog. Therefore, there would be opportunity for dispersal from source areas. If released to water, acrylonitrile will slowly evaporate (half-life 1-6 days) and also biodegrade (complete degradation in approx one week in receiving water in which microorganisms would be acclimated). If spilled on land, it will volatilize rapidly due to its high vapor pressure, although some of it would leach into the ground where its fate is unknown. Humans are exposed to acrylonitrile primarily in the workplace via inhalation and possibly dermal contact. There is a potential for the general public to be exposed to acrylonitrile in the air from its dispersal from source areas, outgassing from acrylic fibers, auto exhaust and cigarette smoke as well as via ingestion from food which is in contact with plastic containers where acrylonitrile monomer has leached out.

Natural Sources: Acrylonitrile is not known to occur as a natural product.

Artificial Sources: Acrylonitrile is a very high production chemical (2.3 billion lb demand in 1985 [5]) which may be released to the environment as fugitive emissions or in wastewater during its production and use in the manufacture of acrylic and modacrylic fibers by copolymerization, ABS and SAN resins, adiponitrile, acrylamide, and other chemicals and resins [5,34]. It is also found in auto exhaust [13]. Residual amounts are reported to be released from clothing, furniture, etc., made with polyacrylic fiber; and to leach from polyacrylonitrile packaging material [34]. Acrylonitrile monomer is released during the burning of polyacrylonitrile plastic [34] and cigarette smoke [34]. Sources of atmospheric emissions from production facilities include absorber vents 0.44 g/kg, flare stacks 0.5 g/kg, product loading 0.14 g/kg, fugitive emissions 0.26 g/kg, storage tanks 0.81 g/kg, deep well ponds 0.10 g/kg, and incinerators, negligible; annual emissions are estimated at 2298 million tons [32].

Terrestrial Fate: If spilled on land, acrylonitrile will evaporate rapidly. Because it is so poorly adsorbed to soil, it may also leach into the ground water.

Aquatic Fate: If released into water, acrylonitrile will probably biodegrade slowly - approx 20 days, or a week or less in waters where the microbes were unacclimated or acclimated, respectively. Volatilization would also occur with expected half-lives of 1-6 days in environmental waters. In humic waters, photooxidation by radicals may occur, but data in natural systems are not available. Adsorption to sediment or particulate matter and bioconcentration in aquatic organisms will not be significant.

Atmospheric Fate: If released to the atmosphere, acrylonitrile will degrade by reaction with hydroxyl radicals with a reaction half-life of 3.5 12-hr sunlit days. This would be reduced considerably under smog conditions. Since it is relatively long-lived in the atmosphere, considerable dispersion would be expected to occur.

Biodegradation: Available data indicates that acrylonitrile degrades in aerobic systems after acclimation. In studies using activated sludge inocula >95% degradation, >70% of theoretical BOD removal was reported after 21 days acclimation in a screening study [20]; 30% of theoretical BOD removal was reported after 10 days in a treatment plant [23]; and >99% degradation, 30% of theoretical BOD removal was reported in a bench-scale continuous flow reactor [31]. Others reported 0 and 38% of theoretical BOD removal after 5 and 20 days, respectively [37], and complete degradation in 7 days [33] in screening studies with sewage seed. Acrylonitrile completely degraded in Mississippi river water in 6 days [12] and degraded completely in 20 days in another study in river water, requiring less time for degradation on redose [6]. Sixty-five percent theoretical BOD was removed from river water in 5 days after 27 days acclimation [24]. In the only data under anaerobic conditions, acrylonitrile was poorly degraded in a reactor with a 2 - 10 day retention time, with only 17% utilization being reported after 110 days acclimation [7].

Abiotic Degradation: Acrylonitrile does not absorb light >290 nm and is therefore not susceptible to direct photolysis [14]. On standing it is reported to slowly develop a yellow color particularly after exposure to light [22]. This may be due to its reaction with photochemically produced free radicals. Acrylonitrile (10 ppm) was stable at pH 4 to 10 for 23 days indicating that hydrolysis is negligible under these conditions [12]. Acrylonitrile reacts with photochemically produced hydroxyl radicals with a reaction half-life of 3.5 days (assuming 12 hr of sunlight) [11]. Reaction with ozone is not fast enough to be a significant sink [16]. In a smog chamber, the formation of ozone is significant with the time to maximum ozone formation averaging 5.3 times faster than propane in five experiments [10,27] and 5.3%/hr of acrylonitrile disappeared [10]. For a first- order reaction, this is equivalent to a half-life of 13 hr. Formaldehyde and PAN-type compounds are formed [29].

Bioconcentration: The whole body bioconcentration factor for a bluegill exposed to acrylonitrile for 28 days, or until equilibrium was obtained in a flowing water system, was 48 [1]. The bioconcentration factor estimated from the water solubility is 1 [18]. The experimental and estimated bioconcentration factors indicate that bioconcentration of acrylonitrile in aquatic organisms is not significant.

Soil Adsorption/Mobility: The Koc for acrylonitrile calculated from the water solubility is 9 [18] indicating that adsorption to soil will be insignificant.

Volatilization from Water/Soil: The overall transfer coefficient for acrylonitrile relative to oxygen is 0.59 [4]. Coupled with the reaeration rates for oxygen in typical bodies of water, one can estimate that the volatilization half-life of acrylonitrile in a typical pond, river, and lake are 6, 1.2, and 4.8 days, respectively [21]. Acrylonitrile is highly volatile and not strongly adsorbed to soil and will therefore volatilize rapidly from soil and other surfaces.

Acrylonitrile

Water Concentrations: DRINKING WATER: Detected, not quantified in one or more water supplies in a five city survey of U.S. drinking water with known sources of contamination [8]. SURFACE WATER: US-914 STORET (EPA Water Quality Database) stations - 5% of samples had detectable quantities of acrylonitrile [30]. GROUND WATER: Screening of 1174 community wells and 617 private wells in Wisconsin during the early 1980s did not detect any acrylonitrile [19].

Effluent Concentrations: 1278 stations in STORET database - 1.6% of samples had detectable quantities of acrylonitrile [30]. Detected in the treated wastewater from the following industries (mean concn): iron and steel manufacturing (1600 ppb), foundries (23 ppb), organic chemicals/plastics manufacture (93 ppb), and rubber processing (10 ppb) [35]. Rio Blanco-CO-Oil shale retort outgas 0.5-2 ppm [28]. U.S. acrylic fiber mfg plant 0.1 g/L [17]. Louisville, KY - chemical and latex manufacturing plant - detected, not quantified [17].

Sediment/Soil Concentrations: Not detected in all but one soil sample at 11 industrial sites or in sediment at four industrial sites in the United States [12]. The one positive soil sample was at the detection limit of 0.5 ppb [12]. United States - 351 samples from STORET database - ND [30].

Atmospheric Concentrations: SOURCE AREAS: 12 sites in United States (43 samples) 970 ppt median, range 46-110000 ppt [3].

Food Survey Values:

Plant Concentrations:

Fish/Seafood Concentrations: STORET database - not detected [30].

Animal Concentrations:

Milk Concentrations:

Other Environmental Concentrations:

Probable Routes of Human Exposure: Besides occupational exposure, the public may be exposed by inhalation of acrylonitrile vapors in source areas, from acrylic fibers and plastics, cigarette smoke, auto exhaust, and from the burning of acrylonitrile-based polymers. It may be ingested in food into which the monomer has been leached from the food packaging material.

Average Daily Intake:

Occupational Exposures: The number of workers exposed to acrylonitrile according to a National Occupational Hazard Survey (NOHS) (1973-1974) was estimated as 374,345 [25]. The more recent National Occupational Exposure Survey statistically estimated that 51,153 workers were exposed to acrylonitrile [26]. Acrylic fiber plant where large fraction of workers complained of symptoms of illness contained 3-20 mg/m^3 [34]. Air in old thermosetting plastics molding plant where workers had adverse symptoms contained 1.4 mg/m^3 [34]. The general population living within 30 km of an industrial plant which manufactures or uses acrylonitrile may be exposed to atmospheric levels as high as 20 ug/m^3 [32].

Body Burdens:

REFERENCES

1. Barrows ME et al; Am Chem Soc Div Environ Chem 18: 345-6 (1978)
2. Bocek K; Experientia Suppl 23: 231-40 (1976)
3. Brodzinsky R, Singh HB; Volatile Organic Chemicals in the Atmosphere An Assess of Available Data p 198 SRI Contract 68-02-3452 (1982)
4. Cadena F et al; J Water Pollut Control Fed 56: 460-3 (1984)
5. Chemical Marketing Reporter; 3/17/86 p 58 (1986)
6. Cherry AB et al; Sewage Indust Wastes 28: 1137-46 (1956)
7. Chou WL et al; Bioeng Symp 8: 391-414 (1979)
8. Coleman WE et al; pp 305-27 in Analysis and Identification of Org Substances in Water Keith L ed Ann Arbor MI Ann Arbor Sci (1976)
9. Daubert TE, Danner RP; Data Compilation Tables of Properties of Pure Compounds. Amer Instit Chemical Engr pp 450 (1985)
10. Dimitriades B, Joshi SB; pp 705-11 in Inter Conf on Photochem Oxidant Pollut and its Control Dimitriades Bed USEPA-600/3-77-001B (1977)
11. Edney E et al; Atmospheric Chem of Several Toxic Compounds USEPA-600/53-82-092 (1983)

Acrylonitrile

12. Going J et al; Environ Monitoring Near Industrial Sites Acrylonitrile USEPA-560/6-79-003 (1979)
13. Graedel TE; Chemical Compounds in the Atmosphere Academic Press NY p 284 (1978)
14. Grasselli J, Ritchey W; Atlas of Spectral Data and Physical Constants for Organic Compounds 2nd ed Vol 6 The Chemical Rubber Co (1975)
15. Hansch C, Leo AJ; Medchem Project Issue No 26. Claremont CA: Pomona College (1985)
16. Harris GW et al; Chem Phys Lett 80: 479-83 (1981)
17. IARC; Some Monomer Plastics and Synthetic Elastomers, and Acrolein 19: 73-86 (1979)
18. Kenaga EE; Ecotox Environ Safety 4: 26-38 (1980)
19. Krill RM, Sonzogni WC; J Water Pollut Control Assoc 78: 70-5 (1986)
20. Ludzack FJ et al; J Water Pollut Control Fed 33: 492-505 (1961)
21. Lyman WJ et al; Handbook of Chem Property Estimation Methods Environ Behavior of Org Compounds McGraw-Hill NY (1982)
22. Merck Index; An Encyclopedia of Chemicals, Drugs and Biologicals 10th ed p 127 (1983)
23. Mills EJ Jr, Stack VT Jr; Proc 8th Industrial Waste Conf Eng Bull Purdue Univ Eng Ext Ser pp 492-517 (1954)
24. Mills EJ et al; 9th Industrial Waste Conf Eng Bull of Purdue U 9: 449-64 (1955)
25. NIOSH; National Occupational Health Survey (1973)
26. NIOSH; National Occupational Exposure Survey (1984)
27. Sickles JE II et al; Proc Annu Meet Air Pollut Control Assoc 73rd Paper 80-50.1 (1980)
28. Sklarew DS, Hayes DJ; Environ Sci Technol 18: 600-3 (1984)
29. Spicer CW et al; Atmospheric Reaction Products from Hazardous Air Pollut Degradation pp 4 USEPA-600/S3-85-028 (1985)
30. Staples CA et al; Environ Toxicol Chem 4: 131-42 (1985)
31. Stover EL, Kincannon DF; J Water Pollut Control Fed 55: 97-109 (1983)
32. Suta BE; in Human Exposure to Atmospheric Concentrations of Selected Chemicals NTIS PB81-193278 (1980)
33. Tabak HH et al; J Water Pollut Control Fed 53: 1503-18 (1981)
34. USEPA; Ambient Water Quality Criteria Document for Acrylonitrile PB81-117285 (1980)
35. USEPA; Treatability Manual pp I.7.7-1 to I.7.7-5 USEPA-600/2-82-001a (1981)
36. Valvani SC et al; J Pharm Sci 70: 502-7 (1981)
37. Young RHF et al; J Water Pollut Control Fed 40: 354-68 (1968)

Adipic Acid

SUBSTANCE IDENTIFICATION

Synonyms: 1,6-Hexanedioic acid

Structure:

CAS Registry Number: 124-04-9

Molecular Formula: $C_6H_{10}O_4$

Wiswesser Line Notation: QV4VQ

CHEMICAL AND PHYSICAL PROPERTIES

Boiling Point: 337.5 °C at 760 mm Hg

Melting Point: 152 °C

Molecular Weight: 146.14

Dissociation Constants: pKa1 = 4.44, pKa2 = 5.4 [16]

Log Octanol/Water Partition Coefficient: 0.08 [9]

Water Solubility: 15,000 mg/L at 15 °C [20]

Vapor Pressure: 0.073 mm Hg at 18.5 °C [4]

Henry's Law Constant: 9.4 x 10^{-7} atm-m^3/mole (approximate-calculated from water solubility and vapor pressure)

Adipic Acid

ENVIRONMENTAL FATE/EXPOSURE POTENTIAL

Summary: Adipic acid is produced in large quantities and may be released into wastewater during its production and use in the manufacture of polymer products. Adipic acid is also released into the atmosphere from motor exhaust and as a result of photooxidation of precursor molecules that are commonly present in urban air. As a constituent of tobacco smoke, it would be emitted by smokers. If released on land, adipic acid will leach into the ground and probably biodegrade. If released into water, adipic acid will readily biodegrade (half-life 3.5 days). Adsorption to sediment, bioconcentration in aquatic organisms, and volatilization should not be significant. It will be primarily associated with aerosols in the atmosphere and be subject to gravitational settling; any vapor phase adipic acid will be subject to degradation by reaction with photochemically produced hydroxyl radicals (vapor phase half-life 4.4 days). The general population is exposed to adipic acid in aerosols from auto exhaust and due to the photooxidation of precursor molecules that are present in urban air. Occupational exposure would be via dermal contact and inhalation of aerosols containing adipic acid.

Natural Sources:

Artificial Sources: Adipic acid is produced in large quantities and may be released into wastewater during its production and use in the manufacture of synthetic fibers, plasticizers, resins, plastics, and foams [2]. Adipic acid is also released into the atmosphere as the result of photooxidation of cycloalkene and diolefin precursors [3]. Motor exhaust is the most important primary source in urban areas, although it is not clear what the relative contribution of photochemical production is compared to primary production in these urban areas [10]. Adipic acid is a chemical constituent of tobacco smoke [8] and this may be an important source indoors.

Terrestrial Fate: If released on land, adipic acid will leach into the ground and probably biodegrade. While adipic acid is readily biodegraded, no degradability data were found for soil systems.

Adipic Acid

Aquatic Fate: If released into water, adipic acid will readily biodegrade (half-life 3.5 days). Adsorption to sediment and volatilization should not be significant.

Atmospheric Fate: Due to its polar nature, adipic acid released into the atmosphere will be primarily associated with aerosols and subject to gravitational settling. Any vapor phase adipic acid will also degrade by reaction with photochemically produced hydroxyl radicals (vapor phase half-life 4.4 days).

Biodegradation: Adipic acid is readily degradable in biodegradability screening tests [5,7,19,22,23]. In four of the tests that were designed as models for degradability in surface water, the results ranged from 92% of theoretical BOD in 14 days to 83% in 30 days [7,23]. A test designed to simulate degradation in polluted river water, the AFNOR test, gave a 5-day BOD of 36% of theoretical [5]. In 5 tests designed to simulate treatment plants, results ranged from 99% DOC removal in one day to 92% of theoretical BOD in 14 days [7,23,22]. A screening procedure that was systematically applied to a large number of organic chemicals typified adipic acid as being "completely biodegraded in a short time by general microorganism" [19]. After a 5-10 hr lag, 50-75% of theoretical BOD was obtained in 90-100 hr [19]. Adipic acid was rapidly degraded in a river die-away test using Main River (Germany) water; 50% and 90% degradation being achieved in 3.5 and 7 days, respectively, at concn levels of 700 mg/L [22]. While no biodegradation studies were found in soil, adipic acid is readily degraded by soil microorganisms [1,17,18]. Under anaerobic conditions, it was 82% degraded after a 10-day lag in a screening test [2]. It is degraded in an anaerobic reactor employing acetate-enriched cultures and a 20-day hydraulic retention time with 67% utilization after 90 days of acclimation [2].

Abiotic Degradation: The pKa values for adipic acid indicate that it will exist largely in the dissociated form in the environment and form salts with cations. The vapor should react with photochemically produced hydroxyl radicals in the atmosphere by H-atom abstraction with a resulting half-life of 4.4 days [6].

34

Adipic Acid

Bioconcentration: The log octanol/water partition coefficient suggests that the potential for bioconcentration in fish is negligible [12].

Soil Adsorption/Mobility: Adipic acid is extremely soluble in water and therefore would not adsorb appreciably to soil [12].

Volatilization from Water/Soil: For molecules with such low Henry's Law constants, volatilization will be slow with the rate being controlled by slow diffusion through air [12]. The half-life for volatilization from a model river 1 m deep with a 1 m/sec current and a 3 m/sec wind is 47 days [12].

Water Concentrations:

Effluent Concentrations: Adipic acid was detected in effluents from advanced treatment plants in Pomona, CA and Lake Tahoe, CA [11]. Motor exhaust of two automobiles contained 1.07 and 4.75 ug/m^3 of adipic acid [10].

Sediment/Soil Concentrations: Soil samples taken at the Univ of California, Los Angeles campus, contained 215 and 568 ppb of adipic acid, whereas bog sediment samples from the Sierra Nevada foothills contained 2050 ppb [10].

Atmospheric Concentrations: The concn of adipic acid in nine 2-hr air samples taken on July 24, 1973, during a smog episode in West Covina, CA (30 km east of Los Angeles) ranged from 1.5 to 8.9 ug/m^3 [3]. The fact that the concn peaked in midafternoon, coincident with the peak in ozone concn, suggested that it was formed by photochemical reaction of cycloalkene or diolefin precursor [3]. The diurnal variation of adipic acid in Pasadena, CA was 0.04 to 0.78 ug/m^3 with the peak concentration occurring shortly before noon [15]. The concn of adipic acid in aerosol samples from West Los Angeles and Los Angeles were 12-484 and 29-92 ng/m^3, respectively [10]. Dust samples from these areas contained 11.4 and 5.90 ppm of adipic acid, respectively [10]. A suburban area of Japan had an adipic acid concn of 10 ng/m^3 [21].

Food Survey Values:

Adipic Acid

Plant Concentrations:

Fish/Seafood Concentrations:

Animal Concentrations:

Milk Concentrations:

Other Environmental Concentrations:

Probable Routes of Human Exposure: The general population is exposed to adipic acid in aerosols from auto exhaust and due to the photooxidation of precursor molecules that are present in urban air. Occupational exposure would be via dermal contact and inhalation of aerosols containing adipic acid.

Average Daily Intake: AIR INTAKE (assume air concn of 0.1 ug/m^3): 2 ug; WATER INTAKE insufficient data; FOOD INTAKE insufficient data.

Occupational Exposures: NIOSH has estimated that 14,851 workers are potentially exposed to adipic acid based on statistical estimates derived from a survey conducted in 1972-1974 in the U.S. [13]. This survey sampled 5000 businesses in 67 metropolitan areas throughout the United States for the manufacture and use of chemicals, trade-name products known to contain the compound, and generic products suspected of containing the compound. NIOSH has estimated that 113,667 workers are potentially exposed to adipic acid based on statistical estimates derived from a survey conducted in 1981-1983 in the U.S. [14].

Body Burdens:

REFERENCES

1. Anderson MS et al; J Gen Microbiol 120: 89-94 (3) (1980)
2. Chou WL et al; Biotechnol Bioeng Symp 8:391-414 (1979)
3. Cronn DR et al; Atmos Environ 11:929-37 (1977)
4. Danly DE, Campbell CR; Adipic Acid IN: Kirk-Othmer Encycl Chem Tech 3rd ed 1:510-31 (1978)
5. Dore M et al; Trib Cebedeau 28:3-11 (1975)

Adipic Acid

6. GEMS; Graphical Exposure Modeling System. FAP. Fate of Atmos Pollut (1986)
7. Gerike P, Fischer WK; Ecotox Environ Safety 3:159-73 (1979)
8. Graedel TE; Chemical Compounds in the Atmosphere pp 213 Academic Press NY (1978)
9. Hansch C, Leo AJ; Medchem Project Issue No 26. Claremont CA: Pomona College (1985)
10. Kawamura K, Kaplan IR; Environ Sci Technol 21:105-10 (1987)
11. Lucas SV; GC/MS Analysis of Organics In Drinking Water Concentrates and Advanced Waste Treatment Concentrates: VOL 1 Analysis Results for 17 Drinking Water, 16 Advanced Waste Treatment and 3 Process Blank Concentrates EPA-600/1-84-020A Columbus OH: Columbus Labs Health Eff Res LAB PP.321 (1984)
12. Lyman WJ et al; Handbook of Chem Property Estimation Methods. Environ Behavior of Organic Compounds. McGraw-Hill, NY (1982)
13. NIOSH; National Occupational Health Survey (1975)
14. NIOSH; National Occupational Exposure Survey (1985)
15. Schuetzie D et al; Environ Sci Technol 9: 838-45 (1975)
16. Serjeant EP, Dempsey B; Ionisation constants of organic acids in Aqueous Solution IUPAC Chemical Data Series No.23 New York NY: Pergamon Press. pp.989 (1979)
17. Tanaka H et al; Hakkokogaku Kaishi 55:57-61 (1977)
18. Trower MK et al; Appl Environ Microbiol 49:1282-9 (1985)
19. Urano K, Kato Z; J Hazardous Materials 13:147-59 (1986)
20. Verschueren K; Handbook of Environ Data on Organic Chemicals 2nd ed Van Nostrand Reinhold, NY p. 165 (1985)
21. Yokouchi Y, Ambe Y; Atmos Environ 20:1727-35 (1986)
22. Zahn R, Wellens H; Z Wasser Abwasser Forsch 13:1-7 (1980)
23. Zahn R, Huber W; Tenside Deterg 12:266-70 (1975)

Allyl Alcohol

SUBSTANCE IDENTIFICATION

Synonyms: 2-Propen-1-ol

Structure:

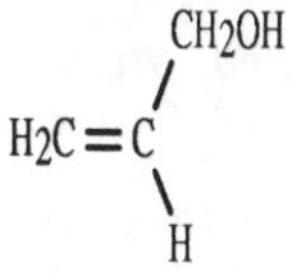

$$H_2C = C \begin{array}{c} CH_2OH \\ \\ H \end{array}$$

CAS Registry Number: 107-18-6

Molecular Formula: C_3H_6O

Wiswesser Line Notation: Q2U1

CHEMICAL AND PHYSICAL PROPERTIES

Boiling Point: 96-97 °C

Melting Point: -129 °C

Molecular Weight: 58.09

Dissociation Constants:

Log Octanol/Water Partition Coefficient: 0.17 [8]

Water Solubility: Miscible [21]

Vapor Pressure: 23.5 mm Hg at 25 °C [4]

Henry's Law Constant: 4.9 x 10^{-6} atm-m^3/mol at 25 °C [10]

ENVIRONMENTAL FATE/EXPOSURE POTENTIAL

Summary: Allyl alcohol is used as an intermediate in the preparation of a variety of substances and has been used as a

38

contact pesticide, although it is not known whether or not this is a current use. When released to water, allyl alcohol is not expected to volatilize, photooxidize, or directly photolyze. Biodegradation is expected to be the predominant fate of allyl alcohol in water. Release of allyl alcohol to soil is expected to result in biodegradation and possible migration to ground water. Volatilization from wet soil, direct photolysis, and bioconcentration are not expected to be significant. Volatilization from dry surfaces or soil should be significant. Release of allyl alcohol to the atmosphere is expected to result mainly in reaction with photochemically generated hydroxyl radicals with estimated half-lives of 6.03-14.7 hr. Direct photolysis is not expected to be significant. Due to the high water solubility of allyl alcohol, rainout may also occur. Allyl alcohol has been detected in human breath. Human exposure is expected to result mainly from the presence of allyl alcohol in exhaust from internal combustion engines, but some may result from use of the compound as a pesticide.

Natural Sources:

Artificial Sources: Allyl alcohol is used in the preparation of esters for use in resins and plasticizers, as an intermediate in the production of pharmaceuticals and other organic chemicals, in the manufacture of glycerol and acrolein, and in the production of military poison gas and herbicides [9,17]. Allyl alcohol has also been used as a contact pesticide for weed seeds and certain fungi [24]. Dow Chemical Company, however, has discontinued production of allyl alcohol, presumably for this pesticide purpose [16]. When used as a herbicide, allyl alcohol is used as 98% active ingredient [19]. Allyl alcohol has been detected, but not quantified, in exhaust gases from internal combustion engines [7].

Terrestrial Fate: Residue disappearance and leaching of ^{14}C-allyl alcohol from different soils was studied in laboratory experiments [23]. Residue disappearance is rapid (approx 90% in 10 days) and leaching of the chemical was fairly rapid (80-90% leached from soil column in 2 days) and was correlated negatively to the organic matter content [23]. Volatilization and direct photolysis are not expected to be significant.

Allyl Alcohol

Aquatic Fate: The most likely fate of allyl alcohol is expected to be biodegradation. Volatilization, direct photolysis, and photooxidation are all expected to be slow processes.

Atmospheric Fate: Using a 2nd-order rate constant of 25.9 cm^3/molecule-sec [2] and an average hydroxyl radical concentration of 5×10^{-5} molecules/cm^3 [20], a pseudo first-order rate constant of 1.3×10^{-5} sec^{-1} can be estimated. From this value, a half-life of 14.7 hr can be estimated. Using the Fate of Atmospheric Pollutants portion of GEMS, a half-life for the reaction of allyl alcohol with photochemically generated hydroxyl radicals of 6.03 hr was estimated [6]. The predominant fate of allyl alcohol in the atmosphere is therefore expected to be reaction with hydroxyl radicals. Due to the miscibility of allyl alcohol with water, rainout may also occur. Direct photolysis is not expected to be important, as allyl alcohol should not absorb significant amounts of radiation at >290 nm.

Biodegradation: Following incubation of 20 °C with settled sewage seed, 2.5 ppm of allyl alcohol had degraded to 9.1%, 55.0%, 78.2%, and 81.8% of the theoretical BOD after 5, 10, 15, and 20 days, respectively [13]. After 10 days exposure to a sewage seed at 20 °C, a BOD of 1.60 (ppm oxygen/ppm allyl alcohol) was observed. The theoretical BOD was 2.2 ppm oxygen/ppm allyl alcohol) [18]. In a 5-day BOD test, 81% of the theoretical oxygen demand was observed following incubation of allyl alcohol at 20 °C with a sewage seed [4]. Allyl alcohol has been confirmed to be easily biodegradable in the 14-day Japanese MITI screening biodegradability test [22].

Abiotic Degradation: When a 13.5 mg/L aqueous solvents of allyl alcohol reacted with hydroxyl radicals formed by the photolysis of hydrogen peroxide by light at >290 nm at a pH of 5.9 for 3 hr, 14.85% of the allyl alcohol degraded [15]. The bimolecular rate constant for the reaction between allyl alcohol and hydroxyl radicals in water is $1.2 \times 10^{+9}$ L/m-sec at pH 7 [1]. Using 1×10^{-17} M for an average aqueous concentration of the hydroxyl radical, a pseudo first-order rate constant of 1.2×10^{-8} sec^{-1} was estimated. A half-life of 1.8 years was calculated from this rate constant. Upon exposure to UV light (<290 nm) for 24 hr, 13.9% of the 171.0 mg allyl alcohol initially present in the vapor phase was degraded [12].

Allyl Alcohol

The allyl alcohol was placed into a 1 liter flask in which it was allowed to volatilize before being passed to a 20 liter flask in which it was irradiated with UV light. The temperature of the water circulating through the system reached 50 °C about six hours after the irradiation began [12]. Using a 2nd-order rate constant measured of 25.9 cm^3/molecule-sec at 167 °C [2], and an average hydroxyl radical concentration of 5 x 10^{-5} molecules/cm^3 [20], a pseudo first-order rate constant of 1.3 x 10^{-5} sec^{-1} was estimated. From this value, a half-life of 14.7 hr was estimated, but the reaction at ambient temperatures would be expected to be slower. Using the Fate of Atmospheric Pollutants portion of GEMS, a half-life of 6.03 hr for the reaction of allyl alcohol with photochemically generated hydroxyl radicals was estimated [6].

Bioconcentration: Allyl alcohol is not expected to bioconcentrate, based on its measured log octanol/water partition coefficient.

Soil Adsorption/Mobility: Using the measured log octanol/water partition coefficient, a soil sorption coefficient of 1.47 can be estimated [14]. A soil sorption coefficient of this magnitude suggests that allyl alcohol will not adsorb strongly to soil and may therefore leach to ground water [11].

Volatilization from Water/Soil: Based on the Henry's Law constant, only slow volatilization of allyl alcohol, if any, is expected to occur from water or moist soil.

Water Concentrations:

Effluent Concentrations: Allyl alcohol has been detected but not quantified in exhaust gases from internal combustion engines [7].

Sediment/Soil Concentrations:

Atmospheric Concentrations:

Food Survey Values:

Plant Concentrations:

Fish/Seafood Concentrations:

Allyl Alcohol

Animal Concentrations:

Milk Concentrations:

Other Environmental Concentrations:

Probable Routes of Human Exposure: The most probable route of exposure to allyl alcohol in humans is the inhalation of the compound released from internal combustion vehicles in the exhaust [7]. Other exposure may result from the use of allyl alcohol as a contact pesticide for weed seeds and certain fungi [24].

Average Daily Intake:

Occupational Exposures:

Body Burdens: Allyl alcohol was found at 0.52 and 9.5 ug/hr in expired air from a smoker and a non-smoker, respectively [5].

REFERENCES

1. Anbar M, Neta P; Int J Appl Rad Isot 18: 493-523 (1967)
2. Atkinson R et al; Adv Photochem 11: 375-488 (1979)
3. Boublik T et al; The Vapor Pressures of Pure Substances Vol 17 Amsterdam, Netherlands: Elsevier Science Publ (1984)
4. Bridie A et al; Water Res 13: 627 (1979)
5. Conkle JP et al; Arch Environ Health 30: 290-5 (1975)
6. Graphical Exposure Modeling System. FAP. Fate of Atmos Pollut (1986)
7. Hampton CV et al; Environ Sci Technol 16: 287-98 (1982)
8. Hansch C, Leo AJ; Medchem Project Issue No 26. Claremont CA: Pomona College (1985)
9. Hawley GG; Condensed Chem Dictionary 10th ed. Van Nostrand Reinhold NY p. 34 (1981)
10. Hine J, Mookerjee PK; J Org Chem 40: 292 (1975)
11. Kenaga EE; Ecotox Environ Safety 4: 26-38 (1980)
12. Knoevenagal K, Himmelreich R; Arch Environ Contam Toxicol 4: 324-33 (1976)
13. Lamb CB, Jenkins GF; Proc 8th Indus Waste Conf Purdue Univ pp. 329-9 (1952)
14. Lyman WJ et al; Handbook of Chemical Property Estimation Methods. Environmental Behavior of Organic Compounds. McGraw-Hill NY (1982)
15. Mansour M; Bull Environ Contam Toxicol 34: 89-95 (1985)
16. Meister RT, ed; Farm Chemicals Handbook p. C-13 (1987)

Allyl Alcohol

17. Merck Index; An Encyclopedia of Chemicals, Drugs and Biologicals 10th ed. p. 44 (1983)
18. Mills EJ, Stack VT; Proc 8th Indus Waste Conf Purdue Univ Extension Series 83: 492-517 (1954)
19. National Pesticide Information Retrieval System, Purdue Univ (1987)
20. Neely WB; Chemicals in the Environment Marcel Dekker, Inc. New York (1980)
21. Riddick JA et al; Organic Solvents John Wiley, New York p. 256 (1986)
22. Sasaki S; pp. 283-98 in Aquatic Pollut Transformation and Biological Effects. Hutzinger O, et al; eds., Oxford Pergamon Press pp. 283-98 (1978)
23. Scheunert I et al; J Environ Sci Health, Part B Pestic Food Contam Agric Wastes 16 (6): 719-42 (1981)
24. Verschueren K; Handbook of Environ Data on Organic Chemicals. 2nd ed Van Nostrand Reinhold NY p. 176 (1983)

Aniline

Synonyms: Benzenamine

Structure:

CAS Registry Number: 62-53-3

Molecular Formula: C_6H_7N

Wiswesser Line Notation: ZR

CHEMICAL AND PHYSICAL PROPERTIES

Boiling Point: 184-186 °C

Melting Point: -6.3 °C

Molecular Weight: 93.12

Dissociation Constants: 4.596 [55]

Log Octanol/Water Partition Coefficient: 0.90 [28]

Water Solubility: 36,070 mg/L at 25 °C [60]

Vapor Pressure: 0.489 mm Hg at 25 °C [13]

Henry's Law Constant: 0.136 atm-m³/mole at pH 7.3 [26]

Aniline

ENVIRONMENTAL FATE/EXPOSURE POTENTIAL

Summary: Aniline is produced in large quantities and will be released to the environment primarily in wastewater from its manufacture and use in the production of polyurethanes, rubber, pesticides, dyes, etc. If released into water, it will primarily be lost due to biodegradation and in surface waters, photooxidation (half-life of the order of days). It will not bioconcentrate in fish. If spilled on land, it will be lost by a combination of biodegradation, oxidation, and chemical binding to components of soil. In cases where it leaches into ground water, it will probably biodegrade, although slowly. If released into air, aniline will photodegrade (estimated half-life 3.3 hr). Human exposure will be primarily in the workplace.

Natural Sources: Aniline has not been reported to occur, as such, in nature [32].

Artificial Sources: Aniline is used in the manufacture of polyurethanes (MDI), rubber processing chemicals, pesticides, fibers, dyes and pigments, hydroquinone, pharmaceuticals, etc. [11]. Wastewater and emissions from its manufacture, transport, storage, and use are likely sources of environmental release.

Terrestrial Fate: If released on land, aniline will exhibit low to moderate sorption to soils with the sorption being stronger at lower pH. Binding to humic materials results in covalent bond formation and slow oxidation. It will also sorb to clay minerals and, again, the sorption will be stronger under acidic conditions. However, its sorption to colloidal organic matter is extremely high and may increase its rate of leaching into ground water. Aniline is readily biodegraded under aerobic conditions, especially after a short acclimation period, and substantial loss can be expected by this means. Aniline has only been reported in ground water associated with wastes.

Aquatic Fate: If released into water, aniline will extensively biodegrade, photodegrade, and to some extent adsorb to sediment and humic materials, especially in more acidic situations. Biodegradation rates most appropriate for natural systems include a half-life of 6 days in a eutrophic pond, as well as 75-90%

mineralization in 21 days in an oligotrophic lake. Photodegradation will occur in surface waters with estimated half-lives ranging from hours to weeks. The only reported half-life in a natural aquatic system which included all loss processes was 2.3 days in an industrial river [76]. Although aniline does not bioconcentrate in fish, it is taken up and metabolized in fish [17,39].

Atmospheric Fate: If released into the atmosphere, aniline will degrade primarily by reaction with photochemically produced hydroxyl radicals (estimated half-life 3.3 hr).

Biodegradation: Aniline is degraded by many common species of bacteria and fungi found in soil [2,36,61] and acetanilide, 2-hydroxyacetanilide, 4-hydroxyaniline and catechol are reported metabolites [61,66,68]. Aniline is a benchmark chemical for aerobic biodegradability tests and there are abundant data on its biodegradation. Degradation is frequently 90-100% in laboratory tests utilizing activated sludge or sewage seed lasting from 3 to 28 days with acclimation not always being required [5,6,9,19,21,23,27,30,34,37,43,44,57,59,72]. It is completely degraded by a soil inoculum in 4 days [1] and by bacteria in river mud in 20 days [10]. In an oligotrophic lake water sample, 75-99% mineralization occurred in 21 days [65], and 40-60% of the aniline degraded in 1 day in river water and seawater when incubated at 30 °C [36]. Degradation occurred in Nile River water after a 3-day lag [15]. The half-life in the Rhine River was 2.30 days as determined by concentration reduction between sampling points [76]. Biodegradation removed 56% of anions in a week and was the most significant removal process for aniline in water from a eutrophic pond, with a substantial portion being mineralized to CO_2 within this 1-week period [9]. Amendment of the pond with sewage sludge greatly increased the rate of degradation. The major pathway of biodegradation involved oxidative deamination to catechol, which was further metabolized to CO_2 whereas minor pathways involved reversible acylation to form acetanilide and anilide [41]. The rate of degradation in four soils reached a maximum value after one week and declined to a low value after 2 weeks [67]. After 10 weeks, 16-26% of the aniline was mineralized [67]. However, approx 60% of the aniline was bound to the soil and probably unavailable [67]. In <1 yr, 39% of aniline in Rhine River water was removed by bank filtration [76], which would involve both biodegradation and adsorption. Although aniline is amenable to degradation in anaerobic reactors [63], only 10% degradation was reported in 53 days, including a 28-day lag with a sewage inoculum forming acetanilide and 2-methylquinoline as products [27], and no degradation occurred in an anaerobic reactor in 110 days with a 2 to 10 day retention time using an inoculum maintained on acetate [12]. Aniline was completely degraded in 7 days using composting [62].

Abiotic Degradation: Aniline has a strong UV adsorption band at 285 nm which extends above 290 nm [42] and is therefore a candidate for direct photolysis by sunlight. When aniline in water is irradiated in July sunlight, 19.3% degradation occurred in 5 hr; however, the rate decreased with increasing concentration [36]. Other investigators report only 0.7% oxidant formation after 4 hr exposure to sunlight [14], and a half-life of 0.2 hr during exposure to light >290 nm in the laboratory [38]. Humic acids photosensitize the reaction of aniline in water. The half-life in distilled water was 1 week, whereas it was 4-8 hr in Georgia's May sunshine in the near-surface black water from the Aucilla River [75]. The presence of various species of algae can increase the photodegradation rates by factors of 4-50 over that in distilled water [74]. When adsorbed on silica gel, 46.5% mineralization occurs in 17 hr upon irradiation with light >290 nm [19]. Aniline is oxidized on exposure to sunlight in air, forming products such as hydrazobenzene, 4-aminodiphenylamine, 2-aminodiphenylamine, benzidine, and azobenzene [73,75]. Variability in results can be a result of aeration of the water as well as concentration since shaking increases the rate of degradation [36]. In the presence of nitrite ions, photolytic products formed are mutagenic using the Ames assay [69]. In the atmosphere, aniline inhibits the conversion of NO to NO_2 and photochemical smog formation [24,50]. The half-life for direct photolysis of aniline in the atmosphere has been estimated to be 2.1 days based on a measured reaction rate constant of 0.32 day^{-1} [46]. The half-life for aniline vapor reacting with photochemically generated hydroxyl radicals in the atmosphere has been estimated to be 3.3 hr, based on a measured reaction rate constant of 1.17×10^{-10} cm^3-molecule/mole at 25 °C and assuming an avg hydroxyl radical concn of $5 \times 10^{+5}$ molecules/cm^3 [3]. Aromatic amines are resistant to hydrolysis [40]; however, they are sensitive to oxidation by air at room temperature [35] and may be oxidized by metal cations and humic acid in the environment. A rapid, reversible equilibrium occurs with humates or humate analogs followed by addition to quinoidal structures and oxidation [53]. Soil catalyzed oxidation occurs in sterilized soil with 9 products being detected after 2 days, including azobenzene, azoxybenzene, phenazine, formanilide and acetanilide [56]. The pKa of aniline is 4.596 [55], which means it may be partially ionized at environmental pHs and some reactions may be sensitive to the pH of the ambient medium.

Bioconcentration: The log BCF for aniline in two species of fish are 0.78 [39] and <1.0 [19], which demonstrates that aniline does not bioconcentrate in fish. The log BCF for algae is 0.60 [19,22].

Soil Adsorption/Mobility: Aniline has a pKa value of 4.596 [55] and, therefore, exists partially as a cation. It is not surprising, therefore, that the soil adsorption is not only a function of percent organic carbon but also the pH of the soil, increasing with the percentage organic carbon and decreasing with pH [47]. Additionally, adsorption is correlated with the clay content of the soil [47]. For 14 soils, the soil/water partition coefficient (Kd) averaged 0.074 and the Koc value averaged 1.86 [47]. The Koc values in two silt loams were 130 and 410, with the higher value occurring in the more acidic soil [56], and averaged 25.5 in seven agricultural soils [7]. Its Koc to colloidal organic carbon from ground water is a relatively high 3900, and this effectively increases aniline solubility and leaching into ground water [43]. The adsorption constant for adsorption to H-montmorillonite (pH 8.35) and Na-montmorillonite (pH 6.8) is 1300 and 130 [4]. Aromatic amines, including aniline, are known to form covalent bonds with humic materials, adding to quinone-like structure followed by slow oxidation [20,53]. In a generic aquatic-terrestrial environment, 0.26% was distributed in soil and 1.23% in sediment [71]. The sediment-water distribution constants ranged from approx 3 to 900, which is higher than predicted [71] and possibly reflects a component of chemical binding.

Volatilization from Water/Soil: Using a measured Henry's Law constant of 1.2×10^{-4} atm-m^3/mole [71], the estimated half-life for evaporation of aniline from a model river 1 m deep with a 1 m/sec current and a 3 m/sec wind is 12 days [40]. When an environmental test system containing soil, plants, simulated photoperiods, rain, and wind was dosed with ^{14}C labeled aniline and left for 30 days, 24% of the radioactivity was found in the air compartment, of which 5% was identified as CO_2 [17]. The remainder was either aniline or volatile metabolic products. In an OECD generic aquatic-terrestrial environment, 14% of applied aniline partitioned to the air phase [71].

Water Concentrations: GROUND WATER: Detected not quantified in St. Louis Park, MN - shallow aquifer contaminated by coal-tar wastes [54]. Hoe Creek underground coal gasification site, Wyoming - 15 mo after gasification completed 0 and 36 ppb in two aquifers [64] SURFACE WATER: Rhine River and two tributaries (The Netherlands) 2.0-12 ppb avg, 5.5-12 ppb max, 100% frequency of detection [70]. Meuse River (The Netherlands) 0.62-0.77 ppb avg, 1.6-2.4 ppb max, 100% frequency of detection [70]. Detected in Rhine delta water but not in surface water from agricultural areas in the Netherlands [25] and the Waal River at Brackel [45], Germany, nine surface waters 0-3.7 ppb [49].

Effluent Concentrations: Shale oil wastewater 0.48-14 [29]. Identified, not quantitated, in trench leachate from Maxey Flats and West Valley radioactive waste disposal sites [18]. Secondary effluents from 10 municipal and industrial wastewater treatment plants discharging into Illinois rivers - detected, not quantified in 1 of 10 plants; a composite sampler from Sauget POTW, which receives wastes from a heavy chemical plant and metal alloy and tubing manufacturing [16].

Sediment/Soil Concentrations: Not detected in three samples of bottom sediment from the Buffalo River, Buffalo, NY [48] or in sediment from an industrial location on an unspecified U.S. river [33]. Microgram quantities in soil samples in the Federal Republic of Germany were attributed to industrial waste [31]. Soil near Buffalo River, NY - 5 ppm, one of two samples positive [48].

Atmospheric Concentrations: URBAN/SUBURBAN: Raleigh, NC (one sample) - 90 ppb, however data quality considered questionable [8]. Detected, not quantified in air 4000 m from a chemical factory [32].

Food Survey Values: Fresh fruits, vegetables, salad, and maize (Germany) - 14 of 16 types positive at 0.6-30.9 ppm [49]. Preserved vegetables (Germany, 10 samples) <0.1 ppm [49]. Animal feed (Germany, rapeseed cake), 120 ppm, two of four samples positive [49]. Volatile component of black tea [31].

Plant Concentrations:

Fish/Seafood Concentrations:

Animal Concentrations:

Milk Concentrations:

Other Environmental Concentrations: Tobacco smoke - 50.6-577 ng/cigarette [31]. Swine manure - 8.2 ug/L in the raw liquid [31].

Probable Routes of Human Exposure:

Average Daily Intake:

Occupational Exposures: Oil shale wastewater facility, indoors 33 ug/m^3, outdoors 5 ug/m^3 [29]. Foundry work atmospheres, 0.004-1.8 mg/m^3 [58]. NIOSH (NOES Survey 1981-1983) has statistically estimated that 19,276 workers are potentially exposed to aniline in the United States [51]. NIOSH (NOHS Survey 1972-1974) has statistically estimated that 852,757 workers are potentially exposed to aniline in the United States [52].

Body Burdens:

REFERENCES

1. Alexander M, Lustigman BK; J Agric Food Chem 14:410-3 (1966)
2. Aoki K et al; Agric Biol Chem 46:2563-71 (1982)
3. Atkinson R; Inter J Chem Kinet 19: 799-828 (1987)
4. Bailey GW et al; Soil Sci Soc Amer Proc 32:222-34 (1968)
5. Baird R et al; J Water Pollut Control Fed 49:1609-15 (1977)

Aniline

6. Bilyk A et al; Pr Nauk Inst Inz Sanit Wodnej Politech Wroclaw pp. 3-32 (1971)
7. Briggs GG: J Agric Food Chem 29:1050-9 (1981)
8. Brodzinsky R, Singh HB; Volatile organic chemicals in the atmosphere: an assessment of available data; pp.198 SRI contract 68-02-3452 (1982)
9. Brown D, Laboureur P; Chemosphere 12:405-14 (1983)
10. Calamari D et al; Chemosphere 9:753-62 (1980)
11. Chemical Marketing Reporter; Chemical Profile; 7 Sept (1985)
12. Chou WL et al; Biotech Bioeng Symp 8:391-414 (1979)
13. Daubert TE, Danner RP; Data compilation tables of properties of pure compounds. Amer Inst Chem Engr pp450 (1985)
14. Draper WM, Crosby DG; Arch Environ Contam Toxicol 12:121-6 (1983)
15. El-Dib MA, Aly OA; Water Res 10:1055-9 (1976)
16. Ellis DD et al; Arch Environ Contam Toxicol 11:373-82 (1982)
17. Figge K et al; Regul Toxicol Pharmacol 3:199-215 (1983)
18. Francis AJ et al; Nuclear Tech 50:158-63 (1980)
19. Freitag D et al; Ecotox Environ Safety 6:60-81 (1982)
20. Furukawa T, Brindley GW; Clays Clay Miner 21:279-88 (1973)
21. Gerike P, Fischer WK; Ecotox Environ Safety 3:159-73 (1979)
22. Geyer H et al; Chemosphere 10:1307-13 (1981)
23. Gilbert PA, Lee CM; pp. 34-45 in Biotransformation and fate of chemicals in the aquatic environment; Maki AW et al eds; Univ Mich, Pellston, MI (1980)
24. Gitchell A et al; J Air Pollut Control Assoc 24:357-61 (1974)
25. Greve PA, Wegman RCC; Schriftenr Ver Wasser, Boden-, Lufthyg. Berlin-Dohlem 46: 59-80 (1975)
26. Hakuta T et al; Desalination 21: 11-21 (1977)
27. Hallas LE, Alexander M; Appl Environ Microbiol 45:1234-41 (1983)
28. Hansch C, Leo AJ; Medchem Project Claremont, CA: Pomona College (1985)
29. Hawthorne SB, Sievers RE; Environ Sci Technol 18:483-90 (1984)
30. Helfgott TB et al; An index of refractory organics USEPA-600/2-77-174 (1977)
31. IARC: Some aromatic amines, anthraquinones and nitroso compounds, and inorganic fluorides used in drinking-water and dental preparations; 27:39-61 (1982)
32. IARC; Some aromatic amines, hydrazine and related substances, N-nitroso compounds and miscellaneous alkylating agents; 4:27-39 (1974)
33. Jungclaus CA et al; Environ Sci Technol 12:88-96 (1978)
34. Kawasaki M; Ecotox Environ Safety 4:444-54 (1980)
35. Kirk-Othmer Encycl Chem Tech, 3rd ed.; 2:309-21 (1978)
36. Kondo M; Simulation studies of degradation of chemicals in the environment: simulation studies of degradation of chemicals in the water and soil; Environment agency. Office of health studies Japan (1978)
37. Korte F, Klein W; Ecotox Environ Safety 6:311-27 (1982)
38. Kotzias D et al; Naturwissenschaften 69:444-5 (1982)
39. Lu PY, Metcalf RL; Environ Health Perspect 10:269-84 (1975)
40. Lyman WJ et al; Handbook of chemical property estimation methods. Environmental behavior of organic compounds. McGraw-Hill New York (1982)
41. Lyons CD et al; Appl Environ Microbiol 48:491-6 (1984)

Aniline

42. Meallier P et al; Ann Chim 4:15-28 (1969)
43. Means JC; Amer Chem Soc 186th Natl Mtg 23:250-1 (1982)
44. Means JL, Anderson SJ; Water Air Soil Pollut 16:301-15 (1981)
45. Meyers AP, Vanderleer RC; Water Res 10: 597-604 (1976)
46. Mill T, Davenport J; ACS Div Environ Chem 192nd Natl Mtg 26: 59-63 (1986)
47. Moreale A, Van Bladel R; J Soil Sci 27:48-57 (1976)
48. Nelson CR, Hites RA; Environ Sci Technol 14:1147-9 (1980)
49. Neurath GB et al; Food Cosmet Toxicol 15:275-82 (1977)
50. NIOSH; National Occupational Exposure Survey (NOES) (1983)
51. NIOSH; National Occupational Hazard Survey (NOHS) (1974)
52. Nguyen YV, Phillips CR; Chemosphere 4:125-30 (1975)
53. Parris GE; Environ Sci Technol 14:1099-105 (1980)
54. Pereira WE et al; Environ Toxicol Chem 2: 283-94 (1983)
55. Perrin DD; Dissociation constants of organic bases in aqueous solution; IUPAC Chemical Data Series; Supplement; London Buttersworth (1972)
56. Pillai P et al; Chemosphere 11:299-317 (1982)
57. Pitter P; Water Res 10:231-5 (1976)
58. Renman L et al; Am Ind Hyg Assoc J 47: 621-8 (1986)
59. Schefer W, Walechli O; Z Wasser Abwasser Forch 13:205-9 (1980)
60. Seidell A; Solubilities of Organic Compounds. NY, NY: Van Nostrand Co Inc (1941)
61. Smith RV, Rosazza JP; Arch Biochem Biophys 161:551-8 (1974)
62. Snell Environmental Group; Rate of biodegradation of toxic organic compounds while in contact with organics which are actually composting; pp. 100 NSF/CEE-82024 (1982)
63. Speece Re; Environ Sci Technol 2:557-8 (1983)
64. Stuermer DH et al; Environ Sci Technol 16: 582-7 (1982)
65. Subba-Rao RV et al; Appl Environ Microbiol 43:1139-50 (1982)
66. Subramanian V et al; Indian Inst Sci J 60:143-78 (1978)
67. Suess A et al; Bayerisches Landwireschaftersches Jahrbuch 55:565-70 (1978)
68. Surovtseva EG et al; Dokl Akad Nauk SSSR 237:220-3 (1977)
69. Suzuki J et al; Bull Environ Contam Toxicol 31:79-84 (1983)
70. Wegman RCC, DeKorte GAL; Water Res 15: 391-4 (1981)
71. Yoshida K et al; Ecotox Environ Safety 7:179-90 (1983)
72. Zahn R, Wellens H; Z Wasser Abwasser Forsch 13:1-7 (1980)
73. Zechner J et al; Z Phys Chem 102:137-50 (1976)
74. Zepp RG, Schlotzhauer PF; Environ Sci Technol 17:462-8 (1983)
75. Zepp RG et al; Chemosphere 10:109-17 (1981)
76. Zoeteman BCF et al; Chemosphere 9:231-49 (1980)

Benzidine

SUBSTANCE IDENTIFICATION

Synonyms: (1,1'-Biphenyl)-4,4'-diamine

Structure:

$$H_2N-\bigcirc-\bigcirc-NH_2$$

CAS Registry Number: 92-87-5

Molecular Formula: $C_{12}H_{12}N_2$

Wiswesser Line Notation: ZR DR DZ

CHEMICAL AND PHYSICAL PROPERTIES

Boiling Point: about 400 °C

Melting Point: 116-129 °C

Molecular Weight: 184.23

Dissociation Constants: 4.66 (pK1); 3.57 (pK2) [21]

Log Octanol/Water Partition Coefficient: 1.34 [9]

Water Solubility: 520 mg/L at 25 °C [25]

Vapor Pressure:

Henry's Law Constant: 0.388 x 10^{-10} atm-m^3/mole (estimated) [10]

Benzidine

ENVIRONMENTAL FATE/EXPOSURE POTENTIAL

Summary: Benzidine may be released as emissions and in wastewater during its production and use in the manufacture of azo dyes or may be formed during the degradation of benzidine-based dyes which have been discharged in waste water. If spilled on soil, it will adsorb to it, especially if the soil is acidic, form complexes with clay particles, and be oxidized by metal cations. The rate of degradation in soil in the few studies reported in the literature; 79% degradation in 4 weeks and 10% mineralization in 1 yr. If released in water, it will rapidly adsorb to suspended clay particles, and be oxidized by naturally occurring metal cations such as Fe(III). It will also be lost by reaction with radicals and photolysis. Its half-life in water is approx 1 day. It will adsorb to sediments and bioconcentrate only moderately in fish. In the atmosphere, it would primarily exist in aerosols, be bound to particulate matter, and be subject to gravitational settling and washout. It may photolyze and would be readily oxidized by reactive species in the atmosphere such as hydroxyl radicals. Human exposure to benzidine would be primarily occupational.

Natural Sources: Benzidine has not been reported to occur in nature [12].

Artificial Sources: Benzidine may be released as emissions and in wastewater during its production and use as an intermediate in the manufacture of direct azo dyes [12]. However, it is now only produced in the United States for captive consumption with strict regulations that it be maintained in isolated or closed systems which would limit its release [12]. Another source is the conversion of benzidine-based dyes to benzidine in streams into which these dye wastes have been discharged [12].

Terrestrial Fate: If released on land, benzidine will both sorb to soil and react with natural cations such as Fe(III) in the soil. Benzidine adsorbs more strongly at lower pH's when larger fractions of the compound are ionized. Its reaction with clay minerals is particularly striking, since a blue complex is formed. Only a few determinations of its persistence in soil could be found, 79% degradation in 4 weeks in a silty clay loam and 10% mineralization in 1 yr.

Benzidine

Aquatic Fate: When released into water, benzidine will completely degrade in approx 1 day due to reaction with radicals, redox reactions with naturally occurring cations, and perhaps photodegradation. Degradation should occur more rapidly in humic waters because of the presence of reactive radicals, cations, and molecules; however, measurements of degradation rates in different water types are lacking.

Atmospheric Fate: Should benzidine be released into the atmosphere, it will most likely be in the form of aerosols or adsorbed to particulate matter and be subject to washout by rain and gravitational settling. Although there are no experimental data on persistence in the atmosphere, it may photolyze and would react in the vapor phase with reactive atmospheric species such as hydroxyl radicals and ozone with an estimated half-life of approx 1 day.

Biodegradation: Although early investigators reported that benzidine was not degradable [16], its resistance to biodegradation was probably due to its toxicity to microorganisms at the high concentrations used. No mineralization occurred when 0.05 ppm of benzidine was incubated for 5 days with activated sludge [5]. Several workers report almost complete degradation in a few weeks in simulated treatment plants using activated sludge or a sewage inoculum, although the percent removal falls off at higher concentrations [1,11,27]. Degradation is due in part to air oxidation [11]. There are conflicting data over the need for acclimation of microorganisms [1,11,27]. No information could be found relating to degradation in natural waters. Benzidine was reported to be stable for 9 weeks to degradation by soil bacteria [31]. In Drummer silty clay loam, 79% of benzidine was lost in 4 weeks [15]. When applied to a silt loam soil (pH 5.4) and incubated for 45 days, 3.3% of the benzidine was mineralized [7]. After 365 days, 8.13-11.6% of benzidine had mineralized in 4 soils, pH 5.2-7.8 [7]. Amendment of the soil with glucose or alfalfa forage had no effect on the decomposition rate [7]. The low decomposition rate, compared with that in simulated treatment plants, suggests that degradation may be controlled by the soil environment or adsorption to soil [7]. However, when benzidine in sludge was applied to a sandy loam soil in a biological soil reactor and worked into the top 20 cm of soil, 42% degraded in 48 days and 55% degraded in 97 days [14]. The overall half-life was 76 days [14].

Abiotic Degradation: Benzidine absorbs light above 290 nm [22] and is therefore a candidate for direct photolysis. Its half-life when irradiated in methanol at 254 nm is 2 hr [15]; however, its quantum yield is the same at 254 and 300 nm in aqueous solution, 0.012 [2]. Its half-life for photodegradation in water was not determined [2]. When benzidine was adsorbed on silica gel and irradiated for 17 hr with radiation >290 nm, 40.8% degradation occurred [5]. Benzidine (1-70 ppb) is completely removed from undisturbed, aerated, or chlorinated lake water in 24 hr [1]. When aerated solutions of benzidine (100 ppb) were exposed to a xenon lamp (300-400 nm), it completely degraded in 12 hr [1]. The principal reactions taking place are believed to be oxidative degradation by free radicals [25]. Benzidine is very rapidly oxidized by Fe(III) and several other natural cations which are found in environmental waters, complexes of fulvic acids, and in clay minerals [4]. Little is known about the products of degradation [4]. Hydrolysis is probably not an important process [4]. Benzidine is predicted to react with ozone and hydroxyl radicals in air with an estimated half-life of 1 day [25].

Bioconcentration: In a 42-day experiment in a flow-through tank in which bluegills were exposed to ^{14}C-benzidine, the log BCF was 1.6 in the edible portion of the fish [29]. The depuration half-life of the ^{14}C-residues was about 7 days [29]. After 3 days in a model ecosystem, the log BCF for fish, mosquitos, snail, and algae were 1.74, 2.66, 2.81, and 3.4, respectively [15].

Soil Adsorption/Mobility: Benzidine exists as the neutral molecule, and as a singly and doubly ionized cation at environmental pH's (ionization constants pK1 and pK2 are 4.3 and 3.3 respectively) [33]. It is not surprising, therefore, that adsorption to soil should be sensitive to the soil pH. In a study of the adsorption of benzidine to 14 soils and sediments, it was found that the Freundlich adsorption constant was not correlated with percentage of organic carbon, but rather with soil pH [33]. The pH controls the amount of benzidine in the ionized form and sorption increases as the pH decreases, that is, as a greater fraction of the total benzidine occurs in the ionic form. The adsorption of the ionized species was highly correlated with soil surface area, while that of the neutral form was correlated with soil organic matter. The Freundlich adsorption constant for the 14 soils and sediments ranged from 50 to 3940 on a molar basis. The adsorption curve is highly non-linear with the average value of $1/n$ in the Freundlich adsorption equation being 0.5 [33]. In another study with 4 soils, the Freundlich adsorption constants ranged from 7600 to 21,000, and the mean $1/n$ was 0.768 [7]. Even though benzidine was largely present as the neutral species and the sorption was correlated with organic carbon (K_{oc} ranged from 227,000 to 882,000), the K_{oc} values are greater than can be accounted for by hydrophobic sorption [7]. Similar results were obtained in a study of the adsorption of benzidine to 3 estuarine sediments containing 0.93, 1.36, and 3.01% organic carbon [13]. K_{oc} values obtained were 4899, 462, and 3307 for the three sediments, in order of organic carbon content [13]. Additionally, the distribution coefficients were inversely related to the salinity of the water [13]. For adsorption to Chesapeake Bay sediment, the Freundlich adsorption constant was 6025, and $1/n$ was 0.75 at pH 7.9 [17]. Adsorption increased as pH decreased, indicating that the protonated form of benzidine was more strongly bound to colloids than the neutral form [17]. A sequential extraction procedure was used to show that benzidine binds to soil in two phases [7]. Initially, a reversible equilibrium is established, followed by covalent bonding to soil organic matter, primarily humic acids [7]. The benzidine concentration in soil solution decreases rapidly over the first 6 hours and then more slowly [19]. After 48 hr the level of benzidine in soil solution remains constant, 6-22 ppb for soil application rates of 30-50 ppm [19]. Benzidine adsorbs to clay minerals forming a blue-colored species, the adsorption increasing with increasing pH [6]. Aromatic amines are known to form covalent bonds with humic materials, adding to quinone-like structures, followed by slow oxidation [20]; however, no data specific to benzidine are available.

Volatilization from Water/Soil: Since benzidine has a high boiling point and a moderate aqueous solubility, volatilization from water should not be a significant transport process [4].

Benzidine

Water Concentrations: Surface Water: Buffalo River, Buffalo, NY - upstream and downstream of Allied Chemical plant where benzidine was believed to have been discharged (42 samples from 7 sites) - not detected [11]. Niagara River near intake of Tonawanda water treatment plant - not detected [11]. Sumida River, Japan - site of several dye and pigment plants - concentration of aromatic amines, including benzidine 0.21-0.56 ppm [28]. Rhine River - detected, not quantified [24]. Benzidine was specifically looked for but not found in Lake Erie or Lake Michigan [8]. Of the 879 stations reporting benzidine in ambient water in EPA's STORET database, 0.1% contained detectable levels of the chemical [26].

Effluent Concentrations: Maximum concentration in wastewater from foundries - 10 ppb; wastewater from nonferrous metals manufacture, 1.2 ppb avg, 6.0 ppb max [30]. Detected in oil refinery, municipal, and industrial effluents [12]. Effluents from textile factories using benzidine-based dyes - 3.5 ppb avg; leather factory - 0.25 ppb; manufacturing plant using benzidine dyes - 3.5 ppb [12]. In a comprehensive survey of wastewater from 4000 industrial and publicly owned treatment works (POTWs) sponsored by the Effluent Guidelines Division of the USEPA, benzidine was identified in 1 discharge of the auto and other laundries industry, 70.0 ppb, and two discharges from an unidentified industry, 215.1 ppb mean [23]. Of the 1235 stations reporting benzidine in effluents in EPA's STORET database, 1.1% contained detectable levels of the chemical [26].

Sediment/Soil Concentrations: Buffalo River watershed - not detected in 21 samples from 7 sites upstream and downstream from Allied Chemical plant, where benzidine was believed to have been discharged [11]. Of the 3240 stations reporting benzidine in sediments in EPA's STORET database, none contained detectable levels of the chemical [26].

Atmospheric Concentrations:

Food Survey Values:

Plant Concentrations:

Fish/Seafood Concentrations: Of the 110 stations reporting benzidine in biota in EPA's STORET database, none contained detectable levels of the chemical [26].

Animal Concentrations:

Milk Concentrations:

Other Environmental Concentrations: Benzidine-based dyes produced in the United States and other countries were found to contain <1-1254 ppm benzidine [12].

Probable Routes of Human Exposure: Exposure to benzidine is primarily occupational via dermal adsorption, inhalation, and ingestion in workers connected with its production and conversion into direct azo dyes [12]. The respiratory route is of major importance under some manufacturing conditions [32]. In the United States, benzidine is only produced for captive consumption which would geographically limit exposure to the chemical [12]. However, occupational exposure can occur while using benzidine-based dyes [12]. Exposure is most likely to be through inhalation of dust or mist, or through dermal adsorption [18].

Average Daily Intake:

Occupational Exposures: Benzidine production plant 10 ug/m^3 avg in air and contamination of workers' clothing to 200 ug/sq dm [3]. Atmospheric concentrations of benzidine at different locations of a benzidine manufacturing plant (location, concn in mg/m^3): reducers and conversion tubs, 0.007; clarification tub, 0.005; filter press, 0.072-0.415; salting-out tub, 0.152; centrifuge, 0.005; location for shoveling benzidine into drums, 17.600 [32]. A 1973 report stated that 17 U.S. companies were using benzidine and that 62 employees were potentially exposed [12]. NIOSH estimates that about 700 people were exposed to benzidine and that approx 79,000 workers in 63 occupations were potentially exposed to benzidine-based dyes [12]. Benzidine manufacturing plant equal or less than 0.007 mg/m^3 at 4 locations, 0.0152 mg/m^3 at a salting-out tub, 0.072-0.415 mg/m^3 at a filter press, and 17.6 mg/m^3 where benzidine was shoveled into drums [12].

Benzidine

Body Burdens: Urine of industrial employees exposed to benzidine-derived dyes 1-112 ppb [18]. The worker who excreted 112 ppb operated a spray dryer, which is operated at 232-260 °C and therefore may have been exposed to the volatilized amine [18]. Urine of workers in benzidine manufacturing plant where the air concn was 10 ug/m^3, and contamination of working clothes was as much as 200 ug/sq dm 5.7 and 9.0 ppb in winter and summer, respectively [3]. Urine of workers in Russian dye manufacturing plant, trace-300 ppb [18].

REFERENCES

1. Baird R et al; J Water Pollut E Control Fed 49: 1609-15 (1977)
2. Banerjee S et al; pp. 113-28 in Conf Environ Chem Hydrazine NTIS AD-A054194 (1978)
3. Bolanowska W et al; Med Pr 23: 129-38 (1972)
4. Callahan MA et al; Water-related environmental fate of 129 priority pollutants vol II; pp.102-1 to 102-7 USEPA-440/4-79-029b (1979)
5. Freitag D; Chemosphere 14: 1589-616 (1985)
6. Furukawa T, Brindley GW; Clays Clay Miner 21: 279-88 (1973)
7. Graveel JG et al; J Environ Qual 15: 53-9 (1986)
8. Great Lakes Water Quality Board; Report to the Great Lakes Water Quality Board. Windsor, Canada (1983)
9. Hansch C, Leo AJ; MEDCHEM Project Claremont CA: Pomona College (1985)
10. Hine J, Mookerjee PK; J Org Chem 40: 292-8 (1975)
11. Howard PH, Saxena J; Persistence and degradability testing of benzidine and other carcinogenic compounds USEPA-560/5-76-005 (1976)
12. IARC; Some Industrial Chemicals and Dyestuffs 29: 149-173 (1982)
13. Johnson WE, Means JC; Am Chem Soc Div Environ Chem 191st Natl Meeting 26: 241 (1986)
14. Kincannon DF, Lin YS; Proc Ind Waste Conf 40: 607-19 (1985)
15. Lu PY et al; Arch Environ Contam Toxicol 6: 129-42 (1977)
16. Lutin PA et al; Purdue Univ Eng Bull Ext Ser 118: 131-45 (1965)
17. Means JC, Wijayaratne RD; Am Chem Soc Div Environ Chem 193rd Natl Meeting 27: 417 (1987)
18. NIOSH; Carcinogenicity and Metabolism of Azo Dyes especially those derived from Benzidine (1980)
19. Ononye AI et al; Bull Environ Contam Toxicol 39: 524-32 (1977)
20. Parris GE; Environ Sci Technol 14: 1099-106 (1980)
21. Perrin DD; Dissociation Constants of Organic Bases in Aqueous Solution. IUPAC Chem Data Series London: Buttersworth (1965)
22. Sadtler; Sadtler Standard Spectra 1830UV Philadelphia PA: Sadtler Res Lab
23. Shackelford WM et al; Analyt Chim Acta 146: 15-27 (1983)
24. Shakelford WM, Keith LH; Frequency of organic compounds identified in water; USEPA-600/4-76-062 (1976)
25. Shriner CR et al; Reviews of the environmental effects of pollutants. II. Benzidine USEPA-600/1-78-024 (1978)

Benzidine

26. Staples CA et al; Environ Toxicol Chem 4: 131-42 (1985)
27. Tabak HH, Barth EF; J Water Pollut Control Fed 50: 552-8 (1978)
28. Takemura N et al; Int J Water Poll 9: 665-701 (1965)
29. USEPA; Ambient Water Qual Criteria for Benzidine; p.B1 to B6 (1980)
30. USEPA; Treatability Manual; p. I.7.4-1 to I.7.4-5 USEPA-600/2-82-001a (1982)
31. Yoshida O et al; Igaku To Seibutsugaku 86: 361-4 (1973)
32. Zavon MR et al; Arch Environ Health 27: 1-7 (1973)
33. Zierath DL et al; Soil Sci 129: 277-81 (1980)

Benzoic Acid

SUBSTANCE IDENTIFICATION

Synonyms: Benzenecarboxylic acid

Structure:

CAS Registry Number: 65-85-0

Molecular Formula: $C_7H_6O_2$

Wiswesser Line Notation: QVR

CHEMICAL AND PHYSICAL PROPERTIES

Boiling Point: 249.2 °C at 760 mm Hg

Melting Point: 122.4 °C

Molecular Weight: 122.13

Dissociation Constants: pKa = 4.205 [50]

Log Octanol/Water Partition Coefficient: 1.87 [23]

Water Solubility: 2700 mg/L at 18 °C [40]

Vapor Pressure: 4.5 x 10^{-3} mm Hg at 20 °C [44]

Henry's Law Constant: 7.0 x 10^{-8} atm-m³/mole [62]

ENVIRONMENTAL FATE/EXPOSURE POTENTIAL

Summary: Benzoic acid may be released into the environment as emissions or, more commonly, in wastewater during its production and use as a chemical intermediate and additive. Benzoic acid and sodium benzoate are commonly added to food products as a preservative and as an antimicrobial agent. Formed in the combustion process, benzoic acid is found in auto exhaust, refuse combustion, and tobacco smoke. Benzoic acid is also widely distributed in nature and naturally occurs in foods such as berries. If released on land, benzoic acid should leach into the ground and biodegrade (half-life <1 week). If released in water, benzoic acid should also readily biodegrade (half-life 0.2- 3.6 days). Adsorption to sediment and volatilization should not be significant. Bioconcentration in fish and algae is not important. In the atmosphere, benzoic acid will be largely associated with aerosols, be subject to gravitational settling, and be scavenged by rain. Exposure of the general population through ingestion of food containing benzoic acid either naturally or as an additive will be considerable. Occupational exposure should be primarily through dermal contact or inhalation of aerosols containing the chemical.

Natural Sources: Benzoic acid in the free state, or in the form of simple derivatives such as salts, esters, and amides, is widely distributed in nature [66]. Gum benzoin may contain as much as 20% benzoic acid and acaroid resin contains 4.5-7% benzoic acid [66]. Natural products containing the free acid include scent glands of beavers, bark of the black cherry tree, cranberries, berries, prunes, ripe cloves, and oil of anise seed, and Tolu balsam [44,66].

Artificial Sources: Benzoic acid may be released into the environment as emissions or, more commonly, in wastewater during its production and use in the manufacture of phenol, benzoate plasticizers, benzoyl chloride, and other chemicals and in medicinals, cosmetics, and industrial preservatives [9,66]. Benzoic acid and sodium benzoate are common food additives, being used as food preservatives at concn of 0.1% and as antimicrobial agents in food at concn levels of 0.29-0.00001% [63]. 0.5 g/ton pulp of benzoic acid is released in the spent chlorination liquor from the bleaching of sulfite pulp [7]. It is formed in combustion processes and found in gasoline and diesel exhaust, refuse combustion, and tobacco smoke [7,13,20].

Terrestrial Fate: If released on land, benzoic acid will leach into the ground and biodegrade (half-life <1 week). After application of contaminated municipal sludge on land in Muskegon County, MI and tilling to 15 cm, the soil contained 461 ppb of benzoic acid. The chemical had disappeared from this layer of soil and the next lower 15 cm within 216 days, when it was next analyzed [11]. After deep well injection with other wastes from a dimethyl terphthalate plant, benzoic acid, which had averaged 54 ppm in the injected waste, appeared only in trace quantities in monitoring wells 427-823 m away [32]. The degradation of the acid in the 2-4 yr residence may have resulted from biodegradation or reaction with subsurface material or other waste components [32].

Aquatic Fate: If released in water, benzoic acid should readily biodegrade (half-life 0.2-3.6 days). Adsorption to sediment or volatilization should not be significant. Benzoic acid was found to be readily degraded in a model ecosystem in which the measure of degradability, the biodegradability index (polar metabolites/nonpolar metabolites) was 2.97 [6,35].

Atmospheric Fate: In the atmosphere, benzoic acid will be largely associated with aerosols, be subject to gravitational settling, and be scavenged by rain. It may photolyze when associated with material such as sand that catalyze this process. The free vapor reacts with photochemically produced hydroxyl radicals with an estimated half-life of 2.0 days.

Biodegradation: Benzoic acid has been studied extensively and shown to be biodegradable in screening tests. Eleven laboratories testing a respiratory biodegradability test utilizing an unacclimated sludge inoculum found benzoic acid to be readily degradable, obtaining a mean oxygen uptake of 84% of theoretical after 10 days, and no lag period before biodegradation commenced [30]. Some results from other investigators are: 99% COD removal in 5 days with acclimated activated sludge [45]; 67% of theoretical BOD removal in 5 days [25]; 97% degradation in 20 days by activated sludge, where 10% of the benzoic acid was replaced every 2 days to acclimate the sludge [36]; 68.2 and 86.9%

mineralization in 5 days by acclimated activated sludge in salt solution and simulated industrial effluent, respectively [5]; 65.4% mineralization in 5 days by activated sludge [16]; complete disappearance in 1 day using an activated sludge inoculum [12]; 73% of theoretical BOD utilized in 6 days using activated sludge from 3 municipal sewage plants [38]; 74% of theoretical BOD utilized in a 5-day test with a sewage seed [68]; >90% degraded in 2 days using activated sludge [69]; 84.1 and 74.9% of theoretical BOD in 5 days by the standard and seawater dilution methods, respectively [59]. Using a Captina silt loam inoculum, the half-life for mineralization of benzoic acid in solution was 4.5 hr after a 30 min lag [10]. Complete degradation occurred in 1 day with a Niagara silt loam inoculum [2]. At concns of 15-18 mg/L, benzoic acid was biodegraded with half-lives of 0.85 and 3.6 days in a polluted river and reservoir, respectively [4]. Low concentrations of benzoic acid are rapidly mineralized in both eutrophic and oligotrophic lake water with the rate of disappearance being proportional to its concentration [47,57]. At 59 ng/L, over 98% mineralization had occurred in 7 days in both eutrophic and oligotrophic water [47]. The half life for mineralization in eutrophic water was 0.22 days over a concn range of 32 ng/L to 50 ug/L [57]. From 63 to 83% of the benzoic acid was lost in 6 hr and >94% in 58 hr [57]. Mineralization was not usually affected by montmorillonite or kaolinite in the water [56]. When benzoic acid was incubated in an acidic loam soil adjusted to 60% of its water holding capacity, 74 and 81% was mineralized in 1 and 12 weeks [28]. The same experiment using a neutral, sandy loam soil resulted in 55 and 71% mineralization in 1 and 12 weeks [28]. After 12 weeks, 3.0 and 4.4% of the chemical was incorporated into the biomass of the two soils [28]. In an alkaline para-brown soil, 40 and 63% mineralization occurred in 3 days and 10 weeks, respectively [21]. An experiment was performed in which ^{14}C-labeled benzoic acid was added to subsurface soil taken from the unsaturated zone beneath the tile of a septic tank and incubated both aerobically and anaerobically [64]. Under aerobic conditions the half-life was 3.9 and 7.3 for carboxyl- and ring-labeled chemical, respectively [64]. The ring-labeled benzoic acid had a mineralization half-life of 18.2 hr when incubated anaerobically [64]. Benzoic acid is biodegradable under anaerobic conditions, indicated by the fact that >75% of theoretical methane production was obtained when incubated for 8 weeks with 10% sludge from a secondary digester [52]. Under anaerobic conditions, 91% of benzoic acid was converted to methane and carbon dioxide in 18 days, including an 8-day lag period [24]. In another experiment, 86-93% conversion to methane and carbon dioxide occurred in 14 days with a sewage sludge inoculum [43]. Benzoic acid was completely mineralized in a week when incubated anaerobically with municipal digested sludge or in anoxic sediment from a hypereutrophic lake in Kalamazoo County, MI [26].

Abiotic Degradation: The pKa for benzoic acid suggests that it exists almost exclusively in the dissociated form at environmental pHs. Benzoic acid absorbs UV radiation up to approx 310 nm [48], and therefore may photolyze. In a photomineralization test in which the chemical is adsorbed on silica gel and irradiated with light >290 nm, 10.2% mineralization occurred in 17 hr [16]. When illuminated with a sunlamp for 24 hr in solution containing zinc oxide, 67% degradation occurred in 24 hr [31]. However it was stable when exposed to sunlight or a sunlamp for 137 hr in aqueous solution [65]. Zinc oxide therefore appears to possess catalytic activity, as does beach sand [31]. In the vapor phase, benzoic acid should react with photochemically produced hydroxyl radicals by aromatic ring addition with an estimated half-life of 2.0 days [17].

Bioconcentration: The BCF of benzoic acid in golden ide and algae (Chlorella fusca) was <10 as determined in a 3- and 1-day static tests, respectively [16]. The BCF for trout muscle calculated by regression analysis from its octanol/water partition coefficient is 14 [6]. While the BCF of mosquito fish, algae, and mosquito larvae after 1 day in an aquatic ecosystem is relatively low (21, 100, and 138, respectively), the BCF in daphnia and snail is high, 1800 and 2800, respectively [35]; however, the investigators measured total radioactive, not benzoic acid, so a metabolite may have been accumulated.

Soil Adsorption/Mobility: Benzoic acid did not adsorb appreciably to two different sandy soils, a clayey subsoil [34], and montmorillonite clay [3].

Benzoic Acid

Volatilization from Water/Soil: Based on the calculated Henry's Law constant for benzoic acid, it would not be expected to volatilize significantly from water [39].

Water Concentrations: DRINKING WATER: In a five-city survey of drinking water, 15 ppm benzoic acid was found in the tap water of Ottumwa, IA but not in that of Miami, Seattle, Philadelphia, or Cincinnati [61]. Another study found it in water from the Torresdale water treatment plant in Philadelphia [58]. Benzoic acid was detected, but not quantified, in treated drinking water in England, whose source was a lowland river containing relatively high levels of wastewater [15]. SURFACE WATER: Benzoic acid was detected, but not quantified, in a Norwegian river downstream from an industrial treatment facility [49]. GROUND WATER: 16-860 ppb of benzoic acid were found in 2 aquifers at the Hoe Creek underground coal gasification site 15 mo after gasification was completed [55]. Concns of benzoic acid in the plumes in shallow, sandy aquifers emanating from landfills in Ontario were 17 to >1000 ppb in one aquifer and ND to 8.8 ppb in another [46]. The concn in background monitoring wells was at trace levels (<0.1 ppb) in the first aquifer and was not determined in the second [46]. Two wells monitoring near-surface ground water adjacent to an unlined surface impoundment at a wood-preserving facility at Pensacola, FL contained 3.1 and 27.5 ppm of benzoic acid, while wells 150 m away contained 0-0.01 ppm of the chemical [18]. It is believed that the benzoic acid was rendered from the wood during treatment or was a degradation product of creosote solutes [18]. Benzoic acid was found in ground water in Australia underlying an area where acid wastes from a manufacturing process of a chemical company were stored in unlined ponds [53]. Since the chemical was only found in the aquifer downgradient from the believed source of pollution and not closer to this source, it was either formed by bacterial action or came from another source [53]. RAIN/SNOW: Benzoic acid was found in the particulate fraction of four samples of rain and snow in Norway [37]. While no concns were indicated, the size of the gas chromatography peaks ranged widely in size [37].

Effluent Concentrations: In a comprehensive survey of wastewater from 4000 industrial and publicly owned treatment works (POTWs) sponsored by the Effluent Guidelines Division of the U.S. EPA,

Benzoic Acid

benzoic acid was identified in discharges of the following industrial category (frequency of occurrence, median concn in ppb): timber products (15, 57.7); leather tanning (7, 89.6); iron and steel mfg (7, 33.4); petroleum refining (1, 503.3); nonferrous metals (19, 62.5); paint and ink (36, 162.1); printing and publishing (18, 228.9); ore mining (13, 32.6); organics and plastics (35, 669.9); inorganic chemicals (9, 56.6); textile mills (12, 46.9); plastics and synthetics (16, 36.2); pulp and paper (49, 133.3); rubber processing (6, 223.3); soaps and detergents (2, 148.3); auto and other laundries (13, 127.8); pesticides manufacture (7, 44.3); photographic industries (2, 69.7); pharmaceuticals (15, 121.6); explosives (4, 20.8); foundries (19, 61.4); porcelain/enameling (4, 176.5); electronics (19, 80.3); electroplating (1, 2.8); oil and gas extraction (24, 23.8); organic chemicals (16, 241.3); mechanical products (34, 104.2); transportation equipment (6, 163.5); synfuels (24, 96.3); publicly owned treatment works (84, 35.9); rum industry (1, 405.3) [51]. The highest effluent concn was 72,124 ppb in the pesticides mfg industry. The paint and ink and organics and plastics industries also had maximum effluents exceeding 10,000 ppb [51]. Benzoic acid appeared in the process exhaust from a phthalic anhydride manufacturing plant without pollution abatement equipment at concn ranging from 5-40 ppm (v/v) [14]. It has been reported in the exhaust gas from diesel-powered vehicles [22] and the concn in the exhaust from a 1982 Toyota Corolla was 0.164 ppb [29]. Extracts of 5 incinerator effluents contained 6-3500 ppm of benzoic acid [27]. Effluent from the Los Angeles County wastewater treatment plant contained 400 ppb of benzoic acid [19]. It was detected, but not quantified, in the effluent of the publicly owned treatment works in Addison, IL that accepts waste from over 300 manufacturing and industrial firms, but not in 9 other treatment facilities sampled in the state [13]. Leachate from a sanitary landfill contained benzoic acid but it was not quantified [1]. Benzoic acid occurred at concn levels of 1-50 ppm in settling basins and other standing water at the Valley of the Drums waste site in Bullitt County, KY [54]. It was a component of spent bleach liquor from a softwood kraft pulp plant [33] and averaged 54 ppm in effluent from a dimethyl terphthalate plant near Wilmington, NC that was disposed of by deep well injection [32].

Sediment/Soil Concentrations:

Atmospheric Concentrations: URBAN/SUBURBAN: Los Angeles 1-26 ppt, 10 ppt mean [29]. An unspecified sample of urban air contained benzoic acid in both the gas and aerosol phases [8]. It was contained in the aerosol fraction of air obtained in a suburban area of Japan 60 km northeast of Tokyo [67]. However, it was not detected in the Allegheny Mountain Tunnel, a tunnel that received considerable traffic [22].

Food Survey Values: Apple wine and apple essence contain 0.329 and 40 ppm of benzoic acid, respectively [63]. Most berries contain about 0.05% of benzoic acid [41].

Plant Concentrations:

Fish/Seafood Concentrations:

Animal Concentrations:

Milk Concentrations:

Other Environmental Concentrations: Used motor oil contained 45.3 umol/L of benzoic acid, although new motor oil did not contain detectable quantities [29]. The chemical was found in fly ash from a municipal waste incinerator in Ontario [60].

Probable Routes of Human Exposure: The general population will be exposed to benzoic acid through the ingestion of foods such as berries and prunes that contain the chemical naturally, as well, from foods in which it is added as a preservative. In addition, exposure would result from inhalation of aerosols from auto exhaust, tobacco smoke, and other combustion sources. Occupational exposure to benzoic acid should primarily be through dermal contact or inhalation of aerosols containing it.

Average Daily Intake: AIR INPUT: insufficient data; WATER INPUT: insufficient data; FOOD INPUT: 312 mg (278 mg as sodium benzoate and 34 mg as benzoic acid) [63].

Occupational Exposures: NIOSH has estimated that 93,571 workers are potentially exposed to benzoic acid based on statistical estimates derived from a survey conducted in 1981-83 in the United States [42]

Body Burdens:

REFERENCES

1. Albaiges J et al; Water Res 20: 1153-9 (1986)
2. Alexander M, Lustigman BK; J Agric Food Chem 14: 410-3 (1966)
3. Bailey GW et al; Soil Sci Soc Of Amer Proc 32: 222-34 (1968)
4. Banerjee S et al; Environ Sci Technol 18: 416-22 (1984)
5. Belly RT, Goodhue CT; pp. 1103-7 in Proc Int Biodegrad Symposium 3rd (1976)
6. Branson DR; in Predicting the Fate of Chemicals in the Aquatic Environment from Laboratory Data ASTM STP 657 Phila PA: American Society For Testing And Materials, pp. 55-70 (1978)
7. Carlberg GE et al; Sci Tot Environ 48: 157-67 (1986)
8. Cautreels W, VanCauwenberghe K; Atmos Environ 12: 1133-41 (1978)
9. Chemical Marketing Reporter; Chemical Profile: Benzoic Acid December 24 (1984)
10. Dao TH, Lavy TL; Soil Sci 143: 66-72 (1987)
11. Demirjian YA et al; J Water Pollut Control Fed 59: 32-8 (1987)
12. Digeronimo MJ et al; in Microbial Degradation Of Pollutants In Marine Environments pp. 154-66 EPA-600/9-79-012 (1979)
13. Ellis DD et al; Arch Environ Contam Toxicol 11: 373-82 (1982)
14. Fawcett RL; J Am Pollut Control Assoc 20: 461-5 (1970)
15. Fielding M et al; Organic Micropollutants In Drinking Water TR-159 Medmenham, England: Water Res Cent pp. 49 (1981)
16. Freitag D et al; Chemosphere 14: 1589-616 (1985)
17. GEMS; Graphical Exposure Modeling System. FAP. Fate of Atmospheric Pollutants (1987)
18. Goerlitz DF et al; Environ Sci Tech 19: 955-61 (1985)
19. Gossett RW et al; Mar Pollut Bull 14: 387-92 (1983)
20. Graedel TE; Chemical Compounds in the Atmosphere p. 220 Academic Press NY (1978)
21. Haider K et al; Arch Microbiol 96: 183-200 (1974)
22. Hampton CV et al; Environ Sci Technol 16: 287-98 (1982)
23. Hansch C, Leo AJ; Medchem Project Issue No 26. Claremont CA: Pomona College (1985)
24. Healy JB Jr, Young LY; Appl Environ Microbiol 38: 84-9 (1979)
25. Heukelekian H, Rand MC; J Water Pollut Contr Assoc 29: 1040-53 (1955)
26. Horowitz A et al; Dev Ind Microbiol 23: 435-44 (1982)
27. James RH et al; in J Proc-APCA 77th Annual Meeting pp. 1-25 (1984)
28. Kassim G et al; Soil Sci Soc Am J 46: 305-9 (1982)
29. Kawamura K et al; Environ Sci Technol 19: 1082-6 (1985)

Benzoic Acid

30. King EF, Painter HA; Ring-test Program 1981-1982 Assessment Of Biodegradability Of Chemicals In Water By Manometric Respirometry Comm Eur Communities, Eur 8631 (1983)
31. Kinney LC et al; Photolysis Mechanisms For Pollution Abatement TWRC-13 Robert A Taft Water Res Cent Rep 1969 (1969)
32. Leenheer JA et al; Environ Sci Technol 10: 445-51 (1976)
33. Lindstrom K, Osterberg F; Environ Sci Technol 20: 133-8 (1986)
34. Loekke H; Water, Air, Soil Pollut 22: 373-87 (1984)
35. Lu PY, Metcalf RL; Environ Health Perspect 10: 269-84 (1975)
36. Lund FA, Rodriguez DS; J Gen Appl Microbiol 30: 53-61 (1984)
37. Lunde G et al; Organic Micropollutants In Precipitation In Norway SNSF Project, FR-9/76, 17 pp. (1977)
38. Lutin PA et al; Purdue Univ Eng Bull, Ext Series 118: 131-45 (1965)
39. Lyman WJ et al; pp. 15-1 to 15-34 in Handbook of Chem Property Estimation Methods. McGraw-Hill NY (1982)
40. MacKay D et al; Chemosphere 9: 701-11 (1980)
41. Merck Index; An Encyclopedia of Chemicals and Drugs 10th ed. p.155 (1983)
42. NIOSH; National Occupational Exposure Survey (1985)
43. Nottingham PM, Hungate RE; J Bacter 98: 1170-2 (1969)
44. Organ Econ Coop Devel; OECD Guidelines for Testing of Chemicals. Berlin: Umweltbundesant 842 pp. (1981)
45. Pitter P; Water Res 10: 231-5 (1976)
46. Reinhard M et al; Environ Sci Technol 18: 953-61 (1984)
47. Rubin HE et al; Appl Environ Microbiol 43: 1133-8 (1982)
48. Sadtler; Sadtler Standard Spectra UV 252 Philadelphia PA: Sadtler Research Lab
49. Schou L et al; Total Environ 20: 277-86 (1981)
50. Serjeant EP, Dempsey B; Ionisation Constants of Organic Acids in Aqueous Solution. IUPAC Chemical Data Series No. 23. New York,NY: Pergamon Press pp. 989 (1979)
51. Shackelford WM et al; Analyt Chim Acta 146: 15-27 (1983)
52. Shelton DR, Tiedje JM; Appl Environ Microbiol 47: 850-7 (1984)
53. Stepan S et al; Austral Water Resources Council Conf Ser 1: 415-24 (1981)
54. Stonebraker RD, Smith AJ Jr; pp. 1-10 in Control Hazard Mater Spills, Proc Natl Conf Nashville, TN (1980)
55. Stuermer DH et al; Environ Sci Technol 16: 582-7 (1982)
56. Subba-Rao RV, Alexander M; Appl Environ Microbiol 44: 659 (1982)
57. Subba-Rao RV et al; Appl Environ Microbiol 43: 1139-50 (1982)
58. Suffet IH et al; Water Res 14: 853-67 (1980)
59. Takemoto S et al; Suishitsu Odaku Kenkyu 4: 80-90 (1981)
60. Tong HY et al; J Chrom 285: 423-41 (1984)
61. USEPA; Interim Report To Congress, June, 1975 Washington, DC (1975)
62. USEPA; Treatability Manual Vol 1 EPA-600/2-82-001a (1981)
63. USEPA; Health and Environmental Effects Document for Benzoic Acid. p.15 ECAO-CIN-G007 (1987)
64. Ward TE; Environ Toxicol Chem 4: 727-37 (1985)
65. Ware GW et al; Arch Environ Contam Toxicol 9: 135-46 (1980)
66. Williams AE; Kirk-Othmer Encycl Chem Tech 3rd ed. 3: 778-92 (1978)
67. Yokouchi Y, Ambe Y; Atmos Environ 20: 1727-35 (1986)

Benzoic Acid

68. Young RHF et al; J Water Pollut Contr Fed 40: 354-68 (1968)
69. Zahn R, Wellens H; Wasser Abwasser Forsch 13: 1-7 (1980)

Benzotrichloride

SUBSTANCE IDENTIFICATION

Synonyms: (Trichloromethyl)benzene

Structure:

CAS Registry Number: 98-07-7

Molecular Formula: $C_7H_5Cl_3$

Wiswesser Line Notation: GXGGR

CHEMICAL AND PHYSICAL PROPERTIES

Boiling Point: 220.8 °C at 760 mm Hg

Melting Point: -5.0 °C

Molecular Weight: 195.48

Dissociation Constants:

Log Octanol/Water Partition Coefficient: 2.92 [3]

Water Solubility: Hydrolyzes rapidly in water [7]

Vapor Pressure: 0.23 mm Hg at 20 °C [12]

Henry's Law Constant:

Benzotrichloride

ENVIRONMENTAL FATE/EXPOSURE POTENTIAL

Summary: Fugitive emissions from the industrial manufacture and use of benzotrichloride may be a potential source of exposure of the chemical to the environment. Still bottoms from the distillation of benzyl chloride contain benzotrichloride, but are disposed of in hazardous waste land disposal sites. If released to soil, benzotrichloride is expected to hydrolyze in the presence of moisture. Evaporation from dry surfaces is likely to occur. If released to water, benzotrichloride hydrolyzes rapidly to form benzoic acid and hydrochloric acid, with a hydrolysis half-life of 11 seconds at 25 °C (pH 7) and 3 minutes at 5.10 °C. Due to the rapid hydrolysis, volatilization from water, bioconcentration, and adsorption to sediments are not expected to be important. If released to air, benzotrichloride will react in vapor phase with photochemically produced hydroxyl radicals with an estimated half-life of 2 days in a typical atmosphere. Atmospheric hydrolysis may occur in the presence of moisture; however, direct photolysis is not expected to occur. Occupational exposure probably occurs during manufacture and use.

Natural Sources: Benzotrichloride has not been reported to occur in nature [4].

Artificial Sources: Still bottoms from the distillation of benzyl chloride contain benzotrichloride (0.0005 kg per 1 kg of benzyl chloride product) and are disposed of in hazardous waste land disposal sites [2]. Fugitive emissions from the industrial manufacture and use of benzotrichloride may be a potential source of exposure of the chemical to the environment.

Terrestrial Fate: Benzotrichloride hydrolyzes rapidly in the presence of moisture; therefore, it is expected to hydrolyze in moist soil. Evaporation from dry surfaces is likely to occur.

Aquatic Fate: Benzotrichloride hydrolyzes rapidly in water to form benzoic and hydrochloric acid with a hydrolysis half-life of 11 seconds at 25 °C (pH 7) and 3 minutes at 5.10 °C. Due to the rapid hydrolysis, volatilization, bioconcentration, and adsorption to sediments are not expected to be significant.

Benzotrichloride

Atmospheric Fate: If released to atmosphere, benzotrichloride will react in vapor phase with photochemically produced hydroxyl radicals with an estimated half-life of 2 days in a typical atmosphere. Atmospheric hydrolysis may occur in the presence of moisture; however, direct photolysis is not expected to occur.

Biodegradation: Degradation observed in the Modified OECD-Screening Test (95% elimination of DOC in 3 days, 100% mineralization of organochlorine in 1 day [11]) is most likely degradation of benzotrichloride's hydrolysis product [11], benzoic acid, since benzotrichloride has an estimated half-life of 11 sec in water at 25 °C and pH 7.

Abiotic Degradation: Benzotrichloride hydrolyzes in the presence of moisture, forming benzoic acid and hydrochloric acid [8]. The hydrolysis rate constant at 25 °C and pH 7 is reported to be 0.063 sec^{-1} [7] which corresponds to a half-life of 11 sec. The hydrolysis rate constant at 5.10 °C was found to be 0.00387 sec^{-1} [5], which corresponds to a half-life of 3.0 minutes. The ultraviolet adsorption spectrum for benzotrichloride in methanol solution shows virtually no absorption above 290 nm [9], therefore direct photolysis in the environment is not expected to occur. The vapor-phase reaction of benzotrichloride with photochemically produced hydroxyl radicals in the atmosphere at 25 °C has an estimated half-life of 2 days at a hydroxyl radical concentration of 8 x 10^{+5} mol/m^3 [1].

Bioconcentration: Based on the log Kow, the BCF for benzotrichloride can be estimated to be 98 [6]. Due to the rapid hydrolysis of benzotrichloride in water, bioconcentration in aquatic organisms is not expected to occur.

Soil Adsorption/Mobility: Due to the rapid hydrolysis of benzotrichloride in water, leaching in moist soils should not be significant, due to its degradation to benzoic and hydrochloric acid.

Volatilization from Water/Soil: Due to the rapid hydrolysis of benzotrichloride in water, volatilization from water is not expected to be important. Benzotrichloride is a liquid which fumes in air [8] and has a relatively high vapor pressure; therefore, evaporation from dry surfaces may be expected to occur.

Water Concentrations:

Effluent Concentrations:

Sediment/Soil Concentrations: Benzotrichloride may have been present in a soil sample collected at an abandoned chemical dump site; however, its presence could not be positively confirmed by the analytical techniques used [10].

Atmospheric Concentrations:

Food Survey Values:

Plant Concentrations:

Fish/Seafood Concentrations:

Animal Concentrations:

Milk Concentrations:

Other Environmental Concentrations:

Probable Routes of Human Exposure: Occupational exposure to benzotrichloride probably occurs during its manufacture and use as an intermediate [4]. In occupational settings, exposure by dermal and inhalation routes may be possible.

Average Daily Intake:

Occupational Exposures:

Body Burdens:

REFERENCES

1. GEMS; Graphical Exposure Modeling System. Fate of atmospheric pollutants (FAP) data base. Office of Toxic Substances. USEPA (1986)

Benzotrichloride

2. Gruber GI; Assessment of Industrial Hazardous Waste Practices - Organic Chemicals, Pesticides and Explosives Industries p. 5-49 USEPA/530/SW-118c (1975)
3. Hansch C, Leo AJ; Medchem Project Issue No 26. Claremont CA: Pomona College (1985)
4. IARC; Some Industrial Chemicals and Dyestuffs 29: 74 (1982)
5. Laughton PM, Robertson RE; Can J Chem 37: 1491 (1959)
6. Lyman WJ et al; Handbook of Chemical Property Estimation Methods. Environmental Behavior of Organic Compounds. McGraw-Hill NY (1982)
7. Mabey W, Mill T; J Phys Chem Ref Data 7: 383 (1978)
8. Merck Index; An Encyclopedia of Chemicals, Drugs and Biologicals 10th ed (1983)
9. Sadtler; 380 UV, Sadtler Research Laboratories (1966)
10. Shafer KH et al; Anal Chem 56: 237 (1984)
11. Steinhaeuser KG et al; Vom Wasser 67: 147-54 (1986)
12. Weber RC et al; Vapor Pressure Distribution of Selected Organic Chemicals USEPA-600/2-81-021 (1981)

Benzyl Chloride

SUBSTANCE IDENTIFICATION

Synonyms: (Chloromethyl)benzene

Structure:

CAS Registry Number: 100-44-7

Molecular Formula: C_7H_7Cl

Wiswesser Line Notation: G1R

CHEMICAL AND PHYSICAL PROPERTIES

Boiling Point: 179 °C

Melting Point: -43 to -48 °C

Molecular Weight: 126.58

Dissociation Constants:

Log Octanol/Water Partition Coefficient: 2.30 [6]

Water Solubility: 493 mg/L at 20 °C [15]

Vapor Pressure: 1 mm Hg at 22 °C [16]

Henry's Law Constant: 3.4 x 10^{-4} atm-m³/mole (calculated from water solubility and vapor pressure)

Benzyl Chloride

ENVIRONMENTAL FATE/EXPOSURE POTENTIAL

Summary: Benzyl chloride is released to the environment by emissions involved with its production and use, by emissions from waste incineration, and by emissions from floor tile manufacturing plants where butyl benzyl phthalate is used. Due to its rapid transformation in environmental media, benzyl chloride is not expected to be a persistent environmental contaminant. If released to water, hydrolysis will be the dominant removal mechanism, as the hydrolysis half-life ranges from 19.1 to 0.58 days at respective temperature ranges of 0.1 to 25 °C. Volatilization from water in shallow, rapidly moving streams may be competitive with hydrolysis. If released to soil, benzyl chloride can be expected to hydrolyze under moist conditions. Leaching is likely to occur, although hydrolytic decomposition may occur at rates sufficient to minimize the importance of leaching. Evaporation from dry surfaces will probably occur. If released to the atmosphere, the dominant removal mechanism will be reaction with hydroxyl radicals. The atmospheric residence time, with respect to hydroxyl radical reactions, has been estimated to be approx 3 days. Monitoring of ambient outdoor air in seven U.S. cities between 1980 and 1981 found benzyl chloride levels generally below detection limits of 5 ppt, although concentrations as high as 111 ppt were detected. Indoor inhalation exposure may occur due to emissions from floor tile made with butyl benzyl phthalate.

Natural Sources: Benzyl chloride has not been detected in nature [8].

Artificial Sources: In the United States, benzyl chloride is emitted to the atmosphere from its commercial production and use at an estimated rate of 45,000 pounds/year, with the emissions originating from scrubber vents, vacuum jets, storage and fugitive emissions, and equipment leaks [2]. Stack emissions from waste incineration have been found to contain benzyl chloride [10]. Emissions of benzyl chloride from floor tile manufactured with butyl benzyl phthalate have been reported [18]. Fractionating column wastes (containing 0.0005 kg benzyl chloride/kg production) generated from the commercial production of benzyl chloride have been landfilled in the past [5].

79

Benzyl Chloride

Terrestrial Fate: Benzyl chloride can be expected to hydrolyze quite rapidly in moist soils based on its relatively rapid hydrolysis in aqueous solutions. Significant leaching in soil is likely to occur; however, hydrolytic decomposition may occur in moist soil at rates sufficient to significantly minimize the importance of leaching. Screening tests have shown that benzyl chloride is readily biodegradable; therefore, biodegradation in various soils may be important. The relatively high vapor pressure of benzyl chloride suggests that evaporation from dry surfaces may be relatively rapid.

Aquatic Fate: In water, the dominant removal mechanism for benzyl chloride will be hydrolysis, which has half-lives ranging from 19.1 to 0.58 days at respective temperature ranges of 0.1 to 25 °C. Volatilization from shallow, rapidly moving streams may be a competitive removal mechanism with hydrolysis; however, in most environmental bodies of water, hydrolysis should be dominant. Screening tests have shown that benzyl chloride is readily biodegradable; therefore, biodegradation may have some importance in various natural waters. Bioconcentration in aquatic organisms and adsorption to sediment and suspended matter is not expected to be important.

Atmospheric Fate: The dominant removal mechanism of benzyl chloride in the atmosphere is its reaction with hydroxyl radicals [22]. The atmospheric residence time of benzyl chloride, with respect to hydroxyl radical reactions, has been estimated to be approx 3 days. The atmospheric half-life with respect to ozone reaction is in excess of 200 days and is therefore not competitive with hydroxyl radical reaction. Physical removal to benzyl chloride from the atmosphere is not likely to be important with respect to hydroxyl reactions [4].

Biodegradation: Biodegradability tests performed under the Japanese MITI protocol have found benzyl chloride to be readily biodegradable [11,20]. Benzyl chloride biodegraded significantly with the formation of dechlorinated products during a 2-day incubation period using raw sewage and raw sewage acclimated to non-chlorinated compounds [9]. None of the above methods have controls for chemical hydrolysis, so these results may be more indicative of the biodegradation of the chemical hydrolysis product, benzyl alcohol.

Benzyl Chloride

Abiotic Degradation: The aqueous hydrolysis rate constant for benzyl chloride has been found to range from 0.042×10^{-5} sec^{-1} at 0.1 °C to 1.38×10^{-5} sec^{-1} at 25 °C [1,12,19], which corresponds to half-lives of 19.1 and 0.58 days, respectively. The rate of aqueous hydrolysis has been found to be independent of pH up to pH 13.0 [24]. The aqueous hydrolysis products of benzyl chloride are benzyl alcohol and hydrogen chloride [1,24]. The atmospheric residence time of benzyl chloride due to vapor-phase reaction with hydroxyl radicals, has been estimated to be approx 3 days in a typical atmosphere containing $1 \times 10^{+6}$ hydroxyl radicals/cm^3 [3]. The rate constant for the reaction between benzyl chloride and ozone in the atmosphere has been determined to be less than 0.04×10^{-8} cm^3/molecule-sec [3], which corresponds to a half-life >200 days in a typical atmosphere.

Bioconcentration: Based upon the log octanol-water partition coefficient and the water solubility, the bioconcentration factor for benzyl chloride can be estimated to be 33 and 16, respectively, utilizing recommended regression equations [13]. These estimations suggest that bioconcentration in aquatic organisms will not be significant.

Soil Adsorption/Mobility: Based upon the log octanol-water partition coefficient and the water solubility, the soil sorption coefficient for benzyl chloride can be estimated to range from 123 to 482 utilizing various regression equations [13]. These estimates indicate that benzyl chloride will have medium to high mobility in soil [23].

Volatilization from Water/Soil: Using the Henry's Law constant, the volatilization half-life of benzyl chloride from a model river 1 m deep flowing 1 m/sec, with a wind velocity of 3 m/sec, can be estimated to be about 0.3 days [13].

Water Concentrations: SURFACE WATER: Benzyl chloride was positively detected in one of 17 samples (no concentration reported) collected from the Delaware River in October, 1976 [21].

Effluent Concentrations: Benzyl chloride has been detected in various unidentified industrial effluents [8]. It has also been detected in stack effluents from waste incinerators [10].

Sediment/Soil Concentrations: Benzyl chloride has been detected (no concentrations reported) in the soil-sediment-water matrix of the Love Canal near Niagara, NY [7].

Atmospheric Concentrations: Monitoring of ambient air in seven U.S. cities (Houston, St. Louis, Denver, Riverside, Staten Is., Pittsburgh, and Chicago) between 1980 and 1981 found benzyl chloride levels generally below analytical detection limits of 5 ppt, although concentrations as high as 111 ppt were detected [22]. The ambient indoor air of various residential homes in North Carolina was found to contain mean benzyl chloride levels of 6.23 ppb during winter time monitoring while the outdoor air contained no detectable levels [17].

Food Survey Values:

Plant Concentrations:

Fish/Seafood Concentrations:

Animal Concentrations:

Milk Concentrations:

Other Environmental Concentrations:

Probable Routes of Human Exposure: Exposure of benzyl chloride to the general population is possible by inhalation of air contaminated by emissions from industrial sources or incinerators. Exposure can also occur through inhalation of indoor air contaminated by emissions from floor tile manufactured with butyl benzyl phthalate.

Average Daily Intake:

Occupational Exposures: It has been estimated that approx 3000 U.S. workers are potentially exposed to benzyl chloride resulting from its manufacture and use [8]. NIOSH has estimated that 11,483 workers are potentially exposed to benzyl chloride based on statistical estimates derived from a survey conducted between 1981-1983 in the United States [14].

Body Burdens:

REFERENCES

1. Albery WJ, Curran JS; J Chem Soc Chem Comm 1972: 425-6 (1972)
2. Anderson GE; Human Exposure to Atmospheric Concentrations of Selected Chemicals. Vol. 2 Air Quality Planning Standards, U.S. EPA NTIS PB83-26529 (1983)
3. Atkinson R et al; Int J Chem Kin 14: 13-8 (1982)
4. Cupitt LT; Fate of Toxic and Hazardous Materials in the Air Environment USEPA-600/S3-80-084 (1980)
5. Gruber GI; Assessment of Industrial Hazardous Waste Practices, Organic Chemicals, Pesticides and Explosives Industries pp. 5-50 USEPA/530/SW-118c NTIS PB-251 307 (1975)
6. Hansch C, Leo AJ; Medchem Project Issue No 26. Claremont CA: Pomona College (1985)
7. Hauser TR, Bromberg SM; Env Monit Assess 2: 249-72 (1982)
8. IARC; Some Industrial Chemicals and Dyestuffs 29: 49-91 (1982)
9. Jacobson SN, Alexander M; Appl Environ Microbiol 42: 1062-6 (1981)
10. James RH et al; J Air Pollut Control Assoc 35: 959-61 (1984)
11. Kitano M; Biodegradation and Bioaccumulation Test on Chemical Substances. OECD Tokyo Meeting Reference Book Tsu-No. 3 (1978)
12. Koskikallio J; Acta Chem Scand 21: 397-407 (1967)
13. Lyman WJ et al; Handbook of Chem Property Estimation Methods. Environ Behavior of Organic Compounds McGraw-Hill NY (1982)
14. NIOSH; National Occupational Exposure Survey (1985)
15. Ohnishi R, Tanabe K; Bull Chem Soc Japan 41: 2647-9 (1971)
16. Perry RH, Green D; Perry's Chemical Engineer's Handbook, 6th ed. McGraw-Hill Book Co., pp. 3-50 (1984)
17. Pleil JD et al; Volatile Organic Compounds in Indoor Air: A Survey of Various Structures USEPA-600/D-85-100 NTIS PB85-198356 (1985)
18. Rittfeldt L et al; Scand J. Work Environ Health 9: 367-8 (1983)
19. Robertson RE, Scott JMW; J Chem Soc 1961: 1596-604 (1961)
20. Sasaki S; pp. 283-98 in Aquatic Pollutants Transformations and Biological Effects Pergamon Press NY (1978)
21. Sheldon LS, Hites RA; Environ Sci Technol 12: 188-94 (1978)
22. Singh HB et al; Environ Sci Technol 16: 872-80 (1982)
23. Swann RL et al; Residue Rev 85: 17-28 (1983)
24. Tanabe K, Sano T; Hokkaido Daigaku 10: 173-82 (1962)

Bis(2-chloroethyl) Ether

SUBSTANCE IDENTIFICATION

Synonyms:

Structure:

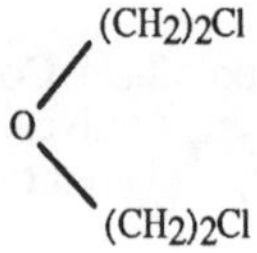

O
/ (CH₂)₂Cl
\ (CH₂)₂Cl

CAS Registry Number: 111-44-4

Molecular Formula: $C_4H_8Cl_2O$

Wiswesser Line Notation: G2O2G

CHEMICAL AND PHYSICAL PROPERTIES

Boiling Point: 178 °C

Melting Point: -24.5 °C

Molecular Weight: 143.02

Dissociation Constants:

Log Octanol/Water Partition Coefficient: 1.29 [11]

Water Solubility: 1,020 mg/L at 20 °C [25]

Vapor Pressure: 1.55 mm Hg at 25 °C [25]

Henry's Law Constant: 2.86 x 10⁻⁴ atm-m³/mole (calculated) [18]

ENVIRONMENTAL FATE/EXPOSURE POTENTIAL

Summary: Bis(2-chloroethyl) ether (BCEE) is most likely released to the environment from the use of products containing the compound. Chlorination of drinking water containing ethyl ether can result in the formation of BCEE. Release of BCEE to water is expected to result in hydrolysis (estimated half-life 40 days) and volatilization. BCEE biodegrades in water following several weeks of acclimation. Aqueous photolysis and photooxidation are not expected to be important processes in the aquatic fate of BCEE. Bioconcentration in aquatic organisms is extremely low. When released to soil, BCEE may hydrolyze and is expected to leach extensively to ground water. A half-life of 13.44 hr was estimated for the reaction of BCEE with photochemically produced hydroxyl radicals in the atmosphere. Direct atmospheric photolysis is not expected to be important, since BCEE should not absorb light of wavelengths above 290 nm. Monitoring studies indicate that BCEE is a contaminant in air, water, sediment, and soil. Human exposure probably results primarily from drinking water contaminated with BCEE.

Natural Sources:

Artificial Sources: The mean loading in treated wastewater from foundries was 0.029 kg/day [29].

Terrestrial Fate: In a 140 cm long column containing Lincoln fine sand collected in Ada, OK, 86% of the applied BCEE reached 140 cm, indicating low adsorption to this soil [30]. Using the water solubility, a log soil sorption coefficient (Koc) of 1.38 was estimated [18]. A Koc of this magnitude indicates that BCEE should be highly mobile in soil [14] and should therefore leach rapidly to ground water. No data concerning the hydrolysis of BCEE in the soil were available. A hydrolysis half-life of 40 days at pH 7 and an unspecified temperature for BCEE was estimated from ethyl chloride data in water, since BCEE is a primary chloride similar to ethyl chloride [3]; thus, some hydrolysis may also occur in soil. No biodegradation data using mixed cultures of soil microorganisms were available. Screening biodegradability test data are not completely consistent, but do suggest that BCEE may biodegrade in soil and that acclimation may be necessary.

Bis(2-chloroethyl) Ether

Aquatic Fate: A hydrolysis half-life for BCEE of 40 days at pH 7 was estimated at an unspecified temperature in water based upon analogy to ethyl chloride, since both these compounds are primary chlorides [3]. No data concerning volatilization of BCEE from natural water bodies were available. Volatilization half-life values for BCEE volatilization were estimated from the Henry's Law constant to be 3.5, 4.4, and 180.5 days for the streams, rivers, and lakes, respectively [18]. No data on the aqueous photolysis of BCEE were available, but photolysis is not expected to be an important fate process. Biodegradation in river water appears to take place (50% theoretical CO_2 observed in 35 days) and acclimation appears to be important [17].

Atmospheric Fate: An atmospheric half-life of 13.44 hr was estimated for the reaction of BCEE with photochemically generated hydroxyl radicals in the atmosphere [10]. Direct photolysis in the atmosphere is not expected to be a major significant degradation process, since no functional groups are present in BCEE that would lead to absorption of sunlight.

Biodegradation: BCEE was added to Ohio River water at pH 7.2 and 22-25 °C [17]. Thirty-five days following the addition of BCEE to the water, 50% of theoretical carbon dioxide was recovered. After a second addition of BCEE, only 9-10 days were required to observe 50% recovery of theoretical carbon dioxide, suggesting acclimation was necessary for optimal biodegradation rates [17]. The absence of evidence of biodegradation in another study [6] may be attributed to an insufficiently long acclimation period (18 days).

Abiotic Degradation: A hydrolysis half-life for BCEE of 40 days at pH 7 was estimated from ethyl chloride data in water at an unspecified temperature [3]. Photolysis is not expected to be significant since BCEE has no conjugated, unsaturated system. An atmospheric half-life of 13.44 hr was estimated for the reaction of BCEE with hydroxyl radicals [10].

Bioconcentration: A bioconcentration factor (BCF) of 10.96 was determined using liquid scintillation analysis of radioactivity for ^{14}C labelled BCEE in bluegills after 14 days [4]. A BCF of 11 was observed in bluegills exposed to BCEE for 14 days [1]. Based on these BCF values, BCEE is not expected to bioconcentrate significantly in aquatic organisms.

Soil Adsorption/Mobility: In a 140 cm long column containing Lincoln fine sand collected in Ada, OK, 86% of the applied BCEE reached 140 cm, indicating minimal adsorption to this soil [30]. Using the water solubility, a log soil sorption coefficient (Koc) of 1.38 was estimated [18]. A Koc of this magnitude indicates that BCEE should be highly mobile in soil [14] and should therefore readily leach to ground water.

Volatilization from Water/Soil: No data concerning volatilization of BCEE from natural water bodies were available. Volatilization half-life values for BCEE volatilization from lakes, rivers, and streams were estimated [18] using the Henry's Law constant. The depths of the lakes were assumed to be 50 m, and that of the rivers and streams, 1 m. A wind velocity of 3 m/sec was assumed and the current velocities of the lakes, rivers, and streams were assumed to be 0.01, 1 and 2, respectively. The half-life values were 3.5, 4.4, and 180.5 days for the streams, rivers, and lakes, respectively. Loss of BCEE due to volatilization was 0% from a microcosm simulating high-rate application of municipal wastewater to land [24].

Water Concentrations: SURFACE WATER: The median BCEE concentration in water samples taken across the United States is <10.0 ug/L based on 808 samples, of which 3 contained detectable BCEE residues. These data are contained in the USEPA STORET database and only an unspecified portion pertains to surface water [27]. Trace concentrations of BCEE were found in five surface water samples taken from sites along the Delaware River in March 1977 [26]. No BCEE was detected in 11 surface water samples taken from the river in August 1976 [26]. The BCEE concentration in three samples of surface water taken from New Orleans and Baton Rouge ranged from 0.040 to 0.16 ug/L with a mean BCEE concentration of 0.11 ug/L [21]. One surface water sample taken in the Houston area contained 1.4 ug/L [21]. Raw water samples collected in unspecified locations in the United Kingdom contained BCEE, but no quantitative data were presented [8]. Identified, not quantified in bankfiltered Rhine water [22].   GROUND WATER:

Bis(2-chloroethyl) Ether

BCEE was detected but not quantified in samples of well water collected near a solid waste landfill located 60 miles southwest of Wilmington, DE [5]. Leachates from wells near low level radioactive waste disposal sites contained BCEE, but no quantitative data were presented [9].   DRINKING WATER: Drinking water samples taken from New Orleans, LA contained 0.04 ug/L BCEE [13]. Samples of drinking water obtained from unspecified locations in the United States and the Netherlands contained 0.42 ug/L and 0.1 ug/L BCEE, respectively [16]. Drinking water samples taken in the Netherlands contained a maximum of 30 ng/L BCEE [23]. BCEE was detected but not quantified in drinking water samples from Evansville, IN [15], Philadelphia, PA [13] and from an unspecified treatment plant in the United Kingdom [8]. Identified, not quantified, in drinking water from the Torresdale Water Treatment Plant in Philadelphia, PA, Feb 1975-Nov 1976, 7 samples, 86% pos [28].

Effluent Concentrations: BCEE residues in treated wastewater effluents from several industries were as follows: foundries - 2 samples, 100% positive, 8-9 ug/L, 8.5 ug/L avg; organic chemicals manufacturing/plastics, 3 of an unspecified number of samples pos, 710 ug/L avg; paint and ink formulation, unspecified number of samples or number pos, >10 ug/L maximum concentration [29]. Synthetic rubber plant effluent, 0.16 mg/L [7].

Sediment/Soil Concentrations: Soil and sediment samples collected at Love Canal, NY, contained BCEE but no quantitative data were presented [12].

Atmospheric Concentrations: BCEE was detected but not quantified in the atmosphere above the Lipari and BFI landfills in New Jersey [2].

Food Survey Values:

Plant Concentrations:

Fish/Seafood Concentrations: Concn (ppb, wet weight) in seafood from areas of Lake Pontchartrain, LA, May-Jun, 1980,

Bis(2-chloroethyl) Ether

Inner Harbor Navigation Canal, oysters, 8 samples, 0.6 avg; Chef Menteur Pass, clams, and The Rigolets, clams, not detected in composite samples [19].

Animal Concentrations:

Milk Concentrations:

Other Environmental Concentrations:

Probable Routes of Human Exposure: The most probable route of human exposure is the ingestion of drinking water contaminated with BCEE.

Average Daily Intake:

Occupational Exposures: NIOSH (NOHS Survey 1972-1974) has statistically estimated that 42 workers are exposed to BCEE in the United States [20].

Body Burdens:

REFERENCES

1. Barrows ME et al; pp 379-92 in: Dynamics, Exposure and Hazard Assessment of Toxic Chemicals. Ann Arbor Science Ann Arbor MI (1980)
2. Bozzelli JW et al; Analysis of Selected Toxic and Carcinogenic Substances in Ambient Air in New Jersey. State of New Jersey Dept of Environ Protection (1980)
3. Brown et al; Research Program on Hazard Priority Ranking of Manufactured Chemicals Stanford Research Institute Menlo Park NTIS PB-263164 (1975)
4. Davies RP, Dobbs AJ; Water Res 18: 1253-62 (1984)
5. DeWalle FB, Chian ESK; J Am Water Works Assn 73: 206-11 (1981)
6. Dojlido JR; Investigations of Biodegradability and Toxicity of Organic Compounds; Final Report 1975-79 USEPA-600/2-79-163 (1979)
7. Durkin PR et al; Investigation of Selected Potential Environmental Contaminants: Haloethers, Final Report USEPA-560/2-75-006 (1975)
8. Fielding M et al; Organic Micropollutants in Drinking Water Medmenham Eng Water Res Center TR-159 (1981)
9. Francis AJ et al; Nucl Technol 50: 158-63 (1980)
10. GEMS; Graphical Exposure Modeling System. Fate of atmospheric pollutants (FA) data base. Office of Toxic Substances. USEPA (1986)
11. Hansch C, Leo AJ; Medchem Project Issue No 26. Claremont CA: Pomona College (1985)

Bis(2-chloroethyl) Ether

12. Hauser TR, Bromberg SM; Environ Monit Assess 2: 249-72 (1982)
13. Keith LH et al; pp. 329-93 in Identification and analysis of organic pollutants in water. Ann Arbor Press Ann Arbor (1976)
14. Kenaga EE; Ecotox Env Safety 4: 26-38 (1980)
15. Kleopfer RD, Fairless BJ; Environ Sci Technol 6: 1036-7 (1972)
16. Kraybill HF; NY Acad Sci Annals 298: 80-9 (1977)
17. Ludzak FJ, Ettinger MB; Proc 18th Indus Waste Conf Purdue Univ Eng Bull Ext Ser 18: 278-82 (1971)
18. Lyman WJ et al; Handbook of Chem Property Estimation Methods. Environ Behavior of Org Compounds McGraw-Hill NY (1982)
19. McFall JA et al; Chemosphere 14: 1561-9 (1985)
20. NIOSH; The National Occupational Hazard Survey (NOHS) (1974)
21. Pellizzari ED et al; Formulation of Preliminary Assessment of Halogenated Organic Compounds in Man and Environmental Media USEPA-560/13-79-006 (1979)
22. Piet et al; pp. 69-80 in Hydrocarbon and Halogenated Hydrocarbon Aquatic Environ; Afghan BK, Mackay D, eds (1980)
23. Piet GJ, Morra CF; pp. 31-42 in Artificial Groundwater Recharge. Huisman L, Olsthorn TN eds; Pitman Pub (1983)
24. Piwoni MD et al; Haz Waste Haz Mat 3: 43-55 (1986)
25. Riddick JA et al; Organic Solvents: Physical Properties and Methods of Purification, 4th Edit. New York: J Wiley & Sons (1986)
26. Sheldon LS, Hites RA; Environ Sci Technol 12: 1188-94 (1978)
27. Staples CA et al; Environ Toxicol Chem 4: 131-42 (1985)
28. Suffet IH et al; Water Res 14: 853-67 (1980)
29. USEPA; Treatability Manual Vol 1 Treatability Data USEPA-600/2-81-001a (1981)
30. Wilson JT et al; J Environ Qual 10: 501-6 (1981)

Bis(chloromethyl) Ether

SUBSTANCE IDENTIFICATION

Synonyms:

Structure:

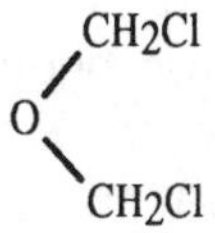

$$O \begin{array}{c} CH_2Cl \\ \\ CH_2Cl \end{array}$$

CAS Registry Number: 542-88-1

Molecular Formula: $C_2H_4Cl_2O$

Wiswesser Line Notation: G1O1G

CHEMICAL AND PHYSICAL PROPERTIES

Boiling Point: 106 °C

Melting Point: -41.5 °C

Molecular Weight: 114.97

Dissociation Constants:

Log Octanol/Water Partition Coefficient: Hydrolyzes rapidly in water [5]

Water Solubility: Hydrolyzes rapidly in water [5]

Vapor Pressure: 30 mm Hg at 22 °C [3]

Henry's Law Constant:

ENVIRONMENTAL FATE/EXPOSURE POTENTIAL

Summary: Bis(chloromethyl) ether (BCME) might possibly enter the atmosphere in exhaust gases associated with the use of chloromethyl methyl ether in which it is a contaminant, as a chemical intermediate, or in situations where it may be formed from hydrochloric acid and formaldehyde in the atmosphere. Due to its rapid rate of hydrolysis (half-life <1 min), BCME would not be expected to persist in water. It is, however, relatively stable in air (estimated half-life <1 day for photooxidation and >18 hr for hydrolysis) and may therefore be an occupational contaminant if it escapes in exhaust gases or is formed in the workplace.

Natural Sources: Does not occur in nature [6].

Artificial Sources: Possibly present in exhaust gases from plants where it is used as a chemical intermediate; however, it would not be expected to occur in wastewater from such plants [6]. There are conflicting data on whether BCME can be formed spontaneously from formaldehyde and HCl in some textile plants [5].

Terrestrial Fate:

Aquatic Fate: BCME will rapidly disappear from any aquatic system by hydrolysis (half-life 10-38 sec).

Atmospheric Fate:

Biodegradation:

Abiotic Degradation: Rapid hydrolysis (half-life 10-38 sec) [3,5,11], slightly faster in base and slower in acid [11]. Hydrolysis half-life in humid air >25 hr [5]. Another report claimed that BCME was stable in air (70% relative humidity) for at least 18 hr [8]. Results suggest that the reaction is surface catalyzed and extrapolation from lab data to the environment is not justified. Therefore the half-life should be considered a lower limit [4]. The decomposition products are HCl and formaldehyde [5]. The estimated half-life for oxidation by photochemically produced hydroxyl radicals is <1 day [3]. The rate constant for reaction with photochemically produced hydroxyl radicals has been estimated to be 3.9×10^{-12} cm^3/molecule-sec at 25 °C, which corresponds to an estimated half-life of 4.1 days based on a hydroxyl radical concn of $5 \times 10^{+5}$ molecules/cm^3 [2].

Bioconcentration: Hydrolysis half-lives are sufficiently fast to preclude any possibility of bioconcentration in the food chain. Reported BCF for bluegill sunfish, 11 [12].

Soil Adsorption/Mobility:

Volatilization from Water/Soil: BCME hydrolyzes so rapidly in water [5] that evaporation is not an important process. Based on the vapor pressure, evaporation from surfaces and dry, near-surface soil may be an important process.

Water Concentrations: DRINKING WATER: Identified, not quantified, in U.S. drinking water [7] and New Orleans, LA, drinking water [1]. SURFACE WATER: USEPA STORET database, 317 surface water data points, 0% pos [10].

Effluent Concentrations: STORET database, 977 effluent data points, 1.4% pos, median < 1.0 ug/L [10]. Municipal solid waste leachate, WI, 5 sites, 20% pos, 250 ppb [9].

Sediment/Soil Concentrations: STORET database, 213 sediment data points, 0% pos [10].

Atmospheric Concentrations:

Food Survey Values:

Plant Concentrations:

Fish/Seafood Concentrations:

Animal Concentrations:

Milk Concentrations:

Bis(chloromethyl) Ether

Other Environmental Concentrations: 1-7% of commercial chloromethyl methyl ether (CMME) [4].

Probable Routes of Human Exposure:

Average Daily Intake:

Occupational Exposures: In two textile mills using formaldehyde resins, magnesium chloride, and zinc nitrate catalysts in the permanent press process, the maximum BCME detected was 2.7 ppb [4]

Body Burdens:

REFERENCES

1. Abrams EF et al; Identification of Organic Compounds in Effluents From Industrial Sources p. D-3 USEPA-560/3-75-002 (1975)
2. Atkinson R; Intern J Chem Kinetics 19: 799-828 (1987)
3. Callahan MA et al; Water Related Environ Fate of 129 Priority Pollut vol.II USEPA-440/4-79-029B (1979)
4. Durkin PR et al; Investigation of Selected Potential Environmental Contaminants: Haloethers; p. 21 USEPA-560/2-75-006 (1975)
5. Fishbein L; Sci Total Environ 11: 223-57 (1979)
6. IARC; Some Aromatic Amines, Hydrazine and Related Substances, N-Nitroso Compounds and Miscellaneous Alkylating Agents; 4: 233 (1974)
7. Kopfler FC et al; Adv Environ Sci Technol 8: 419-33 (1977)
8. Nichols RW, Merritt RF; J Nat Cancer Inst 50:1373-4 (1973)
9. Sabel GV, Clark TP; Waste Management Res 2: 119-30 (1984)
10. Staples CA et al; Environ Toxicol Chem 4: 131-4 (1985)
11. Tou JC et al; J Phys Chem 78: 1096-8 (1974)
12. Vieth GD et al; pp. 116-29 in Aquatic Toxicology; Easton JG et al eds (1980)

Bisphenol A

SUBSTANCE IDENTIFICATION

Synonyms: 2,2-(4,4-Dihydroxydiphenyl)propane

Structure:

CAS Registry Number: 80-05-7

Molecular Formula: $C_{15}H_{16}O_2$

Wiswesser Line Notation: QR DX1&1&R DQ

CHEMICAL AND PHYSICAL PROPERTIES

Boiling Point: 220 °C at 4 mm Hg

Melting Point: 150-155 °C

Molecular Weight: 228.28

Dissociation Constants:

Log Octanol/Water Partition Coefficient: 3.32 [5]

Water Solubility: 120 mg/L at 25 °C [2]

Vapor Pressure: 4×10^{-8} mm Hg at 25 °C [11] (estimated value)

Henry's Law Constant: 1×10^{-10} atm-m³/mole (calculated from water solubility and vapor pressure)

Bisphenol A

ENVIRONMENTAL FATE/EXPOSURE POTENTIAL

Summary: The primary sources of environmental release of bisphenol A are expected to be effluents and emissions from its manufacturing facilities and facilities which manufacture epoxy, polycarbonate, and polysulfone resins. If released to soil, bisphenol A is expected to have moderate to low mobility. This compound may biodegrade under aerobic conditions following acclimation. If released to acclimated water, biodegradation would be the dominant fate process (half-life less than or equal to 4 days). In nonacclimated water, bisphenol A may biodegrade after a sufficient adaption period, it may adsorb extensively to suspended solids and sediments, or it may photolyze. If released to the atmosphere, bisphenol A is expected to exist almost entirely in the particulate phase. Bisphenol A in particulate form may be removed from the atmosphere by dry deposition or photolysis. The small fraction of bisphenol A which would exist in the vapor phase may react with photochemically generated hydroxyl radicals (half-life 4 hours) or it may photolyze. Photodegradation products of bisphenol A vapor are phenol, 4-isopropylphenol, and a semiquinone derivative of bisphenol A. The most probable routes of human exposure to bisphenol A are inhalation and dermal contact of workers involved in the manufacture, use, transport, or packaging of this compound or use of epoxy powder paints.

Natural Sources:

Artificial Sources: The primary sources of environmental release of bisphenol A are expected to be effluents and emissions from manufacturing and use facilities. The primary uses of this compound are in the manufacture of epoxy, polycarbonate, and polysulfone resins [1,17]. Environmental concentrations resulting from treated process effluent discharge are expected to be <0.1 mg/L [2]. Bisphenol A is a thermal degradation product of epoxy resin and powder paint used to paint metal objects [15,16].

Terrestrial Fate: If released to soil, bisphenol A is expected to have moderate to low mobility. Bisphenol A may biodegrade under aerobic conditions following acclimation. This compound is not expected to undergo chemical hydrolysis or volatilize significantly from soil surfaces.

Bisphenol A

Aquatic Fate: If released to acclimated water, biodegradation would be the dominant fate process (half-life less than or equal to 4 days). In nonacclimated waters, bisphenol A may biodegrade after a sufficient adaption period, it may adsorb extensively to suspended solids and sediments, or it may photolyze. This compound is not expected to bioaccumulate significantly in aquatic organisms, volatilize, or undergo chemical hydrolysis.

Atmospheric Fate: Based on the estimated vapor pressure, bisphenol A is expected to exist almost entirely in the particulate phase in the atmosphere [3]. Bisphenol A particles may be removed from the atmosphere by dry deposition or photolysis. The small fraction of bisphenol A which would exist in the vapor phase may react with photochemically generated hydroxyl radicals (half-life 4 hours) or it may photolyze. Photodegradation products of bisphenol A vapor are phenol, 4-isopropylphenol, and a semiquinone derivative of bisphenol A.

Biodegradation: Half-life of 3 mg/L bisphenol A in natural receiving waters: bisphenol A plant discharge, 3 days; Patrick's Bayou water (obtained 200 yards downstream from a bisphenol A plant discharge), 2.5 days; and Houston Ship Channel water, 4 days [2]. Loss was attributed to biodegradation since 3 mg/L bisphenol A in a control sample (deionized water) underwent no observable change in concentration over an 8-day test period. 105 mg/L bisphenol A, acclimated activated sludge inocula from an industrial wastewater treatment plant, 72% COD removal in 24 hours [8]. OECD biodegradation screening test, domestic sewage as seed, <1% degradation in 28 days [2]. Bisphenol A was not oxidized in either the Closed Bottle Test or the Modified Sturm Test [2]. Japanese MITI Test: 100 mg/L substrate, 30 mg/L activated sludge, <30% degradation in 2 weeks [6,19].

Abiotic Degradation: Bisphenol A is not expected to undergo chemical hydrolysis under environmental conditions, since it contains no hydrolyzable functional groups [7]. In neutral and acidic methanol solutions, bisphenol A exhibits slight absorption of UV light wavelengths >290 nm, while in basic methanol solution, bisphenol A exhibits significant absorption of UV >290 nm [18]. These data indicate that bisphenol A has the potential to photolyze in water, and that this potential is somewhat greater under basic conditions. These data also indicate that bisphenol A has potential to undergo photolysis in the atmosphere. Photodecomposition products of bisphenol A vapor are phenol, 4-isopropylphenol, and a semiquinone derivative of bisphenol A [14]. The half-life for bisphenol A vapor reacting with photochemically generated hydroxyl radicals in the atmosphere has been estimated to be 4 hours based on a reaction rate constant of 6.0 x 10^{-11} cm^3/molecule-sec at 25 °C and an average hydroxyl radical concentration of 8.0 x 10^{+5} molecules/cm^3 [4]. However, bisphenol A is expected to exist almost entirely in the particulate phase in the atmosphere, and reaction with hydroxyl radicals is expected to be much slower in particulate form than in vapor form.

Bioconcentration: A bioconcentration factor (BCF) of <100 was measured for bisphenol A in carp [6]. BCF of 42 and 196 were estimated for bisphenol A based on the water solubility and the log Kow, respectively [7]. These BCF values indicate that bisphenol A should not bioaccumulate significantly in aquatic organisms.

Soil Adsorption/Mobility: Soil adsorption coefficients (Koc) of 314 and 1524 have been estimated for bisphenol A based on the water solubility and the log Kow, respectively [7]. These Koc values suggest that mobility of bisphenol A in soil would be moderate to low and that adsorption to suspended solids would be moderate to extensive [21].

Volatilization from Water/Soil: The value of Henry's Law constant suggests that volatilization would be insignificant from all bodies of water [7]. Due to its relatively low vapor pressure and its tendency to adsorb to soil, bisphenol A is not expected to volatilize significantly from wet or dry soil surfaces.

Water Concentrations: SURFACE WATER: Water samples collected from rivers in the Tokyo, Japan area, 1974-1978, concn range of pos. samples 0.06-1.9 ug/L [9].

Effluent Concentrations: Dec. 1974, qualitatively identified in the effluent from a chemical industry in Mt. Vernon, IN [20].

Bisphenol A

Sediment/Soil Concentrations: Not detected in soil samples taken from residential area of Tokyo, Japan, where bisphenol A had been detected in atmospheric fallout [10].

Atmospheric Concentrations: Atmospheric fallout collected from a residential area of Tokyo, Japan between Feb. 1976 and Jan. 1978, 8 samples, 63% pos., deposition rate of bisphenol A in pos. samples 0.04-0.2 ug/sq m-day [10].

Food Survey Values:

Plant Concentrations:

Fish/Seafood Concentrations:

Animal Concentrations:

Milk Concentrations:

Other Environmental Concentrations:

Probable Routes of Human Exposure: The most probable routes of human exposure to bisphenol A are inhalation and dermal contact of workers involved in the manufacture, use, transport, or packaging of this compound or use of epoxy powder paint [16].

Average Daily Intake:

Occupational Exposures: NIOSH has estimated that 23,340 workers are potentially exposed to bisphenol A based on statistical estimates derived from a survey conducted 1981-83 in the U.S. [12]. NIOSH has estimated that 281,011 workers are potentially exposed to bisphenol A based on statistical estimates derived from the survey conducted 1972-74 in the U.S. [13].

Body Burdens:

REFERENCES

1. Chemical Marketing Reporter; Chemical Profile: Bisphenol A NY: Schnell Publishing July 16 (1984)

Bisphenol A

2.   Dorn PB et al; Chemosphere 16: 1501-7 (1987)
3.   Eisenriech SJ et al; Environ Sci Tech 15: 30-8 (1981)
4.   GEMS; Graphical Exposure Modelling System. FAP. Fate of Atmospheric Pollutants. (1987)
5.   Hansch C, Leo AJ; Medchem Project Issue no. 26 Claremont, CA: Pomona College (1985)
6.   Kawasaki M; Ecotoxic Environ Safety 4: 444-54 (1980)
7.   Lyman WJ et al; Handbook of Chemical Property Estimation Methods. McGraw-Hill, NY: (1982)
8.   Matsui S et al; Prog Water Technol 7:645-59 (1975)
9.   Matsumoto G; Water Res 16: 551-7 (1982)
10. Matsumoto G, Hanya T; Atmos Environ 14: 1409-19 (1980)
11. Neely WB, Blau GE; p. 30-2 in Environmental Exposure from Chemicals Vol. 1 Boca Raton, FL: CRC Press (1985)
12. NIOSH; National Occupational Exposure Survey (NOES) (1985)
13. NIOSH; National Occupational Hazard Survey (NOHS) (1974)
14. Peltonen K et al; Photochem Photobiol 43: 481-4 (1986)
15. Peltonen K; J Anal Appl Pyrolysis 10: 51-7 (1986)
16. Peltonen K et al; Am Ind Hyg Assoc J 47: 399-403 (1986)
17. Reed HWB; Kirk-Othmer Encycl Chem Tech 3rd ed NY: Wiley 2: 90 (1978)
18. Sadtler; Standard UV Spectra No. 325 Philadelphia, PA: Sadtler Research Lab (1966)
19. Sasaki S; pp. 283-98 in Aquatic Pollutants: Transformation and Biological Effects Hutzinger O et al eds Oxford: Pergamon Press (1978)
20. Shackelford WM, Keith LH; Frequency of Organic Compounds Identified in Water p.88 USEPA-600/4-76-062 (1976)
21. Swann RL et al; Res Rev 85: 17-28 (1983)

1,3-Butadiene

SUBSTANCE IDENTIFICATION

Synonyms:

Structure:

$$H_2C \diagup\diagdown CH_2$$

CAS Registry Number: 106-99-0

Molecular Formula: C_4H_6

Wiswesser Line Notation: 1U2U1

CHEMICAL AND PHYSICAL PROPERTIES

Boiling Point: -4.5 °C at 760 mm Hg

Melting Point: -108.91 °C

Molecular Weight: 54.09

Dissociation Constants:

Log Octanol/Water Partition Coefficient: 1.99 [6]

Water Solubility: 735 mg/L at 25 °C [14]

Vapor Pressure: 856 mm Hg at -1.5 °C [2]

Henry's Law Constant: 2.57 [8]

ENVIRONMENTAL FATE/EXPOSURE POTENTIAL

Summary: 1,3-Butadiene is a gas at most environmental temperatures and very volatile even at the lower temperatures. Its release to the atmosphere is as emissions from motor vehicles; burning of fossil fuels; manufacture, transport, use, and disposal of petroleum, plastic, and synthetic rubber as well as from its use as a chemical intermediate. Release to the aquatic environment in wastewaters from the above uses may also be significant. Once in the atmosphere, 1,3-butadiene will photooxidize by reaction primarily with hydroxyl radicals as well as other species, with an estimated half-life of several hours (less in smog and polluted air). Amounts released into water or land will rapidly decrease due to evaporation and possibly also due to biodegradation. The compound may leach through soil to ground water. The primary source of human exposure is from the air, especially from urban atmospheres and those around heavy traffic and manufacturing plants using the chemical. Exposure from water is probably minor, although 1,3-butadiene has been detected in drinking water.

Natural Sources: Emissions from forest fires [5].

Artificial Sources: Emissions from motor vehicles; petroleum, plastics, and synthetic rubber manufacture; tobacco smoke [5]. Release associated with its manufacture, transport, storage, and end use in polymers and as a chemical intermediate [23].

Terrestrial Fate: If spilled on land, 1,3-butadiene will predominately volatilize very rapidly due to its very low boiling point. Dissolved in water, it may leach through soil into ground water due to its high water solubility and low estimated soil adsorption coefficient. It will not appreciably hydrolyze but may be subject to biodegradation based on screening tests.

Aquatic Fate: When released into water, 1,3-butadiene will volatilize rapidly with a half-life estimated to be several hours. It will not hydrolyze appreciably, but may be subject to biodegradation, based on screening tests.

Atmospheric Fate: Reaction with hydroxyl radicals is the dominant removal mechanism, with an estimated half-life of several hours. Reaction with ozone and nitrate radicals may also contribute to the degradation of the chemical. Polluted urban atmospheres increase the rate of degradation somewhat during daylight hours as suggested by the detection of the highest atmospheric levels of the chemical in the early morning hours. Acetaldehyde and acrolein have been identified as products of photooxidation. Washout may contribute to removal of 1,3-butadiene from the atmosphere; however, evaporation from the rain may be rapid and the compound returned to the atmosphere relatively quickly unless it leaches into the soil.

Biodegradation: No data concerning the biodegradation of 1,3-butadiene in natural systems could be found in the literature. 1,3-Butadiene was listed in a group of chemicals which should be biodegraded by biological sewage treatment, as long as suitable acclimatization is achieved [22]. Screening tests suggest that 1,3-butadiene may be biodegradable in the environment with 1,2-epoxybutene being a potential product [9].

Abiotic Degradation: Alkenes are known to be resistent to hydrolysis [13]. The estimated half-life for 1,3-butadiene due to photooxidation with hydroxyl radicals is 3.1 hours [13] based on a rate constant of 0.77×10^{-10} cm^3/molecule-sec [5] and hydroxyl radical concn of $8 \times 10^{+5}$ molecule/cm^3 [7]. Within 5 hours, 100% degradation of 1,3-butadiene was observed in a smog chamber artificially irradiated [24] and in 6 hours in Los Angeles air irradiated with sunlight [11]. When the compound was irradiated with black lights in Riverside, CA air, 100% degradation was observed within 8 hours with heavy haze and within 4 hours with light haze [21]. The reaction with nitrate radicals has been recognized as an important nighttime sink for some chemicals. The half-life for the reaction of 1,3-butadiene is 15 hr [1]. Stable reaction products of photooxidation are acetaldehyde and acrolein [5].

Bioconcentration: No information on the bioconcentration factor for 1,3-butadiene could be found in the literature. However, based on its octanol/water partition coefficient, its estimated bioconcentration factor is 19.1 [13] and, therefore, the chemical is not expected to appreciably bioconcentrate.

1,3-Butadiene

Soil Adsorption/Mobility: The range of estimated adsorption coefficients for 1,3-butadiene from the soils and sediments is 72-228 [13] based on its octanol/water partition coefficient or its water solubility and would therefore not be expected to appreciably adsorb in soils and sediments.

Volatilization from Water/Soil: Using the Henry's Law constant, the estimated half-life for evaporation of 1,3-butadiene from a river 1 m deep with a 1 m/sec current and a 3 m/sec wind is 3.8 hours [13]. Due to its low boiling point, 1,3-butadiene would be expected to rapidly evaporate from soils.

Water Concentrations: SURFACE WATER: 14 heavily industrialized river basins, 1975-76, 2 ppb (1 of 204 sites positive) [4].

Effluent Concentrations: Wastewater from synthetic rubber manufacturers did not contain 1,3-butadiene (survey of industry) [16]. Estimated total mass of fugitive emission from petrochemical processes, 1112 metric tons/yr [10].

Sediment/Soil Concentrations:

Atmospheric Concentrations: RURAL/REMOTE: 0.1 to 6.5 ppb C (upper limit as cis-2-butene not separated) in Jones State Forest, TX (3.74 ppb C avg; all 15 samples pos, Jan 1978) [20]. SUBURBAN/URBAN: Riverside, CA - 0 to 0.7 ppb (6 samples afternoons with heavy haze; Aug-Nov 1965); 2.0 ppb (moderately heavy haze and clear sky, March 1966) to 9.0 ppb (light haze and partly cloudy, Dec 1965) early morning sampling [21]. Los Angeles central business district: 0 to 9 ppb (6th floor level) [18]. United States - 1.5 ppb (avg of 498 samples, 1977-78) [3]. SOURCE DOMINATED: Houston area including industrial and tunnels: 0-33.3 ppb (ave 24.8 ppb, 16 of 20 samples pos) [12]. United States - 1.9 ppb (avg of 9 samples, 1977-1980) [3]. Ambient concn - 1-9 ppb [5]. Homebush Bay, Australia (industrial area), June 3, 1975, detected, not quantified [17].

Food Survey Values: Apparent concn in foods packaged in 1,3-butadiene (BD) rubber-based plastic containers: olive oil (bottled in BD rubber-modified acrylonitrile bottles), 6 samples, 100% pos, 8-

1,3-Butadiene

9 ppb; vegetable oil (packaged in BD rubber-modified PVC), 2 samples, 0% pos (limit of detection 1 ppb); yogurt (packaged in polystyrene with BD rubber-modified polystyrene lids), 2 samples, 0% pos [15].

Plant Concentrations:

Fish/Seafood Concentrations:

Animal Concentrations:

Milk Concentrations:

Other Environmental Concentrations:

Probable Routes of Human Exposure: Exposure to 1,3-butadiene would be from inhalation of contaminated ambient air, especially in urban and suburban areas. In addition, air close to and inside of the manufacturing plants which produce or use 1,3-butadiene may contain significant amounts of 1,3-butadiene. Exposure also may occur from ingestion of drinking water [23].

Average Daily Intake: Average exposure for urban/suburban residents: AIR INTAKE: (assume typical concentration 1.5 ppb) - 67.5 ug; WATER INTAKE: insufficient data; FOOD INTAKE: insufficient data.

Occupational Exposures: NIOSH (NOHS Survey 1972-1974) has statistically estimated that 69,555 workers are exposed to 1,3-butadiene in the United States [19]. In a later survey (1981-1983), NIOSH has estimated that 17,033 workers are exposed to 1,3-butadiene. Time-weighted average exposure in rubber production plant: 0.22-59 ppm (over 1 month in 1977); air in petrochemical workplace in USSR: 4.1-4.2 ppm [23].

Body Burdens:

REFERENCES

1. Atkinson R et al; Environ Sci Technol 18: 370-5 (1984)

1,3-Butadiene

2. Boublik T et al; The Vapor Pressures of Pure Substances Vol 17 Amsterdam, Netherlands: Elsevier Science Publ (1984)
3. Brodzinsky R, Singh HB; Volatile organic compounds in the atmosphere: An assessment of available data p. 198 SRI Inter contract 68-02-3452 (1982)
4. Ewing BB et al; Monitoring to detect previously unrecognized pollutants in surface waters. Appendix: organic analysis data pp. 75 USEPA-560/6-77-015 (1977)
5. Graedel TE; Chemical Compounds in the Atmosphere Academic Press New York p. 76 (1978)
6. Hansch C, Leo AJ; Medchem Project Issue No 26. Claremont CA: Pomona College (1985)
7. Hewitt CN, Harrison RM; Atmos Environ 19: 545-54 (1985)
8. Hine J, Mookerjee PK; J Org Chem 40: 292-8 (1975)
9. Hou CT et al; Appl Environ Microbiol 46: 171-7 (1983)
10. Hughes TW et al; Chem Eng Prog 75: 35-9 (1979)
11. Kopcynski SL et al; Environ Sci Technol 6: 342 (1972)
12. Lonneman WA et al; Hydrocarbons in Houston air p. 44 USEPA-600/3-79-018 (1979)
13. Lyman WJ et al; Handbook of Chemical Property Estimation Methods. Environmental behavior of organic compounds McGraw-Hill New York (1982)
14. McAuliffe C; J Phys Chem 70: 1267-75 (1966)
15. McNeal TP, Breder CV; J Assoc Off Anal Chem 70: 18-21 (1987)
16. Miller LM; Investigation of selected potential environmental contaminants: Butadiene and its oligomers p. 64 USEPA-560/2-78-008 (1978)
17. Mulcahy MFR et al; Occurrence Control Photochem Pollut, Proc Symp Workshop Session, Papaer No IV, pp. 17 (1976)
18. Neligan RE; Arch Environ Health 5: 581-91 (1962)
19. NIOSH; National Occupational Hazard Survey (1972-1974), National Occupational Exposure Survey (1981-83)
20. Seila RL; Nonurban hydrocarbon concentrations in ambient air north of Houston p. 38 USEPA-500/3-79-010 (1979)
21. Stephens ER, Burleson FR; J Air Pollut Control Assoc 17: 147-53 (1967)
22. Thom NS, Agg AR; Proc Roy Soc Lond B 189: 347-57 (1975)
23. USEPA; CHIP (Draft) 1,3-Butadiene (1981)
24. Yanagihara S et al; Int Clean Air Cong Proc 4th pp. 472-7 (1977)

Butyl Benzyl Phthalate

SUBSTANCE IDENTIFICATION

Synonyms:

Structure:

CAS Registry Number: 85-68-7

Molecular Formula: $C_{19}H_{20}O_4$

Wiswesser Line Notation: QVR BV01R

CHEMICAL AND PHYSICAL PROPERTIES

Boiling Point: 370 °C

Melting Point: -35 °C

Molecular Weight: 312.39

Dissociation Constants:

Log Octanol/Water Partition Coefficient: 4.91 [9]

Water Solubility: 2.69 mg/L at 25 °C [13]

Vapor Pressure: 8.6 x 10^{-6} mm Hg at 20 °C [24]

Henry's Law Constant: 1.3 x 10^{-6} atm-m³/mole (calculated) [18]

Butyl Benzyl Phthalate

ENVIRONMENTAL FATE/EXPOSURE POTENTIAL

Summary: Over 100 million pounds of butyl benzyl phthalate (BBP) were produced in the United States in 1978. BBP is used as a plasticizer for polyvinyl and cellulosic resins, primarily in poly(vinyl chloride). Possible sources of BBP release to the environment are from its manufacture, distribution, and PVC blending operations; however, release from consumer products is expected to be minimal. Most BBP releases will be to soil and water and not to the air. BBP released to soil is expected to adsorb (Koc 65-350) and not to leach extensively, although it has been detected in ground water. BBP released to aquatic systems will adsorb to sediments and biota but will not volatilize significantly (Henry's Law constant $<1 \times 10^{-6}$ atm/mol m^3) except under windy conditions or from shallow rivers. Biodegradation appears to be the primary fate mechanism for BBP. BBP is readily biodegraded in activated sludge, semicontinuous activated sludge, salt water, lake water, and under anaerobic conditions. For example, at an initial concentration of 1 mg/L in a lake water microcosm, primary degradation accounted for >95% loss of BBP in 7 days; after 28 days, 51-65% of BBP had mineralized (ultimate degradation). U.S. workers are potentially exposed to BBP while a larger population of the United States may be exposed to lower concentrations of BBP found in some drinking waters.

Natural Sources:

Artificial Sources: Over 100 million pounds of BBP were produced in 1978 [8]. It is used as a plasticizer for polyvinyl and cellulosic resins [10], primarily in poly(vinyl chloride) (PVC) [8]. The major use of BBP is in flooring materials with minor amounts used in household products. Diffusion of BBP from consumer products is expected to be minimal [8]. Other possible sources of BBP release to the environment are from its manufacture, distribution, and from PVC blending operations [8].

Terrestrial Fate: A measured soil adsorption constant for BBP is 68-350 [8]; thus if released to land it will sorb to soil and should not leach appreciably, although it has been detected in ground water. The most significant fate process for BBP in soil will be biodegradation [8]. Because of its low volatility, evaporation of

Butyl Benzyl Phthalate

BBP from soil is not expected to be significant.

Aquatic Fate: BBP has a log Kow of 4.77 [8]. Thus, BBP released to waters will partition to solids such as sediment and biota [8]. The primary fate mechanism for BBP will be biodegradation [8]. At an initial concentration of 1 mg/L in a lake water microcosm, primary degradation accounted for >95% loss of BBP in 7 days; after 28 days, 51-65% of it had mineralized (ultimate degradation) [8]. Based on the estimated Henry's Law constant, volatilization of BBP from water will not be significant except from shallow rivers or during high wind activity [18]. Photodegradation and hydrolysis will not be significant, since the half-lives for these processes are >100 days [8].

Atmospheric Fate: BBP released to the atmosphere has an estimated half-life of 1-5 days [35]. However, volatilization of BBP to the atmosphere is not expected to be a significant transport mechanism, since its vapor pressure is only 8.6×10^{-6} mm Hg at 20 °C and its Henry's Law constant is $<1.0 \times 10^{-6}$ atm/mol m^3 [24].

Biodegradation: BBP biodegradation has been studied under a variety of laboratory conditions: activated sludge [8,22,23,24], semicontinuous activated sludge [22,27,28], static flask [33], anaerobic [12,29,30], marine [34], river water [8,27] and lake water [8]. In general, BBP is readily degraded by mixed microbial cultures. In activated sludge, for example, 99% of BBP at a mixed concentration of 3.3 mg/L degraded in 48 hours [22]. Under anaerobic conditions with 10% sludge, >90% of BBP at an initial concentration of 20 mg/mL degraded in about a week [30]. Five bacterial isolates from the ocean were able to grow on and hence degrade BBP [34]. In river water, 100% primary degradation was observed after 9 days while the half-life was 2 days [8]. At an initial concentration of 1 mg/L in a lake water microcosm, primary degradation accounted for >95% loss of the BBP in 7 days; after 28 days, 51-65% of it had mineralized (ultimate degradation) [8].

Abiotic Degradation: In general, water and soil abiotic degradation of BBP will not be significant. Photodegradation and hydrolysis half-lives are >100 days [8]. Atmospheric half-life for BBP is estimated to be 1.5 days [7]; however, not much BBP is expected to enter the atmosphere because of its low vapor pressure and Henry's Law constant.

Bioconcentration: Bluegill sunfish exposed to BBP for 21 days at a mean BBP concentration of 9.73 mg/L had a bioconcentration factor of 663. The tissue half-life for BBP was >1 but less than 2 [1]. In a 3-hour exposure to BBP at concentrations from 20-250 mg/L, English sole gills had an uptake efficiency of 42.2% [2].

Soil Adsorption/Mobility: Measured soil adsorption coefficients for BBP ranged from 68 to 350 [8]; thus it will sorb to soil if released to land. Koc of 17,000 was report with Broome County, NY, soil [26].

Volatilization from Water/Soil: Based on the estimated Henry's Law constant, volatilization from water will not be significant except from shallow rivers or during high wind activity [18]. Because of its low volatility, evaporation of BBP from soil is not expected to be significant.

Water Concentrations: DRINKING WATER: BBP was found but not quantified in Cincinnati drinking water [17]. Some maximum reported concn of BBP in drinking water were: 0.1 ppb in Philadelphia [11], 1.8 ug/L in New Orleans [14], and 3.8 ppb (well water) in New York State [3]. GROUND WATER: BBP was detected in 5 out of 39 public water system wells in New York State; the highest concentration found was 38 ppb [15]. BBP was also detected in ground water near a landfill [6]. SURFACE WATER: BBP was found in the Delaware River at 0.6 ppb [11] and at 2.4 ug/L in Mississippi River near St. Louis [16]. BBP was also detected in Lake Michigan water [8]. The median BBP concentration from 1220 stations in the USEPA STORET database was <10 ug/L for ambient U.S. waters [32]. BBP concentrations in the Illinois River, the Mississippi River (below St. Louis), and Lake Superior averaged 0.7, 0.3, and 0.3 ug/L, respectively [19]. Concn (ppt) in Mississippi River, Lake Itasca, MN (river's source), not detected; Cario, IL, 620; 20 miles south of Memphis, TN, 720; Carrolton St. water intake in New   SEAWATER: Mersey Estuary, UK, 17 sites, 100% pos, dissolved fraction, 100% pos, 6-28.5 ppt; particulate fraction, 100% pos, 3.4-18 ppb (dry weight) [25].

Butyl Benzyl Phthalate

Effluent Concentrations: Effluent from a sewage treatment plant contained 100 ppb BBP [11]. The median BBP concentration from 1337 stations in the STORET database was <6 ug/L for U.S. industrial effluents [32]. BBP concentration was 10 ug/L in urban stormwater from Lake Quinsigamend, MA [4].

Sediment/Soil Concentrations: BBP concentration in a sediment was 567 ng/g; the average for the study was <100 ng/g [8]. Average BBP sediment concentrations in the Kanauha River, Lake Erie, the Mississippi River (Memphis), and the Missouri River were 0.13, 0.41, 0.63, and 0.23 mg/g, respectively [19].

Atmospheric Concentrations: BBP was quantified in indoor air at 20 ng/m^3 [36].

Food Survey Values:

Plant Concentrations:

Fish/Seafood Concentrations:

Animal Concentrations:

Milk Concentrations:

Other Environmental Concentrations:

Probable Routes of Human Exposure: The most likely exposure to BBP is occupational; however, a large population may be exposed to BBP in drinking water and in indoor air where BBP products such as flooring materials are used.

Average Daily Intake:

Occupational Exposures: NIOSH (NOHS Survey 1972-1974) has statistically estimated that 68,488 workers are exposed to BBP in the United States [21]. NIOSH (NOES Survey 1981-1983) has statistically estimated that 89,644 workers are exposed to BBP in the United States [20].

Butyl Benzyl Phthalate

Body Burdens: National Human Adipose Tissue Survey, Fiscal Year 1982, 46 tissue composites, 69% pos, wet tissue concn, not detected (< 0.008-0.012)-1.6 ppm, concn in extractable lipid, not detected (< 0.009-0.019)-1.7 ppm [31].

REFERENCES

1. Barrows ME et al; pp. 379-92 in Dynamics, exposure, and hazard assessment of toxic chemicals Hague R ed Ann Arbor Science Publishers Ann Arbor MI
2. Boese BL; Can J Fish Aquat Sci 41: 1713-18 (1984)
3. Burmaster DE; Environ 24: 6-13, 33-6 (1982)
4. Cole RH et al; J Water Pollut Control Fed 56: 898-908 (1984)
5. DeLeon IR et al; Chemosphere 15: 795-805 (1986)
6. Dunlap WJ et al; Organic pollutants contributed to ground water by a landfill pp.96-110 USEPA-600/9-76-004 (1976)
7. GEMS; Graphical Exposure Modeling System Fate of Atmospheric Pollutants (FAP) Data Base Office of Toxic Substances USEPA (1984)
8. Gledhill WE et al; Env Sci Tech 14: 301-5 (1980)
9. Hansch C, Leo AJ; Medchem Project Issue No 26. Claremont CA: Pomona College (1985)
10. Hawley GG; The Condensed Chemical Dictionary 10th ed Van Nostrand Reinhold New York p.162 (1981)
11. Hites RA; p.107-19 in Natl Conf Munic Sludge Manage 8th (1979)
12. Horowitz A et al; Dev Ind Microb 23: 435-44 (1982)
13. Howard PH et al; Environ Toxicol Chem 4: 653-61 (1985)
14. Keith LH et al; pp 329-73 in Identification and analysis of organic pollutants in water. Keith LH ed Ann Arbor MI Ann Arbor Press (1976)
15. Kim NK, Stone DW; Organic chemicals and drinking water; New York State Dept of Health Albany NY (1980)
16. Konasewich D et al; Status report on organic and heavy metal contaminants in the Lakes Erie, Michigan, Huron and Superior Basins; Great Lake Water Qual Board (1978)
17. Kopfler FC et al; Amer Chem Soc Natl Mtg Div Env Chem Prep 15: 185-7 (1975)
18. Lyman WJ et al; Handbook of chemical property estimation methods. Environmental behavior of organic compounds McGraw-Hill NY (1982)
19. Michael PR et al; Environ Toxicol Chem 3: 377-89 (1984)
20. NIOSH; The National Occupational Exposure Survey (NOES) (1983)
21. NIOSH; The National Occupational Hazard Survey (NOHS) (1974)
22. O'Grady DP et al; Appl Env Microb 49: 443-5 (1985)
23. Patterson JW, Kodukula PS; Chem Eng Prog 77: 48-55 (1981)
24. Petrasek AC et al; J Water Pollut Control Fed 55: 1286-96 (1983)
25. Preston MR, Al-Omran LA; Mar Pollut Bull 17: 548-53 (1986)
26. Russell DJ, McDuffie B; Chemosphere 15: 1003-21 (1986)
27. Saeger VW, Tucker ES; Appl Env Microbiol 31: 29-34 (1976)
28. Saeger VW, Tucker ES; Biodegradation of Phthalate Esters Tech Pap Reg Tech Conf Soc Plast Eng Palisades Sect pp.105-13 (1973)
29. Shelton DR, Tiedje JM; Appl Env Microb 47: 850-7 (1984)

Butyl Benzyl Phthalate

30. Shelton DR et al; Env Sci Tech 18: 93-7 (1984)
31. Stanley JS; Broad Scan Analysis of the FY82 Nat'l Human Adipose Tissue Survey Specimens pp. 36,90-1 USEPA-560/5-86-037 (1986)
32. Staples CA et al; Environ Toxicol Chem 4: 131-42 (1985)
33. Sugatt RH et al; Appl Env Microb 47: 601-6 (1984)
34. Taylor BF et al; Appl Env Microb 42: 590-5 (1981)
35. USEPA; Graphical Exposure Modelling System. Fate of atmospheric pollutants (FAP) data base Office of Toxic Substances (1984)
36. Weschler CJ; Environ Sci Technol 18: 648-52 (1984)

Caprolactam

SUBSTANCE IDENTIFICATION

Synonyms: 2-Azacycloheptanone

Structure:

CAS Registry Number: 105-60-2

Molecular Formula: $C_6H_{11}NO$

Wiswesser Line Notation: T7MVTJ

CHEMICAL AND PHYSICAL PROPERTIES

Boiling Point: 139 °C at 12 mm Hg

Melting Point: 70 °C

Molecular Weight: 113.16

Dissociation Constants:

Log Octanol/Water Partition Coefficient: -0.19 [9]

Water Solubility: 5,250,000 mg/L at 25 °C [5]

Vapor Pressure: 0.0019 mm Hg at 25 °C [11]

Henry's Law Constant:

Caprolactam

ENVIRONMENTAL FATE/EXPOSURE POTENTIAL

Summary: Caprolactam may be released to the environment from its manufacturing and use facilities (primarily manufacturers of nylon 6 resins and plastics). Small amounts may be released from products or combustion of products which contain this compound (e.g., nylon materials, plastics, paints, coatings, and floor polishes). If released to soil, caprolactam is expected to be rapidly degraded by both microbial and chemical degradation processes (half-life on the order of 5 to 14 days). Residual caprolactam may leach into ground water. If released to water, aerobic biodegradation and chemical degradation would be the dominant removal processes (half-life 5-14 days). Volatilization, bioaccumulation in aquatic organisms, and adsorption to suspended solids and sediments are not expected to be significant fate processes. If released to the atmosphere, caprolactam is expected to exist almost entirely in the vapor phase. The dominant removal mechanism is expected to be reaction with photochemically generated hydroxyl radicals (half-life 4.9 hours). This compound may also be removed from the atmosphere by wet deposition. The most probable routes of human exposure to this compound are dermal contact and inhalation by workers involved in the manufacture and use of this compound. Small segments of the general population may be exposed by ingestion of contaminated drinking water or dermal contact with paints, coatings, or floor polishes which contain this compound.

Natural Sources: Caprolactam is not known to occur as a natural product [1].

Artificial Sources: Caprolactam may be released to the environment in emissions and effluents from its manufacturing and use facilities (primarily manufacturers of nylon 6 resins and plastics) [1,23]. Small amounts of caprolactam may be released from products which contain this compound such as nylon materials, plastics, paints, coatings, and floor polishes [6,10]. Combustion of these products may also result in the release of this compound to the environment [18].

Terrestrial Fate: If released to soil, caprolactam is expected to be rapidly degraded by both microbial and chemical degradation processes. The degradation rate of caprolactam in soil is expected

115

to be at least as fast, if not faster, than in water. (The half-life in water has been found to range between 5 and 14 days). Residual caprolactam may leach into ground water. Volatilization is not expected to be an important fate process in soil.

Aquatic Fate: If released to water, aerobic biodegradation and chemical degradation would be the dominant removal processes. The biodegradation rate would be dependent upon size of the microbial population, nutrient availability, concentration of caprolactam, and temperature. Based on results of a study using amended and unamended natural water samples, the half-life of caprolactam in water is expected to range between 5 and 14 days [7]. Volatilization, bioaccumulation in aquatic organisms, and adsorption to suspended solids and sediments are not expected to be significant aquatic fate processes.

Atmospheric Fate: Based on the vapor pressure, caprolactam is expected to exist almost entirely in the vapor phase in the atmosphere [4]. The dominant removal mechanism is expected to be reaction with photochemically generated hydroxyl radicals (half-life 4.9 hours). The complete water solubility of caprolactam suggests that this compound may also be removed from the atmosphere in wet deposition [5].

Biodegradation: GRAB SAMPLES-SURFACE WATER: Grab sample, initial concn 50 ppm, 21 days (% degradation): 50% in sterilized stream water (control); 75% in unsupplemented stream water; 100% in stream water plus sediment; 100% in stream water with yeast extract; 35% in sterilized lake water (control); 72% in unsupplemented lake water; 100% in lake water plus sediment; and 100% in lake water with yeast extract. Grab sample, initial concn 40.4 ppm, 21 days ($\%^{14}CO_2$ evolution): <5% in sterilized stream water (control); <5% in unsupplemented stream water; 8% in stream water plus sediment; 36% in stream water with yeast extract; <5% in sterilized lake water (control); 5% in unsupplemented lake water; 10% in lake water plus sediment; and 50% in lake water with yeast extract. Grab sample, initial concn 50 ppm, 21, 14, and 14 days: 100% degradation in lake water with yeast extract at 10, 20, and 25 °C, respectively. Grab sample, initial concn 40.4 ppm, 21 days: 8, 72, 79% CO_2 evolution in lake water with yeast extract at 10, 20, and 25 °C, respectively. Grab

116

sample, initial concn 2000, 1000, and 100 ppm, 21 days: 36, 85 and 100% degradation, respectively, in lake water supplemented with yeast extract. Grab sample, initial concn 2000, 1000, and 100 ppm, 21 days: 8, 32 and 80% CO_2 evolution, respectively, in lake water supplemented with yeast extract [7]. SCREENING STUDIES: Standard dilute BOD water, 2-day 90% TOC removal, activated sludge inocula [17]. Standard dilute BOD water, initial concn corresponding to 200 mg/L C, 5-day 94.3% COD removal, acclimated activated sludge inocula (vigorous system) [19]. OECD, initial concn corresponding to 40 ppm C, 10 day acclimation, 19-day 93% COD removal [26]. Zahn-Wellens, initial concn equivalent to 1000 mg/L COD, 6 days >90% degradation, 3.5-day lag period, non-adapted acclimated sludge inocula [25]. Zahn-Wellens, 5-day 88% COD removal; Sapromat respirometer, 5 day 82% COD removal; Closed-bottle test, 5-day 10% BODT [24]. A proposed mechanism for metabolism: caprolactam to epsilon-aminocaproic acid to adipic semialdehyde to adipic acid [21].

Abiotic Degradation: Caprolactam, at an initial concentration of 50 ppm, incubated in sterilized lake and stream water for 3 weeks at 20 °C underwent 35 and 50% degradation, respectively [7]. Loss was attributed to primary chemical degradation, since evolution of $^{14}CO_2$ from sterilized water samples was found to be insignificant [7]. Amides are susceptible to chemical hydrolysis under environmental conditions [14]. The half-life for caprolactam vapor reacting with photochemically generated hydroxyl radicals in the atmosphere has been estimated to be 4.9 hours based on an estimated reaction rate constant of 7.9 x 10^{-11} cm³/molecule-sec at 25 °C and an average ambient hydroxyl radical concentration of 5.0 x 10^{+5} molecules/cm³ [3].

Bioconcentration: A bioconcentration factor (BCF) of <1 was estimated for caprolactam based on the measured log Kow [14]. This BCF value and the complete water solubility of caprolactam suggest that this compound will not bioaccumulate significantly in aquatic organisms [5].

Soil Adsorption/Mobility: A soil adsorption coefficient (Koc) of 0.8 was estimated for caprolactam based on the log Kow [14]. This Koc value and the complete water solubility of caprolactam suggest that adsorption to suspended solids and sediments in water would be insignificant and that this compound would be extremely mobile in soil [22].

Volatilization from Water/Soil: The relatively low vapor pressure and complete water solubility of caprolactam suggest that volatilization from water and soil surfaces would not be an important fate process for this compound.

Water Concentrations: DRINKING WATER: July 1975 and Jan. 1976 qualitatively identified in drinking water in the U.S. [20]. Qualitatively identified in drinking water in Germany [12]. GROUND WATER: June 1975 qualitatively identified in well water [20]. April 1974 qualitatively identified in landfill leachate from sites in Dover, DE and Newcastle County, DE [20]. SURFACE WATER: Stream water samples collected Dec. 1983, downstream from tire fire which broke out in Winchester, VA during Oct. 1983, 9700 ug/L - 0.67 km downstream, 250 ug/L - 1.1 km downstream, and 44 ug/L - 9 km downstream [18].

Effluent Concentrations: July 1976, dye manufacturing plant in MA, concn range 36-150 ug/L [8]. Qualitatively identified in water samples taken from advance waste treatment plants in Lake Tahoe, CA Oct. 1974 and Washington, D.C. Sept 1974 [13]. Present in trace amounts in the wastewaters from nylon 6 manufacturing plants in Japan and Russia [10].

Sediment/Soil Concentrations:

Atmospheric Concentrations:

Food Survey Values:

Plant Concentrations:

Fish/Seafood Concentrations:

Animal Concentrations:

Milk Concentrations:

Caprolactam

Other Environmental Concentrations:

Probable Routes of Human Exposure: The most probable routes of human exposure to this compound are dermal contact and inhalation by workers involved in the manufacture and use of this compound [2]. Small segments of the general population may be exposed by ingestion of contaminated drinking water or contact with paints, coatings, or floor polishes which contain this compound.

Average Daily Intake:

Occupational Exposures: NIOSH (NOES Survey 1981-1983) has statistically estimated that 7105 workers are potentially exposed to caprolactam [15]. NIOSH (NOHS Survey 1972-1974) has statistically estimated that 6207 workers are potentially exposed to caprolactam [16].

Body Burdens:

REFERENCES

1. Abrams EF et al; Identification of Organic Compounds in Effluents from Industrial Sources p 29 USEPA-560/3-75-002 (1975)
2. Am Conf Ind Hyg; Appendix: Documentation of the Threshold Limit Values and Biological Exposure Indices 5th ed. pp. 95-6 Cincinnati, OH (1986)
3. Atkinson R; Inter J Chem Kinet 19: 799-828 (1987)
4. Eisenreich SJ et al; Environ Sci Tech 15: 30-8 (1981)
5. Fischer WB, Crescentini L; Kirk-Othmer Encycl Chem Tech 18: 425-36 (1982)
6. Fischer WB, Crescentini L; p 920 in Kirk-Othmer Concise Encycl of Chem Tech NY: Wiley-Interscience (1985)
7. Fortman L, Rosenberg A; Chemosphere 13: 53-65 (1984)
8. Games LM, Hites RA; Anal Chem 49: 1433-40 (1977)
9. Hansch C, Leo AJ; Medchem Project Issue no. 26 Claremont, CA: Pomona College (1985)
10. IARC; Caprolactam and Nylon 6; Inter Agency for Research on Cancer 19: 115-8 (1979)
11. Jones AH; J Chem Eng Data 5:196-200 (1960)
12. Kool HJ; Toxicology Assessment of Organic Compounds in Drinking Water. Crit Rev Env Control 12: 307-57 (1982)
13. Lucas SV; GC/MS Analysis of Organics in Drinking Water Concentrates and Advanced Waste Treatment Concentrates Vol 2 p 145 USEPA-600/1-84-020B NTIS PB85-128239 (1984)

14. Lyman WJ et al; Handbook of Chemical Property Estimation Methods NY: McGraw-Hill (1982)
15. NIOSH; National Occupational Exposure Survey (NOES) (1985)
16. NIOSH; National Occupational Hazard Survey (NOHS) (1974)
17. Pagga U, Guenthner W; pp 498-504 in Comm Eur Communities Eur 7549 Environ Qual Life (1982)
18. Peterson JC et al; Anal Chem 58: 70A-74A (1985)
19. Pitter P; Water Res 10: 231-5 (1976)
20. Shackelford WM, Keith LH; Frequency of Organic Compounds Identified in Water USEPA-600/4/76-062 p.96-7 (1976)
21. Shama G, Wase DAJ; Int Biodeterior Bull 17: 1-9 (1981)
22. Swann RL et al; Res Rev 85: 17-28 (1983)
23. Wilkins GE; Industrial Process Profiles for Environmental Use. Chpt 10. pp 133-5 USEPA-600/2-77-023j (1977)
24. Zahn R, Wellens H; Chemker Z 98: 228-32 (1974)
25. Zahn R, Wellens H; Z Wasser Abwasser Forsch 13: 1-7 (1980)
26. Zahn R; Huber W; Tenside Deterg 12: 266-70 (1975)

2-Chloroaniline

SUBSTANCE IDENTIFICATION

Synonyms: o-Chloroaniline; 1-Amino-2-chlorobenzene

Structure:

CAS Registry Number: 95-51-2

Molecular Formula: C_6H_6ClN

Wiswesser Line Notation: ZR BG

CHEMICAL AND PHYSICAL PROPERTIES

Boiling Point: 208.84 °C

Melting Point: -14 °C

Molecular Weight: 127.57

Dissociation Constants: 2.661 at 25 °C [26]

Log Octanol/Water Partition Coefficient: 1.90 [11]

Water Solubility: 3800 mg/L at 20 °C [5]

Vapor Pressure: 0.17 mm Hg at 20 °C [27]

Henry's Law Constant: 7.5 x 10^{-6} atm-m³/mol (calc from water solubility and vapor pressure)

ENVIRONMENTAL FATE/EXPOSURE POTENTIAL

Summary: 2-Chloroaniline may be released to the environment from process and waste emissions involved in its production or use as a chemical intermediate, or by formation in the environment as a degradation product of various pesticides. If released to soil, 2-chloroaniline will undergo chemical bonding with humic materials, which can result in its chemical alteration and prevent leaching and retard biodegradation. The results of one field study indicate that it is also extremely photosensitive on soil surfaces. If released to water, 2-chloroaniline will bind with humic materials in the water column and sediments. Volatilization is not environmentally rapid (half-life of 64 days from a stagnant pond), but may have some importance in shallow rivers. Various screening tests suggest that 2-chloroaniline is generally resistant to biodegradation or biodegrades slowly. Significant acclimation of microbes may be required for biodegradation to become environmentally important. If released to the atmosphere, 2-chloroaniline will react with sunlight-produced hydroxyl radicals (estimated half-life of 2 days). Humans will be primarily exposed to 2-chloroaniline by dermal contact or inhalation in occupational settings.

Natural Sources:

Artificial Sources: Chloroanilines may be released to the environment as fugitive emissions or in wastewater during their production or use as chemical intermediates. Chloroanilines may also form in the environment as degradation products of various pesticides [37].

Terrestrial Fate: A field study involving the application of 2-chloroaniline (as a component of sludge) to soil plots at varying concentrations measured dissipation rates ranging from 22 to 90% over a 241-day period with dissipation more rapid at lower application concentrations [7]; monitoring of 2,2'-dichloroazobenzene formation levels suggested that degradation of 2-chloroaniline was partially biological in nature [7]; tests involving spraying of 2-chloroaniline on soil surfaces suggested that 2-chloroaniline is extremely photosensitive [7]. When 2-chloroaniline was applied to soil there was an initial, rapid rate of disappearance lasting 2 weeks followed by a much more gradual rate of decline [35]. The percentage of 2-chloroaniline remaining after 2 and 10 weeks were 40 and 20, respectively [35]. While both chemical and biological processes are responsible for the degradation, the former is more important [35]. When released to soil, 2-chloroaniline will undergo chemical bonding with humic materials, which can result in its chemical alteration and prevent leaching.

Aquatic Fate: If released to water, 2-chloroaniline will undergo chemical bonding with humic materials and clay in the water column and in the sediment. 2-Chloroaniline near the water surface is susceptible to degradation by photolysis and photooxidation. Volatilization from stagnant bodies of water does not appear to be important (half-life of 64 days from a representative pond), but may be significant from shallow rivers. Various screening tests suggest that 2-chloroaniline is generally resistant to biodegradation or biodegrades slowly. Significant acclimation of microbes may be required for biodegradation to become environmentally important. Aquatic hydrolysis and bioconcentration are not environmentally important.

Atmospheric Fate: If released to the ambient atmosphere, 2-chloroaniline will exist almost entirely in the vapor phase due to its relatively high vapor pressure. Reaction with sunlight-produced hydroxyl radicals (estimated half-life of 2 days) should rapidly remove 2-chloroaniline from the atmosphere. Direct photolysis may also contribute to its atmospheric removal.

Biodegradation: 2-Chloroaniline was found to be resistant to microbial degradation using the Japanese MITI protocol [12]. A 36% BODT was measured over a 190-hr incubation period with a Warburg respirometer [17]. 100% loss of UV absorbance in a mineral salts solution, with a soil inoculum, required in excess of 64 days [1]. Using an acclimated activated sludge inoculum, 98% of initial 2-chloroaniline was degraded in a biodegradation test [29]. Significant biological transformation (31% degradation in 6 hr) was observed in an aqueous test system receiving activated sludge from two treatment plants [2]. Half-lives greatly in excess of 4 weeks were observed using the modified OECD and Repetitive Die-Away Test [4,34]. Incubation of 2-chloroaniline in soil for 14 days resulted in formation of dichloroazobenzene, but no dichloroazobenzene was formed using sterilized soil [3]. No biodegradation was observed using modified procedures of the OECD and MITI test methods [21]. Results of standard biodegradation tests were reported as follows: Coupled units, 5-6% DOC removal; Zahn-Wellens, 94% DOC removal; MITI, 0% BODT; Sturm, 0% CO_2 evolution, 9% DOC removal; Closed bottle, 0% BODT [10].

Abiotic Degradation: Chloroanilines are expected to be resistant to environmental hydrolysis [37]. 2-Chloroaniline absorbs ultraviolet light above 290 nm indicating that direct environmental photolysis is possible [30]. Irradiation of an aqueous solution of 2-chloroaniline in a quartz tube with a fluorochemical lamp (wavelengths above 300 nm) resulted in a photodegradation half-life of 11.5 hours [13]. The quantum yield was measured to be 0.000062 at 292 nm in hexane solution with photodegradation to aniline [22]. The half-life for the vapor-phase reaction of 2-chloroaniline with sunlight-produced hydroxyl radicals in a typical ambient atmosphere has been estimated to be about 2 days based on an estimated reaction rate constant of 5.1 x 10^{-12} cm^3/molecule-sec at 25 °C [9]. Oxidation of aromatic amines can occur on clay surfaces, but is dependent on the exchangeable cation in the clay and the presence of oxygen [37].

Bioconcentration: Log BCF of 2-chloroaniline in fish were experimentally determined to be less than 2.0 using the Japanese MITI test procedures [12]. 2-Chloroaniline was found to have little or no bioconcentration in carp [31]. The log BCF of 2-chloroaniline has been theoretically estimated to be 1.3 [4].

Soil Adsorption/Mobility: 2-Chloroaniline has been observed to undergo rapid and reversible covalent bonding with humic materials in aqueous solution; the initial bonding reaction is followed by a slower and much less reversible reaction believed to represent the addition of the amine to quinoidal structures followed by oxidation of the product to give an amino-substituted quinone [23]. A similar compound 4-chloroaniline has been found to have Koc values ranging from 96 to over 5000 in various soils and soil materials; the lower Koc values were generally produced in soils with lower pHs, which may be indicative of the importance of covalent binding in soil for all chloroanilines [37].

Volatilization from Water/Soil: The volatilization half-life of 2-chloroaniline from a representative environmental pond (stagnant) has been estimated to be 64 days [36]. Using the Henry's Law constant, the volatilization half-life from a model river (1 m deep) can be estimated to be 5.6 days [16].

Water Concentrations: DRINKING WATER: 2-Chloroaniline has been qualitatively identified in drinking water samples obtained from Cincinnati, OH in 1980 and Seattle, WA in 1976 [15]. Unspecified drinking water in Germany has been reported to contain 2-chloroaniline [14]. Tap water from the Netherlands, using bank filtered Rhine River water as the source, contained 3 ppb 2-chloroaniline in 1977 monitoring [28]. SURFACE WATER: 2-Chloroaniline was detected in 83% of all samples collected at a single site on the Rhine River in 1979 at mean, median, and maximum concn of 0.54, 0.20, and 3.9 ppb, respectively [38]. Mean, median, and maximum concn of 0.45, 0.21-0.42, and 1.3-1.7 ppb, respectively, were found in two tributaries of the Rhine River in 1979 [38]. Mean, median, and maximum concn of 0.06-0.15, 0.02, and 0.23-0.86 ppb, respectively, were found at two locations on the Meuse River in 1979 [38]. Concentrations of 2-chloroaniline in the Rhine River in 1977, 1978, and 1982 have been reported to be 0.27, 0.20, and < 0.1, respectively [18]. It was also detected, but not quantitated, in water taken from the Waal River (Netherlands) in Oct 1974 [19].

Effluent Concentrations: Unspecified isomers of chloroaniline have been detected in the effluent from the publicly owned treatment works (POTW) of Sauget, IL [8]. Wastewater discharges from the commercial manufacture of 4,4'-methylenebis(2-chloroaniline) can contain 2-chloroaniline [24]. 2-Chloroaniline was qualitatively detected in effluent discharges from chemical production facilities along the Upper Catawba River in NC in 1973 and 1974 [33]. Sludge from the Muskegon County (MI) wastewater treatment system was found to contain 48.3 mg/kg 2-chloroaniline [7]. In a comprehensive survey of wastewater from 4000 industrial and publicly owned treatment works (POTWs) sponsored by the Effluent Guidelines Division of the U.S. EPA, 2-

chloroaniline was identified in discharges of the following industrial category (frequency of occurrence, median concn in ppb): pharmaceuticals (4, 588.1); organic chemicals (3, 57.1) [32]. The highest discharge was in pharmaceuticals, 2564 ppb [32].

Sediment/Soil Concentrations: Sediment from a waste treatment lagoon of a small 4,4'-methylenebis(2-chloroaniline) manufacturer in Adrian, MI was found to contain 600 ppm (dry wt basis) 2-chloroaniline [24].

Atmospheric Concentrations: SOURCE AREAS: 8 Industrial sites in the New Jersey area - 33 ng/m^3 of chloroaniline (isomer not specified) near American Cyanamide plant in Bound Brook; not detected in the 7 other areas [25].

Food Survey Values: 2-Chloroaniline has been qualitatively identified as a volatile flavor component of Idaho Russet baked potatoes [6].

Plant Concentrations:

Fish/Seafood Concentrations: Levels of 13-49 ppb 2-chloroaniline were detected in samples of white suckers collected near the discharge point of the Adrian, MI sewage treatment plant [24]

Animal Concentrations:

Milk Concentrations:

Other Environmental Concentrations:

Probable Routes of Human Exposure: Humans will be primarily exposed to 2-chloroaniline by dermal contact or inhalation in occupational settings.

Average Daily Intake:

Occupational Exposures: NIOSH has estimated that 18,138 workers are potentially exposed to 2-chloroaniline based on statistical estimates derived from a survey conducted between 1972-1974 in the United States [20].

2-Chloroaniline

Body Burdens:

REFERENCES

1. Alexander M, Lustigman BK; J Agric Food Chem 14: 410-3 (1966)
2. Baird R et al; J Water Pollut Control Fed 49: 1609-15 (1977)
3. Bartha R et al; Science 16: 582-3 (1968)
4. Canton JH et al; Regul Toxicol Pharmacol 5: 123-31 (1985)
5. Chiou CT et al; Environ Sci Technol 16: 4-10 (1982)
6. Coleman EC et al; J Agric Food Chem 29: 42-8 (1981)
7. Demirjian YA et al; J Water Pollut Control Fed 59: 32-8 (1987)
8. Ellis DD et al; Arch Environ Contam Toxicol 11: 373-82 (1982)
9. GEMS; Graphical Exposure Modeling System. FAP. Fate of Atmospheric Pollutants (1987)
10. Gerike P, Fischer WL; Ecotox Environ Safety 3: 159-73 (1979)
11. Hansch C, Leo AJ; MEDCHEM Project issue 26 Claremont CA: Pomona College (1985)
12. Kawasaki M; Ecotoxic Environ Safety 4: 444-54 (1980)
13. Kondo M; Simulation Studies of Degradation of Chemicals in the Environment, Tokyo, Japan: Office of Health Studies Environ Agency (1978)
14. Kool HJ et al; Crit Reviews Environ Control 12: 307-57 (1982)
15. Lucas SV et al; GC/MS Analysis of Organics in Drinking Water Concentrates and Advanced Waste Treatment Concentrates: Vol 1. USEPA-600/1-84-020a (1984)
16. Lyman WJ et al; Handbook of Chemical Property Estimation Methods New York: McGraw-Hill (1982)
17. Malaney GW; J Water Pollut Control Fed 32: 1300-11 (1960)
18. Malle KG; Z Wasser-Abwasser Forsch 17: 75-81 (1984)
19. Meijers AP, Vanderleer RC; Water Res 10: 597-604 (1976)
20. NIOSH; National Occupational Hazard Survey (1974)
21. Painter HA, King EF; Ecotox Environ Safety 9: 6-16 (1985)
22. Parlar H, Korte F; Chemosphere 10: 797-807 (1979)
23. Parris GE; Environ Sci Technol 14: 1099-1105 (1980)
24. Parris GE et al; Bull Environ Contam Toxicol 24: 497-503 (1980)
25. Pellizzari ED; The Measurement of Carcinogenic Vapors in Ambient Atmospheres USEPA-600/7-77-055 (1977)
26. Perrin DD; Dissociation Constants of Organic Bases in Aqueous Solution: Supplement 1972 London: Butterworth (1972)
27. Piacente V et al; J Chem Eng Data 30: 372-6 (1985)
28. Piet GJ et al; Hydrocarbon, Halogenated Hydrocarbon in the Aquatic Environ; Afghan BK, Mackay D eds NY: Plenum Press p.69-80 (1980)
29. Pitter P; Water Res 10: 231-5 (1976)
30. Sadtler; 254 UV Philadelphia, PA: SP Sadtler & Sons
31. Sasaki S; p.283-98 in Aquatic Pollutants: Transformations and Biological Effects. Hutzinger O et al eds Oxford: Pergamon Press (1978)
32. Shackelford WM et al; Analyt Chim Acta 146: 15-27 (1983)
33. Shackelford WM, Keith LH; Frequency of Organic Compounds Identified in Water. USEPA-600/4-76-062 (1976)
34. Steinhaeuser KG et al; Vom Wasser 67: 147-54 (1986)

2-Chloroaniline

35. Thompson FR, Corke CT; Can J Microbiol 15: 791-6 (1969)
36. USEPA; EXAMS II Computer Modeling System (1987)
37. USEPA; Health and Environmental Effects Document for Chloroanilines ECAO-CIN-G003 (Final Draft) (1987)
38. Wegman RC, DeKorte AL; Water Res 15: 391-4 (1981)

4-Chloroaniline

SUBSTANCE IDENTIFICATION

Synonyms: p-Chloroaniline; 1-Amino-4-chlorobenzene

Structure:

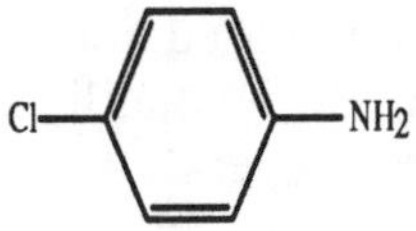

$$Cl - \bigcirc - NH_2$$

CAS Registry Number: 106-47-8

Molecular Formula: C_6H_6ClN

Wiswesser Line Notation: ZR DG

CHEMICAL AND PHYSICAL PROPERTIES

Boiling Point: 232 °C at 760 mm Hg

Melting Point: 72.5 °C

Molecular Weight: 127.58

Dissociation Constants: 3.982 at 25 °C [10]

Log Octanol/Water Partition Coefficient: 1.83 [16]

Water Solubility: 3.9 g/L [18]

Vapor Pressure: 0.025 mm Hg at 25 °C [18]

Henry's Law Constant: 1.07×10^{-5} atm-m^3/mol (calculated from vapor pressure and water solubility)

4-Chloroaniline

ENVIRONMENTAL FATE/EXPOSURE POTENTIAL

Summary: 4-Chloroaniline may be released to the environment during its production or use in the manufacture of dye intermediates, agricultural chemicals, and pharmaceuticals. If released on soil, it will rapidly combine chemically with soil components and partially be mineralized by chemical and biological action. A few percent of the 4-chloroaniline will volatilize from the soil. If released into water, 4-chloroaniline will be primarily lost due to volatilization (half-life 6.4 hr), photooxidation in surface layers (half-life 0.4 hr), and rapid chemical reactions with humic materials and clay in the water column and sediment. Degradation in air will primarily be due to reaction with hydroxyl radicals (half-life 4.6 hr), although direct photolysis is also possible. Human exposure will primarily be in the workplace from inhalation or dermal contact.

Natural Sources:

Artificial Sources: 4-Chloroaniline may be released to the environment as fugitive emissions, in wastewater during its production, or use as a dye intermediate or in the manufacture of pharmaceuticals and agricultural chemicals [38].

Terrestrial Fate: If released on soil, 4-chloroaniline will bind tightly to soil, although in the first few hours after a spill, a small amount will volatilize. The 4-chloroaniline will undergo both biological and chemical transformation. Mineralization occurs most rapidly in the early weeks of incubation with as much as 7.5% degradation to CO_2 occurring in 6 weeks and 17% occurring in 16 weeks. Most (70-90%) of the 4-chloroaniline is transformed into inextractable residues and there is no significant leaching, either vertically or horizontally, into surrounding layers of soil. A three-year field test in which ^{14}C labeled 4-chloroaniline was applied to 60 x 60 x 60 cm box cultivated with barley and later with potatoes and carrots under outdoor conditions resulted in approx 30% of the applied radioactivity being retained at the application site in the upper layer of soil after the first year and 67% being lost to the atmosphere [10]. Uptake by plants, migration into deeper soil layers, or leaching into water was low [10]. From previous experiments, it is known that the bulk of the atmospheric loss is not volatilized 4-chloroaniline or conversion products, but rather CO_2 resulting from mineralization [10]. The situation after the second and third year didn't alter appreciably [10]. No free unchanged 4-chloroaniline could be detected in either soil or plants [10]. Conversion products isolated under these environmental conditions included 4-chloroformanilide, 4-chloroacetanilide, 4-chloronitrobenzene, 4-chloronitrosobenzene, and the condensation products 4,4'-dichloroazoxybenzene and 4,4'-dichloroazobenzene [10]. It is hypothesized that phenolic and hydroxylamine metabolites were not identified because of their chemical instability, although they were found in laboratory experiments [10]. In summary, it is clear that free 4-chloroaniline is not persistent in soil; it is subject to various acylation and oxidation reactions and finally to total biodegradation and incorporation into soil and plant constituents [10]. During bank infiltration of Rhine River water, levels of 4-chloroaniline decreased about 25% [20].

Aquatic Fate: If released into water, 4-chloroaniline will volatilize (half-life 6.4 hr in a model river), photooxidize in surface layers (half-life 0.4 hr), biodegrade (half-life several days in well-acclimated water), and chemically bind to clays and humus in sediment and particulate matter in the water column. When ^{14}C labeled 4-chloroaniline was added to an experimental pond, the ^{14}C label disappeared from the water phase in two stages with half-lives of about 3 and 11 days, respectively [31]. It was assumed that the initial loss resulted from volatilization. After a day, a thin brown film of decomposition products formed. The reduced loss rate after the first few days is probably a result of lower volatility of the decomposition products [31]. The half-life of 4-chloroaniline in a river estimated by sampling at points along the Rhine River in the Netherlands was 0.3 to 3 days [43].

Atmospheric Fate: If released into air, 4-chloroaniline will degrade by reacting with photochemically produced hydroxyl radicals (half-life 4.6 hr) and possibly also by photolysis in the vapor phase or while adsorbed on airborne particulate matter. The rates for vapor phase photolysis are not available. A highly soluble chemical, 4-chloroaniline will probably be scavenged by rain.

Biodegradation: The results of biodegradability screening studies for 4-chloroaniline are conflicting, results range from no degradation to rapid degradation, using soil, sewage, activated sludge, and fresh water inocula. The most frequently reported results are that 4-chloroaniline biodegrades rather slowly with acclimation [1,17,28,30,32,36]. The reason for the conflicting results may be due to toxicity of metabolic intermediates [1], differences in concentrations and inocula used, sensitivity of 4-chloroaniline to chemical oxidation, and lack of sufficient acclimation. Some results are: 28% degradation in 5 days [30], 10% and 18% degradation in 28 days [30,32], no degradation in 28-30 days [17,30], 97% degradation in 10 days including an 8-day lag period [17], 97% removal in 5 days after 20 days acclimation [28], 46 and 100% removal in 8 and 22 days, respectively, after a 14-day acclimation period [32], no degradation in 30 days in the Closed Bottle Test which is supposed to simulate degradation in surface water [17]. In a test which is supposed to simulate degradation in a sewage treatment plant, the Confirmatory Test, 97% degradation was rapidly obtained after a 16-day acclimation period [17]. Biotransformation frequently involves acylation and oxidation to phenols [10]. In the one available river die-away test, 10 ppm 4-chloroaniline degraded slowly in Nile River water over a period of 2 months, but on redose increasingly larger concentrations of the chemical were degraded in shorter and shorter times so that on the 8th redose 100 ppm was degraded in a few days [8]. Several studies were performed where labeled 4-chloroaniline was incubated with soil [4,11,35]. In one study, 12-17% mineralization occurred in 16 weeks with the maximum rate of degradation occurring between 1 and 3 weeks [35]. 86% of the labeled residue was present as unextractable material bound to the soil and 1-4% of the residues were extractable [35]. In another study 7.5% mineralization occurred in 6 weeks, with 72% and 7% unextractable and extractable residue remaining [5]. In comparing three fractions with those of autoclaved controls, it is apparent that part of the extractable fraction was being biologically mineralized [5]. Biological transformation of the unextractable residue, however, also occurs [5]. 8-9%, 69-73%, and 5-7% CO_2, unextractable residue, and extractable residue, respectively, resulted in a third study after 16 weeks of incubation [11]. It has been suggested that the binding of the chloroaniline to soil may extend its life in soil to 10 years [2]. During composting with refuse, 14% of the 4-chloroaniline was metabolized in 21 days [20]. No mineralization occurred when 4-chloroaniline was incubated under anaerobic conditions with digester sludge for 1 month [33].

Abiotic Degradation: 4-Chloroaniline absorbs light >290 nm [34]. On irradiation with light >290 nm in air-saturated water, it photolyzes to form 4-chloronitrobenzene and 4-chloronitrosobenzene [23]. 4-Chloroaniline completely disappears within 6 hours and dark purple condensation products are subsequently formed [23]. The half-life of 4-chloroaniline under illumination conditions typifying those for U.S. surface waters in summer is 0.4 hr [42]. This rate is not increased by the presence of algae in the water [42]. 28% of the 4-chloroaniline adsorbed on silica gel was mineralized to CO_2 in 17 hr [9]. After adsorption onto clay surfaces or humus, irreversible chemical reactions occur. In the atmosphere, vapor phase 4-chloroaniline will react with photochemically produced hydroxyl radicals with an estimated reaction half-life of 4.6 hr assuming a hydroxyl radical concentration of $5 \times 10^{+5}$ radicals/cm^3 [39].

Bioconcentration: Golden orfe - log BCF <1.30 for 3-day exposure [19]. Green algae - 24-hr log BCF 3.08 (dry wt basis) [13], 2.42 (wet weight basis) [19].

Soil Adsorption/Mobility: The Koc for 5 Belgium soils ranged from 230 to 469 [37]. The adsorption isotherm was not linear and the exponent in the Freundlich adsorption isotherm averaged 0.70 [37]. The Koc for 5 German soils ranged from 96 to 1530 with the Freundlich exponent ranging from 0.92 to 1.23 [29]. From adsorption studies on three soils with radically different organic carbon and clay contents, it was shown that the percent ^{14}C labeled 4-chloroaniline that was bound increased with the organic carbon content and decreased with the clay content [41]. Higher binding at low concentrations, particularly on the soil with low organic carbon content suggested a limited number of available binding sites on the soil [41]. At the lower concentration used, 5 ppm, percent binding ranged from 46-78% [41]. The Koc to colloidal organic matter in ground water was high, 5550, suggesting that adsorption onto this microparticulate matter could effectively increase the solubility and leaching of 4-chloroaniline into landfill ground water [22]; however, the bound chemical is chemically altered. The pKa of 4-chloroaniline is 3.98 [26] so it will be partially ionized at acid pH. Binding to clays which may be ionic in character should therefore be stronger at lower pH. Thus the binding of 4-chloroaniline to bentonite was significant at pH 3 but not at pH 6.5 [3]. Adsorption and oxidation of aromatic amines which can occur on clay surfaces are dependent on the exchangeable cation in the clay and the presence of oxygen [6,12] as is partially evidenced by the formation of oligomeric or polymeric complexes when 4-chloroaniline is absorbed on montmorillonite clay [6]. Aromatic amines form covalent bonds with humic materials, adding to model quinoidal structures such as are found in humic materials, followed by slow oxidation to nitrogen-substituted quinoid rings [24]. The reaction half-life of 4-chloroaniline with one test humic constituent was 13 min [24]. Hybrid oligomers were formed between 4-chloroaniline and carboxyphenolic humus constituents [4]. This type of cross-coupling product would explain the fast, strong, and irreversible binding of anilines to soil [4]. In a field experiment in a test pond, no significant leaching occurred either vertically or horizontally into the surrounding soil [31]. After a year, 10-20% of ^{14}C residues remained in the upper 10 cm of soil [31]. Another field experiment using ^{14}C-4-chloroaniline resulted in 30% of the label remaining where it was applied, while little was found at lower depths or in leachate [10].

Volatilization from Water/Soil: Using the Henry's Law constant, one would estimate a half-life of 6.4 hr for a model river 1 m deep with a 1 m/sec current and 3 m/sec wind [21]. In an experimental pond, the concentration of spiked 4-chloroaniline initially declined exponentially (half-life 3 days) which was ascribed to volatilization [31]. When 50 ppm of labeled 4-chloroaniline was put in water, 0.62% of the radioactivity was lost in an hour [18]. When added to 3 soils dampened to 40% total holding capacity with 0.51%, 1.00%, and 2.89% organic carbon, 2.53%, 0.70%, and 0.19% volatilized, respectively, during the first hour per mL of evaporated water [18]. 3.5% of the radioactivity recovered when ^{14}C labeled 4-chloroaniline was incubated in soil was from air. The volatile material was trapped in sulfuric acid solution [5] and, therefore, probably was volatilized 4-chloroaniline.

Water Concentrations: DRINKING WATER: Germany (treated Rhine water source) 7 ppt, annual average [20]. The Netherlands:

maximum from bank-filtered Rhine River water 1 ppb, including m-isomer [27]. SURFACE WATER: Not detected in Lakes Ontario (1 location), Erie (2 locations), Michigan (5 locations), and Superior (1 location) [14}. Rhine River (Germany) - 80 ppt, yearly average [20]. Rhine River and two tributaries (the Netherlands) 130-220 ppt, average; 240-740 ppt maximum with 96-100% frequency of detection [40]. Meuse River (The Netherlands) 20-30 ppt, average, 80-120 ppt maximum, 44-50% frequency of detection [40]. While detected in Rhine Delta water, it was not found in surface waters from agricultural areas in the Netherlands [15].

Effluent Concentrations:

Sediment/Soil Concentrations:

Atmospheric Concentrations: SOURCE AREAS: 8 industrial sites in the New Jersey area - 33 ng/m^3 of chloroaniline (isomer not specified) near American Cyanamide plant in Bound Brook; not detected in the 7 other areas [25].

Food Survey Values: 4-Chloroaniline has been qualitatively identified as a volatile flavor component of Idaho Russet baked potatoes [7].

Plant Concentrations:

Fish/Seafood Concentrations:

Animal Concentrations:

Milk Concentrations:

Other Environmental Concentrations:

Probable Routes of Human Exposure: Humans will be primarily exposed to 4-chloroaniline by dermal contact or inhalation in occupational settings.

Average Daily Intake:

Occupational Exposures:

4-Chloroaniline

Body Burdens:

REFERENCES

1. Baird R et al; J Water Pollut Control Fed 49: 1609-15 (1977)
2. Bartha R; J Agric Food Chem 19: 385-7 (1971)
3. Bengtsson G et al; Water Res 20: 935-7 (1986)
4. Bollag JM et al; Environ Sci Tech 17: 72-80 (1983)
5. Bollag JM et al; J Agric Food Chem 26: 1302-6 (1978)
6. Cloos P et al; Chim Organo Minerale Lab Phys Chim Minerale and Catalyse Grp Catholic Univ Louvain Louvain LA Neuve, Belgium 14: 307-21 (1979)
7. Coleman EC et al; J Agric Food Chem 29: 42-8 (1981)
8. El-Dib MA, Aly OA; Water Res 10: 1055-9 (1976)
9. Freitag D et al; Ecotox Environ Safety 6: 60-81 (1982)
10. Freitag D et al; J Agric Food Chem 32: 203-7 (1984)
11. Fuchsbichler G et al; Z Pflanzenkr Pflanzenschutz 85: 724-34 (1978)
12. Furukawa T, Brindley GW; Clay Minerals 21: 279-88 (1973)
13. Geyer H et al; Chemosphere 13: 269-84 (1984)
14. Great Lakes Water Quality Board; An Inventory of Chemical Sub Identified in the Great Lakes Ecosystem Vol 1 Windsor, Ontario: Great Lakes Water Quality Board (1983)
15. Greve PA, Wegman RCC; Schriftenr Ver Wasser, Boden, Lufthyg Berlin-Dahlem 46: 59-80 (1975)
16. Hansch C, Leo AJ; MEDCHEM Project ser 26 Claremont CA: Pomona College (1985)
17. Janicke W, Hilge G; GWT-Wasser/Abwasser 121: 131-5 (1980)
18. Kilzer L et al; Chemosphere 8: 751-61 (1979)
19. Korte F et al; Chemosphere 1: 79-102 (1978)
20. Kussmaul H; Behavior of persistent organic compounds in bank-filtrated Rhine water Pergamon Ser Environ Sci pp. 265-75 (1978)
21. Lyman WJ et al; Handbook of Chemical Property Estimation Methods McGraw-Hill NY (1982)
22. Means JC; Amer Chem Soc 186th Natl Mtg Washington, DC Preprints Div Environ Chem 23: 250-1 (1983)
23. Miller GC, Crosby DG; Chemosphere 12: 1217-28 (1983)
24. Parris GE; Environ Sci Tech 14: 1099-105 (1980)
25. Pellizzari ED; The Measurement of Carcinogenic Vapors in Ambient Atmospheres USEPA-600/7-77-055 (1977)
26. Perrin DD; Dissociation Constants of Organic Bases in Aqueous Solution: Supplement 1972 London: Butterworth (1972)
27. Piet GJ, Morra CF; Artificial Groundwater Recharge, Water Resources Engineering Series Huisman L, Olsthorn TN eds. Pitman pp. 31-42 (1983)
28. Pitter P; Water Res 10: 231-5 (1976)
29. Rippen G et al; Ecotox Environ Safety 6: 236-45 (1982)
30. Rott B et al; Chemosphere 11: 531-8 (1982)
31. Schauerte W et al; Ecotox Environ Safety 6: 560-9 (1982)
32. Schmidt-Bleek F et al; Chemosphere 11: 383-415 (1982)

4-Chloroaniline

33. Shelton DR, Tiedje JM; Development of Tests for Determining Anaerobic Biodegradation Potential p. 92 USEPA 560/5-81-013 (1981)
34. Sadtler Index; 393 UV Philadelphia, PA: SP Sadtler & Sons
35. Suess A et al; Z Pflanzezernaehr Bodenkd 141: 57-66 (1978)
36. Torgeson DC; Interaction of Herbicides and Soil Microorganisms p 73 16060 DMP 03/71 Water Pollution Control Res Series Washington DC: USEPA (1971)
37. Van Bladel R, Moreale A; J Soil Sci 28: 93-102 (1977)
38. Verschueren K; Handbook of Environmental Data on Organic Chemicals 2nd Ed New York: Von Nostrand Reinhold (1983)
39. Wahner A, Zetzsch C; J Phys Chem 87: 4945-51 (1983)
40. Wegman RCC, Dekorte GAL; Water Res 15: 391-4 (1981)
41. Worobey BL, Webster GRB; J Agric Food Chem 30: 164-9 (1982)
42. Zepp RG, Schlotzhauer PF; Environ Sci Technol 17: 462-8 (1983)
43. Zoeteman BC et al; Chemosphere 9: 231-49 (1980)

Chlorobenzene

SUBSTANCE IDENTIFICATION

Synonyms:

Structure:

CAS Registry Number: 108-90-7

Molecular Formula: C_6H_5Cl

Wiswesser Line Notation: GR

CHEMICAL AND PHYSICAL PROPERTIES

Boiling Point: 132 °C at 760 mm Hg

Melting Point: -45.6 °C

Molecular Weight: 112.56

Dissociation Constants:

Log Octanol/Water Partition Coefficient: 2.84 [17]

Water Solubility: 471.7 mg/L at 25 °C [29]

Vapor Pressure: 11.9 mm Hg at 25 °C [10]

Henry's Law Constant: 3.45 x 10^{-3} atm-m³/mole [30]

Chlorobenzene

ENVIRONMENTAL FATE/EXPOSURE POTENTIAL

Summary: Chlorobenzene will enter the atmosphere from fugitive emissions connected with its use as a solvent in pesticide formulations and as an industrial solvent. Once released it will decrease in concentration due to dilution and photooxidation. Releases into water and onto land will decrease in concentration due to vaporization into the atmosphere and slow biodegradation in the soil or water. Chlorobenzene has potential to percolate into the ground water, particularly if soil is sandy and poor in organic matter. Little bioconcentration is expected into fish and food products. Primary human exposure is expected to be from ambient air, especially near point sources.

Natural Sources:

Artificial Sources: Release to the environment is estimated to be due mostly to volatilization losses associated with its use as a solvent in pesticide formulations and in degreasing and other industrial applications. It is estimated by Dow Chemical to be 30-50% of its production. Disposal of industrial wastes may be another source of release. The U.S. EPA estimates that <0.1% of production would be disposed of in water and <1% on land [52].

Terrestrial Fate: Since chlorobenzene is fairly volatile from water [27], most of any released on moist soil will be lost to the atmosphere. It is relatively mobile in sandy soil and aquifer material and biodegrades very slowly or not at all in these soils [42,45,52,61]. Therefore, it can be expected to leach into the ground water. Degradation will generally be slow, but fairly rapid mineralization (20%/week) has been reported in one study [43]. Acclimation of soil microorganisms is an important factor [16]. Fate of chlorobenzene added to a 140 cm deep soil column packed with sandy soil at a concentration of 1.04 mg/L: 27% volatilized, 23-33% percolated through column, and 40-50% degraded or was not accounted for; at a concentration of 0.18 mg/L, 54% volatilized, 26-34% percolated through the soil column, and 12-20% degraded or was not accounted for [60].

Aquatic Fate: The primary loss process from water will be evaporation. The rate of evaporation will depend on the wind speed and water movement. The half-life for evaporation is estimated to range between 1 and 12 hours in a rapidly flowing stream [6]. Biodegradation will occur during the warmer seasons and will proceed more rapidly in fresh water than in estuarine and marine systems [4,40]. The rate will also depend on the acclimation of microbial communities to chlorobenzene or related chemicals. One reported half-life for an estuarine river with near natural conditions (22 °C) is 75 days [26]. A moderate amount of adsorption will occur onto organic sediments [56].

Atmospheric Fate: If released to the atmosphere, chlorobenzene is expected to exist almost entirely in the vapor phase based upon the vapor pressure [11]. Reaction with hydroxyl radicals is the dominant removal mechanism with an estimated half-life of 17 days with the formation of chlorophenols [2,46]. Reaction in polluted air with nitric oxide is somewhat faster and produces chloronitrobenzene and chloronitrophenols [20]. Photolysis would proceed at a much slower rate, with monochlorobiphenyl being produced [54].

Biodegradation: A large number of bacteria and fungi found in the environment are capable of degrading chlorobenzene and mineralizing it. 2- and 4-Chlorophenol are products of this biodegradation [3,16,48,50]. Degradation is generally slow in water and soil, but may be significant in some situations (i.e., during warmer weather, when microbial activity is increased) [3,16,48,50,57]. Acclimation of the degrading microorganisms is an important factor [3,16,48,50]. Biodegradation half-life of chlorobenzene, reported to be 150 days in river water and 75 days in sediment [26]. No significant mineralization was found for chlorobenzene (27 and 315 ng/g soil) incubated for 8 months [1].

Abiotic Degradation: Chlorobenzene absorbs light in the 290-310 nm region [54], suggesting that this compound may photolyze in the troposphere [51]. This process will occur over the course of a month in the atmosphere. Monochlorobiphenyl has been identified as a photoproduct [41]. Based on experimental data the half-life for direct photolysis of chlorobenzene in surface water at 40 ° latitude during the summer season has been estimated to be 170 years [2]. Indirect photolysis sensitized by dissolved humic acids in natural waters may lead to much faster reactions than will direct photolysis

[2]. Chemical hydrolysis is not expected to be an environmentally relevant fate process [27]. Oxidation by hydroxyl radicals and in air-NO_x systems (smog situations) is relatively fast and results in 5-10% loss/day in the first case. The half-life for chlorobenzene vapor reacting with photochemically generated hydroxyl radicals in the atmosphere has been estimated to be 17 days based on a measured reaction rate constant of 0.94 x 10^{-12} cm^3/molecules-sec at room temperature and an average ambient hydroxyl radical concentration of 5.0 x 10^{+5} molecules/cm^3 [2]. The rate of reaction when NO_x is present is somewhat higher [46]. Products formed include chlorophenols, nitrochlorophenols, and m-chloronitrobenzene [20].

Bioconcentration: Little or no bioconcentration is expected [21,22,23,55]. Log BCF is 1-2 for several species of fish [21,22,23], although a log BCF of 2.65 has been reported for fathead minnows [55].

Soil Adsorption/Mobility: Soil adsorption coefficients of 83-389 have been measured for chlorobenzene [7,27,44,45]. These Koc values suggest that chlorobenzene would have moderate to high mobility in soil and that slight to moderate adsorption to sediments and suspended solids in water may occur [49].

Volatilization from Water/Soil: The value of Henry's Law constant suggests that volatilization of chlorobenzene would be significant from all bodies of water [27]. The half-life of chlorobenzene volatilizing from a rapidly flowing stream has been estimated to range between 1 and 12 hours based on a measured reaeration coefficient ratio of 0.42 and oxygen reaeration coefficients of 0.14-1.96 hr^{-1} [6]. Chlorobenzene applied to soil at a uniform concentration of 1 kg/ha at depths of 1 cm and 10 cm underwent 86.5 and 23.4% loss, respectively, in 1 day. Based on these data the volatilization half-lives from soil depths of 1 and 10 cm were estimated to be 0.3 and 12.6 days, respectively [19]. Volatilization half-lives of 13, 21, and 4.6 days were estimated for chlorobenzene using data obtained from an experimental marine mesocosm under simulated winter, spring, and summer conditions, respectively [57]. Volatilization in a bay or ocean should be significantly faster (perhaps up to an order of magnitude), since turbulence in the mesocosm was substantially less than is found in bays or open oceans [57].

Water Concentrations: DRINKING WATER: 1975 USEPA National Organics Reconnaissance Survey (NORS) finished drinking water samples: Miami, FL - 1 ug/L; Seattle, WA - not detected (detection limit not reported); Cincinnati,OH - 0.1 ug/L; Ottumwa, IA - not detected; and Philadelphia, PA - <0.1 ug/L [9,34]. Finished water in 9 of 10 supplies surveyed by the USEPA contained chlorobenzene with 5.6 and 4.7 ppb in Terrebonne Parrish, LA and New York City, respectively [32]. Chlorobenzene may be formed during chlorination [32]. GROUND WATER: As of June 30, 1984 - Wisconsin, 1174 community wells, 0% pos; 617 private wells, 0% pos, detection limit 1.0-5.0 ug/L [25]. During 1981-1982, 945 wells scattered throughout the United States, 0.1% pos., concn detected 2.7 ug/L, detection limit 0.5 ug/L [59]. 14 ppt (Zurich, 1973) under densely populated, partly industrial area [15]. SURFACE WATER: Ohio River Basin (large survey, 1980-1981) detected in 9.6% of samples, but only 0.01% were >1 ppb. Maximum concentration was >10 ppb [33]. In 14 industrial river basins (1975-76), only 5% of samples in the range 1-4 ppb [13].

Effluent Concentrations: Detected in leachate from the Kin-Buc Landfill (which contains diverse industrial and chemical waste) in Edison, NJ, max concn 4620 ug/L [24]. As of July 31, 1982, results of the Nationwide Urban Runoff Program (NURP), detected in runoff from 3 out of 19 cities, 86 samples, 5% pos, concn range 1-10 ug/L [8].

Sediment/Soil Concentrations: Not detected in sediment in an industrial river location, Lake Ontario, or Raritan Bay (Lower Hudson) [35].

Atmospheric Concentrations: Remote areas <0.02 ppb [39]. Cities in U.S. typical values - 0.0-0.8 ppb, maximum value measured 12 ppb [5,38,46,47]. July - Aug 1981, New Jersey program on Airborne Toxic Elements and Organic Substances (ATEOS), detection limit 0.005 ppb: Newark, NJ, 38 samples, 92% pos, mean concn 0.11 ppb; Elizabeth, NJ, 37 samples, 97% pos, mean concn 0.08 ppb; Camden, NJ, 35 samples, 91% pos, mean concn 0.07

ppb [18]. Jan-Feb 1982, NJ ATEOS program, detection limit 0.005 ppb, Newark, NJ, 30 samples, 100% pos, mean concn 0.22 ppb; Elizabeth, NJ, 38 samples, 100% pos, mean concn 0.21 ppb; Camden, NJ, 37 samples, 97% pos., mean concn 0.18 ppb [18]. USEPA TEAM Study - Bayonne and Elizabeth, NJ, fall 1981, estimated percentage of population (128,000 people) with measurable levels of chlorobenzene in their personal air, daytime - 4%, nighttime - 9%, detection limit 1-5 ug/m^3 [58].

Food Survey Values:

Plant Concentrations:

Fish/Seafood Concentrations: Two studies of chlorobenzenes in fish from the Great Lakes and Japanese coast failed to detect any chlorobenzene [31,35].
Animal Concentrations:

Milk Concentrations:

Other Environmental Concentrations:

Probable Routes of Human Exposure: Populations at special risk of exposure include: urban residents - ambient air; people near manufacturing plants; people near locations where products containing chlorobenzene as a solvent is used; and occupational workers in manufacturing plants or where chlorobenzene is used as a solvent [14].

Average Daily Intake: AIR INTAKE: (Assume avg 0.4 ppb) - 42 ng; WATER INTAKE: (Assume avg 0.1 ppb) - 15 pg; FOOD INTAKE: insufficient data.

Occupational Exposures: 0.0-1.9 mg/m^3 (Dupont - Deepwater, NJ, 10/79) [14]. NIOSH (NOES Survey 1981-1983) has statistically estimated that 15,958 workers are potentially exposed to chlorobenzene in the United States [36]. NIOSH (NOHS Survey 1972-1974) has statistically estimated that 740,615 workers are potentially exposed to chlorobenzene in the United States [37].

Chlorobenzene

Body Burdens: Trace - 10 ppb (0.37 ppb avg) in mothers' milk (42 samples) from subjects living near manufacturing plants or industrial facilities [12]. USEPA TEAM Study - Bayonne and Elizabeth, NJ, Fall 1981, 3% of population (128,000 people) estimated to have measurable levels of chlorobenzene in their breath, detection limit 1-5 ug/m^3 [58].

REFERENCES

1. Aelion CM et al; Appl Environ Microbiol 53: 2212-7 (1987)
2. Atkinson R; Inter J Chem Kinet 19: 799-828 (1987)
3. Ballschmitter K, Scholz C; Chemosphere 9: 457-67 (1980)
4. Bartholomew GW, Pfaender FK; Appl Environ Microbiol 45: 103-9 (1983)
5. Bozzelli JW et al; Analysis of selected toxic and carcinogenic substances in ambient air in New Jersey. State of NJ, Dept Environ Protect (1980)
6. Cadena F et al; J Water Pollut Contrl Fed 460-3 (1984)
7. Chiou CT et al; Environ Sci Tech 17: 227-8 (1983)
8. Cole RH et al; J Water Poll Control Fed 56: 898-908 (1984)
9. Coleman WE et al; pp.305-27 in Analysis and Identification of Organic Substances in Water. L Keith ed Chapter 21 Ann Arbor Sci, Ann Arbor MI (1976)
10. Daubert TE, Danner RP; Data Compilation Tables of Properties of Pure Compounds. American Institute of Chemical Engineers pp 450 (1985)
11. Eisenreich SJ et al; Environ Sci Tech 15: 30-8 (1981)
12. Erickson MD et al; Acquisition and chemical analysis of mothers milk for selected toxic substances pp.164 USEPA 560/13-80-029 (1980)
13. Ewing BB et al; Monitoring to detect previously unrecognized pollutants in surface waters USEPA-560/6-77-015a (1979)
14. Fannick N; Toxic evaluation of inhaled chlorobenzene (monochlorobenzene) NIOSH HE-77-099-726 (1977)
15. Grob K, Grob G; J Chromatog 90: 303-13 (1974)
16. Haider K et al; Arch Microbiol 96: 183-200 (1974)
17. Hansch C, Leo AJ; Medchem Project Claremont, CA: Pomona College (1985)
18. Harkov R et al; Sci Tot Environ 38: 259-74 (1984)
19. Jury WA et al; J Environ Qual 13: 573-9 (1984)
20. Kanno S, Nojima K; Chemosphere 8: 225-32 (1979)
21. Kawasaki M; Ecotox Environ Safety 4: 444-54 (1980)
22. Kenaga EE; Bull Environ Safety 4: 26-38 (1980)
23. Kitano M; OECD Tokyo Mtg Ref Book TSU-No.3 (1978)
24. Kosson DS et al; Proc Ing Waste Conf 39: 329-41 (1985)
25. Krill RM, Sonzogni WC; J Am Water Works Assoc 78: 70-5 (1986)
26. Lee RF, Ryan CC; Microbial degradation of pollutants in marine environments pp.443-50 USEPA-600/9-79-012 (1979)
27. Lyman WJ et al; Handbook of Chemical Property Estimation Methods New York: McGraw-Hill NY (1982)
28. Mackay D, Yeun ATK; Environ Sci Technol 17: 211-17 (1983)
29. Mackay D et al; Environ Sci Technol 13: 333-6 (1979)
30. Mackay D, Shiu WY; J Phys Chem Ref Data 19: 1175-99 (1981)

Chlorobenzene

31. Morita M et al; Environ Pollut 9: 175-9 (1975)
32. NAS; Drinking Water and Health 4: 709 (1977)
33. Ohio River Valley Water Sanit Comm; Assessment of water quality conditions. Ohio River mainstream 1980-81. Cincinnati, OH (1980)
34. Ohio River Valley Water Sanit Comm; Water treatment process modifications for trihalomethane control and organic substances in the Ohio River. Cincinnati, OH (1979)
35. Oliver BG, Nichol KD; Environ Sci Technol 16:532-6 (1982)
36. NIOSH; National Occupational Exposure Survey (NOES) (1983)
37. NIOSH; National Occupational Hazard Survey (NOHS) (1974)
38. Pellizzari ED; Formulation of preliminary assessment of halogenated organic compounds in man and environmental media USEPA560/13-79-006 (1979)
39. Pellizzari ED; Quantification of chlorinated hydrocarbons in previously collected air samples. USEPA 450/3-78-112 (1978)
40. Pfaender FK, Bartholomew GW; Appl Environ Microbiol 44: 159-64 (1982)
41. Pinhey JT, Rigby RDG; Tetra Lett 16: 1267-70 (1969)
42. Rittmann BE et al; Ground Water 18: 236-43 (1980)
43. Roberts PV et al; J Water Pollut Control Fed 52: 161-71 (1980)
44. Sabljic A; J Agric Food Chem 32: 243-6 (1984)
45. Schwarzenbach RP, Westall J; Environ Sci Technol 15: 1360-7 (1981)
46. Singh HB et al; Atmos Environ 15: 601-12 (1981)
47. Singh HB et al; Atmospheric measurements of selected hazardous organic chemicals USEPA-600/3-81-032 (1981)
48. Smith RV, Rosazza JP; Arch Biochem Biophys 161: 551-8 (1974)
49. Swann RL et al; Res Rev 85: 17-28 (1983)
50. Tabak HH et al; J Water Pollut Control Fed 53: 1503 18 (1981)
51. Tissot A et al; Chemosphere 12: 859-72 (1983)
52. Tomson MB et al; Water Res 15: 1109-16 (1981)
53. USEPA; Health Assessment Document of Chlorinated Benzenes; ECAO CIN-301A-DRAFT. (1983)
54. Uyetta M et al; Nature 264: 583-4 (1976)
55. Veith GD et al; J Fish Board Canada 36: 1040-48 (1979)
56. Voice TC et al; Environ Sci Technol 17: 513-8 (1983)
57. Wakeham SG et al; Environ Sci Tech 17: 611-7 (1983)
58. Wallace LA et al; Atmos Environ 19: 1651-61 (1985)
59. Westrick JJ et al; J Am Water Works Assoc 76: 52-9 (1984)
60. Wilson JT et al; J Environ Qual 10: 501-6 (1981)
61. Wilson JT et al; Devel Indust Microbiol Vol.24 (1983)

1-Chloro-2-nitrobenzene

SUBSTANCE IDENTIFICATION

Synonyms: o-Chloronitrobenzene

Structure:

CAS Registry Number: 88-73-3

Molecular Formula: $C_6H_4ClNO_2$

Wiswesser Line Notation: WNR BG

CHEMICAL AND PHYSICAL PROPERTIES

Boiling Point: 245-246 °C

Melting Point: 32-33 °C

Molecular Weight: 157.56

Dissociation Constants:

Log Octanol/Water Partition Coefficient: 2.24 [5]

Water Solubility: 199 mg/L at 25 °C [22]; 440 mg/L at 20 °C [3]

Vapor Pressure: 3.0 x 10^{-2} mm Hg at 20 °C [13] (estimated)

Henry's Law Constant: 3.6 x 10^{-5} atm cu.m/mol [6] (estimated)

1-Chloro-2-nitrobenzene

ENVIRONMENTAL FATE/EXPOSURE POTENTIAL

Summary: The major sources of environmental release of nitroaromatic compounds, such as 1-chloro-2-nitrobenzene, appear to be its production and use plants and by-product manufacturing plants. Minor sources of release to the environment may be loss during transport, storage or land burial, formation in the environment by oxidation of man-made aromatic amines, or by reaction of nitrogen oxides in highly polluted air with chlorinated aromatic hydrocarbons. If released to soil, 1-chloro-2-nitrobenzene should be resistant to oxidation and hydrolysis. Leaching to ground water may be significant since 1-chloro-2-nitrobenzene is predicted to be moderately mobile in soil. Based on Henry's Law constant, volatilization from wet soils may be possible; however, leaching would lessen the significance of volatilization as a removal mechanism. The estimated vapor pressure indicates that volatilization from dry soil surfaces is probably not rapid, although it may be a significant removal mechanism in the absence of other faster process(es). If released to water, 1-chloro-2-nitrobenzene should be resistant to oxidation, hydrolysis, and biodegradation. 1-Chloro-2-nitrobenzene may photolyze in water since it absorbs sunlight and the reaction is catalyzed by TiO_2. Bioconcentration in aquatic organisms and sorption to sediments should not be significant. The volatilization half-life from 1 m deep in surface water with a current velocity of 1 m/sec and a wind speed of 3 m/sec has been estimated to be 33.5 hours. Volatilization from surface waters with slower current speeds is expected to be less rapid. 1-Chloro-2-nitrobenzene has an estimated half-life in river water of 3.2 days based on monitoring data from the Rhine River. In contrast, it was not altered or physically removed during the time period required for transport down 900 miles of the Mississippi River. If released to the atmosphere, 1-chloro-2-nitrobenzene in the vapor phase is predicted to react with photochemically generated hydroxyl radicals with an estimated reaction half-life of 1.97 days at 25 °C. 1-Chloro-2-nitrobenzene has the potential to directly photolyze in the atmosphere since it absorbs sunlight. Chloronitrophenols may form as a result of photochemical reaction of 1-chloro-2-nitrobenzene in air. Exposure to 1-chloro-2-nitrobenzene is probably mainly to occupational groups during production or use in dye manufacturing. A small segment of the general public may be exposed through ingestion of contaminated drinking water or fish.

Natural Sources:

Artificial Sources: The major source of environmental release of nitroaromatic compounds, such as 1-chloro-2-nitrobenzene, appears to be from production plants and from by-product manufacturing plants [7]. 1-Chloro-2-nitrobenzene is used as a chemical intermediate predominantly in dye manufacturing and is known to be inadvertently formed as a by-product of 1-chloro-4-nitrobenzene manufacture [7]. Some loss to the environment may occur during transport, storage, or improper land burial [7]. Potential minor sources of 1-chloro-2-nitrobenzene are formation in the environment by oxidation of man-made aromatic amines and by reaction of nitrogen oxides present in highly polluted air with chlorinated aromatic hydrocarbons [7].

Terrestrial Fate: If released to soil, 1-chloro-2-nitrobenzene should be resistant to oxidation and hydrolysis. Leaching to ground water should be significant since 1-chloro-2-nitrobenzene is predicted to be moderately mobile in soil. Volatilization from wet soil may be significant; however, leaching would lessen the importance of volatilization as a removal mechanism. Volatilization from dry soil surfaces is probably not rapid, although it may be a significant removal mechanism under some circumstances.

Aquatic Fate: If released to water, 1-chloro-2-nitrobenzene should be resistant to oxidation, hydrolysis, and biodegradation. 1-Chloro-2-nitrobenzene may photolyze in water, since it absorbs sunlight and the photolysis is catalyzed by TiO_2 [8]. Bioconcentration in aquatic organisms and adsorption to suspended solids and sediments should not be significant. The volatilization half-life from 1 m deep in surface water with a current velocity of 1 m/sec and a wind speed of 3 m/sec has been estimated to be 33.5 hours. Volatilization from surface waters with slower current speeds should be less rapid. 1-Chloro-2-nitrobenzene has an estimated half-life in river water of 3.2 days based on monitoring data from the Rhine River [24]. However, this compound has been found to travel long distances in surface waters (900 miles in the Mississippi River) at concentrations that could be explained by simple dilution [7]. This observation indicates that 1-chloro-2-nitrobenzene was not altered or physically removed, at least during the time period required for transport down this river [7].

Atmospheric Fate: If released to the atmosphere, 1-chloro-2-nitrobenzene in the vapor phase is predicted to react with photochemically generated hydroxyl radicals with an estimated reaction half-life of 1.97 days at 25 °C. 1-Chloro-2-nitrobenzene has the potential to photolyze in the atmosphere. Chloronitrophenols may form as a result of the photochemical reaction of 1-chloro-2-nitrobenzene in air.

Biodegradation: 1-Chloro-2-nitrobenzene (21.1 ppm) in Ohio River water inoculated weekly with settled sewage underwent no degradation in 175 days [12]. 1-Chloro-2-nitrobenzene (100 ppm) inoculated with 30 ppm activated sludge at 25 °C was less than 30% degraded after 2 weeks [10,19]. Other investigators reported aquatic biodegradation half-life of much >4 weeks with both adapted and unadapted microorganisms [1].

Abiotic Degradation: Based on the molecular structure of 1-chloro-2-nitrobenzene, this compound should be resistant to oxidation and hydrolysis [13]. Weak adsorption of UV light greater than 290 nm by 1-chloro-2-nitrobenzene in methanol [18] indicates potential for photolysis by sunlight in water and air. In the presence of TiO_2, the photolysis half-life of the compound in aqueous solutions was about 2 hr with artificial light of wavelengths greater than 290 nm [8]. In air, 1-chloro-2-nitrobenzene in the vapor phase is predicted to react with photochemically generated hydroxyl radicals with an estimated reaction rate constant of 5.1 x 10^{-12} cm^3/molecule-sec at 25 °C [4]. Assuming an ambient hydroxyl radical concentration of 8.0 x 10^{+5} molecules/cm^3, the reaction half-life has been calculated to be 1.97 days [4]. Chloronitrophenols may form as the result of the photochemical reaction of 1-chloro-2-nitrobenzene in air [9].

Bioconcentration: Results of the Japanese MITI test for bioaccumulation indicate 1-chloro-2-nitrobenzene has a bioconcentration factor (BCF) of less than 100 [19]. Using a recommended value for the log octanol-water partition coefficient of 2.24 [5] and a measured water solubility of 440 mg/L at 20 °C [3], the BCF for 1-chloro-2-nitrobenzene has been estimated from regression equations [13] to be 30 and 20, respectively. Other authors have estimated a BCF value of 28 for this compound [1]. An experimentally measured value in guppy (Piecilia reticulata) was 195 [2]. Based on these BCF values, 1-chloro-2-nitrobenzene should not significantly bioconcentrate in aquatic organisms.

Soil Adsorption/Mobility: 1-Chloro-2-nitrobenzene has been found to travel long distances in surface waters (900 miles in the Mississippi River) at concentrations that could be explained by simple dilution [7]. This observation suggests only minimal adsorption to river sediment and fairly high transportability in the aqueous environment [7]. Based on a recommended value for the log octanol-water partition coefficient of 2.24 [5] and a measured water solubility of 440 mg/L at 20 °C [3], the soil adsorption coefficient (Koc) for 1-chloro-2-nitrobenzene has been estimated from regression equations [13] to be 398 and 155, respectively. These Koc values suggest that this compound should be moderately mobile in soil [20].

Volatilization from Water/Soil: Henry's Law constant for 1-chloro-2-nitrobenzene has been estimated to be 3.6×10^{-5} atm-m^3/mol using a method of group contributions based on molecular structure [6]. Using this value for Henry's Law constant, the volatilization half-life from 1 m deep in surface water with a current velocity of 1 m/sec and wind speed of 3 m/sec has been estimated to be 33.5 hours [13]. Volatilization from waters with slower current speed should be less rapid. Based on the soil adsorption coefficient of 155-398, water solubility of 440 mg/L at 20 °C [3] and estimated vapor pressure of 3.0×10^{-2} mm Hg at 20 °C [13], volatilization from wet soil surfaces may be significant. Based on the vapor pressure of this compound, volatilization from dry soil surfaces would probably not be rapid; although it may be a significant in absence of other more rapid removal process(es).

Water Concentrations: SURFACE WATER: 1-Chloro-2-nitrobenzene was monitored from Sept. 1957 - April 1959 in Mississippi River water samples from Cape Girardeau, MO and New Orleans, LA at concentrations of 0.004-0.037 ppm and 0.001-0.002 ppm, respectively [15]. 1-Chloro-2-nitrobenzene has been detected at levels ranging from less than 0.1 ug/L to 1.0 ug/L in the Rhine River during 1978 to 1982 [14,17]. It was also detected in Dutch coastal waters at a mean concentration of 11.0 ng/L and a maximum concentration of 50.0 ng/L [21]. DRINKING WATER: 1-Chloro-2-nitrobenzene has been identified but not quantified in drinking water from New Orleans LA [11].

Effluent Concentrations: 1-Chloro-2-nitrobenzene has been identified in treated effluents from an advanced wastewater treatment plant in Orange County, CA [11]. 1,2-, 1,3- and 1,4-chloronitrobenzene have been identified in the effluent from 1-chloro-3-nitrobenzene production at a concentration of 1500-1800 mg/L [7].

Sediment/Soil Concentrations:

Atmospheric Concentrations:

Food Survey Values:

Plant Concentrations:

Fish/Seafood Concentrations: 1-Chloro-2-nitrobenzene (0.006-0.24 ppm) was found in the edible portion of various species of fish taken from the Mississippi River 0.60 and 150 miles south of St. Louis, MO. 1-Chloro-2-nitrobenzene was not detected in fish taken from the Mississippi River 100 miles north of St. Louis (2 samples), 260-400 miles south of St. Louis (3 samples), or taken from the Missouri River (6 samples) [23].

Animal Concentrations:

Milk Concentrations:

Other Environmental Concentrations:

Probable Routes of Human Exposure: Exposure to 1-chloro-2-nitrobenzene is probably due mainly to occupational exposure during manufacture or use as a dye intermediate. A small segment of the general public may be exposed through ingestion of contaminated drinking water or fish.

1-Chloro-2-nitrobenzene

Average Daily Intake:

Occupational Exposures: NIOSH has statistically estimated that 2215 workers are exposed to 1-chloro-2-nitrobenzene in the United States [16].

Body Burdens:

REFERENCES

1. Canton JH et al; Reg Toxicol Pharmacol 5:123-31 (1985)
2. Deneer JW et al; Aquatic Toxicol 10:115-29 (1987)
3. Eckert JW; Phytopathol 52: 642-9 (1962)
4. GEMS; Graphical Exposure Modeling System. Fate of Atmospheric Pollutants, Office of Toxic Substances. USEPA (1986)
5. Hansch C, Leo AJ; Medchem Project Issue No. 26 Pomona College Claremont CA (1985)
6. Hine J, Mookerjee PK; J Org Chem 40: 292-8 (1975)
7. Howard PH et al; Investigation of Selected Environmental Contaminants : Nitroaromatics USEPA 560/2-76-010 Research Triangle Park NC (1976)
8. Hustert K et al; Chemosphere 16:809-12 (1987)
9. Kanno S, Nojima K; Chemosphere 8: 225-32 (1979)
10. Kitano M; Biodegradation and Bioaccumulation Test on Chemical Substances OECD Tokyo Meeting Reference Book TSU-NO 3 (1978)
11. Lucas SV; GC/MS Analysis of Organics in Drinking Water Concentrates and Advanced Waste Treatment Concentrates 2: Computer-Printed Tabulations of Compound Identification Results from Large Volume Concentrates, Columbus OH Health Eff Res Lab (1984)
12. Ludzack FJ, Ettinger MB; Eng Bull Ext Ser No 115: 278-82 (1963)
13. Lyman WJ et al; Handbook of Chemical Property Estimation Methods. Environmental Behavior of Organic Compounds. McGraw-Hill p 4-9,5-5,7-4,15-21 (1982)
14. Malle KG; Z Wasser-Abwasser Forsch 17: 75-81 (1984)
15. Middleton FM, Lichtenberg JJ; Ind Eng Chem 52: 99A-102A (1960)
16. NIOSH National Occupational Exposure Survey (NOES) Sept (1985)
17. Piet GJ, Zoeteman BCJ; J Am Water Works Assoc 72: 400-4 (1980)
18. Sadtler; Sadtler Standard Spectra Philadelphia PA
19. Sasaki S; pp 283-98 in Aquatic Pollutants Transformation and Biological Effects Hutzinger O et al eds; Oxford Pergamon Press (1978)
20. Swann RL et al; Res Rev 85: 17-28 (1983)
21. van de Meent D. et al; Wat Res Tech 18:73-81 (1986)
22. Yalkowsky SH et al; Arizona Database of Aqueous Solubility, University of Arizona (1987)
23. Yurawecz MP, Puma BJ; J Assoc Off Anal Chem 66: 1345-52 (1983)
24. Zoeteman BCJ et al; Chemosphere 9: 231-49 (1980)

1-Chloro-4-nitrobenzene

SUBSTANCE IDENTIFICATION

Synonyms: p-Chloronitrobenzene

Structure:

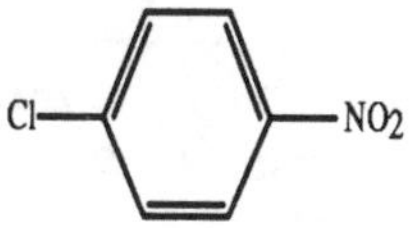

Cl—⟨benzene ring⟩—NO₂

CAS Registry Number: 100-00-5

Molecular Formula: $C_6H_4ClNO_2$

Wiswesser Line Notation: WNR DG

CHEMICAL AND PHYSICAL PROPERTIES

Boiling Point: 242 °C at 760 mm Hg

Melting Point: 82-84 °C

Molecular Weight: 157.56

Dissociation Constants:

Log Octanol/Water Partition Coefficient: 2.39 [8]

Water Solubility: 225 mg/L at 20 °C [24]; 453 mg/L at 20 °C [6]

Vapor Pressure: 1.04 x 10^{-2} mm Hg at 20 °C (estimated) [16]

Henry's Law Constant: 3.6 x 10^{-5} atm-m³/mole (estimated) [9]

ENVIRONMENTAL FATE/EXPOSURE POTENTIAL

Summary: The major source of environmental release of nitroaromatic compounds, such as 1-chloro-4-nitrobenzene, appears to be from production and use plants and by-product manufacturing plants. Minor sources of release to the environment may occur during transport or storage. If released to soil, 1-chloro-4-nitrobenzene should be resistant to hydrolysis, oxidation, and biodegradation. Biological reduction of 1-chloro-4-nitrobenzene under aerobic conditions may result in p-chloroaniline, 4-chloroacetaniline and 4-chloro-2-hydroxyaniline, as well as other metabolites. Leaching to ground water may be significant since moderate mobility in soil is predicted. Volatilization from wet soil surfaces may be significant, but should be considerably slower from dry soil surfaces. If released to water, 1-chloro-4-nitrobenzene should be resistant to hydrolysis, oxidation, and biodegradation. Potential biodegradation products in water are the same as for soil. 1-Chloro-4-nitrobenzene could potentially photolyze, although in air-saturated water this compound has been shown to be fairly stable to photolysis (5% degradation in 84 hours). Bioconcentration in aquatic organisms and adsorption to suspended solids and sediments should not be significant. The volatilization half-life from 1 m deep in surface water with a wind speed of 3 m/sec has been estimated to be 33.5 hours. Volatilization from surface waters with slower current speeds is expected to be less rapid. Based on monitoring data, 1-chloro-4-nitrobenzene has an estimated half-life of 0.3-3 days in rivers and 3-30 days in ground water. If released to the atmosphere, 1-chloro-4-nitrobenzene vapor is predicted to react with photochemically generated hydroxyl radicals (half-life 1.97 days). Reaction with ozone is not significant. 1-Chloro-4-nitrobenzene has the potential to directly photolyze in the atmosphere and 4-chloro-2-nitrophenol may be produced, although no experimental evidence is available. Exposure to 1-chloro-4-nitrobenzene is probably due mainly to occupational exposure during production or use as an intermediate. A small segment of the general population may be exposed through ingestion of contaminated drinking water or fish.

Natural Sources:

Artificial Sources: The major source of environmental release of nitroaromatic compounds, such as 1-chloro-4-nitrobenzene, appears to be from production and use plants and from by-product manufacturing plants [10]. 1-Chloro-4-nitrobenzene is used to manufacture p-nitrophenol, p-nitroaniline, p-aminophenol, phenacetin, and agriculture and rubber chemicals and has other minor uses. Some loss to the environment may occur during transport or storage [10].

Terrestrial Fate: If released to soil, 1-chloro-4-nitrobenzene should be resistant to oxidation, hydrolysis, and biodegradation. Reduction of 1-chloro-4-nitrobenzene by microorganisms under aerobic conditions may result in p-chloroaniline, 4-chloroacetanilide, and 4-chloro-2-hydroxyacetanilde as well as other metabolites. Leaching to ground water may be significant since 1-chloro-4-nitrobenzene is predicted to be moderately mobile in soil. Volatilization from wet soil surfaces may be significant, and should be considerably slower from dry soil surfaces. 1-Chloro-4-nitrobenzene at 0.3 ug/L was detected in Rhine River water before infiltration and less than 0.01 ug/L was detected after bank infiltration (infiltration time - 1-12 months) and after dune infiltration (infiltration time - 2-3 months) [26].

Aquatic Fate: If released to water, 1-chloro-4-nitrobenzene should be resistant to oxidation, hydrolysis, and biodegradation. Reduction of 1-chloro-4-nitrobenzene by microorganisms under aerobic conditions may result in p-chloroaniline, 4-chloroacetanilide, and 4-chloro-2-hydroxyacetanilide, as well as other metabolites. 1-Chloro-4-nitrobenzene may photolytically decompose in water in the presence of strong oxidizing agent, such as TiO_2; although, in air-saturated waters this compound may be stable towards photolytic dechlorination [11,18]. Bioconcentration in aquatic organisms and absorption to suspended solids and sediments should not be significant. The volatilization half-life from 1 m deep in surface water with a current velocity of 1 m/sec and a wind speed of 3 m/sec has been estimated to be 33.5 hours. Volatilization from surface waters with slower current speeds should be less rapid. Based on monitoring data, 1-chloro-4-nitrobenzene has an estimated half-life of 0.3 - 3 days in rivers and 3-30 days in ground water [26].

Atmospheric Fate: If released to the atmosphere, 1-chloro-4-nitrobenzene in the vapor phase is predicted to react with photochemically generated hydroxyl radicals with an estimated half-life of 1.97 days at 25 °C. Reaction with ozone is not likely. 1-Chloro-4-nitrobenzene has the potential to photolyze in the atmosphere. 4-Chloro-2-nitrophenol may be produced by the photochemical reaction of 1-chloro-4-nitrobenzene in air.

Biodegradation: 1-Chloro-4-nitrobenzene (100 ppm) inoculated with 30 ppm activated sludge at 25 °C was less than 30% degraded after 2 weeks [13,21]. 1-Chloro-4-nitrobenzene (10 ug/mL) inoculated with a mixed culture of microorganisms in soil was observed to be resistant to biodegradation (significant ring cleavage, as measured by UV absorbance, was not detected after 64 days) [1]. From 61 to 70% reduction of 1-chloro-4-nitrobenzene after 8-13 days was observed when a mixture of 1-chloro-4-nitrobenzene and 2,4-dinitrochlorobenzene under continuous flow conditions involving feeding, aeration, settling, and reflux was inoculated with a species of Arthrobacter simplex isolated from industrial waste. When two aeration columns were used, one with A. simplex and the other with A. simplex, Streptomyces coelicolor, Fusarium sp. and Trichoderma viridis isolated from soil, a 90% reduction of the nitro compounds was observed after 10 days. The reduction of 1-chloro-4-nitrobenzene produced p-chloroaniline and some undefined products [2]. The yeast Rhodosporidiam sp. reduced 1-chloro-4-nitrobenzene under aerobic conditions to give 4-chloroacetanilide and 4-chloro-2-hydroxyacetanilide as final major metabolites [4]. Other investigators have reported aquatic biodegradation half-life of much greater than 4 weeks with both adapted and unadapted microorganisms [3] and the chemical was found to remain unchanged under common biodegradation procedures used in wastewater treatment plants [14].

Abiotic Degradation: Based on the molecular structure of 1-chloro-4-nitrobenzene, this compound should be resistant to oxidation and hydrolysis [16]. When an aqueous solution of 1-chloro-4-nitrobenzene (1.45 x 10^{-6} M) was irradiated with light of wavelengths greater than 290 nm for 84 hours, less than 5% of the starting material underwent reductive dechlorination. The stability of 1-chloro-4-nitrobenzene towards reductive dechlorination was due to oxygenated conditions of this experiment [18]. However, 1-chloro-4-nitrobenzene does absorb UV light in the environmentally significant range, (>290 nm) [20], which indicates that potential exists for photolysis in water and air. In the presence of TiO_2, the photolysis half-life of this compound in aqueous solution was less than an hour with artificial light of wavelength greater than 290 nm [11]. In air, vapor-phase 1-chloro-4-nitrobenzene is predicted to react with a hydroxyl radical with an estimated rate constant of 5.0 x 10^{-12} cm^3/molecule-sec at 25 °C [7]. Assuming an ambient hydroxyl radical concentration of 8.0 x 10^{+5} radicals/cm^3, the reaction half-life has been calculated to be 1.97 days. 4-Chloro-2-nitrophenol may be produced by the photochemical reaction of 1-chloro-4-nitrobenzene in air [12].

Bioconcentration: Results of the MITI test for bioaccumulation indicate 1-chloro-4-nitrobenzene has a bioconcentration factor (BCF) of less than 100 [21]. Using a recommended value for the log octanol-water partition coefficient of 2.39 [8] and a measured water solubility of 453 mg/L at 20 °C [6], the BCF for 1-chloro-4-nitrobenzene has been estimated to be 39 and 20, respectively [16]. Other authors have estimated a BCF value of 39 for this compound [3]. The experimentally measured value in guppy (Poecilia reticulata) on the basis of fat weight was 288 [5]. Based on these BCF values, 1-chloro-4-nitrobenzene should not significantly bioconcentrate in aquatic organisms.

Soil Adsorption/Mobility: Based on a recommended value for the log octanol-water partition coefficient of 2.39 [8] and a water solubility of 453 mg/L at 20 °C [6], the soil absorption coefficient (Koc) for 1-chloro-4-nitrobenzene has been estimated to be 476 and 151, respectively [16]. These Koc values suggest that this compound should be moderately mobile in soil [22].

Volatilization from Water/Soil: Henry's Law constant for 1-chloro-4-nitrobenzene has been estimated to be 3.6 x 10^{-5} atm-m^3/mole using a method of group contributions based on molecular structure [9]. Using this value for Henry's Law constant, the volatilization half-life from 1 m deep in surface water with a current velocity of 1 m/sec and wind speed of 3 m/sec has been estimated to be 33.5 hours [16]. Volatilization from surface waters with slower current speed should be less rapid. Based on a soil absorption coefficient for 1-chloro-4-nitrobenzene of 151-476 and an estimated Henry's Law constant of 3.6 x 10^{-5} atm-m^3/mole [9], volatilization from wet soil surfaces may be significant. Based on the estimated vapor pressure of 1.04 x 10^{-2} mm Hg at 20 °C [16], volatilization from dry soil surfaces should be considerably slower than from wet soil surfaces.

1-Chloro-4-nitrobenzene

Water Concentrations: SURFACE WATER: From 1977 to 1982, 1-chloro-4-nitrobenzene was detected in the Rhine River at levels ranging from less than 0.1 ug/L to 0.11 ug/L. 1-Chloro-4-nitrobenzene was monitored in the Rhine River at a concentration of 1 ug/L, but was not detected in related tap water [19]. 0.3 ug/L 1-chloro-4-nitrobenzene was detected in Rhine River water before bank infiltration and <0.01 ug/L was detected after bank infiltration (infiltration time - 1-12 months) and after dune infiltration (infiltration time - 2-3 months) [26]. It was also detected in Dutch coastal waters at a mean concentration of 6.9 ng/L and a maximum concentration of 31 ng/L [23]. DRINKING WATER: 1-Chloro-4-nitrobenzene was positively identified but not quantified in drinking water from New Orleans, LA [15].

Effluent Concentrations: 1-Chloro-2-, 3-, and 4-nitrobenzene were identified in 1-chloro-3-nitrobenzene wastewater at a concentration of 1.5-1.8 g/L [10].

Sediment/Soil Concentrations:

Atmospheric Concentrations:

Food Survey Values:

Plant Concentrations:

Fish/Seafood Concentrations: This compound was found in the edible portion of various species of fish taken from the Mississippi River 0, 60, and 150 miles south of St. Louis, MO at concentrations ranging from 0.008 to 0.63 ppm [25].

Animal Concentrations:

Milk Concentrations:

Other Environmental Concentrations:

Probable Routes of Human Exposure: Exposure to 1-chloro-4-nitrobenzene is probably due mainly to occupational exposure during production or use as an intermediate. A small segment of

1-Chloro-4-nitrobenzene

the general population may be exposed through ingestion of contaminated drinking water or fish.

Average Daily Intake:

Occupational Exposures:

Body Burdens:

REFERENCES

1. Alexander M, Lustigman BK; J Agric Food Chem 14: 410-3 (1966)
2. Bielaszczyk E et al; Acta Microbiol Pol 16:243-8 (1967)
3. Canton JH et al; Reg Toxicol Pharmacol 5: 123-31 (1985)
4. Corbett MD, Corbett BR; Appl Env Microbiol 41: 942-9 (1981)
5. Deneer JW et al; Aquatic Toxicol 10: 115-29 (1987)
6. Eckert JW; Phytopathol 52: 642-9 (1962)
7. GEMS; Graphical Exposure Modeling System. Fate of Atmospheric Pollutants (FAP) Data Base. Office of Toxic Substances. USEPA (1986)
8. Hansch C, Leo AJ; Medchem Project Issue No 26. Claremont CA: Pomona College (1985)
9. Hine J, Mookerjee PK; J Org Chem 40: 292-8 (1975)
10. Howard et al; Investigation of Selected Environmental Contaminants: Nitroaromatics USEPA-560/2-76-010 (1976)
11. Hustert K et al; Chemosphere 16:809-12 (1987)
12. Kanno S, Nojima K; Chemosphere 8: 225-32 (1979)
13. Kitano M; Biodegradation and Bioaccumulation Test on Chemical Substances OECD Tokyo Meeting Reference Book TSU-No 3 (1978)
14. Lindgaard-Joergensen P, Jacobsen BN; In Comm Eur Communities, Eur 10388. Org Micropollut Aquat Environ p 429-39 (1986)
15. Lucas SV; GC/MS Analysis of Organics in Drinking Water Concentrates and Advanced Waste Treatment Concentrates: Vol 2. Computer-Printed Tabulations of Compound Identification Results from Large Volume Concentrates. Columbus, OH, Health Effects Research Laboratory (1984)
16. Lyman WJ et al; Handbook of Chemical Property Estimation Methods. Environmental Behavior of Organic Compounds. McGraw-Hill NY p4-9,14-8,15-21 (1982)
17. Malle KG; Z Wasser-Abwasser Forsch 17: 75-81 (1984)
18. Miller GC, Crosby DG; Chemosphere 12: 1217-27 (1983)
19. Piet GJ, Morra CF; p 31-42 in Artificial Groundwater Recharge (Water Resources Eng. Series) Pitman Pub (1983)
20. Sadtler; Sadtler Standard Spectra Phil PA UV (No.) 1290
21. Sasaki S; p 283-98 in Aquatic Pollutants: Transformation and Biological Effects, Hutzinger O et al eds; Oxford Pergamon Press (1978)
22. Swann RL et al; Res Rev 85: 17-28 (1983)
23. Van de Meent D et al; Wat Sci Tech 18: 78-81 (1986)

1-Chloro-4-nitrobenzene

24. Yalkowsky SH et al; Arizona Database of Aqueous Solubility, University of Arizona (1987)
25. Yurawecz MP, Puma BJ; J Assoc Off Anal Chem 66: 1345-52 (1983)
26. Zoeteman BCJ et al; Chemosphere 9: 231-49 (1980)

2-Chlorophenol

SUBSTANCE IDENTIFICATION

Synonyms: o-Chlorophenol

Structure:

CAS Registry Number: 95-57-8

Molecular Formula: C_6H_5ClO

Wiswesser Line Notation: QR BG

CHEMICAL AND PHYSICAL PROPERTIES

Boiling Point: 174.9 °C at 760 mm Hg

Melting Point: 9.3 °C

Molecular Weight: 128.56

Dissociation Constants: pKa = 8.52 [10]

Log Octanol/Water Partition Coefficient: 2.15 [16]

Water Solubility: 28,000 ppm at 25 °C [32]

Vapor Pressure: 1.42 mm Hg at 25 °C [13]

Henry's Law Constant: 5.6 x 10^{-7} atm-m^3/mole [19]

2-Chlorophenol

ENVIRONMENTAL FATE/EXPOSURE POTENTIAL

Summary: Release of 2-chlorophenol (2-CP) to the environment will occur through its use as a synthetic intermediate primarily for dyes and higher chlorinated phenols. Since the pKa of 2-CP is 8.52 it will exist in water and moist soils in a partially dissociated state which may affect its transport and reactivity in the environment. If it is released to the soil it will be expected to show low to moderate adsorption to the soil based on estimated Koc's and may leach to the ground water. Hydrolysis in soil cannot be important. Biodegradation in soils may be important with loss of 94% reported for 2-CP incubated in non-sterile clay loam soil at 4 °C in 6.5 hr vs 1% loss in sterile soil in 12 days. If 2-CP is released to water it may adsorb to sediments. It will not be expected to bioconcentrate in aquatic organisms and will not chemically hydrolyze. It will be susceptible to photolysis near the surface of waters, and biodegradation should be an important fate process with complete removal reported in 13 and 36 days in die-away tests using 2 raw river waters, 15 days in 2 river waters seeded with water from previous die-away tests, and 15 days in acclimated river water. Evaporation from water should be a slow transport process with a half-life of 73 days estimated for evaporation from a river 1 m deep, flowing at 1 m/sec with a wind velocity of 3 m/sec. If 2-CP is released to the atmosphere it may be susceptible to photolysis and reaction with NO in polluted air. The estimated vapor phase half-life in the atmosphere is 1.96 days mainly as a result of addition of ozone to the aromatic ring. Exposure to 2-CP from environmental sources may occur through the ingestion of contaminated drinking water where 2-CP is formed by chlorination of water containing phenol. Occupational exposure may occur from its use as a synthetic intermediate primarily for dyes and higher chlorinated phenols.

Natural Sources: 2-CP is a synthetic organic compound and has no known natural sources.

Artificial Sources: Release of 2-CP to the environment will occur through its use as a synthetic intermediate primarily for dyes and higher chlorinated phenols [18].

2-Chlorophenol

Terrestrial Fate: If 2-CP is released to the soil, it will be expected to show low to moderate adsorption to the soil based on estimated Koc's and may leach to the ground water. Hydrolysis in soil will not be important. Biodegradation in soils may be important with loss of 94% reported for 2-CP incubated in non-sterile clay loam soil at 4 °C in 6.5 hr vs 1% loss in sterile soil in 12 days [2]. Since the pKa of 2-CP is 8.52 [10], it will exist in moist soils in a partially dissociated state which may affect its transport and reactivity.

Aquatic Fate: If 2-CP is released to water it may adsorb to sediments based on experimental Koc's, although estimated Koc's predict this adsorption will be low to moderate. It will not be expected to bioconcentrate in aquatic organisms and will not hydrolyze. It will be susceptible to photolysis near the surface of waters, and biodegradation should be an important fate process with complete removal reported in 13 and 36 days in die-away tests using two raw river waters [11] and 15 days in acclimated river water [28]. Addition of seed water from prior die-away testing led to faster complete removal of 2-CP in one of the two raw river waters to which seed had been added (15 days with seed vs 36 days without) [11]. Evaporation from water is expected to be a slow transport process with a half-life of 73 days estimated for evaporation from a river 1 m deep, flowing at 1 m/sec with a wind velocity of 3 m/sec. Since the pKa of 2-CP is 8.52 [10], it will exist in water and sediment in a partially dissociated state which may affect its transport and reactivity in the environment.

Atmospheric Fate: If 2-CP is released to the atmosphere it may be susceptible to photolysis and reaction with NO in polluted air. The estimated vapor phase half-life in the atmosphere is 1.96 days as a result of addition of ozone to the aromatic ring. Washout may be an important transport removal process.

Biodegradation: Complete removal of 2-CP at 1 ppm concn in 15 days reported in acclimated river water. Recovery of residual 2-CP (days) in river die-away tests using Great Miami River water was 980 ppb (0 days), 810 ppb (6 days), 710 ppb (23 days), and 0 ppb (36 days); and in the Little Miami River was 890 ppb (0 days), 870 ppb (6 days), 480 ppb (13 days), 0 ppb (15 days) [11]. Addition of seed water from prior die-away testing led to faster complete removal in seeded vs unseeded Great Miami River water (15 days vs 36 days), whereas time for complete removal in seeded Little Miami River water was nearly the same as that in unseeded water (15 days vs 13 days) [11]. Loss of 94% reported for 2-CP incubated in non-sterile clay loam soil at 4 °C in 6.5 hr vs 1% loss in sterile soil in 12 days [2]. Complete removal reported in 14 and 47 days in Dunkirk and Mardin silt loam suspensions, respectively [1]. Degradation (degr) of 67% 2-CP reported in 10 days upon percolation through Rothamsted clay; degr was faster upon redose [44]. Evolution of 13% and 25% theoretical CO_2 in 1 and 10 weeks in para-brown soil reported [15]. Complete disappearance of 2-CP was reported in aerobic columns of aquifer material contaminated by landfill leachate in 6 days for acclimated columns and 9 days in unacclimated columns; under anaerobic conditions degr occurred only with aquifer material from an actively methanogenic site [42]. Complete loss of 2-CP reported in sediment from farm stream at 20 °C in 10-15 days vs 19% in 30 days in sterile sediments [2]. Complete degradation reported in 3 days in acclimated sludge [21]; 0% theoretical BOD in 11-19 days with wastewater seed [9]; 100% deg (5 ppm) and 0% deg (10 ppm) with activated sludge [9]; 95.6% removal in 6 hr of 200 ppm with acclimated activated sludge [35]. Half-lives (concn of 2-CP tested) for batch cultures experiments using unacclimated anaerobic sewage sludge were: 1 week (3 or 10 ppm), 2 weeks (30 ppm), 8 weeks (97 ppm), and 33 weeks (285 ppm); lag time was <1 week for all concn tested [20]. 2-CP was readily degraded in anaerobic microcosm using subsurface soils from both the unsaturated and saturated zones [39].

Abiotic Degradation: Hydrolysis will not be important under normal environmental conditions [43]. Since the pKa of 2-CP is 8.52 [10], 2-CP will exist in a partially dissociated state in water and moist soils and therefore, its transport should be affected by pH. The photolysis also should be affected by pH, as demonstrated by the fact that the quantum yields for the disappearance of the undissociated form was 10 times less (quantum yield [qy], 0.03-0.04) than that for the dissociated form (qy, 0.30) when 2-CP was irradiated at 296 nm in aqueous solution at pH 8-13; the main product of photolysis from the undissociated form was pyrocatechol and cyclopentadienic acid from the dissociated form [4]. 2-CP (4.46 x 10^{-5} mol) reacted with 1 mL NO in 1 liter air for 5 hr to give the 4- and 6-nitro-2-chlorophenol adducts in 36% and 30% yields, respectively [24]. The estimated vapor phase half-life in the atmosphere as a result of addition of ozone to the aromatic ring is 1.96 days [12].

Bioconcentration: BCF: bluegill sunfish, 214 [3]; goldfish, 7.1 [25]. Using the water solubility and the log octanol/water partition coefficient, BCF's of 1.9 and 25 were estimated, respectively [29]. Based on these experimental and estimated BCFs, 2-CP will not be expected to bioconcentrate in aquatic organisms.

Soil Adsorption/Mobility: Koc: Sediments, fine, 4890; coarse, 3990 [22]; clay loam soil, 51.1 [5]. Adsorption to the organo-clay Bentone 24 has been shown to be pH sensitive for 2-CP; 2-CP was 76.6% adsorbed by Bentone 24 in aqueous solution at pH 7.8 and 15% adsorbed by Bentone 18c at pH 7.6 [31]. Using the water solubility and the log octanol/water partition coefficient, Koc's of 16 and 352 were estimated, respectively [29].

Volatilization from Water/Soil: Half-lives for evaporation of 2-CP from stirred and static water at a depth of 0.38 cm at 23.8 °C were 1.35 and 1.60 hr, respectively [7]. Using the estimated Henry's Law constant, a half-life of 73 days was estimated for evaporation from a river 1 m deep, flowing at 1 m/sec with a wind velocity of 3 m/sec [29].

Water Concentrations: DRINKING WATER: Identified, not quantified: in drinking water [26]; in finished drinking water [6]; in finished drinking water in United States when water chlorination was used [27]. Detected at 39 ppt in the finished drinking water of 1 of 6 Canadian cities in Feb 1985, not detected (nd) in raw water [38]. Frequency of detection/maximum level (ppt) in samples from 40 potable water treatment plants in Canada: treated water, 1/40 (Feb-Mar 1985); 2/65 (May-July 1985); nd (Oct-Dec 1984); nd in any of the samples of raw water from the 40 treatment plants [37]. GROUND WATER: Melbourne, Australia, 1973 and 1975, 4 bore-hole samples from aquifer polluted by chemical company waste ponds, 3 pos, identified, not quantified [41]. SURFACE WATER: USEPA STORET database, 814 samples, 0.2% pos, median, <10 ppb [40]. Great Lakes basin ecosystem, Lakes Erie and Michigan basins, 0% pos [14]. Netherlands: Rhine River, 1974, avg 3-20 ppb

[36], 1976, 206 grab samples, 2% pos, max 2.3 ppb, 1977, not detected(nd) [46]; River Meuse at Eijsden, 1974, avg 2-20 ppb [36], 1976 and 1977, nd [45]. Netherlands, March 1979-March 1980, Ijssel River near Kampen, 13 samples, 7.7% pos, max 0.6 ppb [45]. Rhein River, km 865, 1978, <0.1 ppb 1979, <1 ppb [30].

Effluent Concentrations: USEPA STORET database: 1312 samples, 1.5% pos, median, <10 ppb [40]. U.S. National Urban Runoff Program, through July 1982, 15 cities, 6.7% pos, 86 samples, 1% pos samples, 2 ppb [8]. Secondary sewage effluent, 1.7 ppb, herbicide production waste, 2.88 ppb [36]. Detected, not quantified in Love Canal water, sediment and/or soil [17]. Extract from effluent from test hazardous waste incinerators, 5 extracts, 40% pos, 6.9 and 13 ppb [23]. Municipal treatment plants, July 1978, identified at <10 ppb in primary effluents from Los Angeles (LA) and San Diego, and secondary effluents from Orange County and LA [47]. Identified, not quantified, industrial effluents [6].

Sediment/Soil Concentrations: SEDIMENTS: USEPA STORET database; 308 samples, 1% pos, median, <1000 ppm [40].

Atmospheric Concentrations:

Food Survey Values:

Plant Concentrations:

Animal Concentrations:

Milk Concentrations:

Other Environmental Concentrations:

Probable Routes of Human Exposure: General exposure to 2-CP may occur through the ingestion of contaminated drinking water where 2-CP is formed during the chlorination treatment. Occupational exposure may also occur with its use as a synthetic intermediate primarily for dyes and higher chlorinated phenols.

2-Chlorophenol

Occupational Exposures: NIOSH (NOES Survey 1981-1983) has statistically estimated that 934 employees are exposed to 2-CP in the United States [33].

Body Burdens:

REFERENCES

1. Alexander M, Aleem MIH; J Agric Food Chem 9: 44-7 (1961)
2. Baker MD et al; Water Res 14: 1765-71 (1980)
3. Barrows ME et al; pp 379-92 in Dyn Exposure Hazard Assess Toxicol Chem Ann Arbor, MI, Ann Arbor Sci (1980)
4. Boule P et al; Chemosphere 11: 1179-88 (1982)
5. Boyd SA; Soil Sci 134: 337-43 (1982)
6. Callahan MA et al; Water-Related Environ Fate of 129 Priority Pollut Vol.2 USEPA-440/4-79-029b (1979)
7. Chiou CT et al; Environ Inter 3: 231-6 (1980)
8. Cole RH et al; J Water Pollut Control Fed 56: 898-908 (1984)
9. Dojlido JR; Investigations of Biodegradability and Toxicity Organic Compounds Final Report 1975-79 USEPA-600/2-79-163 (1979)
10. Drahonovsky J, Vacek Z; Collect Czech Chem Commun 36: 3431-40 (1971)
11. Ettinger MB, Ruchhoft CC; Sew Indust Waste 22: 1214-7 (1950)
12. GEMS; Graphical Exposure Modeling System. Fate of Atmos Pollut. FAP. (1986)
13. GEMS; Graphical Exposure Modeling System. PCCHEM. USEPA (1987)
14. Great Lakes Quality Rev Board; An Inventory of Chem Identified in the Great Lakes Ecosystem Vol.1 Summary, Report to the Great Lakes Quality Rev Board, Windsor Ontario, Canada pp 195 (1983)
15. Haider K et al; Arch Microbiol 96: 183-200 (1974)
16. Hansch C, Leo AJ; Medchem Project Issue No 26. Claremont CA: Pomona College (1985)
17. Hauser TR, Bromberg SM; Environ Monit Assess 2: 249-71 (1982)
18. Hawley GG; Condensed Chemical Dictionary 10th ed Von Nostrand Reinhold NY p 240 (1981)
19. Hine J, Mookerjee PK; J Org Chem 40: 292-8 (1975)
20. Hrudey SE et al; Environ Tech Lett 8: 65-76 (1987)
21. Ingols RS et al; J Water Pollut Control Fed 38: 629-35 (1966)
22. Isaacson PJ, Frink CR; Environ Sci Technol 18: 43-8 (1984)
23. James RH et al; J Proc Air Pollut Control Assoc Ann Mtg 77th 1: 84-18.5 (1984)
24. Kanno S, Nojima K; Chemosphere 8: 225-32 (1979)
25. Kobayashi K et al; Bull Japan Soc Sci Fish 45: 173-5 (1979)
26. Kool HJ et al; Crit Rev Environ Control 12: 307-57 (1982)
27. Krijgsheld KR, Vamdergen A; Chemosphere 15: 825-60 (1986)
28. Ludzack FJ, Ettinger MB; J Water Pollut Control Fed 32: 1173-1200 (1960)
29. Lyman WJ et al; Handbook of Chemical Property Estimation Methods. Environmental Behavior of Organic Compounds. McGraw-Hill NY (1982)
30. Malle KG; Z Wasser Abwasser Forsch 17: 75-81 (1984)
31. Miller RW, Faust SD; Environ Lett 4: 211-23 (1973)

32. Nathan MF; Chem Eng 85: 93-100 (1978)
33. NIOSH; The National Occupational Hazard Survey (NOHS) (1974)
34. NIOSH; The National Occupational Exposure Survey (NOES) (1983)
35. Pitter P; Water Res 10: 231-5 (1976)
36. Scow K et al; pp 4-22 to 23 in An Exposure and Risk Assessment for Chlorinated Phenols NTIS PB85-21195 (1982)
37. Sithole BB, Williams DT; J Assoc Off Anal Chem 69: 807-10 (1986)
38. Sithole BB et al; J Assoc Off Anal Chem 69: 466-73 (1986)
39. Smith JA, Novak JT; Water Air Soil Pollut 33: 29-42 (1987)
40. Staples CA et al; Environ Toxicol Chem 4: 131-42 (1985)
41. Stepan S et al; Australian Water Res Council Conf Ser 1: 415-24 (1981)
42. Sulfita JM, Miller GD; Environ Toxicol Chem 4: 751-8 (1985)
43. USEPA; Treatability Manual pp 1.8.2-1 USEPA-600/2-82-001A (1981)
44. Walker N; Plant Soil 5: 194-204 (1954)
45. Wegman RCC, VanDenBroek HH; Water Res 13: 651-7 (1983)
46. Wegman RCC, Hofstee AWM; Water Res 17: 227-30 (1983)
47. Young DR; Ann Rep South Calif Coastal Water Res Proj 1978: 103-12 (1978)

3-Chlorophenol

SUBSTANCE IDENTIFICATION

Synonyms: m-Chlorophenol

Structure:

CAS Registry Number: 108-43-0

Molecular Formula: C_6H_5ClO

Wiswesser Line Notation: QR CG

CHEMICAL AND PHYSICAL PROPERTIES

Boiling Point: 214 °C at 760 mm Hg

Melting Point: 33 °C

Molecular Weight: 128.56

Dissociation Constants: pKa = 9.12 [9]

Log Octanol/Water Partition Coefficient: log Kow = 2.50 [13]

Water Solubility: 26,000 mg/L at 25 °C [20]

Vapor Pressure: 0.119 mm Hg at 25 °C [11]

Henry's Law Constant: 5.6 x 10^{-7} atm-m³/mole [16]

ENVIRONMENTAL FATE/EXPOSURE POTENTIAL

Summary: Release of 3-chlorophenol (3-CP) to the environment may occur through its use as a synthetic intermediate. Since the pKa of 3-CP is 9.12, it will exist in water and moist soils at pH >8.0 in a partially dissociated state which may affect its transport and reactivity in the environment. If 3-CP is released to the soil, it will be expected to show low to moderate adsorption to the soil based on estimated Koc's and may leach to the ground water. Hydrolysis in soil should not be important. Biodegradation in soils may be important with loss of 87% reported for 2-CP incubated in non-sterile clay loam soil at 4 °C in 6.5 hr vs 31% loss in sterile soil in 160 days. If 3-CP is released to water, it may slightly to moderately sorb to sediments based on estimated Koc's. It will not be expected to bioconcentrate in aquatic organisms and will not hydrolyze. It will be susceptible to photolysis near the surface of waters and biodegradation may be an important fate process with degradation of 6% reported in 40 days in water from a farm stream at 20 °C. Evaporation from water should be a slow transport process with a half-life of 73 days estimated for evaporation from a river 1 m deep, flowing at 1 m/sec with a wind velocity of 3 m/sec. If 3-CP is released to the atmosphere, it may be susceptible to photolysis and reaction with NO in polluted air. The estimated vapor phase half-life in the atmosphere is 1.96 days as a result of addition of ozone to the aromatic ring. Washout may be an important transport removal process. Occupational exposure may also occur from its use as a synthetic intermediate.

Natural Sources:

Artificial Sources: Release of 3-CP to the environment will occur through its use as a synthetic intermediate [15]. 3-CP was reported to be formed during the chlorination of sewage [24].

Terrestrial Fate: If 3-CP is released to the soil, it will be expected to show low to moderate adsorption to soil based on estimated Koc's and may leach to the ground water. Hydrolysis in soil will not be important. Biodegradation in soils may be important with loss of 87% reported for 3-CP incubated in non-sterile clay loam soil at 23 °C in 160 days vs 31% loss in sterile soil in 160 days [2]. Since the pKa of 3-CP is 9.12 [9], it will exist in moist soils in a partially dissociated state which may affect its transport and reactivity.

Aquatic Fate: If 3-CP is released to water, it may be slightly to moderately sorb to sediments based on estimated Koc's. It will not be expected to bioconcentrate in aquatic organisms and will not hydrolyze. It will be susceptible to photolysis near the surface of waters and biodegradation may be an important fate process with degradation of 6% reported in 40 days in water from a farm stream at 20 °C; loss of 56% reported in sediment from a farm stream at 20 °C in 30 days vs 14% in 30 days in sterile sediments [2]. Evaporation from water should be a slow process with a half-life of 73 days estimated for evaporation from a river 1 m deep, flowing at 1 m/sec with a wind velocity of 3 m/sec. Since the pKa of 3-CP is 9.12 [9], it will exist in water and sediments at pH >8.0 in a partially dissociated state which may affect its transport and reactivity.

Atmospheric Fate: If 3-CP is released to the atmosphere, it may be susceptible to photolysis and reaction with NO in polluted air. The estimated vapor phase half-life in the atmosphere is 1.96 days as a result of addition of ozone to the aromatic ring. Washout may be an important transport removal process.

Biodegradation: Degradation of 6% 3-CP reported in 40 days in water from a farm stream at 20 °C; loss of 56% 3-CP reported in sediment from farm stream at 20 °C in 30 days vs 14% in 30 days in sterile sediments [2]. Degradation in clay loam soil under aerobic conditions of 87% 3-CP reported in 160 days vs 31% degradation in sterile control; degradation under anaerobic conditions was 37% in 160 days vs 15% in sterile control [3]. 3-CP was not completely removed in 72 and 47 days in nutrient media seeded with Dunkirk and Mardin silt loam suspensions, respectively [1]. 3-CP was degraded by anaerobic columns of actively methanogenic aquifer material contaminated by landfill leachate but not by nonmethanogenic aquifer material [23]. Complete degradation in 7 weeks with incubation using municipal sewage sludge under anaerobic conditions [8]. Degradation of 30% 3-CP in 4 weeks and 100% in 6 weeks reported using fresh unacclimated sludge with a lag of approx 3 weeks; 90% degradation in 3 days and 100% degradation in 7 days reported using acclimated sludge [6]. No degradation reported using settled wastewater solids seed incubated for 7 days followed by 3 weekly subcultures [4].

Abiotic Degradation: Hydrolysis will not be an important degradation process based on the resistance of phenols and halogenated aromatics to hydrolysis [18]. Since the pKa of 3-CP is 9.12 [9], 3-CP will exist in a partially dissociated state in most waters in the environment at pH >8.0 which may affect its reactivity in the environment. The quantum yields (qy) for the disappearance of the undissociated form was 0.09 and for the dissociated form, 0.13, when 3-CP was irradiated at 296 nm in aqueous solution at pH 8-13; the main product of photolysis is resorcinol (>80% of converted 3-CP) [5]. Irradiation of 3-CP in an oxygen-free aqueous solution at pH >13 with 313 nm light resulted in the formation of chloride (qy,0.14) [12]. 3-CP (4.46×10^{-5} mol) reacted with 1 mL NO in 1 L air for 5 hr to give the 2-, 4-, and 6-nitro-3-chlorophenol adducts in 6.3%, 17%, and 11% yields, respectively [17]. The estimated vapor phase half-life in the atmosphere is 1.96 days as a result of addition of ozone and hydroxyl radicals to the aromatic ring [10].

Bioconcentration: Using the water solubility and the log octanol/water coefficient, BCF's of 15 and 352 were estimated, respectively [18]. Based on these estimated BCF's, 3-CP will not be expected to bioconcentrate in aquatic organisms.

Soil Adsorption/Mobility: Koc, clay loam soil, 51.1 [7]. Adsorption to the organo-clay Bentone 24 has been shown to be pH sensitive for 3-CP; 3-CP was 81% adsorbed by Bentone 24 in aqueous solution at pH 7.7 and 13% adsorbed by synthetic clay Bentone-18c at pH 7.6 [19]. Using the water solubility and the log octanol/water partition coefficient, Koc's of 16 and 546 were estimated, respectively [18].

Volatilization from Water/Soil: Using the estimated Henry's Law constant, a half-life of 73 days was estimated for evaporation from a river 1 m deep, flowing at 1 m/sec with a wind velocity of 3 m/sec [18].

3-Chlorophenol

Water Concentrations: GROUND WATER: Melbourne, Australia, 1973 and 1975, 4 bore hole samples from aquifer polluted by chemical company waste ponds, 75% pos, identified, not quantified [22]. SURFACE WATER: Netherlands: Rhine River, 1976, 1 sample date, 100% pos, max 6 ppb; 1977, 1 sample date, 0% pos [26]. River Meuse at Eijsden, 1976 and 1977, 1 sample date each, not detected [26]. Netherlands, March 1979-March, 1980, Ijssel River near Kampen, 13 samples, 31% pos, max 3.4% ppb [25].

Effluent Concentrations: Detected, not quantified in Love Canal water, sediment, and/or soil [14].

Sediment/Soil Concentrations:

Atmospheric Concentrations:

Food Survey Values:

Plant Concentrations:

Fish/Seafood Concentrations:

Animal Concentrations:

Milk Concentrations:

Other Environmental Concentrations:

Probable Routes of Human Exposure: Occupational exposure to 3-CP may occur that is related to its use as an intermediate in organic synthesis.

Average Daily Intake:

Occupational Exposures: NIOSH (NOES Survey 1981-1983) has statistically estimated that 919 employees are exposed to 3-CP in the United States [21].

Body Burdens:

3-Chlorophenol

REFERENCES

1. Alexander M, Aleem MIH; J Agric Food Chem 9: 44-7 (1961)
2. Baker MD et al; Water Res 14: 1765-71 (1980)
3. Baker MD, Mayfield CI; Water Air Soil Pollut 13: 411-24 (1980)
4. Barth EF, Bunch RL; Biodegradation and Treatability of Specific Pollut USEPA-600/9-79-034 (1979)
5. Boule P et al; Chemosphere 11: 1179-88 (1982)
6. Boyd SA, Shelton DR; Appl Environ Microbiol 47: 272-7 (1984)
7. Boyd SA; Soil Sci 134: 337-43 (1982)
8. Boyd SA et al; Appl Environ Microbiol 46: 50-4 (1983)
9. Drahonovsky J, Vasek Z; Collect Czech Chem Commun 36: 3431-40 (1971)
10. GEMS; Graphical Exposure Modeling System. FAP. Fate of Atmos Pollut (1986)
11. GEMS; Graphical Exposure Modeling System. PCCHEM. USEPA (1987)
12. Grabowski ZR; Z Physik Chem 27: 239-52 (1961)
13. Hansch C, Leo AJ; Medchem Project Issue No 26. Claremont CA: Pomona College (1985)
14. Hauser TR, Bromberg SM; Environ Monit Assess 2: 249-71 (1982)
15. Hawley GG; Condensed Chemical Dictionary 10th ed Von Nostrand Reinhold NY p 240 (1982)
16. Hine J, Mookerjee PK; J Org Chem 40: 292-8 (1975)
17. Kanno S, Nojima K; Chemosphere 8: 225-32 (1979)
18. Lyman WJ et al; Handbook of Chemical Property Estimation Methods. Environmental Behavior of Organic Compounds. McGraw-Hill NY (1982)
19. Miller RW, Faust SD; Environ Lett 4: 211-23 (1973)
20. Nathan MF; Chem Eng 85: 93-100 (1978)
21. NIOSH; The National Occupational Exposure Survey (NOES) (1983)
22. Stepan S et al; Australian Water Resources Council Conf Ser 1: 415-24 (1981)
23. Suflita JM, Miller GD; Environ Toxicol Chem 4: 751-8 (1985)
24. USEPA; Ambient Water Quality Criteria Doc Chlorinated Phenols p C-3 USEPA-440/5-80-032 (1980)
25. Wegman RCC, VanDenBroek HH; Water Res 13: 651-7 (1983)
26. Wegman RCC, Hofstee AWM; Water Res 17: 227-30 (1983)

4-Chlorophenol

SUBSTANCE IDENTIFICATION

Synonyms: p-Chlorophenol

Structure:

CAS Registry Number: 106-48-9

Molecular Formula: C_6H_5ClO

Wiswesser Line Notation: QR DG

CHEMICAL AND PHYSICAL PROPERTIES

Boiling Point: 220 °C at 760 mm Hg

Melting Point: 43 °C

Molecular Weight: 128.56

Dissociation Constants: pKa = 9.41 [34]

Log Octanol/Water Partition Coefficient: 2.39 [15]

Water Solubility: 27,000 mg/L at 25 °C [30]

Vapor Pressure: 0.087 mm Hg at 25 °C [11]

Henry's Law Constant: 5.6 x 10^{-7} atm-m^3/mole [19]

4-Chlorophenol

ENVIRONMENTAL FATE/EXPOSURE POTENTIAL

Summary: Release of 4-chlorophenol (4-CP) to the environment may occur due to its industrial use as a synthetic intermediate primarily for dyes and higher chlorinated phenols and its use as a denaturant for alcohol. Since the pKa of 4-CP is 9.41, it will exist in water and moist soils in a partially dissociated state which may affect its transport and reactivity in the environment. If 4-CP is released to the soil, it will be expected to show low to moderate adsorption to the soil based on estimated Koc's and it may leach to the ground water. Hydrolysis in soil will not be important. Biodegradation in soils may be important with loss of 84% reported for 4-CP incubated in non-sterile clay loam soil at 4 °C in 6.5 hr vs 0% loss in sterile soil in 12 days. If 4-CP is released to water, it may adsorb to sediments based on experimental Koc's, although estimated Koc's predict this adsorption will be low to moderate. It will not be expected to bioconcentrate in aquatic organisms and will not hydrolyze. It will be susceptible to photolysis near the surface of waters and biodegradation may be an important fate process with complete removal reported in 15 days and 13 days in unacclimated and acclimated river water, respectively. Half-lives of 20 days were reported in water, and 3 days in sediment and seawater, from the estuarine Skidway River, GA at 22 °C. Evaporation from water should be a slow transport process with a half-life of 73 days estimated for evaporation from a river 1 m deep, flowing at 1 m/sec with a wind velocity of 3 m/sec. If 4-CP is released to the atmosphere it may be susceptible to photolysis and reaction with NO in polluted air. The estimated vapor phase half-life in the atmosphere is 1.96 days as a result of addition of O_3 to the aromatic ring. Washout may be an important transport removal process. Exposure to 4-CP may occur through the ingestion of contaminated drinking water. Occupational exposure may occur due to its industrial uses.

Natural Sources: 4-CP is a synthetic organic compound and has no known natural sources.

Artificial Sources: Release of 4-CP to the environment may occur due to its industrial use as a synthetic intermediate primarily for dyes and higher chlorinated phenols and its use as a denaturant for alcohol [17]. It is also used as an antiseptic and selective solvent for refining mineral oils [17]. It is formed inadvertently through the chlorination of phenol containing effluents and drinking water sources [38].

Terrestrial Fate: If 4-CP is released to the soil, it will be expected to show low to moderate adsorption based on estimated Koc's and it may leach to the ground water. Hydrolysis in soil will not be important. Biodegradation in soils may be important with loss of 84% reported for 4-CP incubated in non-sterile clay loam soil at 4 °C in 6.5 hr vs 0% loss in sterile soil in 12 days [4]. Since the pKa of 4-CP is 9.41 [34], it will exist in moist soils in a partially dissociated state which may affect its transport and reactivity in soil and ground water.

Aquatic Fate: If 4-CP is released to water, it may adsorb to sediments based on experimental Koc's, although estimated Koc's predict this adsorption will be low to moderate. It will not be expected to bioconcentrate in aquatic organisms and will not hydrolyze. It will be susceptible to photolysis near the surface of waters and biodegradation may be an important fate process with complete removal reported in 15 days and 13 days in unacclimated [10] and acclimated [27] river water, respectively. Half-lives of 20 days were reported in water, and 3 days in sediment and seawater, from the estuarine Skidway River, Ga at 22 °C [26]. Evaporation from water is expected to be a slow transport process with a half-life of 73 days estimated for evaporation from a river 1 m deep, flowing at 1 m/sec with a wind velocity of 3 m/sec. Since the pKa of 4-CP is 9.41 [34], it will exist in water and sediment in a partially dissociated state which may affect its transport and reactivity in water and sediment.

Atmospheric Fate: If 4-CP is released to the atmosphere, it may be susceptible to photolysis and reaction with NO in polluted air. The estimated vapor phase half-life in the atmosphere is 1.96 days as a result of addition of O_3 to the aromatic ring. Washout may be an important transport removal process.

Biodegradation: Complete removal of 4-CP at 1 ppm concn in 13 days reported in acclimated river water and 33% removal in 25 days with acclimated sewage seed [27]. Recovery of residual 4-CP (days in Great Miami River water) was 1000 ppb (0 days); 980

ppb (6 days); 80 ppb (13 days); 0 ppb (15 days); and in the Little Miami River was 1000 ppb (0 days); 1000 ppb (6 days); 0 ppb (13 days) [10]. Half-lives of 20 days were reported in water, and 3 days in sediment and seawater, from the estuarine Skidaway River, GA at 22 °C [26]. 100% deg in 30 days reported using sediment from farm stream vs 15% degradation using sterile controls [4]. The rate of degradation of 4-CP was significantly lower using microbes from a mesotrophic reservoir which were adapted to natural humic acids from a highly colored lake than the rate using unadapted microbes [35]. Loss of 84% reported for 4-CP incubated in non-sterile clay loam soil at 4 °C in 12 days vs 0% loss in sterile soil [4]. Complete removal reported in 9 and 3 days in Dunkirk and Mardin silt loam suspensions, respectively [2]. Degradation of 50% 4-CP reported in 21 days upon percolation through phenol pretreated Rothamsted clay; no change in degradation rate upon redose [39]. Evolution of 22.2% and 35% theoretical CO_2 in 1 and 10 weeks in para-brown soil was reported [14]. Degradation of 100% in 16 days in suspension of Niagara silt loam soil was reported [3]. Complete degradation reported in 3 days in acclimated sludge [22]; 96% max removal by acclimated sludge reported [33]. No degradation reported using a Warburg respirometer with acclimated sewage seed [18]. Incubation for 16 weeks with fresh sludge resulted in 61% degradation with a 3 week lag period; incubation with sludge acclimated to 2-, 3-, or 4-CP for 4 weeks resulted in 100% degradation in 8 days [7]. Half-lives for biotransformation and biomineralization in the dark in estuarine water from the Skidaway River, GA, in summer (25 °C)/winter (14 °C) were 11/116 hr and 2/231 days, respectively [21]. Rapid and extensive (approx 30%) mineralization reported in aerobically incubated ground water from Lula, OK [1].

Abiotic Degradation: Halogenated aromatics and phenols are generally resistant to hydrolysis [28]. Since the pKa of 4-CP is 9.41 [34], 4-CP will exist in a partially dissociated state in water and moist soils which may affect its reactivity. The quantum yield (qy) for the disappearance of 4-CP was 0.25 when it was irradiated at 296 nm in aqueous solution at pH 1-13; products of photolysis included hydroquinone [5]. Irradiation of 4-CP in an oxygen-free aqueous solution at pH >13 with 313 nm light resulted in the formation of chloride (qy, 0.13-0.16), hydroquinone (qy, 0.02-0.04), and hydrogen peroxide (qy, 0.02-0.05) [13]. 4-CP (4.46×10^{-5} mol) reacted with 1 mL NO in 1 L air for 5 hr to give the 2-nitro-4-chlorophenol adduct in 64% yield [23]. The estimated vapor phase half-life in the atmosphere is 1.96 days as a result of addition of O_3 to the aromatic ring [12]. Half-lives for phototransformation by sunlight in summer (25 °C)/ winter (14 °C) were: distilled water, 63/99 hr; poisoned estuarine water, 46/63 hr; half-lives for photomineralization were: distilled water, 58/224 days; poisoned estuarine water, 53/334 days [21].

Bioconcentration: BCF of 15 reported for goldfish [24]. Using the water solubility and the log octanol/water partition coefficient, BCF's of 2 and 39 were estimated, respectively [28]. Based on these experimental and estimated BCF, 4-CP will not be expected to bioconcentrate in aquatic organisms.

Soil Adsorption/Mobility: Koc of 70 reported for Brookstone clay loam soil [6]. Adsorption to the organo-clay Bentone 24 has been shown to be pH sensitive for 4-CP; 4-CP was 48.5% adsorbed by Bentone 24 in aqueous solution at pH 8.0 and 7.7% adsorbed by Bentone 18c at pH 7.7 [29]. 4-CP did not appear to be sorbed in an experiment in a sandy aquifer [37]. Using the solubility and the log octanol/water partition coefficient, Koc's of 16 and 476 were estimated, respectively [28]. Based on these experimental and estimated Koc's, high to moderate mobility of 4-CP in soil or sediment is expected.

Volatilization from Water/Soil: Using the estimated Henry's Law constant, a half-life of 73 days was estimated for evaporation from a river 1 m deep, flowing at 1 m/sec with a wind velocity of 3 m/sec [28]. Half-lives for evaporation from stirred and static water at a depth of 0.38 cm at 23.6 °C were 12.8 and 17.4 hr, respectively [8].

Water Concentrations: DRINKING WATER: Identified, not quantified, in drinking water [25]. Frequency of detection/maximum level found (ppt) in samples from 40 potable water treatment plants in Canada: raw water, 2/103; treated water, 6/86 (Oct-Dec 1984); raw water, 1/5.9; treated water, 6/33 (Feb-Mar 1985); raw water, not detected; treated water, 10/127 (May-June 1985) [36].
SURFACE WATER: Netherlands: Rhine River (River Meuse at Eijsden), 1976, 206 grab samples, 16% pos (20% pos), max 3.9

ppb; 1977, not detected (not detected) [41]. Netherlands, March 1979-March, 1980, Ijssel River near Kampen, 13 samples, 2.1% pos, max 2.1 ppb [40].

Effluent Concentrations: Jacksonville, AR, industrial waste, 1970, Jan, 1.7 ppm; March, 14.3 ppm; Apr, 22.9 ppm; May, 22.4 ppm; Aug, 2.1 ppm [9]. Detected, not quantified in Love Canal water, sediment, and/or soil [16].

Sediment/Soil Concentrations: SEDIMENT: Netherlands, 19 sites in lakes, rivers, and canals, not detected [40].

Atmospheric Concentrations:

Food Survey Values: Identified, not quantified, as a volatile flavor component in fried bacon [20].

Plant Concentrations:

Fish/Seafood Concentrations:

Animal Concentrations:

Milk Concentrations:

Other Environmental Concentrations:

Probable Routes of Human Exposure: Exposure to 4-CP may occur through the ingestion of contaminated drinking water where 4-CP is formed during the chlorination treatment. Occupational exposure may occur that is related to its use as a synthetic intermediate primarily for dyes and higher chlorinated phenols and its use as a denaturant for alcohol.

Average Daily Intake:

Occupational Exposures: NIOSH (NOES Survey 1981-1983) has statistically estimated that 2487 employees are exposed to 4-CP in the United States [31]. NIOSH (NOHS Survey 1972-1974) has statistically estimated that >4847 employees are exposed to 4-CP in the United States [32].

4-Chlorophenol

Body Burdens:

REFERENCES

1. Aelion CM et al; Am Soc Microbiol Abstr 87th Ann Mtg, Atlanta, GA (1987)
2. Alexander M, Aleem MIH; J Agric Food Chem 9: 44-7 (1961)
3. Alexander M, Lustigman BK; J Agric Food Chem 14: 410-3 (1966)
4. Baker MD et al; Water Res 14: 1765-71 (1980)
5. Boule P et al; Chemosphere 11: 1179-88 (1982)
6. Boyd SA; Soil Sci 134: 337-43 (1982)
7. Boyd SA, Shelton DR; Appl Environ Microbiol 47: 272-7 (1984)
8. Chiou CT et al; Environ Internat 3: 231-6 (1980)
9. City of Jacksonville; Biological Treatment of Chlorophenolic Wastes, The Demonstration of a Facility for the Biological Treatment of a Complex Chlorophenolic Waste, NTIS PB-206813 (1971)
10. Ettinger MB, Ruchhoft CC; Sew Indust Waste 22: 1214-7 (1950)
11. GEMS; Graphical Exposure Modeling System. PCCHEM. USEPA (1987)
12. GEMS; Graphical Exposure Modeling System. FAP. Fate of Atmos Pollut (1986)
13. Grabowski ZR; Z Physik Chem 27: 239-52 (1961)
14. Haider K et al; Arch Microbiol 96: 183-200 (1974)
15. Hansch C, Leo AJ; Medchem Project Issue No 26. Claremont CA: Pomona College (1985)
16. Hauser TR, Bromberg SM; Environ Monit Assess 2: 249-71 (1982)
17. Hawley GG; Condensed Chemical Dictionary 10th ed Van Nostrand Reinhold NY p 240 (1981)
18. Helfgott TB et al; An Index of Refractory Organics USEPA-600/2-77-174 (1977)
19. Hine J, Mookerjee PK; J Org Chem 40: 292-8 (1975)
20. Ho CT et al; J Agric Food Chem 31: 336-42 (1983)
21. Hwang H et al; Environ Sci Technol 20: 1002-7 (1986)
22. Ingols RS et al; J Water Pollut Control Fed 38: 629-35 (1966)
23. Kanno S, Nojima K; Chemosphere 8: 225-32 (1979)
24. Kobayashi K et al; Bull Jap Soc Sci Fish 45: 173-5 (1979)
25. Kool HJ et al; Crit Rev Environ Control 12: 307-57 (1982)
26. Lee RF, Ryan C; pp 443-50 in Microbial Degradation of Pollut in Marine Environments USEPA-600/9-79-012 (1979)
27. Ludzack FJ, Ettinger MB; J Water Pollut Control Fed 32: 1173-200 (1960)
28. Lyman WJ et al; Handbook of Chemical Property Estimation Methods. Environmental Behavior of Organic Compounds. McGraw-Hill NY (1982)
29. Miller RW, Faust SD; Environ Lett 4: 211-23 (1973)
30. Nathan MF; Chem Eng 85: 93-100 (1978)
31. NIOSH; The National Occupational Exposure Survey (NOES) (1983)
32. NIOSH; The National Occupational Hazard Survey (NOHS) (1974)
33. Pitter P; Water Res 10: 231-5 (1976)
34. Serjeant EP, Dempsey B; Ionization Constants of Organic Acids in Aqueous Solution, IUPAC Chemical Data Series No. 23 New York, NY (1979)

4-Chlorophenol

35. Shimp R, Pfaender FK; Appl Environ Microbiol 49: 402-7 (1985)
36. Sithole BB, Williams DT; J Assoc Off Anal Chem 69: 807-10 (1986)
37. Sutton PA, Barker JF; Groundwater 23: 10-6 (1985)
38. USEPA; Ambient Water Quality Criteria Doc, Chlorinated Phenols pp C-3 USEPA-440/5-80-032 (1980)
39. Walker N; Plant Soil 5: 194-204 (1954)
40. Wegman RCC, VanDenBroek HH; Water Res 13: 651-7 (1983)
41. Wegman RCC, Hofstee AWM; Water Res 17: 227-30 (1983)

Chloroprene

SUBSTANCE IDENTIFICATION

Synonyms: 2-Chloro-1,3-butadiene

Structure:

CAS Registry Number: 126-99-8

Molecular Formula: C_4H_5Cl

Wiswesser Line Notation: 1UYG1U1

CHEMICAL AND PHYSICAL PROPERTIES

Boiling Point: 59.4 °C

Melting Point: -130 °C(freezing point)

Molecular Weight: 88.54

Dissociation Constants:

Log Octanol/Water Partition Coefficient: 2.03 [6]

Water Solubility:

Vapor Pressure: 174 mm Hg at 20 °C [1]

Henry's Law Constant: 3.2 x 10^{-2} atm-m³/mole [10] (estimated by bond contribution method)

Chloroprene

ENVIRONMENTAL FATE/EXPOSURE POTENTIAL

Summary: The main sources of environmental release of chloroprene are probably the effluent and emissions from plants which use this compound to make polychloroprene elastomers. If released to soil, chloroprene should be susceptible to removal by rapid volatilization and transport by leaching into ground water. Chemical hydrolysis is not expected to occur. If released to water, volatilization is predicted to be the dominant removal mechanism (half-life 3 hours from a model river 1 m deep with a current speed of 1 m/sec and wind speed of 3 m/sec). In water, this compound is not expected to chemically hydrolyze, adsorb significantly to suspended solids or sediments, or bioaccumulate in aquatic organisms. If released to the atmosphere, chloroprene is expected to exist almost entirely in the vapor phase. The primary removal mechanism should be reaction with photochemically generated hydroxyl radicals with small amounts of chloroprene being removed by reaction with ozone. The overall reaction half-life has been estimated to be 1.6 hours. Anticipated reaction products include H_2CO, $H_2C=CClCHO$, $OHCCCHO$, $ClCOCHO$, $H_2CCHCClO$, chlorohydroxy acids, and aldehydes. The most probable route of human exposure to chloroprene is inhalation by workers involved in the production or use of this compound.

Natural Sources: Chloroprene is not known to occur as a natural product [11].

Artificial Sources: This compound is used almost exclusively, without isolation, in the production of polychloroprene elastomers [11]. The primary source of environmental release of chloroprene may be in the effluent or emissions from its manufacturing/use facilities.

Terrestrial Fate: If released to soil, chloroprene is expected to be susceptible to rapid volatilization and extensive leaching into ground water. Chloroprene will not be susceptible to chemical hydrolysis.

Aquatic Fate: If released to water, volatilization is expected to be the dominant removal mechanism. The volatilization half-life from water 1 m deep with a current speed of 1 m/sec and a wind speed

184

of 3 m/sec has been estimated to be approx 3 hours. Chloroprene is not expected to undergo chemical hydrolysis, adsorb significantly to suspended solids or sediments, or bioaccumulate in aquatic organisms.

Atmospheric Fate: Based on the vapor pressure, chloroprene is expected to exist almost entirely in the vapor phase in the atmosphere [4]. Chloroprene is predicted to be removed from the atmosphere by reaction with photochemically generated hydroxyl radicals (half-life 1.8 hours) and ozone (half-life 12 hours). The overall reaction half-life of the compound in the atmosphere has been estimated to be 1.6 hours. Anticipated reaction products include H_2CO, $H_2C=CClCHO$, $OHCCHO$, $ClCOCHO$, $H_2CCHCClO$, chlorohydroxy acids, and aldehydes. Removal by wet or dry deposition is unlikely [3].

Biodegradation:

Abiotic Degradation: Chloroprene is not expected to undergo chemical hydrolysis under environmental conditions since it contains no hydrolyzable functional groups [12]. The half-life for the reaction of chloroprene vapor with photochemically generated hydroxyl radicals in the atmosphere has been estimated to be 1.8 hours based on an estimated reaction rate constant of 1.36×10^{-10} cm^3/molecule-sec at 25 °C and an ambient hydroxyl radical concentration of $8.0 \times 10^{+5}$ molecules/cm^3 [7]. The half-life for the reaction of chloroprene vapor with ozone has been estimated to be 12 hours based on an estimated reaction rate constant of 2.6×10^{-17} cm^3/molecule-sec and an ambient ozone concentration of $6 \times 10^{+11}$ molecules/sec [7]. Based on these data, the overall reaction half-life of chloroprene vapor in the atmosphere has been estimated to be 1.6 hours. Anticipated reaction products include H_2CO, $H_2C=CClCHO$, $OHCCHO$, $ClCOCHO$, $H_2CCHCClO$, chlorohydroxy acids, and aldehydes [3].

Bioconcentration: The bioconcentration factor (BCF) of chloroprene has been estimated to be 22 based on an estimated log octanol/water partition coefficient [13]. This BCF value suggests that chloroprene will not bioaccumulate significantly in aquatic organisms.

Chloroprene

Soil Adsorption/Mobility: Based on a method of molecular topology and quantitative structure-activity analysis, a soil adsorption coefficient of 50 (Koc) was estimated for chloroprene [17]. Using a linear regression equation based on an estimated log octanol/water partition coefficient, a Koc of 315 was estimated [13]. These Koc values suggest that adsorption to suspended solids and sediments in water would not be significant. Chloroprene should be moderately to highly mobile in soil [18].

Volatilization from Water/Soil: Using the estimated value for the Henry's Law constant, the volatilization half-life from a model river 1 m deep with a current velocity of 1 m/sec and a wind speed of 3 m/sec is estimated to be approx 3 hours [13]. The relatively high vapor pressure suggests that chloroprene would volatilize rapidly from dry soil surfaces. Chloroprene is also expected to volatilize fairly rapidly from wet surfaces since its Koc values are relatively low and its estimated value for Henry's Law constant is relatively high.

Water Concentrations: SURFACE WATER: Chloroprene was detected in 1 out of 204 samples of surface water taken from sites near heavily industrialized areas across the United States during 1975-76 [5].

Effluent Concentrations: Chloroprene was identified in 2 out of 63 industrial effluents at a concentration of <10 ug/L [16].

Sediment/Soil Concentrations:

Atmospheric Concentrations: Chloroprene was detected in the ambient air in 6/6 NJ cities: avg concn 0.097 ppbv, max concn 4.0 ppbv [9]. Not detected in air above 6 abandoned hazardous wastes sites and one active sanitary landfill in NJ during 1983/84, detection limit 0.01 ppbv [8]. Chloroprene was detected in 2/2 samples of ambient air in Houston, TX during July 1976 at an average concentration of 0.59 ppbv [2]. Chloroprene was not detected in 17 samples of ambient air in Baton Rouge, LA during March 1977, detection limit not reported [2].

Food Survey Values:

186

Chloroprene

Plant Concentrations:

Fish/Seafood Concentrations:

Animal Concentrations:

Milk Concentrations:

Other Environmental Concentrations:

Probable Routes of Human Exposure: The most probable route of human exposure to chloroprene is inhalation by workers involved in the manufacture or use of this compound.

Average Daily Intake:

Occupational Exposures: During 1977, airborne concentrations of chloroprene of up to 0.2 ppm were reported in a roll building area in a metal fabricating plant where polychloroprene was applied extensively to metal cylinders prior to vulcanization [11]. During 1973, at a U.S. chloroprene polymerization plant, airborne concentrations of chloroprene were found to range from 14-1420 ppm in the make-up area, from 130-6760 ppm in the reactor area, from 6-440 in the monomer recovery area and from 113-252 ppm in the latex area [11]. NIOSH has estimated that 7604 workers are potentially exposed to chloroprene based on statistical estimates derived from a survey conducted in 1981-83 in the United States [15]. NIOSH has estimated that 230,555 workers are potentially exposed to chloroprene based on statistical estimates derived from a survey conducted in 1972-74 in the United States [14].

Body Burdens:

REFERENCES

1. Boublik T et al; The Vapor Pressure of Pure Substances Elsevier Sci Pub. Amsterdam p. 229 (1984)
2. Brodzinsky R, Singh HB; Volatile Organic Chemicals in the Atmosphere: An Assessment of Available Data. SRI International. Menlo Park CA (1982)
3. Cupitt LT; Fate of Toxic and Hazardous Materials in the Air Environment USEPA 600/3-80-084 (1980)
4. Eisenreich SJ; Environ Sci Tech 15: 30-8 (1981)

5. Ewing BB et al; Monitoring to Detect Previously Unrecognized Pollutants in Surface Waters USEPA 560/6-77-015 (1977)
6. GEMS; Graphical Exposure Modeling System. CLOG3 (1982)
7. GEMS; Graphical Exposure Modeling System. FAP. Fate of Atmospheric Pollutants (1986)
8. Harkov R et al; J Environ Sci Health 20: 491-501 (1985)
9. Harkov R et al; pp. 104-19 in Proc Int Tech Conf Toxic Air Contam McGovern JJ ed. APCA. Pittsburgh, PA (1981)
10. Hine J, Mookerjee PK; J Org Chem 40: 292-8 (1975)
11. IARC; Chloroprene and Polychloroprene 19: 131-5 (1979)
12. Jaber HM et al; Data Acquisition for Environmental Transport and Fate Screening p. 312 USEPA-600/6-84/009, NTIS PB84-243906, PB84-243955 (1984)
13. Lyman WJ et al; Handbook of Chem Property Estimation Methods McGraw-Hill NY (1982)
14. NIOSH; National Occupational Hazard Survey (NOHS) (1974)
15. NIOSH; National Occupational Exposure Survey (NOES) (1983)
16. Perry DL et al; Identification of Organic Compounds in Industrial Effluent Discharges USEPA 600/4-79-016, NTIS PB-294794 (1979)
17. Sabljic A; J Agric Food Chem 32: 243-6 (1984)
18. Swann RL et al; Res Rev 85: 17-28 (1983)

2-Cresol

SUBSTANCE IDENTIFICATION

Synonyms: o-Cresol; 2-Hydroxytoluene; 2-Methylphenol; o-Cresylic acid; o-Hydroxytoluene

Structure:

CAS Registry Number: 95-48-7

Molecular Formula: C_7H_8O

Wiswesser Line Notation: QR B1

CHEMICAL AND PHYSICAL PROPERTIES

Boiling Point: 190.95 °C at 760 mm Hg

Melting Point: 30.9 °C

Molecular Weight: 108.15

Dissociation Constants: 10.287 [44]

Log Octanol/Water Partition Coefficient: 1.95 [23]

Water Solubility: 30.8 g/L at 40 °C [44]

Vapor Pressure: 0.31 mm Hg at 25 °C [44]

Henry's Law Constant: 1.6 x 10^{-6} atm-m³/mol [31] (calc)

ENVIRONMENTAL FATE/EXPOSURE POTENTIAL

Summary: 2-Cresol is released to the atmosphere in auto and diesel exhaust, during coal tar and petroleum refining, and wood pulping, and during its use in manufacturing and metal refining. Wastewater from these industries as well as from municipal wastewater treatment plants contain 2-cresol. When released to the atmosphere 2-cresol will react with photochemically produced hydroxyl radicals (half-life 9.6 hr) during the day and react with nitrate radicals at night (half-life 2 min). In addition it will be scavenged by rain and oxidized by metal cations in rainwater and fogwater. When released into water, biodegradation will generally occur within days. However, in surface layers of oligotrophic waters, photolysis may be important. Its fate in soil has not been well characterized; it is relatively mobile in most soils and will readily biodegrade (100% in 8 days). Human exposure will primarily be via inhalation and dermal contact in the workplace or inhalation in source areas.

Natural Sources: 2-Cresol occurs in small quantities in petroleum, coal, and wood [50,58]. 2-Cresol was found in surface water in the blast zone of the Mount St. Helens eruption and is believed to have been formed during the destruction of the plant and soil material by the volcanic action [37].

Artificial Sources: 2-Cresol may be released into the environment in emissions and wastewaters during its production and use as a disinfectant and solvent [38]. It may also be released in emissions or wastewater during coal tar and petroleum refining, organic chemicals, plastics, and resins mfg and in its use as an ore floatation and textile scouring agent [58]. It is found in sewage [58]. It is also found in emissions from wood pulping, tobacco smoke, and auto and diesel exhaust [18,22]. It is a product of the photooxidation of toluene [47].

Terrestrial Fate: 2-Cresol should be relatively mobile in most soils. An exception appears to be where iron oxide and pH levels are high. 2-Cresol biodegrades rapidly in water and there is evidence that it also biodegrades in soil. However, information on the biodegradation rate is lacking.

2-Cresol

Aquatic Fate: The primary removal mechanism in most surface waters will be biodegradation. The half-life for biodegradation in river water is reported to be 2 days, which is similar to the half-lives for other cresol isomers. Although this half-life is the result of only one study, results of aerobic screening studies for cresol isomers are similar and this would suggest that the biodegradation half-life in river water is fairly rapid. In an oligotrophic lake, photolysis may contribute to the disappearance of 2-cresol, but experimental data are lacking. While photolysis of 2-cresol is catalyzed by the presence of humic acids and it is oxidized by metal ions such as Fe(III) and Mn(III/IV), it is not clear that these processes contribute significantly to the degradation of 2-cresol in eutrophic waters. There is no evidence that biodegradation occurs under anaerobic conditions, but only a few reports are available. Evaporation and volatilization will not be important.

Atmospheric Fate: The estimated photodegradation half-life of 2-cresol resulting from its reaction with photochemically generated hydroxyl radicals is 9.6 hr. Its lifetime will be less under photochemical smog conditions. In the nighttime, its half-life will be extremely short, 2 min, due to its reaction with NO_3 radicals. In addition oxidation by metal ions such as Fe(III) and Mn(III/IV) may rapidly occur (hours or less) in rainwater and fogwater. 2-Cresol is highly soluble in water and it will be scavenged from the atmosphere by rain [31].

Biodegradation: 2-Cresol biodegrades rapidly in screening studies using soil, sewage, activated sludge, or municipal wastewater inocula [1,8,25,43,48,55,57,59]. Acclimation is frequently not necessary [1,8,25,55,57,59]. It is completely degraded in river water in 2 and 7 days at 20 and 4 °C, respectively [34]. In a field study, water from a downgradient well below a coal distillation plant showed a 76% concentration reduction (after corrections for dispersion were made) compared with near surface water [17]. No mineralization was observed when 2-cresol was incubated with two digester sludge samples for 8 weeks under anaerobic conditions [27], nor was there any mineralization in 29 weeks when it was incubated with anaerobic freshwater sediment [27]. Similarly, no mineralization occurred when it was incubated anaerobically for 40 days at 37 °C with a sludge inoculum [15]. 2-Cresol completely degraded in soil in 8 days when applied at 500 ppm [29]. The disproportionate decrease in concentration of 2-cresol in ground water downgradient from a wood preserving facility has been ascribed to biodegradation [19].

Abiotic Degradation: In the atmosphere during the daytime, 2-cresol reacts principally with photochemically-generated hydroxyl radicals primarily by ring addition but also by H abstraction with a resulting half-life of 9.6 hr [7,42]. However in the nighttime, especially in moderately polluted atmospheres where concentrations of O_3 and NO_2 are high, reaction with NO_3 radicals is a major sink for 2-cresol (half-life 2 min [7]) with the formation of nitrocresols [12,21]. Under photochemical smog conditions, half-lives have been measured to be 1-5 hr with nitrocresol formation [6,20,41]. No data could be found on the abiotic degradation of 2-cresol in aqueous systems. Since 2-cresol has no hydrolyzable groups, hydrolysis will not be significant. The photolysis half-life of 4-cresol in pure water by sunlight is 35 days and it will presumably be similar for the 2-isomer [49]. Since absorptivity increases with pH from pH 5.1 to 8.9, so may the photolysis rate [49]. The rate of photolysis increases by a factor of 12 when humic acids are added to the water [49]. The depth-averaged half-life of 2-cresol in the upper 1 m of a polluted eutrophic pre-alpine Swiss lake is 11 days [14]. This is a result of sensitized photolysis possibly by alkoxy radicals that are photochemically produced from dissolved organic matter [53]. Phenols are oxidized by ions such as Fe(III) and Mn(III/IV) that are widely distributed in suspended particules in surface waters and in soils, especially at low pH [53]. These reactions can also occur in urban rainwater and fogwater since these droplets may contain manganese oxides and have low pH values [14]. Manganese oxides derived from dust and ash may bring about these oxidative degradations in time scales of a few hours or less [53].

Bioconcentration: Using the log octanol-water partition coefficient of 1.95 [23], a bioconcentration factor of 18 is estimated [35]. Therefore 2-cresol would not be expected to bioconcentrate significantly in fish.

Soil Adsorption/Mobility: The value for Koc measured on Brookstone clay loam soil (pH 5.7) is 22 and that predicted from the water solubility is 18 [9]. Koc's for phenols predicted from water solubilities are only good for soils with organic carbon contents greater than approx 0.5% [51]. Adsorptivities are greater than predicted for soils with low organic carbon content because interactions such as H-bonding are dominant [51]. On five fine-textured B horizon clay soils (pH 4.5-7.8), the distribution between soil and water ranged from 2.8 to 500 [5]. For these subsurface soils, the levels of free iron oxide and pH were the key factors in determining adsorption capacity [5].

Volatilization from Water/Soil: 2-Cresol has a low volatility from water, having a calculated Henry's Law constant of 1.6×10^{-6} atm-m^3/mol [32]. Such chemicals will partition into the water phase and volatilization will not be an important loss process [35].

Water Concentrations: DRINKING WATER: Identified, not quantified in drinking water in the United States [30,33,46]. SURFACE WATER: Detected, not quantified in the Danube River in Germany [46]. GROUND WATER: Ground water in a sand aquifer at a wood preserving facility in Pensacola, FL (5 sites, 5 depths) 0-7.10 mg/L [19]. Two aquifers under the Hoe Creek underground coal gasification site in Wyoming - 15 mo after gasification complete 63 and 6600 ppb [54]. Detected in 1 well water in United States [46]. Concentrations of 2-cresol in ground water at an abandoned phenolic waste site (pinetar manufacturer) in Gainesville FL were 1100-5200 ppb at the original site; 920-2,900 ppb at downgradient sites; and <0.3-22 ppb at upgradient sites [36]. RAIN/SNOW: Portland, OR - seven rain events 240-2800 ppt, >1000 ppt mean dissolved in rain [31]. In Switzerland, the concn of 2-cresol dissolved in rain was 1.5 ppb [56].

Effluent Concentrations: 71% frequency of appearance in 28 samples representing effluents from refineries, petrochemical and metallurgical industries, municipal wastewater plants, and polluted fjords in Norway [52]. Identified in finished water from advanced waste treatment plants [33]. 2-Cresol was confirmed in 10 of 4000 effluent samples in a broad survey covering 46 industrial categories [11]. It was found in effluents from timber products, iron and steel manufacturing, coal mining, organics and plastics, inorganic chemicals, textile mills, plastics and synthetics, organic chemicals, and publicly-owned treatment works [11]. Wastewater from the gasification of Indian Head lignite contained 640 ppm of 2-cresol [16]. In a comprehensive survey of wastewater from 4000 industrial and publicly-owned treatment works (POTWs) sponsored by the Effluent Guidelines Division of the U.S. EPA, 2-cresol was identified in discharges of the following industrial category (frequency of occurrence, median concn in ppb): timber products (14, 105.5); iron and steel mfg (9, 22.8); petroleum refining (10, 123.5); nonferrous metals (5, 36.4); paint and ink (1, 30.9); printing and publishing (4, 11.4); coal mining (3, 23.2); organics and plastics (24, 503.5); inorganic chemicals (3, 14.3); textile mills (4, 72.5); plastics and synthetics (1, 1685.9); pulp and paper (4, 59.9); rubber processing (7, 435.3); soaps and detergents (2, 44.2); auto and other laundries (3, 460.9); gum and wood industries (4, 3.3); pharmaceuticals (7, 82.3); foundries (40, 49.7); aluminum (4, 2.4); electronics (5, 237.5); electroplating (2, 4.8); oil and gas extraction (10, 4.1); organic chemicals (11, 1217.3); transportation equipment (4, 8.3); synfuels (2, 290.2); publicly-owned treatment works (79, 51.6); rum industry (3, 208.2) [45]. Both the petroleum products and organics and plastics industries had maximum effluent concentrations exceeding 10 ppm, namely, 10.1 and 18,321.9 ppm, respectively [45].

Sediment/Soil Concentrations: A max concn of 55 ppm of 2-cresol was detected in subsurface soil at the site of an abandoned phenolic waste site in Gainesville, FL [36]. In addition, subsurface soil at downgradient sites contained maximum concns of 12-34 ppm [36].

Atmospheric Concentrations: RURAL/REMOTE: Various sites in United States - not detected [10]. URBAN/SUBURBAN: Various sites in United States - not detected [10]. Portland, OR - during seven rain events 10-29 ppt, 15.6 ppt, mean in gas phase [31]. Particulate matter associated cresol was generally <1% gas-phase value [31]. SOURCE AREAS: 6 areas (54 samples) in United States 1.6 ppb median, 29 ppb max [10]. Oil shale wastewater facility 0.44 ppb [24]. Phenolic resin factory in Japan 40 ppb [28].

Food Survey Values: 2-Cresol was identified as a volatile component of fried bacon [26].

2-Cresol

Plant Concentrations:

Fish/Seafood Concentrations:

Animal Concentrations:

Milk Concentrations:

Other Environmental Concentrations:

Probable Routes of Human Exposure: The most probable route of human exposure to 2-cresol is via inhalation of air, especially in source areas or in occupational settings. Occupational exposure may also occur by dermal contact.

Average Daily Intake: AIR INTAKE: (assume typical concn 15.6 ppt, 1.4 ug); WATER INTAKE: insufficient data; FOOD INTAKE: insufficient data.

Occupational Exposures: Based on 1978 emissions data, it has been estimated that the total annual U.S. dosage (number of exposed persons times annual average atmospheric concn to which they were exposed) of 2-cresol is 5820 mg/m^3-persons [2]. This is a result of inhalation by occupationally exposed persons. NIOSH (NOHS Survey, 1972-74) has statistically estimated that 21,156 workers are exposed to 2-cresol in the United States [40]. NIOSH (NOES Survey, 1981-83) has statistically estimated that 1339 workers are exposed to 2-cresol in the United States [39]. Indoor air at oil shale wastewater facility 1.8 ppb [24].

Body Burdens: Varnish worker exposed to alkyl benzenes including a very low concn of toluene - 0.2 mg/L in urine [3]. Urine of 10 men exposed to approx 200 ppm toluene in air for 4 hr contained 1400 mg/L, mean, of 2-cresol 4 hr after exposure, whereas the mean level prior to exposure was 0.159 mg/L [13]. Printing workers exposed to an average of 139.8 ppm toluene - 3.11 mg/L mean in urine [4].

REFERENCES

1. Alexander M, Lustigman BK; J Agric Food Chem 14: 410-3 (1966)
2. Anderson GE; Human Exposure to Atmos Concentrations of Selected Chemicals Vol 1 (NTIS PB84-102540) USEPA Office of Air Quality Planning (1983)
3. Angerer J, Wulf H; Int Arch Occup Environ Health 56: 307-21 (1985)
4. Angerer J; Int Arch Occup Environ Health 56: 322-8 (1985)
5. Artiola-Fortuny J, Fuller WH; Soil Sci 133: 18-26 (1982)
6. Atkinson R et al; Int J Chem Kinet 12: 779-836 (1980)
7. Atkinson R; Chem Rev 85:69-201 (1985)
8. Baird RB et al; Arch Environ Contam Toxicol 2: 165-78 (1974)
9. Boyd SA; Soil Sci 134: 337-43 (1982)
10. Brodzinsky R, Singh HB; Volatile Org Chem The Atmos An Assess Of Avail Data p 198 SRI 68-02-3452 (1982)
11. Bursey JT, Pellizzari ED; Anal of Industrial Wastewater for Org Pollut in Consent Decree Survey Athens GA USEPA 68-03-2867 (1982)
12. Carter WPL et al; Environ Sci Technol 15: 829-31 (1981)
13. Fatiadi AJ; environ Int 10: 175-205 (1984)
14. Faust BC, Holgne J; Environ Sci Technol 21: 957-64 (1987)
15. Fedorak PM, Hrudey SE; Water Res 18: 361-7 (1984)
16. Giabbai MF et al; Intern J Environ Anal Chem 20: 113-29 (1985)
17. Godsy EM et al; Bull Environ Contam Toxicol 30: 261-8 (1983)
18. Graedel TE; Chemical Compounds in the Atmos Academic Press NY p 256 (1978)
19. Goerlitz DF et al; Environ Sci Technol 19: 955-61 (1985)
20. Grosjean D; Environ Sci Technol 19: 968-74 (1985)
21. Grosjean D; Atmos Environ 18: 1641-52 (1984)
22. Hampton CV; Environ Sci Technol 16: 287-98 (1982)
23. Hansch C, Leo AJ; MEDCHEM Project Claremont CA: Pomona College (1985)
24. Hawthorne SB; Sievers RE; Environ Sci Technol 18: 483-90 (1984)
25. Heukelekian H, Rand MC; J Water Pollut Contr Assoc 29: 1040-53 (1955)
26. Ho CT et al; J Agric Food Chem 31: 336-42 (1983)
27. Horowitz A et al; Dev Ind Microbiol 23: 435-44 (1982)
28. Hoshika Y, Muto G; J Chromatogr 157: 277-84 (1978)
29. Huddleston RL et al; Land treatment biological degradation processes pp.41-61 in Water Resource Symposium 13 (1986)
30. Kopfler FC et al; Adv Environ Sci Technol 8(Fate Pollut Air Water Environ): 419-33 (1977)
31. Leuenberger C et al; Environ Sci Technol 19: 1053-8 (1985)
32. Leuenberger C et al; Water Res 19: 885-94 (1985)
33. Lucas SV; GC/MS Anal of Org in Drinking Water and Advanced Waste Treatment concn Vol 2 p 397 USEPA-600/1-84-020B (1984)
34. Ludzack FJ, Ettinger MB; J Water Pollut Contrl Fed 32: 1173-200 (1960)
35. Lyman WJ et al; Handbook of Chem Property Estimation Methods Environ Behavior of Org Compounds McGraw-Hill NY (1982)
36. McCreary JJ et al; Chemosphere 12: 1619-32 (1983)
37. McKnight et al; Org Geochem 4: 85-92 (1982)

2-Cresol

38. Merck Index; An Encyclopedia of Chemicals, Drugs and Biologicals 10th ed p 2568 (1983)
39. NIOSH; National Occupational Exposure Survey (1985)
40. NIOSH; National Occupational Health Survey (1975)
41. Nojima K, Kanno S; Chemosphere 6: 371-6 (1977)
42. Perry RA et al; J Phys Chem 81: 1607-11 (1977)
43. Pitter P; Water Res 10: 231-5 (1976)
44. Riddick JA et al; Organic Solvents New York: Wiley Interscience (1986)
45. Shackelford WM et al; Analyt Chim Acta 146: 15-27 (1983)
46. Shackelford WM, Keith LH; Freq of Org Compounds Identified in Water USEPA-600/4-76-062 (1976)
47. Shepson PB et al; Environ Sci Technol 19: 249-55 (1985)
48. Singer PC et al; Treatability and Assess of Coal Conversion Waste Waters Phase I p 178 USEPA-600/7-79-248 (1979)
49. Smith JH et al; Environ Pathways of Selected Chemicals in Freshwater Systems part II Lab Studies USEPA-600/7-78-074 (1978)
50. Snyder LR; Anal Chem 41: 314-23 (1969)
51. Southworth GR, Keller JL; Water Air Soil Pollut 28: 239-48 (1986)
52. Sporstoel S et al; Int J Environ Anal Chem 21: 129-38 (1985)
53. Stone AT; Environ Sci Technol 21: 979-88 (1987)
54. Stuermer DH et al; Environ Sci Technol 16: 582-7 (1982)
55. Takemoto S et al; Suishitsu Odaku Kenkyu 4: 80-90 (1981)
56. Tremp J et al; ACS Div Environ Chem 192nd Natl Mtg 26: 142-3 (1986)
57. Urushigawa Y et al; Kogai Shigen Kenkyusho Iho 12: 37-46 (1983)
58. Verschueren K; Handbook of Environ Data on Org Chemicals 2nd ed Von Nostrand Reinhold NY pp 403-6 (1983)
59. Young RHF et al; J Water Pollut Contr Fed 40: 354-68 (1968)

3-Cresol

SUBSTANCE IDENTIFICATION

Synonyms: 1-Hydroxy-3-methylbenzene; m-Cresol; 3-Hydroxytoluene; 3-Methylphenol; m-Cresylic acid; m-Hydroxytoluene; m-Methylphenol

Structure:

CAS Registry Number: 108-39-4

Molecular Formula: C_7H_8O

Wiswesser Line Notation: QR C1

CHEMICAL AND PHYSICAL PROPERTIES

Boiling Point: 202 °C

Melting Point: 11-12 °C

Molecular Weight: 108.15

Dissociation Constants: 10.09 [41]

Log Octanol/Water Partition Coefficient: 1.96 [16]

Water Solubility: 23 g/L at 25 °C [26]

Vapor Pressure: 0.143 mm Hg at 25 °C [41]

Henry's Law Constant: 8.7 x 10^{-7} atm-m³/mol [26] (calc)

3-Cresol

ENVIRONMENTAL FATE/EXPOSURE POTENTIAL

Summary: 3-Cresol is released to the atmosphere in auto and diesel exhaust, during coal tar refining and wood pulping, and during its use in manufacturing and metal refining. Wastewater from these industries as well as from municipal wastewater plants contain 3-cresol. When released to the atmosphere, 3-cresol will react with photochemically produced hydroxyl radicals during the day (half-life 8 hr) and react with nitrate radicals at night (half-life 5 min). It will also be scavenged by rain and oxidized by metal cations in rainwater and fogwater. Biodegradation will generally be the dominant loss mechanism when 3-cresol is released into water, and half-lives in most surface waters would range from hours to days. Longer half-lives would occur in oligotrophic lakes and marine waters. Volatilization, bioconcentration in fish, and adsorption to sediment will be unimportant and photolysis is only expected to be important in surface waters of oligotrophic lakes. Its fate in soil has not been well characterized. It is relatively mobile in most soils and will biodegrade (100% in 11 days). Human exposure will primarily be via inhalation and dermal contact in the workplace and inhalation in source areas.

Natural Sources: Cresols occur in small quantities in petroleum [51]. 3-Cresol was found in surface water in the blast zone of the Mount St. Helen eruption and is believed to have been formed during the destruction of plant and soil material by the volcanic action [33].

Artificial Sources: 3-Cresol is released into the environment in emissions and effluents during its production by coal tar refining and use in disinfectants and fumigants as well as photographic developers, explosives, and polyester solvents, and in wood pulping and metal refining [14,24,34,45]. Emissions have also been detected in tobacco smoke, and in auto and diesel exhaust [14,57]. It is found in sewage plant effluent [57]. 3-Cresol is emitted during the incineration of fruits and vegetables [27]. It is formed during the pyrolysis of some epoxy resins and may be emitted in situations where these resins are exposed to temperatures above 350 °C [38]. It is also a product of the photooxidation of toluene [46].

199

3-Cresol

Terrestrial Fate: 3-Cresol is mobile in soil and therefore it has the potential to leach into ground water. It biodegrades rapidly in soil, with one investigator reporting complete degradation in 11 days. In the subsurface environment, 3-cresol biodegrades on the time scale of weeks to months [1].

Aquatic Fate: The primary removal mechanism for 3-cresol in most surface waters is biodegradation with half-lives in estuarine waters of several days. The half-lives are shorter in freshwater and longer in marine waters. In oligotrophic lakes, photolysis may contribute to the removal process but experimental data are lacking. Also, while oxidation by metal ions such as Fe(III) and Mn(III/IV) occurs, it is not clear whether these processes contribute significantly to the degradation of 3-cresol in eutrophic waters. Degradation under anaerobic conditions is much slower, weeks rather than hours in screening studies, but there are insufficient data from natural systems to estimate degradation rates.

Atmospheric Fate: The photochemical half-life of 3-cresol during the daytime is 8 hr while at night it is 5 min. The dominant reactions are with hydroxyl radicals during the day and with NO_3 radical at night. Daytime half-lives will be reduced under smog conditions. In addition, oxidation by metal ions such as Fe(III) and Mn(III/IV) may occur rapidly (hours or less) in rainwater and fogwater. 3-Cresol is highly soluble in water and it will be scavenged from the atmosphere by rain [25].

Biodegradation: 3-Cresol generally biodegrades rapidly in screening studies using soil, sewage, activated sludge, or freshwater inocula [2,7,11,18,40,47,48,49,56]. Acclimation is frequently not necessary [2,7,18,47,48,56]. Die-away studies were performed in freshwater, estuarine water, and marine water from sites near Newport, NC throughout the year [39]. In estuarine water, the half-life ranged from 1 to 6 days [39]. Rates were faster in freshwater and slower in marine water [39]. While rates were highest during the summer in fresh and estuarine waters, the rate of degradation in marine water was almost independent of the season [39]. In an older die-away study in river water, 99-100% degradation was reported in 7 and 2 days at 4 and 20 °C, respectively [29]. When 3-cresol was incubated with two digester sludges under anaerobic conditions, 92 and 90% mineralization was reported in 4 and 5 weeks, respectively [20]. No mineralization occurred in 29 weeks, however, when it was incubated with anaerobic freshwater sediment [20]. Similarly, no mineralization occurred when 3-cresol was incubated anaerobically for 40 days at 37 °C with a sludge inoculum [10]. In experiments using labeled methyl groups, it was shown that a methanogenic consortia oxidized the methyl group of 3-cresol primarily to methane, whose production began after most of the parent compound was depleted from the media [42]. 3-Cresol completely degraded in soil in 11 days at an application rate of 500 ppm [21]. Low concns (39 ppb) of 3-cresol degraded in subsurface material taken from an uncontaminated aquifer [1]. There was a linear increase in biodegradation with time over the 160-day experiment, with a 0.07% per day average mineralization rate [1]. When the concentration of 3-cresol was increased to 788 ppb, no mineralization was observed [1]. The disproportionate decrease in concentration of 3-cresol in ground water downgradient from a wood-preserving facility has been ascribed to biodegradation [13].

Abiotic Degradation: In the atmosphere during the daytime, 3-cresol reacts principally with photochemically generated hydroxyl radicals with a resulting half-life of 8.0 hr [6]. However in the nighttime, especially in moderately polluted atmospheres where concentrations of O_3 and NO_2 are high, reaction with NO_3 radicals becomes the predominant sink for 3-cresol (half-life 5 min) with the formation of nitrocresols [4,37]. Under photochemical smog conditions 3-cresol is the most reactive cresol isomer and a half-life of 2 hr has been reported with the formation of nitrocresols [5,36]. When 3-cresol sorbed on silica gel was irradiated for 3 days with light >290 nm, 33.5% of the chemical was mineralized to CO_2 [11]. The photolysis half-life for 4-cresol by sunlight in pure water is 35 days; the half-life was a factor of 12 less when humic acids are added [50]. The estimated photolysis half-life of 4-cresol in a river, eutrophic lake or pond, and oligotrophic lake are 400, 830, and 200 summer days (12 hr of sunlight), respectively [50]. These photolysis half-lives may be similar for 3-cresol. 3-Cresol is oxidized by ions such as Fe(III) and Mn(III/IV) that are widely distributed in suspended particles in surface waters and in soils, especially at low pH [54]. These reactions can also occur in urban rainwater and fogwater since these droplets may contain manganese oxides and have low pH values [54]. Manganese oxides derived from dust and ash may bring about these oxidative degradations in time scales of a few hours or less [54]. Since 3-cresol has no hydrolyzable groups, hydrolysis will not be a significant loss process.

Bioconcentration: The bioconcentration of 3-cresol in fish (golden ide) after 3 days was 20 [11] indicating that 3-cresol does not bioconcentrate significantly in fish. After 1 day the bioconcentration in algae was found to be 4900 [11].

Soil Adsorption/Mobility: The value for Koc measured on Brookstone clay loam soil is 35 whereas that predicted from the water solubility is 18 [8]. However, Koc's for phenols predicted from water solubilities are only good for soils with organic carbon contents greater than approx 0.5% [52]. Adsorptivities are greater than predicted for soils with low organic carbon content because interactions such as H-bonding are dominant [52]. There is no adsorption to sodium montmorillonite or sodium kaolinite clay between pH 2 and 10 [30]. Its low adsorptivity to subsurface soil is evidenced by its elution in one pore volume in a 20 cm column filled with aquifer sediment collected at a 30 m depth [13].

Volatilization from Water/Soil: 3-Cresol has a low potential to volatilize from water, having a calculated Henry's Law constant of 8.7×10^{-7} atm-m^3/mol [26]. As such it will partition into the aqueous phase and evaporation will not be an important loss process [31].

Water Concentrations: SURFACE WATER: Detected, not quantitated in Lake Michigan basin - St Joseph River [15]. GROUND WATER: Ground water in a sand aquifer at a wood-preserving facility in Pensacola, FL (5 sites, 5 depths) 0-13.73 mg/L [13]. Southington, CT landfill site 0.6 mg/L [43]. Two monitoring wells under rapid infiltration site at Fort Devens, MA where the 3-cresol concentration is 66 ug/L in the basin, 0.020 ug/L and ND [22]. Concentrations of 3- and 4-cresol combined in ground water at an abandoned phenolic waste site (pinetar manufacturer) in Gainesville FL were 2300-11100 ppb at the original site, 3100-6200 ppb at downgradient sites, and <0.3-95 ppb at upgradient sites [32]. RAIN/SNOW: Portland, OR - seven rain events 380-2000 ppt, >1100 ppt, mean including 4-isomer dissolved in rain [25]. In Switzerland, the concn of 3- and 4-cresol dissolved in rain was 4.5 ppb [55].

Effluent Concentrations: 71% frequency of appearance in effluents from refineries, petrochemical and metallurgical industries, municipal wastewater plants, and polluted fjords in Norway [53]. Identified in finished water from advanced waste treatment plants [28]. Wastewater from the gasification of Indian Head lignite contained 1840 ppm of 3- and 4-cresol, combined [12]. In a comprehensive survey of wastewater from 4000 industrial and publicly owned treatment works (POTWs) sponsored by the Effluent Guidelines Division of the U.S. EPA, 4-cresol was identified in discharges of the following industrial category (frequency of occurrence, median concn in ppb): timber products (8, 209.6); leather tanning (1, 48.4); iron and steel mfg (11, 40.0); petroleum refining (6, 44.8); nonferrous metals (9, 106.1); paving and roofing (1, 15.6); paint and ink (3, 3.6); printing and publishing (1, 21.5); coal mining (1, 26.0); organics and plastics (17, 1075.2); inorganic chemicals (4, 33.1); textile mills (6, 42.3); plastics and synthetics (1, 745.4); pulp and paper (1, 140.7); rubber processing (6, 93.2); auto and other laundries (2, 374.2); pesticides manufacture (1, 27.0); photographic industries (2, 3035.0); gum and wood industries (5, 4.1); pharmaceuticals (7, 19.1); explosives (1, 1.3); foundries (29, 62.8); aluminum (1, 9.8); electronics (5, 48.4); oil and gas extraction (5, 24.9); organic chemicals (21, 13.8); mechanical products (5, 32.3); transportation equipment (3, 37.3); synfuels (7, 456.5); publicly owned treatment works (75, 77.0); rum industry (3,425.4) [44]. Both the timber products industry and organics and plastics industries had maximum effluent concentrations exceeding 10 ppm, namely, 23.1 and 52.6 ppm, respectively [44].

Sediment/Soil Concentrations: 3-Cresol was detected, but not quantified, in soil at the site of an abandoned phenolic waste site in Gainesville, FL [32].

Atmospheric Concentrations: RURAL: Not detected in rural samples from western Colorado and Utah [17]. URBAN: Not detected in urban air samples from western Colorado and Utah [17]. Portland, OR - during seven rain events 19-51 ppt, 28.6 ppt, mean including 4-isomer [25]. Particulate matter associated cresol

3-Cresol

was generally <1% of the gaseous phase chemical [25]. SOURCE AREA: Oil shale wastewater facility 0.88 ppb, including 4-cresol [17].

Food Survey Values: Identified as a volatile component of roasted filberts [23] and fried bacon [19].

Plant Concentrations:

Fish/Seafood Concentrations:

Animal Concentrations:

Milk Concentrations:

Other Environmental Concentrations:

Probable Routes of Human Exposure: Humans will be exposed to 3-cresol via inhalation and dermal contact in the workplace as well as via inhalation in source areas.

Average Daily Intake: AIR INTAKE: (assume typical concn 28.6 ppt for the 3- plus 4- isomers being divided equally between the isomers) 1.3 ug; WATER INTAKE: insufficient data; FOOD INTAKE: insufficient data.

Occupational Exposures: NIOSH (NOHS Survey, 1972-74) has statistically estimated that 11,162 workers are exposed to 3-cresol in the United States [36]. NIOSH (NOES Survey, 1981-83) has statistically estimated that 5573 workers are exposed to 3-cresol in the United States [35]. Based on 1978 emissions data, it has been estimated that the total annual U.S. dosage of 3-cresol is 14,030 mg/m^3 persons [3]. This is a result of inhalation by occupationally exposed persons. Indoor air at a oil shale wastewater facility 5.1 ppb (including 4-isomer) [17].

Body Burdens: Urine of 10 men exposed to approx 200 ppm toluene in the air for 4 hr contained 0.599 mg/L of 3-cresol, mean, 4 hr after exposure compared with <0.2 mg/L, mean, before exposure [9].

3-Cresol

REFERENCES

1. Aelion CM et al; Appl Environ Microbiol 53: 2212-7 (1987)
2. Alexander M, Lustigman, BK; J Agric Food Chem 14: 410-3 (1966)
3. Anderson GE; Human Exposure to Atmos Concn of Selected Chem Vol 1 p 230 (NTIS PB84-102540) (1983)
4. Atkinson R et al; Int J Chem Kinetics 16: 887-98 (1984)
5. Atkinson R et al; Int J Chem Kinetics 12: 779-836 (1980)
6. Atkinson R et al; Adv Photochem 11: 375-488 (1979)
7. Baird RB et al; Arch Environ Contam Toxicol 2: 165-78 (1974)
8. Boyd SA; Soil Sci 134: 337-43 (1982)
9. Fatiadi AJ; Environ Int 10: 175-205 (1984)
10. Fedorak PM, Hrudey SE; Water Res 18: 361-7 (1984)
11. Freitag D et al; Chemosphere 14: 1589-616 (1985)
12. Giabbai MF et al; Intern J Environ Anal Chem 20: 113-29 (1985)
13. Goerlitz DF et al; Environ Sci Tech 19: 955-61 (1985)
14. Graedel TE; Chem Compounds in the Atmos Academic Press, NY p 256 (1978)
15. Great Lakes Water Quality Board; An Inventory of Chem Substances Identified in the Great Lakes Ecosystem Vol 1 Summary Report Windsor Ontario Canada p 195 (1983)
16. Hansch C, Leo AJ; MEDCHEM Project Claremont CA: Pomona College (1985)
17. Hawthorne SB, Sievers RE; Environ Sci Tech 18: 483-90 (1984)
18. Heukelekian H, Rand MC; J Water Pollut Control Assoc 29: 1040-53 (1955)
19. Ho CT et al; J Agric Food Chem 31: 336-42 (1983)
20. Horowitz A et al; Dev Ind Microbiol 23: 435-44 (1982)
21. Huddleston RL et al; Land treatment biological degradation processes pp.41-61 in Water Resource Symposium 13 (1986)
22. Hutchins SR et al; Water Res 18: 1025-36 (1984)
23. Kinlin TE et al; J Agric Food Chem 20: 1021 (1972)
24. Kirk-Othmer; Encycl Chem Tech 3rd Wiley Interscience, NY 18: 557 (1982)
25. Leuenberger C et al; Environ Sci Tech 19: 1053-8 (1985)
26. Leuenberger C et al; Water Res 19: 885-94 (1985)
27. Liberti A et al; Chemosphere 12: 661-3 (1983)
28. Lucas SV; GC/MS Analysis of Org in Drinking Water Concn and Advanced Waste Treatment Concns Vol 2 Computer-Printed Tabulations of Compound Identification Results For Large Volume Concentrates p 397 USEPA-600/1-84-020B (1984)
29. Ludzack FJ, Ettinger MB; J Water Pollut Control Fed 32: 1173-1200 (1960)
30. Luh MD, Baker RA; Proc of the 25th Industrial Waste Conf Purdue Univ, Eng Bull Ext Series 25: 534-42 (1970)
31. Lyman WJ et al; Handbook of Chem Property Estimation Methods Environ Behavior of Org Compounds McGraw-Hill NY pp 15-1 to 15-33 (1982)
32. McCreary JJ et al; Chemosphere 12: 1619-32 (1983)
33. McKnight et al; Org Geochem 4: 85-92 (1982)
34. Merck Index; An Encyclopedia of Chemicals, Drugs and Biologicals 10th ed p 2568 (1983)
35. NIOSH; National Occupational Exposure Survey (1985)

205

36. NIOSH; National Occupational Health Survey (1975)
37. Nojima K, Kanno S; Chemosphere 6: 371-6 (1977)
38. Peltonen K; J Analyt Appl Pyrolysis 10: 51-7 (1986)
39. Pfaender FK, Bartholomew GW; Appl Environ Microbiol 44: 159-64 (1982)
40. Pitter P; Water Res 10: 231-5 (1976)
41. Riddick JA et al; Organic Solvents New York: Wiley Interscience (1986)
42. Roberts DJ et al; Can J Microbiol 33: 335-8 (1987)
43. Sawhney BL, Kozloski RP; J Environ Qual 13: 349-52 (1984)
44. Shackelford WM et al; Analyt Chim Acta 146: 15-27 (1983)
45. Shareef GS et al; Hazardous/Toxic Air Pollut Control Technol A Literature Review p 275 USEPA-600/2-84-194 (1984)
46. Shepson PB et al; Environ Sci Technol 19: 249-55 (1985)
47. Shimp R, Pfaender FK; Appl Environ Microbiol 49: 402-7 (1985)
48. Shimp R, Pfaender FK; Appl Environ Microbiol 49: 394-401 (1985)
49. Singer PC et al; Treatability and Assess of Coal Conversion Waste Waters Phase I p 178 USEPA-600/7-79-248 (1979)
50. Smith JH et al; Environ Pathways of Selected Chemicals in Freshwater Systems Part II Lab Studies USEPA/7-78-074 (1978)
51. Snyder LR; Anal Chem 41: 314-23 (1969)
52. Southworth GR, Keller JL; Water Air Soil Pollut 28: 239-48 (1986)
53. Sporstoel S et al; Int J Environ Anal Chem 21: 129-38 (1985)
54. Stone AT; Environ Sci Technol 21: 979-88 (1987)
55. Tremp J et al; ACS Div Environ Chem 192nd Natl Mtg 26: 142-3 (1986)
56. Urushigawa Y et al; Kogai, Shigen Kenkyusho IHO 12: 37-46 (1983)
57. Verschueren K; Handbook of Environ Data on Org Chemicals 2nd ed Von Nostrand Reinhold NY pp 406-8 (1983)

4-Cresol

Synonyms: 1-Hydroxy-4-methylbenzene; p-Cresol; 4-Hydroxytoluene; 4-Methylphenol; p-Cresylic acid; p-Hydroxytoluene; p-Methylphenol

Structure:

HO—⟨benzene ring⟩—CH₃

CAS Registry Number: 106-44-5

Molecular Formula: C_7H_8O

Wiswesser Line Notation: QR D

CHEMICAL AND PHYSICAL PROPERTIES

Boiling Point: 201.9 °C at 760 mm Hg

Melting Point: 34.8 °C

Molecular Weight: 108.13

Dissociation Constants: 10.26 [42]

Log Octanol/Water Partition Coefficient: 1.94 [20]

Water Solubility: 22.6 g/L at 40 °C [42]

Vapor Pressure: 0.13 mm Hg at 25 °C [42]

Henry's Law Constant: 9.6 x 10^{-7} atm-m³/mol [30] (calc)

207

4-Cresol

ENVIRONMENTAL FATE/EXPOSURE POTENTIAL

Summary: 4-Cresol is released to the atmosphere in auto and diesel exhaust, during coal tar refining, wood pulping, and during its use in manufacturing and metal refining. Wastewater from these industries as well as from municipal wastewater plants contain 4-cresol. When released to the atmosphere, 4-cresol will react with photochemically produced hydroxyl radicals during the day (half-life 10 hr) and react with nitrate radicals at night (half-life 4 min). It will also be scavenged by rain and oxidized by metal cations in rainwater and fogwater. Biodegradation is expected to be the dominant loss mechanism when 4-cresol is released into water. Volatilization, bioconcentration in fish, and adsorption to sediment will be unimportant and photolysis is only expected to be significant in surface waters of oligotrophic lakes. Experimental half-lives are only a few hours in eutrophic lakes and ponds but this may be preceded by an acclimation period ranging from hours to days. Half-lives in an oligotrophic lake, marine waters, and in water/sediment ecocores were 6, <4, and <2 days, respectively. Its fate in soil has not been extensively studied; it is relatively mobile in some soils and readily biodegrades (100% biodegradation in 7 days). Human exposure will primarily be via inhalation and dermal contact in the workplace or via inhalation in source areas.

Natural Sources: Plant volatile [18]. Cresols occur in petroleum [56].

Artificial Sources: 4-Cresol is released into the environment in emissions and wastewater from its production in coal tar refining and use as a disinfectant, in metal refining, and chemical manufacturing [37,49]. It is also found in emissions from autos and diesel engines, wood pulping, brewing, glass fibre manufacture, and in tobacco smoke [18]. It is also a product of the photooxidation of toluene [51].

Terrestrial Fate: 4-Cresol is relatively mobile in some soils and, therefore, may leach into the ground water. It biodegrades rapidly in soil, with one investigator reporting complete degradation in 7 days.

4-Cresol

Aquatic Fate: A one-compartment computer model for 4-cresol predicts that following an acute discharge, the half-life would be 0.55 hr in a river, 12 hr in a pond or eutrophic lake, and 2400 hr in an oligotrophic lake [53]. Biodegradation is predicted to be the dominant transformation process in eutrophic waters, but four times slower than photolysis in oligotrophic waters. Dilution is more important than biodegradation in rivers. Sorption and volatilization are unimportant in all waters [53]. While photolysis is catalyzed by the presence of humic acids, there are no experiments that evaluate whether this process is important in some eutrophic waters. Also, while oxidation by metal ions such as Fe(III) and Mn(III/IV) occurs, it is not clear whether these processes contribute significantly to the degradation of 4-cresol in eutrophic waters. The model results are not entirely in agreement with experiments, where degradation half-lives were only a few hours in eutrophic waters but may be preceded by a lag period ranging from hours to days and where degradation occurred in an oligotrophic lakes in 6 days. In addition, half-lives in marine waters were 9-43 hr. In marine or freshwater/sediment ecocores, degradation is complete in a few days. Degradation is much slower under anaerobic conditions - weeks instead of hours in screening studies. However in the only study in a natural system, no mineralization occurred in 29 weeks in anaerobic lake sediment.

Atmospheric Fate: The photochemical half-life of 4-cresol during the daytime is 10 hr while at night it is 4 min. The dominant reactions are with hydroxyl radical during daylight hours and with nitrate radicals at night. Daytime half-lives will be reduced under smog conditions. 4-Cresol is highly soluble in water and it will be scavenged from the atmosphere by rain [29].

Biodegradation: 4-Cresol has been shown to biodegrade rapidly in screening studies using soil, sewage, activated sludge, and freshwater inocula [1,7,23,28,31,33,35,41,52,62]. Acclimation is frequently not necessary [1,7,23,28,31,35,62]. Complete COD removal was obtained in a simulated biological treatment plant in 16-74 hr [11]. 4-Cresol biodegrades rapidly in environmental waters including oligotrophic lakes, eutrophic ponds, rivers, creeks, and bays [44,54,58,63]. Total degradation occurred in Lake Tahoe water within 6 days and 4 days, respectively, depending on whether the water contained sediment or not [54]. In a eutrophic pond, degradation was complete in 7-8.5 hr; however, this included a 5.5 to 6 hr lag period [54]. In another study, the half-life in two fish ponds ranged from 2-7 hr after a 30 to 56 hr lag [44]. In 3 river waters of the Pacific Northwest, the half-life ranged from 1-10 hr after a 2 day lag period [44]. At 3 sites in Pensacola Bay, FL the half-life ranged from 9 to 43 hr in the saltwater and 6 to 11 hr when anaerobic sediment was added to the water [63]. In marine water/sediment ecocores from 3 Pensacola Bay sites, biodegradation half-lives ranged from 3 to 16 hr [63]. 95% of the 4-cresol added to freshwater/sediment ecocores from the Escamba River, FL degraded in 75 hr and 22% was mineralized in that period [58]. Under anaerobic conditions, 4-cresol was mineralized in 3 and 8 weeks in two screening studies using digester sludge inocula [9,50]. In a third study using domestic anaerobic sludge inocula, mineralization occurred after 15 days at a 4-cresol concn of 100 ppm and increased to >39 days at a concn of 400 ppm [15]. However, no mineralization was observed when incubated for 29 weeks in anaerobic lake sediment [25]. Phenol is produced as an intermediate during anaerobic biodegradation [65]. In experiments using labeled methyl groups, it was shown that a methanogenic consortia oxidized the methyl group of 4-cresol primarily to carbon dioxide [43]. 4-Cresol completely degraded in soil in 7 days at an application rate of 500 ppm [26]. The disproportionate decrease in the concentration of 4-cresol in ground water downgradient from a wood preserving facility has been ascribed to biodegradation [16].

Abiotic Degradation: In the atmosphere during daylight hours, 4-cresol reacts principally with photochemically generated hydroxyl radicals with a resulting half-life of 10 hr [5]. However in the nighttime, especially in moderately polluted atmospheres where concentrations of O_3 and NO_2 are high, reaction with NO_3 radicals becomes the predominant sink for 4-cresol (half-life 4 min) with the formation of nitrocresols [6]. Under photochemical smog conditions, a half-life of 3 hr has been reported and nitrocresols have been formed [4,40]. 4-Cresol in pure water photolyzed in the presence of sunlight (half-life 35 days) in the laboratory [54]. The addition of humic acid to the water increases this rate by a factor of 12 (half-life 3 days) [54]. The resulting photolysis half-life in a river, eutrophic lake or pond, and oligotrophic lake are 400, 830, and 200 summer days (12 hr of sunlight), respectively [54]. The absorptivity of 4-cresol increases markedly as the pH increases from 5.1 to 8.9 [54] and photolysis may therefore be much more important in alkaline lakes. The depth-averaged half-life of 4-cresol in the upper 1 m of a polluted eutrophic pre-alpine Swiss lake is 4.4 days [14]. This is a result of sensitized photolysis, possibly by alkoxy radicals that are photochemically produced from dissolved organic matter [14]. Singlet oxygen is also obtained when sunlight is absorbed by humic substances in water and this species exhibits a high reactivity towards phenols at high pH when they are present as phenoxide anions [46]. The half-life of 4-cresol with 4×10^{-14} moles/L of singlet oxygen at pH 8 is 500 hr [46]. 4-Cresol is oxidized by ions such as Fe(III) and Mn(III/IV) that are widely distributed in suspended particules in surface waters and in soils, especially at low pH [60]. These reactions can also occur in urban rainwater and fogwater since these droplets may contain manganese oxides and have low pH values [60]. Manganese oxides derived from dust and ash may bring about these oxidative degradations in time scales of a few hours or less [60].

Bioconcentration: Using the log octanol-water partition coefficient of 1.94 [20], a bioconcentration factor of 18 is estimated [34]. Therefore 4-cresol would not be expected to bioconcentrate significantly in fish.

Soil Adsorption/Mobility: The value of Koc measured on Brookstone clay loam soil is 49, whereas that predicted from water solubility is 0.9 [8]. Coyote Creek sediment Koc 650 [54]. Based on this value, only 0.05, 0.3, and 0.1% would be sorbed in a one-compartment model of a oligotrophic lake, eutrophic pond, and river, respectively [54]. On five fine-textured B horizon clay soils, the distribution between soil and water ranged from 5 to 50 [3]. For these subsurface soils, the levels of free iron oxide and pH were the key factors in determining adsorption capacity [3]. The Koc of 4-cresol to 3 subsoils (% organic carbon) was: Apison (0.11) - 3420; Fullerton (0.05) - 3350; and Dormont (1.2) - 115 [57]. Koc's for phenols predicted from water solubilities were only good for soils with organic carbon contents >0.5% [57]. Adsorptivities were much greater than predicted for soils with low organic carbon content because interactions such as H-bonding were dominant [57].

4-Cresol

Volatilization from Water/Soil: 4-Cresol has a low volatility from water, having a calculated Henry's Law constant of 9.6 x 10^{-7} atm-m^3/mol [30]. Based on laboratory data on its rate of evaporation from water, 4-cresol would have an estimated half-life of 1.4 yr and 290 yrs in a typical river and lake, respectively [34,53].

Water Concentrations: SURFACE WATER: Lower Tennessee River below Calvert City, KY - 200 ppb in water/sediment sample [17]. Hayashida River in Tatsumo City, Japan - (site of leather industry) 204 ppb [64]. GROUND WATER: Hoe Creek underground coal gasification site, WY - 2 aquifers 15 mo after gasification complete 9.6-16000 ppb [61]. Landfill ground water, Norman, OK 14.6 ppb [12]. Detected, not quantified in ground water [47]. Ground water in a sand aquifer at a wood-preserving facility in Pensacola, Florida (5 sites, 5 depths) 0-6.17 ppm [16]. Southington, CT landfill site 1.5 ppm [45]. Concentrations of 3- and 4-cresol combined in ground water at an abandoned phenolic waste site (pinetar manufacturer) in Gainesville, FL were 2300-11,100 ppb at the original site, 3100-6200 ppb at downgradient sites, and <0.3-95 ppb at upgradient sites [36]. RAIN/SNOW: Portland, OR - seven rain events 380-2000 ppt, >1100 ppt, mean (including 3-isomer) dissolved in rain [29].

Effluent Concentrations: Detected, not quantified in air emanating from sedimentation tank of water treatment plant [19]. Effluents from refineries, petrochemical and metallurgical industries, municipal wastewater plants, and polluted fjords in Norway (28 samples) 71% frequency of appearance [59]. Identified in finished water from advanced waste treatment plants [32]. 4-Cresol was confirmed in 10 of 4000 effluent samples in a broad survey covering 46 industrial categories [10]. It was found in effluents from timber products, leather tanning, organics and plastics, textile mills, rubber processing, auto and other laundries, electronics, mechanical products, synfuels, and publicly owned treatment works [10]. The concn of 4-cresol in wastewater from a petroleum refinery was 50 ppb, but this was removed in the activated sludge unit [55]. In a comprehensive survey of wastewater from 4000 industrial and publicly owned treatment works (POTWs) sponsored by the Effluent Guidelines Division of the U.S. EPA, 4-cresol was identified in discharges of the following industrial category (frequency of occurrence, median concn in ppb): timber products (15, 166.6); leather tanning (7, 31.9); iron and steel mfg (8, 33.2); petroleum refining (1, 10298.3); nonferrous metals (3, 18.4); paving and roofing (1, 18.5); organics and plastics (24, 477.2); inorganic chemicals (3, 37.7); textile mills (9, 50.9); pulp and paper (5, 36.7); rubber processing (2, 17659.4); soaps and detergents (2, 94.2); auto and other laundries (2, 31.6); photographic industries (3, 327.7); gum and wood industries (3, 3.6); pharmaceuticals (4, 89.2); explosives (1, 3.7); foundries (16, 47.0); aluminum (13, 17.3); electronics (7, 66.6); oil and gas extraction (10, 6.4); organic chemicals (13, 95.7); mechanical products (8, 398.5); transportation equipment (1, 0.3); synfuels (15, 181.1); publicly-owned treatment works (134, 44.7); rum industry (3, 87.5) [48]. The highest effluent was 10,298 ppb in the petroleum refining industry [48].

Sediment/Soil Concentrations: 4-Cresol was detected, but not quantified, in soil at the site of an abandoned phenolic waste site in Gainesville, FL [36].

Atmospheric Concentrations: RURAL: not detected in rural air samples from western Colorado and Utah [22]. URBAN: not detected in urban air samples from western Colorado and Utah [22]. Portland, OR - during seven rain events 19-51 ppt, 28.6 ppt, mean (including 3-isomer) [29]. Particulate matter associated cresol was generally <1% of the gas phase chemical [29]. SOURCE AREA: Oil shale wastewater facility 88 ppb, including 3-cresol [22].

Food Survey Values: Identified as a volatile component of roasted filberts [27] and fried bacon [24]. Synthetic food flavor [21].

Plant Concentrations:

Fish/Seafood Concentrations:

Animal Concentrations:

Milk Concentrations:

Other Environmental Concentrations:

4-Cresol

Probable Routes of Human Exposure: Humans are exposed to 4-cresol by inhalation and dermal contact, especially in source areas or occupational settings.

Average Daily Intake: AIR INTAKE: (assume typical concn for 3- plus 4-isomers 28.6 ppt being divided equally between the two isomers), 1.3 ug; WATER INTAKE: insufficient data; FOOD INTAKE: insufficient data.

Occupational Exposures: NIOSH (NOHS Survey, 1972-1974) has statistically estimated that 33,257 workers are exposed to 4-cresol in the United States [38]. NIOSH (NOES Survey, 1981-1983) has statistically estimated that 3269 workers are exposed to 4-cresol in the United States [39]. It has been estimated based on 1978 emissions data that the total annual U.S. dosage of 4-cresol is 11,630 mg/m³ persons [2]. This is a result of inhalation by occupationally exposed persons. Indoor air at an oil shale wastewater facility 5.1 ppb, including 3-isomer [22].

Body Burdens: Urine of 10 men exposed to approx 200 ppm toluene for 4 hr contained 40.968 mg/L 4-cresol, mean, 4 hr after exposure and 31.206 mg/L prior to exposure [13].

REFERENCES

1. Alexander M, Lustigman BK; J Agric Food Chem 14: 410-3 (1966)
2. Anderson GE; Human Exposure to Atmos Concn of Selected Chem Vol I (NTIS PB84-102540) USEPA Office of Air Quality Planning (1984)
3. Artiola-Fortuny J, Fuller WH; Soil Sci 133: 18-26 (1982)
4. Atkinson R et al; Int J Chem Kinet 12: 779-836 (1980)
5. Atkinson R et al; Adv Photochem 11: 375-488 (1979)
6. Atkinson R et al; Int J Chem Kinetics 16: 887-98 (1984)
7. Baird RB et al; Arch Environ Contam Toxicol 2: 165-78 (1974)
8. Boyd SA; Soil Sci 134: 337-43 (1982)
9. Boyd SA et al; Appl Environ Microbiol 46: 50-4 (1983)
10. Bursey JE, Pellizzari ED; Anal of Industrial Wastewater for Org Pollut in Consent Decree Survey Athens GA USEPA 68-03-2867 (1982)
11. Chudoba J et al; SB.VYS Sk Chem Technol Praze Technol Vody 13: 45-63 (1968)
12. Dunlap WJ et al; pp 96-110 in Org Pollut contributed to Ground Water by a Landfill USEPA-600/9-76-004 (1976)
13. Fatiadi AJ; Environ Int 10: 175-205 (1984)
14. Faust BC, Holgne J; Environ Sci Technol 21: 957-64 (1987)
15. Fedorak PM, Hrudey SE; Water Res 18: 361-7 (1984)
16. Goerlitz DF et al; Environ Sci Technol 19: 955-61 (1985)

4-Cresol

17. Goodley PC, Gorden M; Kentucky Academy of Sci 37: 11-5 (1976)
18. Graedel, TE Chemical Compounds in the Atmosphere New York Academic Press p 257 (1978)
19. Hangartner M; Intern J Environ Anal Chem 6: 161-9 (1979)
20. Hansch C, Leo AJ; MEDCHEM Project Claremont CA: Pomona College (1985)
21. Hawley GG; Condensed Chem Dictionary 10th ed Von Nostrand Reinhold NY p 285 (1981)
22. Hawthorne SB, Sievers RE; Environ Sci Technol 18: 483-90 (1984)
23. Heukelekian H, Rand MC; J Water Pollut Control Assoc 29: 1040-53 (1955)
24. Ho CT et al; J Agric Food Chem 31: 336-42 (1983)
25. Horowitz A et al; Dev Ind Microbiol 23: 435-44 (1982)
26. Huddleston RL et al; Land treatment biological degradation processes pp.41-61 in Water Resource Symposium 13 (1986)
27. Kinlin TE et al; J Agric Food Chem 20: 1021 (1972)
28. Kitano M; Biodegradation and bioaccumulation Test on Chem substances OECD Tokyo Mtg Ref TSU-NO3 (1978)
29. Leuenberger C et al; Environ Sci Technol 19: 1053-8 (1985)
30. Leuenberger C et al; Water Res 19: 885-94 (1985)
31. Lewis DL et al; Environ Toxicol chem 3: 563-74 (1984)
32. Lucas SV; GC/MS Anal of Org in Drinking Water Concentrates and Advanced Waste Treatment Concentrates Vol 2 USEPA-600/1-84-020B (1984)
33. Lund FA, Rodriguez DS; J Gen Appl Microbiol 30: 53-61 (1984)
34. Lyman WJ et al; Handbook of Chem Property Estimation Methods Environ Behavior of Org Compounds McGraw-Hill NY (1982)
35. Mackay D et al; Chemosphere 14: 335-74 (1985)
36. McCreary JJ et al; Chemosphere 12: 1619-32 (1983)
37. Merck Index; An Encyclopedia of Chemicals, Drugs and Biologicals 10th ed pp 369 (1983)
38. NIOSH; National Occupational Hazard Survey (1975)
39. NIOSH; National Occupational Exposure Survey (1985)
40. Nojima K, Kanno S; Chemosphere 6: 371-6 (1977)
41. Pitter P; Water Res 10: 231-5 (1976)
42. Riddick JA et al; Organic Solvents New York: Wiley Interscience (1986)
43. Roberts DJ etal; Can J Microbiol 33: 335-8 (1987)
44. Rogers JE et al; Microbial Transformation Kinetics of Xenobiotics in Aquatic Environ USEPA-600/3-84-043 (1984)
45. Sawhney BL, Kozloski RP; J Environ Qual 13: 349-52 (1984)
46. Scully FE Jr, Hoigne J; Chemosphere 16: 681-94 (1987)
47. Shackelford WM, Keith LH; Frequency of Org Compounds Identified in Water USEPA-600/4-76-062 (1976)
48. Shackelford WM et al; Analyt Chim Acta 146: 15-27 (1983)
49. Shareef GS et al; Hazardous/Toxic Air Pollut Control Technology pp 275 USEPA-600/2-84-194 (1984)
50. Shelton DR, Tiedje JM; Appl Environ Microbiol 47: 850-7 (1984)
51. Shepson PB et al; Environ Sci Technol 19: 249-255 (1985)
52. Singer PC et al; Treatability and Assess of Coal Conversion Wastewaters Phase I p 178 USEPA-600/7-79-248 (1979)

4-Cresol

53. Smith JH, Bomberger DC Jr; pp.445-51 in Hydrocarbons and halogenated hydrocarbons in the aquatic environment Afgan, Mackay eds Plenum NY (1980)
54. Smith JH et al; Environ Pathways of Selected Chemicals in Freshwater Systems Part II Lab Studies USEPA-600/7-78-074 (1978)
55. Snider EH, Manning FS; Environ Internat 7: 237-58 (1982)
56. Snyder LR; Anal Chem 41: 314-23 (1969)
57. Southworth GR, Keller JL; Water Air Soil Pollut 28: 239-48 (1986)
58. Spain JC, Van Veld PA; Appl Environ Microbiol 45: 428-35 (1983)
59. Sporstoel S et al; Int J Environ Anal Chem 21: 129-38 (1985)
60. Stone AT; Environ Sci Technol 21: 979-88 (1987)
61. Stuermar DH et al; Environ Sci Technol 16: 582-7 (1982)
62. Urushigawa Y et al; Kogai Shigen Kenyusho Iho 12: 37-46 (1983)
63. Van Veld PA, Spain JC; Chemosphere 12: 1291-305 (1983)
64. Yasuhara A et al; Environ Sci Technol 15: 570-3 (1981)
65. Young LY, Rivera MD; Water Res 19: 1325-32 (1985)

Dibutyl Phthalate

SUBSTANCE IDENTIFICATION

Synonyms:

Structure:

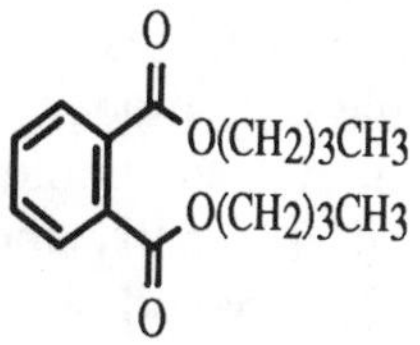

O(CH₂)₃CH₃
O(CH₂)₃CH₃

CAS Registry Number: 84-74-2

Molecular Formula: $C_{16}H_{22}O_4$

Wiswesser Line Notation: 4OVR BVO2

CHEMICAL AND PHYSICAL PROPERTIES

Boiling Point: 340 °C

Melting Point: -35 °C

Molecular Weight: 278.34

Dissociation Constants:

Log Octanol/Water Partition Coefficient: 4.72 [23]

Water Solubility: 11.2 mg/L at 25 °C [29]

Vapor Pressure: 1.4 x 10⁻⁵ mm Hg at 25 °C [20]

Henry's Law Constant: 4.6 x 10⁻⁷ atm-m³/mole (calculated) [43]

ENVIRONMENTAL FATE/EXPOSURE POTENTIAL

Summary: Di-n-butyl phthalate (DBP) is a ubiquitous pollutant due to its widespread use primarily as a plasticizer in plastics which are used throughout society. DBP may be released into the environment as emissions and in wastewater during its production and use, incineration of plastics, and migration of the plasticizer from materials containing it. If released into water, it will adsorb moderately to sediment and particulates in the water column. The DBP will disappear in 3-5 days in moderately polluted waters and generally within 3 weeks in cleaner bodies of water. It will not bioconcentrate in fish since it is readily metabolized. If spilled on land, it will adsorb moderately to soil and slowly biodegrade (66 and 98% in 26 weeks in two cases). DBP is found in ground water under rapid infiltration sites and elsewhere. It has been suggested that its tendency to form complexes with water-soluble fulvic acids, a component of soils, may aid its transport into ground water. Although it degrades under anaerobic conditions, its fate in ground water is unknown. If released into air, DBP is generally associated with the particulate fraction and will be subject to gravitational settling. Vapor phase DBP will degrade by reaction with photochemically produced hydroxyl radicals (estimated half-life 18 hr). Human exposure is from air, drinking water, and food, in addition to in the workplace.

Natural Sources: Natural occurrence in soils by microbial biosynthesis appears to be a possible source [57].

Artificial Sources: In wastewater and airborne particulate matter associated with its manufacture, storage, transportation and use as a plasticizer in nitrocellulose and other lacquers, rubbers, polyvinyl acetate, and polymethyl methacrylate resins and plastics, and as an insect repellant, and as a solvent in perfumes, inks, etc. [25,57,83].

Terrestrial Fate: When spilled on soil, DBP will be adsorbed to a moderate extent and will slowly biodegrade. In two representative soils, 98 and 66% degradation occurred in 26 weeks. Removal rates are increased by acclimation of the microbial populations. Two laboratory models of rapid infiltration sites gave radically different percent removals for DBP; 86% in one case and none in

218

the other. However, DBP has been found in ground water at high concentrations under rapid infiltration sites, which demonstrates that with high input levels it can leach into ground water.

Aquatic Fate: If released into water, DBP will adsorb moderately to sediment and complex with humic material in the water column. Biodegradation rates are rapid with 90-100% degradation being reported in 3-5 days in industrial rivers and pond water and 2-17 days in water from a variety of freshwater and estuarine sites. Biodegradation is slower in seawater, with 33% degradation occurring in 14 days and 100% in 5 days in clean and polluted waters, respectively. In one case where the total loss rate in an industrial river (the Rhine in the Netherlands) was determined by measuring the concentration reduction between fixed points, the half-life was 0.40 days [89]. However, DBP was regularly detected 6 km downcurrent from a kraft pulp mill on a Finnish Lake, making it one of the most persistent compounds discharged by the mill [5]. Although biodegradation occurs (98% in 30 days) in anaerobic sediment/pond water, no rates in ground water could be found. Volatilization will make a small contribution to loss in natural bodies of water. Photooxidation and hydrolysis would not make a significant contribution to DBP's loss in the water, with the possible exception of oligotrophic alkaline waters where hydrolysis may be significant (estimated half-life 76 days at pH 9).

Atmospheric Fate: If released to the atmosphere, DBP will adsorb on particulate matter and be subject to gravitational settling. Vapor phase DBP will photodegrade by reaction with hydroxyl radicals (estimated half-life 18 hr).

Biodegradation: DBP is significantly biodegraded in biodegradation tests utilizing sewage [75,78] and activated sludge [75,82] inoculum, as well as inoculum composed of sewage, soil, and natural waters [67]. In a shake flask biodegradation test, after 28 days 68 to >99% of the DBP had disappeared and 80.6 to >99% was converted to CO_2 [75]. The lag period averaged 4.5 days [75]. 60-70% removal were reported in three treatment plants using activated sludge [58]. 100% degradation occurred in 4 days in water from an urban river in Japan and utilizing water from the Rhine, Meuse, and Ijssel Rivers [24] in the Netherlands, 90% degradation occurred in three days [68]. In an aerobic pond water-sediment mixture, 97% degradation was noted in 5 days [35]. The intermediate products of degradation were the mono-n-butyl ester and phthalic acid [35]. In clean and polluted seawater, 33% degradation in 14 days and 100% in 5 days was observed, respectively [24]. 50% of the DBP biodegraded in 1-5 days and complete disappearance was obtained in 2-13 days in sediment-water systems obtained from 6 geographically different estuarine and freshwater sites bordering on the Gulf of Mexico [84]. Biodegradation in the water alone proceeded somewhat slower (half-life 2-12 days, complete disappearance in 2-17 days) with rapid degradation occurring after a lag period that varied from site to site [84]. Biodegradation under anaerobic conditions was slower with 41% and 98% degradation occurring after 7 and 30 days, respectively, in a sediment-pond water mixture [35]. In Davidson clay loam and Lakeland sand, 98 and 66% occurred in 26 weeks, respectively, as a result of biodegradation [57]. While 86% removal of DBP in secondary sewage occurred in a well acclimated 2.5 m loamy sand soil column [4], no removal occurred in another laboratory model of a rapid infiltration site employing 1 m columns [31]. The feed rate was greater in the first case and the acclimation period may have been longer. Nevertheless, the reason for the disparity in results is not evident. DBP is completely mineralized in digested sludge in 2 weeks under anaerobic conditions [71] and 28% was lost after 7 days in a composting mixture [73].

Abiotic Degradation: DBP has a strong UV absorption band at 274 nm which extends beyond 290 nm [65] and is therefore a candidate for photolysis. However, the estimated photolysis half-life in natural waters is 144 days [87]. No data for the photolysis rate in air could be found. The estimated half-life due to reaction with alkoxy radicals in natural waters is 456 days [87]. Photochemically produced hydroxyl radicals in the atmosphere will react with vapor phase DBP by aromatic ring addition and hydrogen atom abstraction with an estimated half-life of 18.4 hr [18]. The hydrolysis of DBP at neutral pH has a half-life of 10 yr [7]. The rate increases with pH and at pH 9 the estimated half-life is 76 days [87].

Bioconcentration: DBP is rapidly metabolized in fish [34] within 4 hr, 75% of the residue from a channel catfish was in the form of monobutyl phthalate [34]. Therefore log BCF of 3.15 [47] and 3.83 [66] measured in fish using ^{14}C labeled chemicals have no meaning. DBP was confirmed to be at most slightly bioconcentrated in fish in testing by the Japanese Ministry of International Trade and Industry (MITI) [67]. The log BCF in American oyster, Brown shrimp, and Sheepshead minnow were 1.50, 1.22 and 1.07, respectively [86], in agreement with the calculated log BCF of 1.32 [39]. In a study in which clams, Neanthes virens, and sediment samples were measured for DBP residues at two sites in Portland Harbor, the bioconcentration from the sediment was 0.59-1.1 and 0.14-0.25 at the two sites [63].

Soil Adsorption/Mobility: The Koc value for DBP estimated from its water solubility is 160 [39] and 6400 [87]. The partition coefficient of DBP between montmorillonite, kaolinite and calcium montmorillonite and seawater was 40, 20, and 4, respectively, at high concentrations (3-4 ppm) and 2, 4, and 36, respectively, at low concentrations (ca 20 ppb) [76]. The partition coefficient between marine sediment (ca 1% organic carbon) and seawater was 149 [4]. Despite its moderate adsorption to soil, DBP has been found in ground water underlying rapid infiltration sites at high concentrations [80]. DBP forms a 1:1 complex with fulvic acid, which is a water-soluble humic material formed from the decomposition of plants [46]. The water soluble complex so formed may act as a vehicle for the mobilization and transport of DBP as well as altering its reactivity [46]. Koc 1386 in Broome County, NY soil [64].

Volatilization from Water/Soil: The volatilization half-life for DBP from a stirred seawater solution 1 m deep is 28 days [3]. Using the calculated Henry's Law constant, it is estimated that the half-life for DBP in a river 1 m deep with a 1 m/sec current and a 3 m/sec wind is 47 days, with the rate being controlled by the diffusion through the air [43]. Approx 1-5% of the DBP was lost in a sewage treatment plant by air stripping [61]. Due to its low vapor pressure and moderate adsorption to soil, volatilization from soil would not be expected to be a significant loss process. Evaporation rates from the leaves of Sinapis alba in a closed terrestrial simulation chamber were 17 and 0.5 ng/sq cm-hr during time intervals of 0-1 and 8-15 days, respectively, after application of 2.54 ug/sq cm onto the leaves [41].

Dibutyl Phthalate

Water Concentrations: DRINKING WATER: 6 U.S. cities 0.01-5.0 ppb, 0.02 ppb median [38]. 3 New Orleans' drinking water plants 0.1-0.36 ppb [38]. Contaminated drinking water well in N.Y. 470 ppb [6]. Detected, not quantified at 8 water works in Japan [28]. 6 cities in Japan (tap water) 190-240 ppb [1]. 2 drinking water wells in vicinity of landfills 0.5 ppb [12]. Detected in raw and treated drinking water from 12 of 14 sites in England [17].   GROUND WATER: Ground water underlying 2 rapid infiltration sites 0.73-2.38 ppb, not detected at a 3rd site and detected, not quantified at a fourth [30]. Norman, OK (landfill ground water) detected, not quantified [13]. Recovery well removing contaminated water under landfill 1 ppb [12]. SURFACE WATER: 14 heavily industrialized river basins in the United States (204 sites) 87 sites had concentrations >1 ppb, 60 ppb max [16]. Delaware River (river mile 78-132, 16 samples) 0.1-0.6 ppb, all samples positive [70]. Lower Tennessee River below Culvert City, KY (water and sediment) 42 ppb [22]. St. Clair River 1-2 ppb [40]. Missouri River 0.09 ppb [48]. Monatiquote River, MA 1-30 ppb [26]. Lake Erie (2 sites) 1 ppb [40], Lake Michigan (9 sites) 1-4 ppb [40], Lake Huron (2 sites) 0.04-2 ppb [40,48], and Lake Superior (1 site) not detected [48]. Rhine, Ijssel, and Meuse Rivers in the Netherlands (21 sites) 0-2.8 ppb [68].   Tama River, Japan 0.71-3.14 ppb [52]. Shizuoka Prefecture, Japan 1.39 ppb avg, 4.3 ppb max in river water with 22 of 23 samples positive [72].   Lake Saimaa, Finland - site of pulp mill - 5-230 ppb [56]. Concn (ppt) in Mississippi River, Lake Itasca, MN (river's source) 150; Cario, IL 140; 20 miles south of Memphis, TN 150; Carrolton St. water intake in New Orleans, LA 139 [11]. SEAWATER: Gulf of Mexico: Mississippi Delta 6.5 ppb avg; Gulf Coast 3.4-265 ppt, 74 ppt avg; Open Gulf 3.0-133 ppt, 93 ppt avg [19]. Kiel Bight - 15 stations - 46.4-193.8 ppt at 1 m depth [14]. Mersey Estuary, UK, 17 sites, 100% pos, dissolved fraction, 100% pos, 114-2116 ppt, 58 ppt avg pos; particulate fraction, 100% pos, 20-698 ppb (dry wt) [62]. RAIN: Ewiwetak Atoll (North Pacific) 2.6-72.5 ppt, 31 ppt [2]. Detected in water and particulate fraction of rain and snow in Norway [42] and rain in Los Angeles [37]. Urban College Station, TX, Feb-May, 1979, unfiltered rain, 56 ppt avg concn [49].

Dibutyl Phthalate

Effluent Concentrations: Industries whose mean effluent levels of DBP exceed 100 ppb include: aluminum forming (3900 ppb raw, <5400 ppb treated); foundries (280 ppb raw, 440 ppb treated); metal finishing (140 ppb); paint and ink formulations (2300 ppb); and petroleum refining (4100 ppb) [81]. Industries whose maximum effluent levels of DBP exceed 1000 ppb include: aluminum forming (19,000 ppb raw, 90,000 ppb treated); foundries (5400 ppb raw, 9300 ppb treated); paint and ink formulations (69,000 ppb raw, 1300 ppb treated); photographic equipment/supplies (1400 ppb);and metal finishing (3100 ppb) [81]. DBP was detected at concentration levels of 0.5-11 ppb and 5% frequency of detection in urban runoff in Denver and Rapid City, two of the nineteen cities (86 samples) across the U.S. in the Nationwide Urban Runoff Program [10]. Phoenix, AZ secondary sewage effluent 0.25 ppb [80] and detected, not quantified in sewage effluent [1]. Detected in ng quantities in air samples in Hamilton, Ontario, where waste plastics are burned [1]. Ground water from four U.S. rapid-infiltration sites for municipal primary and secondary effluents, 75% pos, 0.73-2.38 ppb range of 2 sites, 1 site not quantified [31].

Sediment/Soil Concentrations: Chesapeake Bay sediment - 2 sites, 0-10 cm layer 89 and 27 ppb dry wt, with site closest to Baltimore Harbor showing the higher concentration [60]. Core samples down to 100 cm have comparable amounts of DBP and this fact along with given deposition rates at the sites show leaching of the DBP, input over many decades, or both. Portland, ME sediment (8 sites) 40-280 ppb, 160 ppb average [63]. Gulf of Mexico (core section, top 10 cm): Mississippi Delta 0-52.1 ppb, 13 ppb average; Gulf Coast 0-15.3 ppb, 7.6 ppb average; and Open Gulf 1.6-5.6 ppb, 3.4 ppb average [19]. Lake Superior (Black Bay) 100 ppb [48]. Lake Erie (Detroit River to Stony Point, 18 sites) 3-6 ppb; Lake Huron (Saginaw River) 290 ppb [40]. San Lurs Pass to Gulf of Mexico, TX 60 ppb [53]. Lake Constance, Switzerland (surface sediment) 0.1-0.3 ppm [21]. Delaware River estuary (surface sediment) 4.5 ppb [27]. Japan - river sediment 0-0.96 ppb, 0.17 ppb average with 12 of 22 samples positive [72]. Rhine and Ijssel Rivers in the Netherlands 0-2 ppm [69]. Rhine and Nelker Rivers, Germany (9 sites) 0-400 ppb [44].

Dibutyl Phthalate

Atmospheric Concentrations: Average concentration in particulate matter in air collected in Queens, Brooklyn, and Staten Island - 3.3-5.7 ng/m³ [5]. Air over the Great Lakes 0.5-5 ng/m³ [15]. Houston (17 samples) 0-567 ng/m³, only one sample positive [59]. Hamilton, Ontario 700 ng/m³ [79]. Gulf of Mexico, 6 sites, 0.65-3.71 ng/m³, 1.30 ng/m³ avg in vapor particulates [20]. North Atlantic 0.4-2.3 ng/m³, 1.0 ng/m³ avg [19]. Eniwetak Atoll in Pacific Ocean 0.9 ng/m³ average, College Station, TX 3.8 ng/m³, Pigeon Key, FL 18.5 ng/m³ [2]. Bolivia 19-36 ng/m³, Antwerp 24-74 ng/m³ in particulate matter [8]. Urban air, Belgium, avg concn (ug/1000 m³): particulates, 101, gas phase, 353 [9].

Food Survey Values: Canned tuna (Canada) 0-78 ppb, canned salmon (Canada) <37 ppb [85]. Egg white (Japan - retail stores) <150 ppb [32]. Fresh and processed food in Japan: meat 100 ppb avg, fish 180 ppb avg, eggs 80 ppb avg with 70% of samples positive [33]. DBP is used as a plasticizer in food wrappings and food containers and it can migrate from the plastic packaging into foods [57]. It has been estimated that 150 mg of DBP will migrate into 1 kg cheese with 15% fat content [57]. Cereal, gelatin, corn starch, and casein components of commercial fish food 20-30 ppb [45].

Plant Concentrations:

Fish/Seafood Concentrations: Clams - 2 sites Portland, ME 40 and 100 ppb; Neanthes virens - 2 sites Portland, Me 70 and 180 ppb [63]. Not detected (<.1 ppb) in 18 species of marine organisms from 14 locations in Mississippi Delta and coastal areas in NW part of the Gulf of Mexico [19]. Detected, not quantified in white sucker, longnose sucker, and yellow perch from Nepugin Bay, Lake Superior [36], in burbot from 2 sites in Lake Huron [40]. Lake Superior (adjacent to Isle Royale, MI): fat siscowet trout trace; lean lake trout 200 ppb; white fish 70 ppb [77]. Selected areas of North America: channel catfish 0-200 ppb; dragonfly niads 200 ppb; tadpoles 500 ppb [48].

Animal Concentrations: Double crested cormorants and herring gulls 11-19 ug/g lipid [88].

Milk Concentrations: Not detected in milk in Japan [33].

Dibutyl Phthalate

Other Environmental Concentrations:

Probable Routes of Human Exposure: Humans may be exposed to DBP in the workplace where it is manufactured and used. Exposure may also occur from particulate matter and vapor in air, drinking water, and food products.

Average Daily Intake: AIR INTAKE: (assume 0-20 ng/m^3) 0-400 ng; WATER INTAKE: (assume 0.01-5.0 ppb) 20 ng-10,000 ng; FOOD INTAKE: insufficient data.

Occupational Exposures: NIOSH (NOHS Survey 1972-1974) has statistically estimated that 905,227 workers are exposed to DBP in the United States [55]. NIOSH (NOES Survey 1981-1983) has statistically estimated that 229,345 workers are exposed to DBP in the United States [54].

Body Burdens: Human adipose tissue 0.10-0.30 ppm [50], 0.57-0.79 ppm [51]. Detected in human tissue and blood [57]. National Human Adipose Tissue Survey, Fiscal Year 1982, 46 tissue composites, 44% pos, wet tissue concn, not detected (<0.009-0.13)-1.5 ppm, concn in extractable lipid, not detected (<0.011-0.18)-2.4 ppm [74].

REFERENCES

1. Akiyama T et al; J UOEH 2: 285-300 (1980)
2. Atlas E, Giam CS; Science 211: 163-5 (1981)
3. Atlas E et al; Environ Sci Technol 16: 283-6 (1982)
4. Bouwer EJ et al; Water Res 15: 151-9 (1981)
5. Bove JL et al; Int J Environ Anal Chem 5: 189-94 (1978)
6. Burmaster DE; Environ 24: 6-13,33-6 (1982)
7. Callahan MA et al; Water-related fate of 129 priority-pollutants vol II p.94-1 to 94-28 USEPA-440/4-79-029b (1979)
8. Cautreels W et al; Sci Total Environ 8: 79-88 (1977)
9. Cautreels W, VanCauwenberghe K; Atmos Environ 12: 1133-41 (1978)
10. Cole RH; J Water Pollut Control Fed 56: 898-908 (1984)
11. DeLeon IR et al; Chemosphere 15: 795-805 (1986)
12. DeWalle FB, Chian ESK; J Amer Water Works Assoc 73: 206-11 (1981)

Dibutyl Phthalate

13. Dunlap WJ et al; Organic pollutants contributed to ground water by a landfill pp.96-110 USEPA-600-9-76-004 (1976)
14. Ehrhardt M, Derenbach J; Mar Chem 8: 339-46 (1980)
15. Eisenreich SJ et al; Environ Sci Technol 15: 30-8 (1981)
16. Ewing BB et al; Monitoring to detect previously unrecognized pollutants in surface waters USEPA-560/6-77-015 (1977)
17. Fielding M et al; Organic micropollutants in drinking water Res Ctr Medmenham Eng p.49 TR-159 (1981)
18. GEMS; Graphical Exposure Modeling System. Fate of atmospheric pollutants (FAP) data base. Office of Toxic Substances USEPA (1987)
19. Giam CS et al; Science 199: 419-21 (1978)
20. Giam CS et al; Atmos Environ 14: 65-9 (1980)
21. Giam CS, Atlas E; Naturwissenschaften 67: 508-10 (1980)
22. Goodley PC, Gordon M; NY Acad Sci 37: 11-5 (1976)
23. Hansch C, Leo AJ; Medchem Project Issue No 26. Claremont CA: Pomona College (1985)
24. Hattori Y et al; Pollut Control Cent Ooaki Prefect Mizu Shori Gigutsu 16: 951-4 (1975)
25. Hawley GG; Condensed Chemical Dictionary 10th ed Van Nostrand Reinhold New York p.330 (1981)
26. Hites RA; J Chromatogr Sci 11: 570-4 (1973)
27. Hochreiter JJ Jr; Chemical quality reconnaissance on the water and surficial bed material in the Delaware River estuary and adjacent New Jersey tributaries, 1980-81 USGS/WRI/NTIS 82-36 (1982)
28. Horibe N et al; Gifu-Ken Eisi Kenkyusho Ho 19: 20-5 (1974)
29. Howard PH et al; Environ Toxicol Chem 4: 653-61 (1985)
30. Hutchins SR et al; Environ Toxicol Chem 2: 195-216 (1983)
31. Hutchins SR, Ward CH; J Hydrol 67: 223-33 (1984)
32. Ishida M et al; J Agric Food Chem 29: 72 (1981)
33. Ishikawa K et al; Miyagi-Ken Eisi Kenkyusho Nempo p.105-11 (1975)
34. Johnson BT et al; in Fate of pollutants in the air and water environment Suffet SH ed John Wiley & Sons NY p.422 (1977)
35. Johnson BT, Lulves W; J Fish Res Board Can 32: 333-9 (1975)
36. Kaiser KLE; J Fish Res Board Can 34: 850-5 (1977)
37. Kawamura K, Kaplan IR; Environ Sci Technol 17: 497-501 (1983)
38. Keith LH et al; pp.329-73 in Identification and analysis of organic pollutants in water Keith LH ed Ann Arbor Press Ann Arbor MI (1976)
39. Kenaga EE; Ecotox Environ Safety 4: 26-38 (1980)
40. Konasewich D et al; Status report on organic and heavy metal contaminants in the Lake Erie, Michigan, Huron and Superior Basins. Great Lakes Qual Rev Board (1978)
41. Loekke H, Bro-Rasmussen F; Chemosphere 10: 1223-35 (1981)
42. Lunde G et al; Organic micropollutants in precipitation in Norway p.17 SNSF Project FR-9176 (1977)
43. Lyman WJ et al; Handbook of chemical property estimation methods. Environmental behavior of organic compounds McGraw-Hill NY (1982)
44. Malesch R et al; Chem Ztg 105: 187-94 (1981)
45. Mathur SP; J Environ Qual 3: 189-97 (1974)

Dibutyl Phthalate

46. Matsuda K, Schnitzer M; Bull Environ Contam Toxicol 6: 200-4 (1971)
47. Mayer FL Jr, Sanders HO; Environ Health Perspect 3: 153-7 (1973)
48. Mayer FL et al; Nature 238: 411-3 (1972)
49. Mazurek MA, Simoneit BRT; Crit Rev Environ Control 16: 69 (1986)
50. Mes J et al; Bull Environ Contam Toxicol 12: 721-5 (1974)
51. Mes J, Campbell DS; Bull Environ Contam Toxicol 16: 53-60 (1976)
52. Morita M et al; Tokyo Toritsu Eisi Kenkyusho Kenkyo Nempo 24: 357-62 (1973)
53. Murray HE et al; Chemosphere 10: 1327-34 (1981)
54. NIOSH; The National Occupational Exposure Survey (NOES) (1983)
55. NIOSH; The National Occupational Hazard Survey (NOHS) (1974)
56. Oikari A et al; Pap Puu 62: 193-6, 199-200 (1980)
57. Overcash MR et al; Behavior of organic priority pollutants in the terrestrial system: di-n-butyl phthalate ester, toluene and 2,4-dinitrophenol p.104 NTIS PB82-224-544 (1982)
58. Patterson JW, Kodukola PS; Chem Ind 15: 609-10 (1981)
59. Pellizzari ED; Quantification of chlorinated hydrocarbons in previously collected air samples USEPA-450/3-78-112 (1978)
60. Peterson JC, Freeman DH; Environ Sci Technol 16: 464-9 (1982)
61. Petrasek AC et al; J Water Pollut Control Fed 55: 1286-96 (1983)
62. Preston MR, Al-Omran LA; Mar Pollut Bull 17: 548-53 (1986)
63. Ray LE et al; Chemosphere 12: 1031-8 (1983)
64. Russell DJ, McDuffie B; Chemosphere 15: 1003-21 (1986)
65. Sadtler 529 UV (1988)
66. Sanders HO et al; Environ Res 6: 84-90 (1973)
67. Sasaki S; pp.283-98 in Aquatic pollutants: Transformation and biological effects Hutzinger O et al eds Pergammon Press Oxford (1978)
68. Schouter MJ et al; Inter J Environ Anal Chem 7: 13-23 (1979)
69. Schwartz HE et al; Int J Environ Anal Chem 6: 133-44 (1979)
70. Sheldon LS, Hites RA; Environ Sci Technol 12: 1188-94 (1978)
71. Shelton DR et al; Environ Sci Technol 18: 93-7 (1984)
72. Shibuya S; Numazi Kogyo Koti Semmon Gakko Kenkyu Hokoku p.63-72 (1979)
73. Snell Environmental Group Inc; Rate of biodegradation of toxic organic compounds while in contact with organics which are actively composting p.100 NSF/CEE 82024 (1982)
74. Stanley JS; Broad Scan Analysis of the FY82 Nat'l Human Adipose Tissue Survey Specimens pp.36,90-1 USEPA-560/5-86-037 (1986)
75. Sugatt RH et al; Appl Environ Microbiol 47: 601-6 (1984)
76. Sullivan KF et al; Anal Chem 53: 1718-9 (1981)
77. Swain WR; J Great Lakes Res 4: 398-407 (1978)
78. Tabak HH et al; J Water Pollut Control Fed 53: 1503-18 (1981)
79. Thomas GH; Environ Health Perspect 3: 23-8 (1973)
80. Tomson MB et al; Water Res 15: 1109-16 (1981)
81. U.S.EPA; Treatability Manual p.1.6.3-1 to 1.6.3-4 USEPA-600/2-82-001a (1981)
82. Urushigawa Y, Yonezawa Y; Chemosphere 5: 317-20 (1979)
83. Verscheuren K; Handbook of environmental data on organic chemicals 2nd ed Van Nostrand Reinhold NY p.468 (1983)
84. Walker WW et al; Chemosphere 13: 1283-94 (1984)

Dibutyl Phthalate

85. Williams DT; J Agric Food Chem 21: 1128-9 (1973)
86. Wofford HW et al; Ecotox Environ Safety 5: 202-10 (1981)
87. Wolfe NL et al; Chemosphere 9: 393-402 (1980)
88. Zitko V; Int J Environ Anal Chem 2: 241-52 (1973)
89. Zoeteman BW et al; Chemosphere 9: 231-49 (1980)

1,2-Dichlorobenzene

SUBSTANCE IDENTIFICATION

Synonyms: o-Dichlorobenzene

Structure:

CAS Registry Number: 95-50-1

Molecular Formula: $C_6H_4Cl_2$

Wiswesser Line Notation: GR BG

CHEMICAL AND PHYSICAL PROPERTIES

Boiling Point: 180.5 °C at 760 mm Hg

Melting Point: -17.0 °C

Molecular Weight: 147.01

Dissociation Constants:

Log Octanol/Water Partition Coefficient: 3.38 [20]

Water Solubility: 156 mg/L at 25 °C [54]

Vapor Pressure: 1.47 mm Hg at 25 °C [38]

Henry's Law Constant: 0.0012 atm-m^3/mole at 20 °C [49]

1,2-Dichlorobenzene

ENVIRONMENTAL FATE/EXPOSURE POTENTIAL

Summary: Chemical waste dump leachates and direct manufacturing effluents are reported to be the major source of pollution of the chlorobenzenes (including the dichlorobenzenes) to Lake Ontario. The major source of 1,2-dichlorobenzene (1,2-DCB) emission to the atmosphere has been reported to be solvent applications which may emit 25% of annual production to the atmosphere. If released to soil, 1,2-DCB can be moderately to tightly adsorbed. Leaching from hazardous waste disposal areas has occurred and the detection of 1,2-DCB in various ground waters indicates that leaching can occur. Volatilization from soil surfaces may be an important transport mechanism. It is possible that 1,2-DCB will be slowly biodegraded in soil under aerobic conditions. Chemical transformation by hydrolysis, oxidation, or direct photolysis are not expected to occur in soil. If released to water, adsorption to sediment will be a major environmental fate process based upon extensive monitoring data in the Great Lakes area and Koc values. Analysis of Lake Ontario sediment cores has indicated the presence and persistence of 1,2-DCB since before 1940. 1,2-DCB is volatile from the water column, with an estimated half-life of 4.4 hours from a river one meter deep flowing 1 m/sec with a wind velocity of 3 m/sec at 20 °C; adsorption to sediment will attenuate volatilization. Aerobic biodegradation in water may be possible; however, anaerobic biodegradation is not expected to occur. Experimental BCF values of 66-560 have been reported and 1,2-DCB has been detected in trout from Lake Ontario. Aquatic hydrolysis, oxidation, and direct photolysis are not expected to be important. If released to air, 1,2-DCB will exist predominantly in the vapor phase and will react with photochemically produced hydroxyl radicals at an estimated half-life rate of 24 days in a typical atmosphere. Direct photolysis in the troposphere is not expected to be important. The detection of 1,2-DCB in rainwater suggests that atmospheric removal via wash-out is possible. General population exposure to 1,2-DCB may occur through oral consumption of contaminated drinking water and food (particularly fish) and through inhalation of contaminated air, since 1,2-DCB has been detected in widespread ambient air.

Natural Sources:

230

1,2-Dichlorobenzene

Artificial Sources: For the chlorobenzenes in general, the major source of pollution to Lake Ontario has been reported to be chemical waste dump leachates and direct chemical manufacturing effluents located around Niagara Falls, NY [42]. 1,2-DCB's use in manufacturing and solvents may be significant sources of discharges into water [25]. The major emission source of 1,2-DCB to the atmosphere has been reported to be solvent applications [22]; approx 25% of production is used in solvent applications, which may eventually be emitted into the atmosphere [57]. Stack effluents from refuse-fired steam boilers and power plants can emit 1,2-DCB into the atmosphere.

Terrestrial Fate: Based on experimental adsorption data, 1,2-DCB can be moderately to tightly adsorbed in soil. Leaching from hazardous waste disposal areas in Niagara Falls to adjacent surface waters has been reported [15] and the detection of 1,2-DCB in various ground waters indicates that leaching can occur. Volatilization from soil surfaces may be an important transport mechanism; however, volatilization may be attenuated by tight adsorption or leaching. It is possible that 1,2-DCB will be slowly biodegraded in soil under aerobic conditions. Chemical transformation processes such as hydrolysis, oxidation, or direct photolysis (on soil surfaces) are not expected to occur.

Aquatic Fate: Based on suspended solid monitored Koc values of about 39,800 and extensive sediment monitoring data in the Great Lakes area, adsorption to sediment is a major environmental fate process for 1,2-DCB. Its detection in Lake Ontario sediment cores indicates that 1,2-DCB has persisted in these sediments since before 1940 [42]. 1,2-DCB is volatile from water, with an estimated half-life of 4.4 hours from a river one meter deep flowing 1 m/sec with a wind velocity of 3 m/sec at 20 °C; adsorption to sediment in water will attenuate volatilization. 1,2-DCB may biodegrade in aerobic water after microbial adaptation; however, it is not expected to biodegrade under anaerobic conditions which may exist in lake sediments or various ground waters. Experimental BCF values of 66-560 have been reported; the detection of 1,2-DCB in trout from Lake Ontario has confirmed that bioaccumulation is important. Aquatic hydrolysis, oxidation, and direct photolysis are not expected to be important. The persistence half-life of 1,2-DCB in the water column has been estimated to be 0.3-3 days in rivers, 3-30 days in lakes, and 30-300 days in ground waters [70].

Atmospheric Fate: 1,2-DCB will exist predominantly in the vapor phase in the atmosphere. The half-life for the vapor phase reaction of 1,2-DCB with photochemically produced hydroxyl radicals in the atmosphere has been estimated to be 24 days. Direct photolysis is not expected to be important. The detection of 1,2-DCB in rainwater suggests that atmospheric removal via wash-out is possible.

Biodegradation: Using a static-culture screening procedure (5 or 10 mg/L test compound, a 7-day static incubation followed by three weekly subcultures and a settled domestic wastewater as microbial inoculum), 1,2-DCB was biodegraded 20-45%, 59-66%, 32-4%, and 18-29% after the original culture, first, second, and third subculture, respectively [61]. 1,2-DCB was reduced from 50 mg/L to 2-4 mg/L in 7 days of incubation using industrial or municipal sludge seeds and a shaker flask procedure [14]. 1,2-DCB was found to be degradation resistant using the Japanese MITI test [31]. In a continuous flow activated sludge system, virtually 100% removal of 1,2-DCB (78% biodegradation, 22% stripping) was observed [30]. In a laboratory aquifer column simulating saturated-flow conditions typical of a river/ground water infiltration system, the dichlorobenzenes were biotransformed under aerobic conditions but not under anaerobic conditions [32]. 1,2-DCB was significantly removed in an aerobic biofilm column (9.6 ppb concn) but was not removed in a methanogenic biofilm column (15 ppb) [5,6]. The dichlorobenzenes are not expected to be biotransformed in anaerobic water conditions found in aquifers [68]. 1,2-DCB had BOD's of 0, 41, and 51% of the theoretical BOD over 5-, 10-, and 20-day periods, respectively [2]. An analysis of monitoring data from sediment cores indicated insufficient evidence to show occurrence of a aerobic dehalogenation of the chlorobenzenes in Lake Ontario sediment [48]. The dichlorobenzenes were slowly biodegraded (6.3% of theoretical CO_2 evolution in 10 weeks) in an alkaline soil sample [19]. It has been suggested that the three dichlorobenzene isomers may undergo slow biodegradation in natural water [9]. Not significantly degraded (half-life of much greater than 4 weeks) in the repetitive die-away test using inoculum which was not adapted [10].

Abiotic Degradation: 1,2-DCB is not expected to undergo significant hydrolysis in environmental waters [9]. It is reported to be resistant towards oxidation by peroxy radicals in aquatic media [9]. In an isooctane solvent, 1,2-DCB absorbs virtually no irradiation above 300 nm [55]; therefore, direct photolysis in the environment should not be significant. The experimentally determined rate constant for the vapor-phase reaction between 1,2-DCB and photochemically produced hydroxyl radicals has been reported to be 0.42 x 10^{-12} cm³/molecule-sec at 22 °C [1]; assuming an average atmospheric hydroxyl radical concn of 8 x 10^{+5} molecules/cm³, the half-life for the reaction is estimated to be 24 days.

Bioconcentration: Mean 1,2-DCB BCF values of 270 to 560 were experimentally determined for rainbow trout exposed up to 119 days in laboratory aquariums [43]. A whole body BCF of 66 was determined for bluegill sunfish exposed to 1,2-DCB over a 28-day period in a continuous flow system [4]. Using the water solubility and the log Kow, BCF values of 36 and 218 are estimated, respectively, from recommended equations [37].

Soil Adsorption/Mobility: An experimental Koc value of 280 (Kom=162) determined for 1,2-DCB in a silt loam soil containing 1.6 % organic matter [11]. An experimental Koc value of 320 (Kom = 186) determined in a silt loam soil containing 1.9% organic matter [12]. In equilibrium batch studies, a relatively strong adsorption of 1,2-DCB to collected aquifer material was observed [34]. Using the water solubility and the log Kow, Koc values of 272 and 1480 are estimated, respectively, from regression-derived equations [37].

Volatilization from Water/Soil: The Henry's Law constant for 1,2-DCB indicates that volatilization will be significant from all waters and should be rapid in most instances [37]. The volatilization half-life from a river one meter deep flowing 1 m/sec with a wind velocity of 3 m/sec is estimated to be 4.4 hours at 20

233

°C [37]. Based on its vapor pressure, 1,2-DCB is expected to evaporate at a relatively significant rate from dry surfaces.

Water Concentrations: DRINKING WATER: A mean 1,2-DCB concn of 0.003 ppb was detected in drinking water samples from 3 cities near Lake Ontario in 1980 [42]. Concn of 1 ppb identified in Miami, FL drinking water and qualitative detections reported for Philadelphia, PA and Cincinnati, OH [64]. 1,2-DCB was found in 2 of 945 finished water supplies throughout the United States that use ground water sources at concn of 2.2 and 2.7 ppb [67]. Qualitative detection reported for Cleveland, OH tap water [56]. Qualitative detection reported for two drinking water supply sources in the United Kingdom [16]. GROUND WATER: 1,2-DCB was positively detected in 20 of 685 ground waters analyzed in NJ during 1977-1979, with 6800 ppb the highest concn found [51]. SURFACE WATERS: 1,2-DCB was detected in 15 of 463 surface waters analyzed in NJ during 1977-1979, with 8.2 ppb the highest concn found [51]. Mean concentrations of 5 and 6 ppt detected in Lake Ontario and Grand River water, respectively, during 1980 near Niagara Falls; concn of 0-56 ppt found in the Niagara River [42]. Concn of 3.9-240 ppt (mean concn of 23 ppt) detected in Niagara River at Niagara-On-The-Lake between 1981 and 1983 [47]. Concn of 5.6-190 ppt (mean concn of 18 ppt) detected in the Niagara River between 1981 and 1983 [46]. An average concn of 20 ppt was found in the Niagara River near Niagara-On-The-Lake between Sept and Oct 1982 [45]. Positive detection of 1,2-DCB was reported by 0.6% of 1077 USEPA STORET stations [60]. Qualitative detection reported for the Delaware and Raritan Canal in NJ [17]. Concn below 0.5 ppb detected in the Rhine River between 1978 and 1982 [39]. An average concn of 0.32 ppb found in the Rhine River near Dusseldorf in 1984 [58]. Mean concn (ppt) from single sampling cruises: Lake Ontario, 1.0; Niagara River, 4.5; Detroit River, 0.4 [44]. SEA WATER: 1,2-DCB was detected in the water column of the Narragansett Bay near Rhode Island [66]. Concn (ppt): Southern North Sea, 108 samples in 9 locations, 2-127, 19 avg, 13 median; Rhine River, 1 sample, 400 [65]. RAIN/SNOW: A mean 1,2-DCB concn of 0.49 ppt was detected in Portland, OR rainwater during March-April 1982 [52]; Concn of not detected to 0.62 ppt found in Portland, OR rainwater during 1984 [35].

1,2-Dichlorobenzene

Effluent Concentrations: The wastewater effluents from four water treatment plants discharging into the Grand River and Lake Ontario contained a mean 1,2-DCB concn of 13 ppt during 1980 sampling [42]. Positive detection of 1,2-DCB was reported by 2.5% of 1311 USEPA STORET stations [60]. Concn of 0.3-41 ppb were detected in Southern California municipal wastewater effluents in 1976 [69]. Concn of 2.8-13 ppb found in leachate, associated ground water beneath a municipal landfill in Ontario, Canada [53]. The dichlorobenzene isomers were qualitatively detected in waters adjacent to hazardous waste disposal areas in Niagara Falls, NY as a result of leaching [15]. Ambient mean air concn of 0.04-0.86 ppb were detected above six abandoned hazardous waste sites in NJ as a result of volatile emanations [21]. Flue gas effluents from a municipal refuse-fired steam boiler in Virginia contained 4.4 ug/m^3 of the dichlorobenzene isomers [63]; flue gas effluents from a refuse-derived-fired power plant in Ohio contained 7.8 ng/m^3 of the dichlorobenzene isomers [63]. Coal-fired power plants in United States, 0.5 ng/m^3 [29]. NJ, concn (avg all samples, 0.01 ppb detection limit added for zero concn)/max concn in ppb by vol) at 5 hazardous waste sites (82 total samples), 0.35/2.59; 0.11/0.87; 0.77/8.4; 0.06/0.69; 0.11/0.7; at one landfill site (15 samples), 0.06/0.53 [33].

Sediment/Soil Concentrations: Mean 1,2-DCB concentrations of 1, 8, 2, and 11 ppb were detected in the superficial sediments from Lakes Superior, Huron, Erie, and Ontario, respectively [42]. Concentrations of 2 to 330 ppb found in a Lake Ontario sediment core (0 to 8 cm in depth) with the higher concn in the upper 2 cm [42]. Mean concn of 10 ng/g detected on settling particulates at a station on Lake Ontario [46]. Concn ranging from 32 to 44 ng/g detected in suspended sediment of Lake Ontario at depths from 20 to 68 meters with an average concn of 10 ng/g found on bottom sediments [45]. Qualitative detection reported for both soil and sediment collected near the Love Canal [23]. 1,2-DCB was qualitatively detected in sediments off the California coast near an effluent discharge point of a wastewater treatment plant [69]. Trace concn were detected in surficial bed material of the Racoon Creek in Bridgeport, NJ [24].

Atmospheric Concentrations: The mean 1,2-DCB concentrations from 226 source-dominant points, 674 urban/suburban points, and 9

235

rural/remote points from the United States have been reported to be 200, 56, and 1.8 ppt, respectively, with an overall mean concn of 90 ppt [8]. Average concn of 0.61 ppb near industrial areas in NJ and 0.06 ppb near residential areas in NJ in 1978 [7]. Mean concn of 12.5, 22.6 and 4.0 ppt detected in the ambient air of Los Angeles, CA, Phoenix, AZ, and Oakland, CA,respectively, during Apr-May 1979 [57]. Mean concn of 0.01-0.03 ppb detected in the ambient air of three NJ cities during July-Aug 1981 [22]. Mean concn of 5.8 ng/m³ detected in the ambient air of Portland, OR during 1984 [35]. Airborne Toxic Element and Organic Substance project, NJ, geometric mean, max concn [ppb by vol] for Summer 1981/Winter 1982: Camden, 0.01, 0.12/0.03, 0.22; Elizabeth, 0.02, 0.65/0.05, 0.80; Newark, 0.03, 0.90/0.02, 9.61 [36].

Food Survey Values: 1,2-DCB concn of 1.0 ng/g were found in market meat samples in Yugoslavia [27]. Identified, not quantified, in volatile flavor compounds from fried chicken in United States [62].

Plant Concentrations: 1,2-DCB has been detected in the roots of wheat plants grown from lindane-treated seeds [25].

Fish/Seafood Concentrations: 1,2-DCB concn 0.3, 1, 1, and 1 ppb were detected in trout taken from Lake Superior, Lake Huron, Lake Erie, and Lake Ontario, respectively, during 1980 [42]. Concn of 0-4.0 ug/wet kg were found in Flatfish off the California coast near Los Angeles [69]; mean concn of less than 0.031 mg/wet kg was found in the muscle tissue of 8 seafood species caught off the California coast [69]. Fish and mussels taken from rivers in Slovenia and the Gulf of Trieste (Yugoslavia) were found to contain traces to 1.2 ug/g 1,2-DCB (on a fat basis) [28].

Animal Concentrations:

Milk Concentrations: Concentrations of 2.6 ng/g 1,2-DCB were found in market milk samples in Yugoslavia [27].

Other Environmental Concentrations:

Probable Routes of Human Exposure: Occupational exposure to 1,2-DCB occurs during its manufacture and its use as a chemical intermediate and solvent [25]; probable routes of occupational exposure are inhalation of contaminated air and dermal contact. General population exposure may occur through oral consumption of contaminated drinking water and food (particularly fish), especially in the vicinity of effluent discharges (e.g., on Lake Ontario). General population exposure may also occur through inhalation of contaminated ambient air since 1,2-DCB has widespread monitoring detection in many areas of the United States.

Average Daily Intake: WATER INTAKE: Insufficient data. AIR INTAKE: Based on monitoring data at three U.S. urban sites (Los Angeles, Phoenix, Oakland), the AVDI for 1,2-DCB has been estimated to be 0.5-2.8 ng/day [57]. Netherlands: Dichlorobenzenes (mixture of 1,2-, 1,3-, and 1,4-), 7.0 ug/day (estimated from 20 m^3 air/day), equivalent to 2.5 mg/year [18]. FOOD INTAKE: Insufficient data.

Occupational Exposures: NIOSH (NOHS Survey 1972-1974) has statistically estimated that 697,803 workers are exposed to 1,2-DCB in the United States [41]. A U.S. EPA report has estimated that 10,000 workers are potentially exposed during production, processing, and industrial solvent use and 2 million workers are potentially exposed for all occupational activities [25]. It has been estimated that 200 workers may be exposed to 1,2-DCB fumes in transmission shops from its use as a cleaner [25]. 1,2-DCB levels up to 8.5 ppm (51 mg/m^3) were detected in the air of a chlorobenzene factory [25]. Indoor air in mobile homes (Texas), mean 0.1 ppb, range 0.006-0.21 ppb, 5% of homes [13]. NIOSH (NOES Survey 1981-1983) has statistically estimated that 41,459 workers are exposed to 1,2-DCB in the United States [40].

Body Burdens: Mean 1,2-DCB of 13 ug/kg (fat basis) detected in Yugoslavian human adipose tissue and 9 ug/kg (as-is basis) or 230 ug/kg (fat basis) in human milk [26]. Unidentified dichlorobenzene isomers were found in human blood samples at levels of 0.76-68 ng/mL taken from residents at the Love Canal [3]. Incidence of detection in the National Human Adipose Tissue Survey fiscal year 1982 composite specimens, 63% pos, not detected to 2 ppb (wet tissue concn) [59].

1,2-Dichlorobenzene

REFERENCES

1. Atkinson R; Chem Rev 85: 69 (1985)
2. Bailey RE; Environ Sci Technol 17: 504 (1983)
3. Barkley J et al; Biomed Mass Spect 7: 139 (1980)
4. Barrows ME et al; pp 379-92 in Dyn Exposure Hazard Assess Toxic Chem Ann Arbor, MI Ann Arbor Sci (1980)
5. Bouwer EJ, McCarty PL; Ground Water 22: 433 (1984)
6. Bouwer EJ; Environ Prog 4: 43 (1985)
7. Bozzilli JW, Kebbekus BB; J Environ Sci Health 17: 693 (1982)
8. Brodzinsky R, Singh HB; Volatile Org Chem in the Atmosphere: An Assess of Available Data Menlo Park, CA Atmospheric Sci Center, SRI Internatl pp 198 (1982)
9. Callahan MA et al; Water-Related Environ Fate of 129 Priority Pollut vol. II USEPA-440/4-79-029B (1979)
10. Canton JH et al; Regulatory Toxicol Pharmacol 5: 123-31 (1985)
11. Chiou CT et al; Sci 206: 831 (1979)
12. Chiou CT et al; Environ Sci Technol 17: 227 (1983)
13. Connor TH et al; Toxicol Lett 25:33-40 (1985)
14. Davis EM et al; Water Res 15: 1125 (1981)
15. Elder VA et al; Environ Sci Technol 15: 1237 (1981)
16. Fielding M et al; Organic Micropollut in Drinking Water Medmenham, Eng Water Res Cent TR-159 (1981)
17. Granstrom ML et al; Water Sci Technol 16: 375 (1984)
18. Guichert R, Schulting; Sci Total Environ 43: 193-219 (1985)
19. Haider K et al; Arch Microbiol 9: 183 (1974)
20. Hansch C, Leo AJ; Medchem Project Issue NO. 26 Claremont, CA: Pomona College (1985)
21. Harkov R et al; J Environ Sci Health 20: 491 (1985)
22. Harkov R et al; J Air Pollut Control Assoc 33: 1177 (1983)
23. Hauser TR, Bromberg SM; Environ Monit Assess 2: 249 (1982)
24. Hochreiter JJ; Chem-Quality Reconnaissance of the Water and Surficial Bed Material in the Delaware River Estuary and Adjacent NJ Tributaries 1980-81 pp 82-36 USGS/WRI/NTIS 82-36 (1982)
25. IARC; Some Industrial Chemicals and Dyestuffs 29: 214 (1982)
26. Jan J; Bull Environ Contam Toxicol 30: 595 (1983)
27. Jan J; Mitt Geb Lebensmittelunters Hyg 74: 420 (1983)
28. Jan J, Malnersic S; Bull Environ Contam Toxicol 24: 824 (1980)
29. Junk GA et al; ACS Symp Ser 319: 109-23 (1986)
30. Kincannon DF et al; J Water Pollut Control Fed 55: 157 (1983)
31. Kitano M; Biodegradation and Bioaccumulation Test on Chem Sub OECD Tokyo Meeting Reference Book TSU-No.3 (1978)
32. Kuhn EP et al; Environ Sci Technol 19: 961 (1985)
33. LaRegina J et al; Environ Progress 5: 18-27 (1986)
34. Lewis TE et al; p 7 in New Persp Env Tox Chem Soc Environ Tox Chem Abstr 6th Meet (1985)
35. Ligocki MP et al; Atmos Environ 19: 1609 (1985)
36. Lioy PJ et al; pp. 3-43 in Toxic Air Pollution; Lioy PJ, Daisey JM, eds (1987)

1,2-Dichlorobenzene

37. Lyman WJ et al; Handbook of Chemical Property Estimation Methods Environ Behavior of Organic Compounds McGraw-Hill NY (1982)
38. Mackay D, Shiu WY; J Phys Ref Data 10: 1193 (1981)
39. Malle KG; Z Wasser-Abwasser Forsch 17: 75 (1984)
40. NIOSH; The National Occupational Exposure Survey (NOES) (1983)
41. NIOSH; The National Occupational Hazard Survey (NOHS) (1974)
42. Oliver BG, Nicol KD; Environ Sci Technol 16: 532 (1982)
43. Oliver BG, Niimi AJ; Environ Sci Technol 17: 287 (1983)
44. Oliver BG, Nicol KD; Intern J Environ Anal Chem 25: 275-85 (1986)
45. Oliver BG, Charlton MN; Environ Sci Technol 18: 903 (1984)
46. Oliver BG; Symp Amer Chem Soc, Div Environ Chem 186th Natl Mtg 23: 421 (1983)
47. Oliver BG, Nicol KD; Sci Tot Env 39: 57 (1984)
48. Oliver BG, Nicol D; Environ Sci Technol 17: 505 (1983)
49. Oliver BG; Chemosphere 14: 1087 (1985)
50. Oliver BG; Symp Amer Chem Soc, Div Environ Chem 186th Natl Mtg Preprint (1983)
51. Page GW; Environ Sci Technol 15: 1475 (1981)
52. Pankow JF et al; Environ Sci Technol 18: 310 (1984)
53. Reinhard M et al; Environ Sci Technol 18: 953 (1984)
54. Riddick JA et al; Organic Solvents: Physical Properties and Methods of Purification, 4th Edit. New York: J Wiley & Sons (1986)
55. Sadtler 303 UV (1988)
56. Sanjivamurthy VA; Water Res 12: 31 (1978)
57. Singh HB et al; Atmos Environ 15: 601 (1981)
58. Sontheimer H et al; Sci Tot Env 47: 27 (1985)
59. Stanley JS; Broad Scan Analysis of the FY82 National Human Adipose Tissue Survey Specimens. Vol 1. Executive Summary. p.1 USEPA-560/5-86-036 (1986)
60. Staples CA et al; Environ Toxicol Chem 4: 131 (1985)
61. Tabak HH et al; J Water Pollut Control Fed 53: 1503 (1981)
62. Tang J et al; J Agric Food Chem 31: 1287-92 (1983)
63. Tiernan TO et al; Environ Health Pers 59: 145 (1985)
64. USEPA; Preliminary Assessment of Suspected Carcinogens in Drinking Water An Interim Report to Congress (1975)
65. vandeMeent D et al; Wat Sci Tech 18: 73-81 (1986)
66. Wakeham SG et al; Can J Fish Aquat Sci 40: 304 (1983)
67. Westrick JJ et al; J Amer Water Works Assoc 76: 52 (1984)
68. Wilson JT, McNabb JF; EOS Transactions, Amer Geophys Union 64: 505-7 (1983)
69. Young DR et al; Water Chlorination Environ Impact Health Eff 3: 471-86 (1980)
70. Zoeteman BCJ et al; Chemosphere 9: 231 (1980)

1,3-Dichlorobenzene

SUBSTANCE IDENTIFICATION

Synonyms: m-Dichlorobenzene

Structure:

CAS Registry Number: 541-73-1

Molecular Formula: $C_6H_4Cl_2$

Wiswesser Line Notation: GR CG

CHEMICAL AND PHYSICAL PROPERTIES

Boiling Point: 173 °C

Melting Point: -24.7 °C

Molecular Weight: 147.01

Dissociation Constants:

Log Octanol/Water Partition Coefficient: 3.60 [19]

Water Solubility: 111 mg/L at 20 °C [44]

Vapor Pressure: 2.3 mm Hg at 25 °C [31]

Henry's Law Constant: 0.0018 atm-m³/mole at 20 °C [33]

1,3-Dichlorobenzene

ENVIRONMENTAL FATE/EXPOSURE POTENTIAL

Summary: Chemical waste dump leachates and direct manufacturing effluents are reported to be the major source of pollution of the chlorobenzenes (including the dichlorobenzenes) to Lake Ontario. Use of 1,3-dichlorobenzene (1,3-DCB) as a fumigant will release it directly to the atmosphere. If released to soil, 1,3-DCB can be moderately to tightly adsorbed. Leaching from hazardous waste disposal areas has occurred and the detection of 1,3-DCB in various ground waters indicates that leaching can occur. Volatilization from soil surfaces may be an important transport mechanism. It is possible that 1,3-DCB will be slowly biodegraded in soil under aerobic conditions. Chemical transformation by hydrolysis, oxidation, or direct photolysis are not expected to occur in soil. If released to water, adsorption to sediment will be a major environmental fate process based upon extensive monitoring data in the Great Lakes area and Koc values. Analysis of Lake Ontario sediment cores has indicated the presence and persistence of 1,3-DCB since before 1940. 1,3-DCB is volatile from the water column, with an estimated half-life of 4.1 hours from a river one meter deep flowing 1 m/sec with a wind velocity of 3 m/sec at 20 °C; adsorption to sediment will attenuate volatilization. Aerobic biodegradation in water may be possible; however, anaerobic biodegradation is not expected to occur. Experimental BCF values of 89-740 have been reported and 1,3-DCB has been detected in trout from Lake Ontario. Hydrolysis, oxidation, and direct photolysis in aquatic environment are not expected to be important. If released to air, 1,3-DCB will exist predominantly in the vapor phase and will react with photochemically produced hydroxyl radicals at an estimated half-life rate of 14 days in a typical atmosphere. Direct photolysis in the troposphere is not expected to be important. The detection of 1,3-DCB in rainwater suggests that atmospheric removal via wash-out is possible. General population exposure to 1,3-DCB may occur through oral consumption of contaminated drinking water and food (particularly fish) and through inhalation of contaminated air since 1,3-DCB has been widely detected in ambient air.

Natural Sources: Dichlorobenzenes are not known to occur in nature [24].

1,3-Dichlorobenzene

Artificial Sources: For the chlorobenzenes in general, the major source of pollution to Lake Ontario has been reported to be chemical waste dump leachates and direct chemical manufacturing effluents located around Niagara Falls, NY [36]. Use of 1,3-DCB as a fumigant will release it directly to the atmosphere; use as a chemical intermediate or solvent are also sources of environmental release [21].

Terrestrial Fate: Based on experimental adsorption data, 1,3-DCB can be moderately to tightly adsorbed in soil. Leaching from hazardous waste disposal areas in Niagara Falls to adjacent surface waters has been reported [13] and the detection of 1,3-DCB in various ground waters indicates that leaching can occur. Volatilization from soil surfaces may be an important transport mechanism; however, volatilization may be attenuated by tight adsorption or leaching. It is possible that 1,3-DCB will be slowly biodegraded in soil under aerobic conditions. Chemical transformation processes such as hydrolysis, oxidation, or direct photolysis (on soil surfaces) are not expected to occur.

Aquatic Fate: Based on Koc values of 12,600-31,600 calculated from sediment/water monitoring data and extensive sediment monitoring data in the Great Lakes area, adsorption to sediment is a major environmental fate process for 1,3-DCB. Its detection in Lake Ontario sediment cores indicates that 1,3-DCB has persisted in these sediments since before 1940 [36]. 1,3-DCB is volatile from water, with an estimated half-life of 4.1 hours from a river one meter deep flowing 1 m/sec with a wind velocity of 3 m/sec at 20 °C; adsorption to sediment in water will attenuate volatilization. 1,3-DCB may biodegrade in aerobic water after microbial adaptation; however, it is not expected to biodegrade under anaerobic conditions which may exist in lake sediments or various ground waters. Experimental BCF values of 89-740 have been reported; the detection of 1,3-DCB in trout from Lake Ontario has confirmed this level of bioaccumulation. Hydrolysis, oxidation, and direct photolysis in aquatic environment are not expected to be important .

Atmospheric Fate: 1,3-Dichlorobenzene will exist predominantly in the vapor phase in the atmosphere. The half-life for the vapor-phase reaction of 1,3-dichlorobenzene with photochemically produced hydroxyl radicals in the atmosphere has been estimated to be 14 days. Direct photolysis is not expected to be important. The detection of 1,3-dichlorobenzene in rainwater suggests that atmospheric removal via wash-out is possible.

Biodegradation: Using a static-culture flask-screening procedure (5 or 10 mg/L test compound, a 7-day static incubation followed by three weekly subcultures and a settled domestic wastewater as microbial inoculum), 1,3-DCB was biodegraded 58-59%, 67-69%, 31-39%, and 33-35% after the original culture, first, second, and third subculture, respectively [50]. In a continuous-flow activated sludge system, nearly 100% of influent 1,3-DCB was removed by an apparent combination of biodegradation and stripping [26]. In a laboratory aquifer column simulating saturated-flow conditions typical of a river/ground water infiltration system, the dichlorobenzenes were biotransformed under aerobic conditions but not under anaerobic conditions [28]. 1,3-DCB was significantly removed in an aerobic biofilm column (9.8 ppb concn) but was not removed in a methanogenic biofilm column (10 ppb) [5,6]. The dichlorobenzenes are not expected to be biotransformed in anaerobic water conditions found in aquifers [56]. An analysis of monitoring data from sediment cores indicated insufficient evidence to show occurrence of anaerobic dehalogenation of the chlorobenzenes in Lake Ontario sediment [37]. The dichlorobenzenes were slowly biodegraded (6.3% of theoretical CO_2 evolution in 10 weeks) in an alkaline soil sample [18]. Microbial decomposition of 1,3-DCB by a Pseudomonas sp. or by a mixed culture of soil bacteria yielded dichlorophenols as products [2]. It has been suggested that the three dichlorobenzene isomers may undergo slow biodegradation in natural water [8]. Not significantly degraded (half-life of much greater than 4 weeks) in the repetitive die-away test using inoculum which was not adapted [9].

Abiotic Degradation: 1,3-DCB is not expected to undergo significant hydrolysis in environmental waters [8]. It is reported to be resistant towards oxidation by peroxy radicals in aquatic media [8]. In a methanol solvent, 1,3-DCB absorbs virtually no irradiation above 290 nm [45]; therefore, direct photolysis in the environment should not be significant. The experimentally determined rate constant for the vapor phase reaction between 1,3-DCB and photochemically produced hydroxyl radicals has been reported to be 0.72 x 10^{-12} cm³/molecule-sec at 22 °C [1]; assuming an average atmospheric hydroxyl radical concn of 8 x 10^{+5} molecules/cm³, the half-life for the reaction is estimated to be 14 days.

Bioconcentration: Mean 1,3-DCB BCF values of 420-740 were experimentally determined for rainbow trout exposed up to 119 days in laboratory aquariums [40]. A whole body BCF of 89 was determined for bluegill sunfish exposed to 1,3-DCB over a 28-day period in a continuous flow system [4]. Using the water solubility and the log Kow, BCF values of 39 and 320 are estimated, respectively, from recommended equations [30]. BCF of 97 was experimentally determined for fathead minnows under flowing water conditions [10].

Soil Adsorption/Mobility: An experimental Koc value of 293 (Kom = 170) was determined for 1,3-DCB in a silt loam soil containing 1.9% organic matter [11]. Koc values of 31,600 and 12,600 were determined for suspended sediment-water samples collected from the Niagara River during 1982 and 1981 [35]. In equilibrium batch studies, a relatively strong adsorption of 1,2-dichlorobenzene to collected aquifer material was observed [29]. Using the water solubility and the log Kow, Koc values of 296 and 2450 are estimated, respectively, from regression-derived equations [30].

Volatilization from Water/Soil: The Henry's Law constant for 1,3-DCB indicates that volatilization will be significant from all waters and potentially rapid [30]. The volatilization half-life from a river one meter deep flowing 1 m/sec with a wind velocity of 3 m/sec is estimated to be 4.1 hours at 20 °C [30]. Based on the vapor pressure, 1,3-DCB is expected to evaporate at a relatively significant rate from dry surfaces.

Water Concentrations: DRINKING WATER: A mean 1,3-DCB concn of 0.001 ppb was detected in drinking water samples from 3 cities near Lake Ontario in 1980 [36]. Concn of 0.5 ppb identified in Miami, FL drinking water and qualitative detections were reported for Philadelphia, PA and Cincinnati, OH [52]. 1,3-DCB was not detected in 945 finished water supplies throughout the U.S. that use ground water sources [55]. Qualitative detection was reported for Cleveland, OH tap water [46]. Qualitative detection was reported for two drinking water supply sources in the United Kingdom [15]. In an analysis of 30 potable Canadian water sources, 1,3-DCB was detected at an average concn below 1 ppb [41]. GROUND WATER: 1,3-DCB was detected in 19 of 685 ground waters analyzed in NJ during 1977-1979 with 236.8 ppb the highest concn found [42]. 1,3-DCB was identified in ground waters at locations using rapid infiltration for wastewater treatment near Ft. Devens, MA, Boulder, CO, and Phoenix, AZ at concn of 0.05-0.56 ppb [23]. SURFACE WATERS: 1,3-DCB was detected in 19 of 463 surface waters analyzed in NJ during 1977-1979 with 241.5 ppb the highest concn found [42]. Mean concn of 1 ppt detected in the Grand River during 1980 near Niagara Falls; concn of 0-18 ppt found in the Niagara River [36]. Concn of 2.1-110 ppt (mean concn of 11 ppt) detected in Niagara River at Niagara-On-The-Lake between 1981 and 1983 [39]. Concn of 2.4-85 ppt (mean concn of 7.8 ppt) detected in the Niagara River between 1981 and 1983 [34]. An average concn of 15 ppt was found in the Niagara River near Niagara-On-The-Lake between Sept. and Oct. 1982 [35]. Positive detection of 1,3-DCB was reported by 0.3% of 986 USEPA STORET stations [49]. Qualitative detection reported for the Delaware and Raritan Canal in NJ [16]. Concn below 0.5 ppb detected in the Rhine River between 1978-1982 [32]. An average concn of 0.05 ppb found in the Rhine River near Dusseldorf [48]. Mean concn (ppt) from single sampling cruises: Lake Ontario, 1.7; Niagara River, 5.7; and Detroit River, 7.1 [38]. SEAWATER: 1,3-DCB was detected in the water column of the Narragansett Bay near Rhode Island [54]. Concn (ppt): Southern North Sea, 108 samples in 9 locations, 2-31, 9 avg, 7 median; Rhine River, 1 sample, 150 [53]. RAIN/SNOW: A mean 1,3-DCB concn of 0.002 ppt was detected in Portland, OR rainwater during March-April 1982 [43].

Effluent Concentrations: The wastewater effluents from four water treatment plants discharging into the Grand River and Lake Ontario contained a mean 1,3-DCB concn of 14 ppt during 1980 sampling [36]. Positive detection of 1,3-DCB was reported by 1.5% of 1301 USEPA STORET stations [49]. Dichlorobenzene isomers were detected in wastewater effluents from 7 treatment facilities in Illinois [14]. The dichlorobenzene isomers were qualitatively detected in waters adjacent to hazardous waste disposal areas in Niagara Falls, NY as a result of leaching [13]. Flue gas effluents from a municipal refuse-fired steam boiler in Virginia contained 4.4 ug/m^3 of the dichlorobenzene isomers [51]; flue gas effluents from a refuse-derived-fired power plant in Ohio contained 7.8 ng/m^3 of the dichlorobenzene isomers [51].

Sediment/Soil Concentrations: Mean 1,3-DCB concentrations of 2, 2, 4, and 74 ppb were detected in the superficial sediments from Lake Superior, Huron, Erie, and Ontario, respectively [36]. Concentrations of 4 to 330 ppb found in a Lake Ontario sediment core (0 to 8 cm in depth) with the higher concn in the upper 2 cm [36]. Mean concn of 13 ng/g detected on settling particulates at a station on Lake Ontario [34]. Concn ranging from 16 to 19 ng/g detected in suspended sediment of Lake Ontario at depths from 20 to 68 meters with an average concn of 15 ng/g found on bottom sediments [35]. Qualitative detection reported for both soil and sediment collected near the Love Canal [20]. Trace concn were detected in surficial bed material of the Racoon Creek in Bridgeport, NJ [22].

Atmospheric Concentrations: The mean 1,3-DCB concentrations from 138 source-dominant points, 652 urban/suburban points and 2 rural/remote points from the U.S. have been reported to be 150, 83, and 6.7 ppt, respectively, with an overall mean concn of 95 ppt [7]. Mean concn of 7.7, 8.7, and 6.5 ppt detected in the ambient air of Los Angeles, CA, Phoenix, AZ, and Oakland, CA, respectively, during Apr-May 1979 [47]. Concn of 0.15-0.82 ug/m^3 detected in the ambient air of Bound Brook, NJ [27].

Food Survey Values:

Plant Concentrations:

Fish/Seafood Concentrations: 1,3-DCB concn of 0.6, 0.6, 0.3, and 2-3 ppb were detected in trout taken from Lake Superior, Lake Huron, Lake Erie, and Lake Ontario, respectively, during 1980 [36].

Animal Concentrations:

Milk Concentrations:

1,3-Dichlorobenzene

Other Environmental Concentrations:

Probable Routes of Human Exposure: General population exposure to 1,3-DCB may occur through oral consumption of contaminated drinking water and food (particularly fish), especially in the vicinity of effluent discharges such as on the Lake Ontario. General population exposure may also occur through inhalation of contaminated ambient air, since 1,3-DCB has been detected in many areas of the U.S.. Occupational exposure probably occurs during the manufacture of dichlorobenzenes and the uses of 1,3-DCB as a fumigant, chemical intermediate and solvent; probable routes of occupational exposure are inhalation of contaminated air and dermal contact.

Average Daily Intake: WATER INTAKE: Insufficient data. AIR INTAKE: Based on monitoring data at three U.S. urban sites (Los Angeles, Phoenix, Oakland), the average daily intake for 1,3-DCB has been estimated to be 0.8-1.1 ug/day [47]. Netherlands: Dichlorobenzenes (mixture of 1,2-, 1,3-, and 1,4-), 7.0 ug/day (estimated from 20 m^3/day), equivalent to 2.5 mg/year [17]. FOOD INTAKE: Insufficient data.

Occupational Exposures: Indoor air in mobile homes (Texas), mean 0.27 ppb, range 0.025-1.4 ppb, 16% of homes [12].

Body Burdens: Trace concn (less than 5 ug/kg) of 1,3-DCB were detected in Yugoslavian human milk [25]. Unidentified dichlorobenzene isomers were found in human blood samples at levels of 0.76-68 ng/mL taken from residents at the Love Canal [3].

REFERENCES

1.   Atkinson R; Chem Rev 85: 69 (1985)
2.   Ballschmiter K et al; Angew Chem Int Ed English 16: 645 (1977)
3.   Barkley J et al; Biomed Mass Spect 7: 139 (1980)
4.   Barrows ME et al; pp 379-92 in Dyn Exposure Hazard Assess Toxic Chem Ann Arbor, MI Ann Arbor Sci (1980)
5.   Bouwer EJ; Environ Prog 4: 43 (1985)
6.   Bouwer EJ, McCarty PL; Ground Water 22: 433 (1984)

1,3-Dichlorobenzene

7. Brodzinsky R, Singh HB; Volatile Org Chem in the Atmosphere An Assess of Available Data Menlo Park, CA Atmospheric Sci Cntr SRI International pp 198 (1982)
8. Callahan MA et al; Water-Related Environ Fate of 129 Priority Pollut vol.II USEPA-440/4-79-029B (1979)
9. Canton JH et al; Regulatory Toxicol Pharmacol 5: 123-31 (1985)
10. Carlson AR, Kosian PA; Arch Environ Contam Toxicol 16: 129-35 (1987)
11. Chiou CT et al; Environ Sci Technol 17: 227 (1983)
12. Connor TH et al; Toxicol Lett 25: 33-40 (1985)
13. Elder VA et al; Environ Sci Technol 15: 1237 (1981)
14. Ellis DD et al; Arch Environ Contam Toxicol 11: 373 (1982)
15. Fielding M et al; Organic Micropollutants in Drinking Water Medmenham, Eng Water Res Cent TR-159 (1981)
16. Granstrom ML et al; Water Sci Tech 16: 375 (1984)
17. Guicherit R, Schulting FL; Sci Total Environ 43: 193-219 (1985)
18. Haider K et al; Arch Microbiol 96: 183 (1974)
19. Hansch C, Leo AJ; Medchem Project Issue No.26 Claremont, CA: Pomona College (1985)
20. Hauser TR, Bromberg SM; Environ Monit Assess 2: 249 (1982)
21. Hawley GG; The Condensed Chemical Dictionary. New York: Van Nostrand Reinbold Co (1981)
22. Hochreiter JJ Jr; Chem-Quality Reconnaissance of the Water and Surficial Bed Material in the Delaware River Estuary and Adjacent NJ Tributaries USGS/WRI/NTIS 82-36 (1981)
23. Hutchins SR et al; Environ Toxicol Chem 2: 195 (1983)
24. IARC; Some Industrial Chemicals and Dyestuffs 29: 213 (1982)
25. Jan J; Bull Environ Contam Toxicol 30: 595 (1983)
26. Kincannon DF et al; J Water Pollut Control Fed 55: 157 (1983)
27. Krost KJ et al; Anal Chem 54: 810 (1982)
28. Kuhn EP et al; Environ Sci Technol 19: 961 (1985)
29. Lewis TE et al; pp 7 in New Persp Env Tox Chem Soc Environ Tox Chem Abstr 6th meet (1985)
30. Lyman WJ et al; Handbook of Chem Property Estimation Methods Environ Behavior of Organic Compounds McGraw-Hill NY (1982)
31. Mackay D, Shiu WY; J Phys Ref Data 10: 1193 (1981)
32. Malle KG; Z Wasser-Abwasser Forsch 17: 75 (1984)
33. Oliver BG; Chemosphere 14: 1087 (1985)
34. Oliver BG; Symp Amer Chem Soc Div Environ Chem 186th Natl Mtg 23: 421 (1983)
35. Oliver BG, Charlton MN; Environ Sci Technol 18: 903 (1984)
36. Oliver BG, Nicol KD; Environ Sci Technol 16: 532 (1982)
37. Oliver BG, Nicol KD; Environ Sci Technol 17: 505 (1983)
38. Oliver BG, Nicol KD; Intern J Environ Anal Chem 25: 275-85 (1986)
39. Oliver BG, Nicol KD; Sci Tot Environ 39: 57 (1984)
40. Oliver BG, Niimi AJ; Environ Sci Technol 17: 287 (1983)
41. Otson R et al; J Assoc Off Analyt Chem 65: 1370 (1982)
42. Page GW; Environ Sci Technol 15: 1475 (1981)
43. Pankow JF et al; Environ Sci Technol 18: 310 (1984)
44. Riddick JA et al; Organic Solvents: Physical Properties and Methods of Purification, 4th Edit. New York: J Wiley & Sons (1986)

1,3-Dichlorobenzene

45. Sadtler 1671 UV (1988)
46. Sanjivamurthy VA; Water Res 12: 31 (1978)
47. Singh HB et al; Atmos Environ 15: 601 (1981)
48. Sontheimer H et al; Sci Tot Environ 47: 27 (1985)
49. Staples CA et al; Environ Toxicol Chem 4: 131 (1985)
50. Tabak HH et al; J Water Pollut Control Fed 53: 1503 (1981)
51. Tiernan TO et al; Environ Health Pers 59: 145 (1985)
52. USEPA; Preliminary Assessment of Suspected Carcinogens in Drinking Water An Interim Report to Congress (1975)
53. vandeMeent D et al; Wat Sci Tech 18: 73-81 (1986)
54. Wakeham JG et al; Can J Fish Aquat Sci 40: 304 (1983)
55. Westrick JJ et al; J Amer Water Works Assoc 76: 52 (1984)
56. Wilson JT, McNabb JF; Bio Transformation of Org Pollut in Groundwater EOS Transactions, Amer Geophys Union 64: 505-7 (1983)

1,4-Dichlorobenzene

SUBSTANCE IDENTIFICATION

Synonyms: p-Dichlorobenzene

Structure:

CAS Registry Number: 106-46-7

Molecular Formula: $C_6H_4Cl_2$

Wiswesser Line Notation: GR DG

CHEMICAL AND PHYSICAL PROPERTIES

Boiling Point: 174 °C at 760 mm Hg

Melting Point: 53.1 °C

Molecular Weight: 147.01

Dissociation Constants:

Log Octanol/Water Partition Coefficient: 3.52 [24]

Water Solubility: 87 mg/L at 25 °C [66]; 74 mg/l at 25 °C [4]

Vapor Pressure: 1.76 mm Hg at 25 °C [66]

Henry's Law Constant: 0.0015 atm-m³/mole at 20 °C [53]

1,4-Dichlorobenzene

ENVIRONMENTAL FATE/EXPOSURE POTENTIAL

Summary: Chemical waste dump leachates and direct manufacturing effluents are reported to be the major source of pollution of the chlorobenzenes (including the dichlorobenzenes) to Lake Ontario. The major source of 1,4-dichlorobenzene (1,4-DCB) emission to the atmosphere is volatilization from use in toilet bowl deodorants, garbage deodorants, and moth flakes. If released to soil, 1,4-DCB can be moderately to tightly adsorbed. Leaching from hazardous waste disposal areas has occurred and the detection of 1,4-DCB in various ground waters indicates that leaching can occur. Volatilization from soil surfaces may be an important transport mechanism. It is possible that 1,4-DCB will be slowly biodegraded in soil under aerobic conditions. Chemical transformation by hydrolysis, oxidation, or direct photolysis are not expected to occur in soil. If released to water, volatilization may be the dominant removal process. The volatilization half-life from a river one meter deep flowing one meter/sec with a wind velocity of 3 m/sec is estimated to be 4.3 hours at 20 °C. Adsorption to sediment will be a major environmental fate process based upon extensive monitoring data in the Great Lakes area and Koc values based upon monitoring samples. Analysis of Lake Ontario sediment cores has indicated the presence and persistence of 1,4-DCB since before 1940. Adsorption to sediment will attenuate volatilization. Aerobic biodegradation in water may be possible; however, anaerobic biodegradation is not expected to occur. For the most part, experimental BCF values reported in the literature are less than 1000, which suggests that significant bioconcentration will not occur; however, a BCF of 1800 was determined for guppies in one study. Aquatic hydrolysis, oxidation, and direct photolysis are not expected to be important. If released to air, 1,4-DCB will exist predominantly in the vapor phase and will react with photochemically produced hydroxyl radicals at an estimated half-life rate of 31 days in a typical atmosphere. Direct photolysis in the troposphere is not expected to be important. The detection of 1,4-DCB in rain water suggests that atmospheric removal via wash-out is possible. General population exposure to 1,4-DCB may occur through oral consumption of contaminated drinking water and food (particularly fish) and through inhalation of contaminated air.

1,4-Dichlorobenzene

Natural Sources: Dichlorobenzenes are not known to occur as such in nature [31].

Artificial Sources: For the chlorobenzenes in general, the major source of pollution to Lake Ontario has been reported to be chemical waste dump leachates and direct chemical manufacturing effluents located around Niagara Falls, NY [57]. In 1972, 70-90% of the annual U.S. production of 1,4-DCB was estimated to have been released into the atmosphere primarily as a result of use in toilet bowl and garbage deodorants and use in moth control as a fumigant [31]. In 1984, it was reported that 67% of the 1,4-DCB consumed in the United States is used for space deodorants and moth control, with 33% used as an intermediate for polyphenylene sulfide resin production [16]; volatilization from the deodorants and moth flakes will therefore be the major emission source to the atmosphere.

Terrestrial Fate: Based on experimental adsorption data, 1,4-DCB can be moderately to tightly adsorbed in soil. Leaching from hazardous waste disposal areas in Niagara Falls to adjacent surface waters has been reported [19] and the detection of 1,4-DCB in various ground waters indicates that leaching can occur. Volatilization from soil surfaces may be an important transport mechanism; however, volatilization may be attenuated by tight adsorption or leaching. It is possible that 1,4-DCB will be slowly biodegraded in soil under aerobic conditions. Chemical transformation processes such as hydrolysis, oxidation, or direct photolysis (on soil surfaces) are not expected to occur.

Aquatic Fate: 1,4-DCB is volatile from water, with an estimated half-life of 4.3 hours from a model river one meter deep flowing 1 m/sec with a wind velocity of 3 m/sec at 20 °C. Studies carried out in an experimental marine ecosystem found volatilization to be the major removal process for 1,4-DCB [81]. Volatilization was also found to be the predominant elimination mechanism of 1,4-DCB from Lake Zurich in Switzerland based on one-year monitoring studies and laboratory studies [70]. Monitoring data based Koc values of 63,000-100,000 and extensive sediment monitoring data in the Great Lakes area indicate that adsorption to sediment is a major environmental fate process for 1,4-DCB. Its detection in Lake Ontario sediment cores indicates that 1,4-DCB

1,4-Dichlorobenzene

has persisted in these sediments since before 1940 [57]. Adsorption to sediment in water will attenuate volatilization. 1,4-DCB may biodegrade in aerobic water after microbial adaptation; however, it is not expected to biodegrade under anaerobic conditions which may exist in lake sediments or various ground waters. Experimental BCF values reported in the literature are less than 1000, for the most part, which suggests that significant bioconcentration will not occur; however, a BCF of 1800 was determined for guppies during one study. Aquatic hydrolysis, oxidation, and direct photolysis are not expected to be important.

Atmospheric Fate: 1,4-DCB will exist predominantly in the vapor-phase in the atmosphere. The half-life for the vapor phase reaction of 1,4-DCB with photochemically produced hydroxyl radicals in the atmosphere has been estimated to be 31 days. Direct photolysis is not expected to be important. The detection of 1,4-DCB in rainwater suggests that atmospheric removal via wash-out is possible.

Biodegradation: Using a static-culture flask-screening procedure (5 or 10 mg/L test compound, a 7-day static incubation followed by three weekly subcultures and a settled domestic wastewater as microbial inoculum), 1,4-DCB was biodegraded 37-55%, 54-61%, 29-34%, and 0-16% after the original culture, and first, second, and third subculture, respectively [77]. 1,4-DCB was found to be degradation resistant using the Japanese MITI test [37]. In a laboratory aquifer column simulating saturated-flow conditions typical of a river/ground water infiltration system, the dichlorobenzenes were biotransformed under aerobic conditions but not under anaerobic conditions [39]. 1,4-DCB was significantly removed in anaerobic biofilm column (10.8 ppb concn) but was not removed in a methanogenic biofilm column (10 ppb) [8,9]. Using the Sapromat respirometer test (similiar to the MITI test except incubation period is two weeks longer) and concn of 8 and 40 mg/L 1,4-DCB, degradation was found to be slow (as it started after 14 days) and dependent on concn with degradation occurring only at lower concn [12]. At concn of 1000 mg/L, 1,4-DCB was not degraded nor was it toxic to benzene-acclimated activated sludge during 192-hr Warburg respirometer tests [45]. The dichlorobenzenes are not expected to be biotransformed in anaerobic water conditions found in aquifers [85]. 1,4-DCB was

253

reported to have BOD's of 65, 77 and 77% of the theoretical BOD, over 5-, 10-, and 20-day periods, respectively [2]. An analysis of monitoring data from sediment cores indicated insufficient evidence to show occurrence of anaerobic dehalogenation of the chlorobenzenes in Lake Ontario sediment [58]. The dichlorobenzenes were slowly biodegraded (6.3% of theoretical CO_2 evolution on 10 weeks) in an alkaline soil sample [23]. Microbial decomposition of 1,4-DCB by a <u>Pseudomonas</u> sp. or by a mixed culture of soil bacteria yielded dichlorophenol as a product [3]. It has been suggested that the three dichlorobenzene isomers may undergo slow biodegradation in natural water [13]. Half-life of 2-3 weeks in the repetitive die-away test using inoculum which was not adapted [14].

Abiotic Degradation: 1,4-DCB is not expected to undergo significant hydrolysis in environmental waters [13]. It is reported to be resistant towards oxidation by peroxy radicals in aquatic media [13]. 1,4-DCB absorbs almost no UV radiation above 300 nm in either a methanol or hexane solvent [35,68]. No significant dechlorination was detected when 1,4-DCB in hexane, methanol, or methanol-water was irradiated for 50 hours with light above 290 nm [35]; direct photolysis is not expected to be important in the environmental disappearance of 1,4-DCB [35]. Although the possibility of sensitized photolysis has been suggested [35], photolysis experiments from rivers flowing into the Great Lakes were found to have no accelerating affect on photodegradation [54]. The experimentally determined rate constant for the vapor-phase reaction between 1,4-DCB and photochemically produced hydroxyl radicals has been reported to be 0.32×10^{-12} cm^3/molecule-sec at 22 °C [1]; assuming an average atmospheric hydroxyl radical concn of $8 \times 10^{+5}$ molecules/cm^3, the half-life for the reaction is estimated to be 31 days. Smog chamber tests have found dichloronitrophenol, dichloronitrobenzene, and dichlorophenol as products of 1,4-DCB irradiation [52,74].

Bioconcentration: Mean 1,4-DCB BCF values of 370 to 720 were experimentally determined for rainbow trout exposed up to 119 days in laboratory aquariums [61]. A whole body BCF of 60 was determined for bluegill sunfish exposed to 1,4-DCB over a 28-day period in a continuous-flow system [6]. A BCF of 214 was determined for rainbow trout during a 4-day exposure under static test conditions [49]. During a 60-day exposure of 1,4-DCB to rainbow trout (from egg to alevin), BCF's were generally in the 100 to 250 range, although a BCF of 1400 was observed at the hatching stage [12]. An experimental BCF of 1800 (on a fat basis) was determined for guppies during a 19-day constant flow test [38]. Based on the water solubility and the log Kow, BCF value of 55 and 222 are estimated, respectively, from recommended equations [44]. BCF of 110 was experimentally determined for fathead minnows under flowing water conditions [15].

Soil Adsorption/Mobility: An experimental Koc value of 273 (Kom=158) was determined for 1,4-DCB in a silt loam soil containing 1.9% organic matter [17]. An experimental Koc value of about 390 was determined for a Lincoln fine sand containing 0.087 to 0.13% organic content [84]. Koc values ranging from 603 to 1833 were measured in laboratory batch and column experiments using very low organic carbon sorbents [71]. A measured Koc value of 700 was determined from adsorption tests with 5 different types of soil [30]. In equilibrium batch studies, a relatively strong adsorption of 1,4-DCB to collected aquifer material was observed [41]. Based on the water solubility and a log Kow, Koc values of 409 and 1514 are estimated, respectively, from regression-derived equations [44].

Volatilization from Water/Soil: The Henry's Law constant indicates that volatilization will be significant from all waters and potentially rapid [44]. The volatilization half-life from a river one meter deep flowing 1 m/sec with a wind velocity of 3 m/sec is estimated to be 4.3 hours at 20 °C [44]. 1,4-DCB is a white crystalline solid at ambient temperatures that is volatile (sublimes readily) [28]; therefore, volatilization from dry surfaces can be expected to occur.

Water Concentrations: DRINKING WATER: A mean 1,4-DCB concn of 0.013 ppb was detected in drinking water samples from 3 cities near Lake Ontario in 1980 [57]. Concn of 0.5 ppb was identified in Miami, FL drinking water and qualitative detections were reported for Philadelphia, PA and Cincinnati, OH [79]. 1,4-DCB was found in 9 of 945 finished water supplies throughout the United States that use ground water sources at mean concn of 0.60-0.74 ppb [83]. Qualitative detection was reported for

1,4-Dichlorobenzene

Cleveland, OH tap water in 1975 [69]. Dichlorobenzene isomers were found in drinking water in the vicinity of the Love Canal at levels of 10-800 ng/L [5]. 1,4-DCB was detected at levels below 1 ppb in an analysis of 30 potable Canadian water sources [62]. Qualitative detection reported for 14 drinking water supply sources in the United Kingdom [20].   GROUND WATER: 1,4-DCB was positively detected in 19 of 685 ground waters analyzed in NJ during 1977-1979 with 995 ppb the highest concn found [63]. 1,4-DCB was detected in ground water near Boulder, CO (0.5 ppb) and near Phoenix, AZ (0.07 ppb) at land application sites using rapid infiltration treatment of wastewaters [29]. Concn ranging from not detected to 32.6 ppb were found in ground water in Texas [7]. 1,4-DCB was detected in ground waters in the Netherlands at maximum concn of 3.0 ppb [88].   SURFACE WATERS: 1,4-DCB was detected in 26 of 463 surface waters analyzed in NJ during 1977-1979 with 30.5 ppb the highest concn found [63]. Mean concentrations of 45, 4, and 10 ppt detected in Lake Ontario, Lake Huron, and Grand River water, respectively, during 1980; concn of 1-94 ppt found in the Niagara River [57]. Concn of 9.0-310 ppt (mean concn of 36 ppt) detected in Niagara River at Niagara-On-The-Lake between 1981 and 1983 [60]. Concn of 8.7-110 ppt (mean concn of 24 ppt) detected in the Niagara River between 1981 and 1983 [55]. An average concn of 48 ppt was found in the Niagara River near Niagara-On-The-Lake between Sept and Oct 1982 [56]. Positive detection of 1,4-DCB was reported by 3.0% of 8576 USEPA STORET stations [76]. Qualitative detection reported for Delaware and Raritan Canal in NJ [21]. Concn below 0.5 ppb detected in the Rhine River between 1978-1982 [46]. An average concn of 0.19 ppb found in the Rhine River near Dusseldorf in 1984 [73]. Mean concn (ppt) form single sampling cruises: Lake Ontario, 0.07; Niagara River, 2.0; Detroit River, 0.4 [59]. SEAWATER: 1,4-DCB was detected in the water column of the Narragansett Bay near Rhode Island [82]. Concn (ppt): Southern North Sea, 108 samples in 9 locations, 5-321, 117 avg, 125 median; Rhine River, 1 sample, 300 [80]. RAIN/SNOW: A mean 1,4-DCB concn of 0.66 ppt was detected in Portland, OR rainwater during March-Apr 1982 [64]; a mean concn of 5.5 ppt detected in Portland, OR rainwater during Oct-Dec 1982 [64]. Average concn of 4.1 ppt found in Portland, OR rainwater during 1984 [42].

1,4-Dichlorobenzene

Effluent Concentrations: The wastewater effluents from four activated sludge wastewater treatment plants discharging into the Grand River and Lake Ontario contained a mean 1,4-DCB concn of 660 ppt during 1980 sampling [57]. Positive detection of 1,4-DCB was reported by 1.7% of 1306 USEPA STORET stations [76]. Concn of 0.4-100 ppb were detected in Southern California municipal wastewater effluents in 1976 [86]. Mean concn of 3.0 and 3.1 ppb were found in the final effluents of the Los Angeles County municipal wastewater treatment plant in July 1978 and Nov 1980, respectively [87]. Qualitative detection reported for chemical plant and sewage treatment plant effluents at unspecified facilities in the United States [72]. Concn of 7.7-16 ppb were detected in leachates from municipal solid wasteland fills in Minnesota [67]. Concn of 7-40 ppb found in leachate associated ground water beneath a municipal landfill in Ontario, Canada [65]. The dichlorobenzene isomers were qualitatively detected in waters adjacent to hazardous waste disposal areas in Niagara Falls, NY as a result of leaching [19]. Ambient mean air concn of 0.03-4.19 ppb were detected above six abandoned hazardous waste sites in NJ as a result of volatile emanations [25]. Flue gas effluents from a municipal refuse-fired steam boiler in Virginia contained 4.4 ug/m^3 of the dichlorobenzene isomers [78]; flue gas effluents from a refuse-derived-fired power plant in Ohio contained 7.8 ng/m^3 of the dichlorobenzene isomers [78]. Coal-fired power plants in United States, 80 ng/m^3 [36]. NJ, concn (avg all samples (0.01 ppb detection limit added for zero concn)/max concn in ppb by vol) at 5 hazardous waste sites (82 total samples), 0.24/1.21, 0.1/0.77, 0.51/4.19, 0.04/0.38, and 0.08/0.45; at one landfill site (15 samples), 0.06/0.44 [40].

Sediment/Soil Concentrations: Mean 1,4-DCB concentrations of 5, 16, 9, and 94 ppb were detected in the superficial sediments from Lakes Superior, Huron, Erie, and Ontario, respectively [57]. Concentrations of 17 to 230 ppb found in a Lake Ontario sediment core (0 to 8 cm in depth) with the higher concn in the upper 4 cm [57]. Mean concn of 49 ng/g detected on settling particulates at a station on Lake Ontario [55]. Concn ranging from 110 to 150 ng/g detected in suspended sediment of Lake Ontario at depths from 20 to 68 meters with an average concn of 63 ng/g found on bottom sediments [56]. Qualitative detection reported for both soil and sediment collected near the Love Canal [27]. 1,4-DCB was qualitatively detected in sediments off the California coast near an effluent discharge point of a wastewater treatment plant [86].

Atmospheric Concentrations: The mean 1,4-DCB concentrations from 36 source-dominant points, 392 urban/suburban points, and 2 rural/remote points from the U.S. have been reported to be 2.6, 290, and 0 ppt, respectively, with an overall mean concn of 260 ppt [11]. Average concn of 0.48 ppb were found near industrial areas in NJ and 0.04-0.10 ppb near residential areas in NJ in 1978 [10]. Mean concn of 0.04-0.07 ppb detected in the ambient air of three NJ cities during July-Aug 1981 [26]. Mean concn of 120 ng/m^3 was detected in the ambient air of Portland, OR during 1984 [42]. A geometric mean concn of 0.06 ppb was detected in the ambient air of Newark, NJ in the winter of 1982 [25]. The ambient air of central Tokyo was reported in 1975 to contain 2.7-4.2 ug/m^3 1,4-DCB and the air in suburban Tokyo contained 1.5-2.4 ug/m^3 [48]. Airborne Toxic Element and Organic Substance project, NJ, geometric mean, max concn (ppb by vol) for Summer 1981/Winter 1982: Camden, 0.04, 0.20/0.02, 0.20; Elizabeth, 0.07, 0.46/0.02, 0.26; Newark, 0.05, 0.73/0.06, 0.12 [43].

Food Survey Values: 1,4-DCB concn of 5.0 ng/g were found in market meat samples in Yugoslavia [33]. Pork has reportedly been tainted with a disagreeable odor and taste as a result of the use of 1,4-DCB in pig stalls as an odor-control agent [31].

Plant Concentrations: 1,4-DCB has been detected in the roots of wheat plants grown from lindane-treated seeds [31].

Fish/Seafood Concentrations: 1,4-DCB concn 1, 4, and 2-4 ppb were detected in trout taken from Lake Huron, Lake Erie, and Lake Ontario, respectively, during 1980 [57]. Concn of 0-67.0 ug/wet kg were found in the livers of flatfish off the California coast near Los Angeles [86]; mean concn of less than 0.0016 mg/wet kg was found in the muscle tissue of 8 seafood species caught off the California coast [86]. Fish and mussels taken from rivers in Slovenia and the Gulf of Trieste (Yugoslavia) were found to contain traces to 0.45 ug/g 1,4-DCB (on a fat basis) [34]. A species of mackerel from Japanese coastal waters was found to contain 1,4-DCB at a level of 0.05 mg/kg wet weight [48].

1,4-Dichlorobenzene

Animal Concentrations: Bovine tissue with an unusual smell was reported to contain 1,4-DCB concn of 4.4-55.9 mg/kg in muscle, 165 mg/kg in perirenal fat, 11.3 mg/kg in pancreas, 1.9 mg/kg in lung, 3.4 mg/kg in liver, and 2.8 mg/kg in spleen [31]. Samples of adipose tissue from pigeons captured in central and suburban Tokyo contained mean concn of 1.35-2.43 mg/kg [31].

Milk Concentrations: Concentrations of 5.3 ng/g 1,4-DCB were found in market milk samples in Yugoslavia [33].

Other Environmental Concentrations:

Probable Routes of Human Exposure: Occupational exposure to 1,4-DCB by inhalation and dermal routes probably occurs during its manufacture and its uses as a space deodorant, moth control agent, and chemical intermediate. General population exposure may occur through oral consumption of contaminated drinking water and food (particularly fish), especially in areas near effluent discharges, for example, certain areas in Lake Ontario. General population exposure may also occur through inhalation of contaminated ambient air, since 1,4-DCB has wide-spread monitoring detection in many areas of the United States. Inhalation exposure will occur in the near vicinity of toilet bowl and garbage deodorants and moth flakes containing 1,4-DCB.

Average Daily Intake: WATER INTAKE: insufficient data. AIR INTAKE: Assume a mean atmospheric concn of 20-200 ppt (120-1200 ng/m^3): 2.4-24 ng/day. Netherlands: Dichlorobenzenes (mixture of 1,2-, 1,3-, and 1,4-), 7.0 ug/day (estimated from 20 m^3 air/day), equivalent to 2.5 mg/year [22]. FOOD INTAKE: Insufficient data.

Occupational Exposures: NIOSH (NOHS Survey 1972-1974) has statistically estimated that 697,803 workers are exposed to 1,4-DCB in the United States [50]. 1,4-DCB concn of 32.5-52.1 mg/m^3 were found in the workplace air of a monochlorobenzene manufacturing plant [31]. A chlorobenzene factory was found to contain levels of 144-204 mg/m^3 [31]. The air in a factory where moth cakes were made contained 54-150 mg/m^3 and the air in an abrasive wheel facility using 1,4-DCB in the manufacturing process contained concn of 48-99 mg/m^3 [31]. Indoor air in mobile homes (Texas)

259

1,4-Dichlorobenzene

mean 1.3 ppb, range 0.006-9.6 ppb, 63% of homes [18]. NIOSH (NOES Survey 1981-1983) has statistically estimated that 27,242 workers are exposed to 1,4-DCB in the United States [51].

Body Burdens: Mean 1,4-DCB concentrations of 146 ug/kg (fat basis) detected in Yugoslavian human adipose tissue and 25 ug/kg (as-is basis) or 640 ug/kg (fat basis) in human milk [32]. Unidentified dichlorobenzene isomers were found in human blood samples at levels of 0.76-68 ng/mL taken from residents at the Love Canal [5]. Concn of 0.2-11.6 ng/g (fat basis) detected in human adipose tissue and 8-12 ng/mL in human blood samples in Japan [47]. Concn of 0.2-9.9 ug/g found in Japanese human adipose tissue [48]. Incidence of detection in the U.S. National Human Adipose Tissue Survey fiscal year 1982 composite specimens, 100% pos, 18-1400 ppb (wet tissue concn) [75].

REFERENCES

1. Atkinson R; Chem Rev 85: 69 (1985)
2. Bailey RE; Environ Sci Technol 17: 504 (1983)
3. Ballschmiter K et al; Angew Chem Int Ed English 16: 645 (1977)
4. Banerjee S et al; Environ Sci Technol 14: 1227 (1980)
5. Barkley J et al; Biomed Mass Spect 7: 139 (1980)
6. Barrows ME et al; pp 379-92 in Dyn Exposure Has Assess Toxic Chem Ann Arbor, MI Ann Arbor Sci (1980)
7. Bedient PB et al; Groundwater 22: 318 (1984)
8. Bouwer EJ; Environ Prog 4: 43 (1985)
9. Bouwer EJ, McCarty PL; Groundwater 22: 433 (1984)
10. Bozzelli JW, Kebbekus BB; J Environ Sci Health 17: 693 (1982)
11. Brodzinsky R, Singh HB; Volatile Organic Chem in the Atmos An Assess of Available Data Menlo Park, CA Atmos Sci Cntr SRI Internatl pp 198 (1982)
12. Calamari D et al; Ecotox Environ Safety 6: 369 (1982)
13. Callahan MA et al; Water-Related Environ Fate of 129 Priority Pollut vol.II USEPA-440/4-79-029B (1979)
14. Canton JH et al; Regulatory Toxicol Pharmacol 5: 123-31 (1985)
15. Carlson AR, Kosian PA; Arch Environ Contam Toxicol 16: 129-35 (1987)
16. Chemical Marketing Reporter Sept 10, p 54 (1984)
17. Chiou CT et al; Environ Sci Technol 17: 227 (1983)
18. Connor TH et al; Toxicol Lett 25: 33-40 (1985)
19. Elder VA et al; Environ Sci Technol 15: 1237 (1981)
20. Fielding M et al; Org Micropollut in Drinking Water Medmenham, Eng Water Res Cent TR-159 (1981)
21. Granstrom ML et al; Water Sci Technol 16: 375 (1984)
22. Guicherit R, Schulting FL; Sci Total Environ 43: 193-219 (1985)
23. Haider K et al; Arch Microbiol 96: 183 (1974)

1,4-Dichlorobenzene

24. Hansch C, Leo AJ; Medchem Project Issue No.26 Claremont, CA: Pomona College (1985)
25. Harkov R et al; J Environ Sci Health 20: 491 (1985)
26. Harkov R et al; J Air Pollut Control Assoc 33: 1177 (1983)
27. Hauser TR, Bromberg SM; Environ Monit Assess 2: 249 (1982)
28. Hawley GG; Condensed Chem Dictionary 10th ed von Nostrand Reinhold NY p 333 (1981)
29. Hutchins SR et al; Environ Toxicol Chem 2: 195 (1983)
30. Hutzler NJ et al; Amer Chem Soc 186th Natl Mtg Div Environ Chem 23: 499-502 (1983)
31. IARC; Some Industrial Chemicals and Dyestuffs 29: 214 (1982)
32. Jan J; Bull Environ Contam Toxicol 30: 595 (1983)
33. Jan J; Mitt Geb Lebensmittelunters Hyg 74: 420 (1983)
34. Jan J, Malnersic S; Bull Environ Contam Toxicol 24: 824 (1980)
35. Jori A et al; Ecotox Environ Safety 6: 413 (1982)
36. Junk GA et al; ACS Symp Ser 319: 109-23 (1986)
37. Kitano M; Biodegradation and Bioaccumulation Test on Chem Sub OECD Tokyo Meeting Ref Book TSU-No.3 (1978)
38. Konemann H, Van Leeuwen K; Chemosphere 9: 3 (1980)
39. Kuhn EP et al; Environ Sci Technol 19: 961 (1985)
40. LaRegina J et al; Environ Progress 5: 18-27 (1986)
41. Lewis TE et al; p 7 Soil in New Persp Env Tox Chem Soc Environ Tox Chem Abstr 6th Meet p 7 (1983)
42. Ligocki MP et al; Atmos Environ 19: 1609 (1985)
43. Lioy PJ et al; pp. 3-43 in Toxic Air Pollution; Lioy PJ, Daisey JM, eds (1987)
44. Lyman WJ et al; Handbook of Chem Property Estimation Methods Environ Behavior of Org Compounds McGraw-Hill NY (1982)
45. Malaney GW, McKinney RE; Water Sewage Works 113: 302 (1966)
46. Malle KG; Z Wasser-Abwasser Forsch 17: 75 (1984)
47. Morita M, Ohi G; Environ Pollut 8: 269 (1975)
48. Morita M et al; Environ Pollut 9: 175 (1975)
49. Neely WB et al; Environ Sci Technol 8: 1113 (1974)
50. NIOSH; The National Occupational Hazard Survey (NOHS) (1974)
51. NIOSH; The National Occupational Exposure Survey (NOES) (1983)
52. Nojimak, Kanna S; Chemosphere 9: 437 (1980)
53. Oliver BG, Chemosphere 14: 1087 (1985)
54. Oliver BG et al; Environ Sci Technol 13: 1075 (1979)
55. Oliver BG; Symp Amer Chem Soc Div Environ Chem 186th Natl Mtg 23: 421 (1983)
56. Oliver BG, Charlton MN; Environ Sci Technol 18: 903 (1984)
57. Oliver BG, Nicol KD; Environ Sci Technol 16: 532 (1982)
58. Oliver BG, Nicol KD; Environ Sci Technol 17: 505 (1983)
59. Oliver BG, Nicol KD; Intern J Environ Anal Chem 25: 275-85 (1986)
60. Oliver BG, Nicol KD; Sci Toxicol Env 39: 57 (1984)
61. Oliver BG, Niimi AJ; Environ Sci Technol 17: 287 (1983)
62. Otson R et al; J Assoc Off Analyt Otson 65: 1370 (1982)
63. Page GW; Environ Sci Technol 15: 1475 (1981)
64. Pankow JF et al; Environ Sci Technol 18: 310 (1984)
65. Reinhard M et al; Environ Sci Technol 18: 953 (1984)

1,4-Dichlorobenzene

66. Riddick JA et al; Organic Solvents: Physical Properties and Methods of Purification, 4th Edit. New York: J Wiley & Sons (1986)
67. Sabel BV, Clark TP; Waste Manag Res 2: 119 (1984)
68. Sadtler 55 UV (1988)
69. Sanjivamurthy VA; Water Res 12: 31 (1978)
70. Schwarzenbach RP et al; Environ Sci Technol 13: 1367 (1979)
71. Schwarzenbach RP, Westall J; Environ Sci Technol 15: 1360 (1981)
72. Shackelford WM, Keith LH; Frequency of Organic Compounds Identified in Water p 72 USEPA-600/4-76-062 (1976)
73. Sontheimer H et al; Sci Tot Env 47: 27 (1985)
74. Spicer CW et al; Atmos Reaction Prod From Hazardous Air Pollut Degradation p 4 USEPA-600/S3-85-028 (1985)
75. Stanley JS; Broad Scan Analysis of the FY82 National Human Adipose Tissue Survey Specimens. Vol 1. Executive Summary. p.1 USEPA-560/5-86-036 (1986)
76. Staples CA et al; Environ Toxicol Chem 4: 131 (1985)
77. Tabak HH et al; J Water Pollut Control Fed 53: 1503 (1981)
78. Tiernan TO et al; Environ Health Pers 59: 145 (1985)
79. USEPA; Preliminary Assess of Suspected Carcinogens in Drinking Water An Interim Report to Congress (1975)
80. vandeMeent D et al; Wat Sci Tech 18: 73-81 (1986)
81. Wakeham SG et al; Environ Sci Technol 17: 611 (1983)
82. Wakeham SG et al; Can J Fish Aquat Sci 40: 304 (1983)
83. Westrick JJ et al; J Amer Water Works Assoc 76: 52 (1984)
84. Wilson JT et al; J Environ Qual 10: 501 (1981)
85. Wilson JT, McNabb JF; Bio Transformation of Org Pollut in Groundwater EOS Transactions, Amer Geophys Union 64: 505-7 (1983)
86. Young DR et al; Water Chlorination Environ Impact Health Eff 3: 471-86 (1980)
87. Young DR et al; Water Chlorination Environ Impact Health Eff 4: 841-84 (1983)
88. Zoeteman BCJ et al; Chemosphere 9: 231 (1980)

3,3'-Dichlorobenzidine

SUBSTANCE IDENTIFICATION

Synonyms: 3,3'-Dichloro-4,4'-diaminodiphenyl

Structure:

CAS Registry Number: 91-94-1

Molecular Formula: $C_{12}H_{10}Cl_2N_2$

Wiswesser Line Notation: ZR BG DR DZ CG

CHEMICAL AND PHYSICAL PROPERTIES

Boiling Point:

Melting Point: 132-133 °C

Molecular Weight: 253.13

Dissociation Constants:

Log Octanol/Water Partition Coefficient: 3.51 [12]

Water Solubility: 3.1 mg/L at 25 °C [4]

Vapor Pressure: 4.2 x 10^{-7} mm Hg at 25 °C (estimated) [15]

Henry's Law Constant: 4.5 x 10^{-8} atm m³/mole at 25 °C (estimated from vapor pressure and water solubility)

3,3'-Dichlorobenzidine

ENVIRONMENTAL FATE/EXPOSURE POTENTIAL

Summary: 3,3'-Dichlorobenzidine (DCB) may be released as emissions or in wastewater during its production or use as an intermediate in the manufacture of pigments. Strict regulations requiring its use in closed systems should limit its environmental release. If released into water, it will rapidly and strongly adsorb to sediment and particulate matter and at least a part of the compound will be irreversibly sorbed. It will undergo very rapid photooxidation in surface layers of water (half-life 90 sec) forming 3-chlorobenzidine, benzidine, and other water insoluble colored substances. Redox reactions and reactions involving free radicals may also be important. It will bioconcentrate in fish. When released on land, it will tightly bind to the soil and possibly undergo chemical reactions with soil components. Very slow mineralization is expected to occur in soil (2% in 32 weeks). If released to the atmosphere it will most likely be adsorbed to particulate matter and rapidly photodegrade. Exposure to DCB will be primarily occupational.

Natural Sources: DCB has not been reported to occur naturally [14].

Artificial Sources: DCB may be released as emissions and in wastewater during its production, use as an intermediate in the manufacture of pigments, and use as curing agents in polyurethane elastomers [14]. Regulations in the United States requiring that DCB and its salts be used in isolated or closed systems would reduce its release [14].

Terrestrial Fate: When released on land, DCB will adsorb tightly to soil and possibly react chemically with soil components. Very slow mineralization (2% in 32 weeks) can occur under aerobic conditions. Loss of DCB due to oxidation by certain (Fe^{+3}, Cu^{+2} etc.) naturally occurring soil cations may be important.

Aquatic Fate: When released in water, DCB will rapidly sorb to sediment and particulate matter in the water column. It will rapidly photodegrade in the surface layers of water (half-life 90 sec) forming 3-chlorobenzidine, benzidine, and other water-insoluble colored products. DCB is relatively resistant to biodegradation and this would be a minor loss process. Chemical oxidative processes involving metal cations occur for the parent molecule, benzidine, and may also occur for DCB.

Atmospheric Fate: Should DCB be released into the atmosphere, it would most likely be associated with particulate matter or aerosols and be subject to gravitational settling. While its degradation in the atmosphere has not been studied, it would probably rapidly photodegrade and this process is likely to be faster than dry deposition.

Biodegradation: In laboratory biodegradability tests using sewage seed, 9-99% of DCB degraded in 28 days when yeast extract was present at concentrations of 50 to 400 mg/L. However, no degradation occurred without this additional nutrient. Therefore, DCB was considered inherently biodegradable although the precise role of the yeast extract was unexplained [6]. When incubated with natural aquatic communities from eutrophic and mesotrophic lakes, 25% of the DCB degraded in a month [1]. When incubated in soil under aerobic conditions, only 2% mineralization occurred in 32 weeks and no degradation intermediates were detected [5]. Under anaerobic conditions, no mineralization occurred in a year [5].

Abiotic Degradation: DCB has a strong absorption band at 282 nm [3] and degrades rapidly in dilute aqueous solutions (half-life 90 sec) when exposed to noonday summer sunlight [3]. The photodegradation products are 3-chlorobenzidine, benzidine, and other water-insoluble colored materials [3]. Short-lived intermediates are also observed when chlorine-water is added to a dilute aqueous solution of DCB [3]. Half-lives of 3-4 minutes were determined in laboratory irradiation experiments [3], which also revealed that the photodegradation was acid catalyzed [3]. The photolability of the compound is much lower in organic solvents, which may lead to an enhanced stability in water contaminated with hydrocarbons [3]. Hydrolysis is not expected to be an important process for DCB [7]. While no data could be found for DCB, unsubstituted benzidine is very rapidly oxidized by Fe(III) and certain other naturally occurring cations. While the chlorosubstituted benzidine would have less of a tendency to oxidize, this type of chemical reaction could be very important environmentally both in natural water and in soil [7,9].

Bioconcentration: Fish rapidly bioconcentrate DCB. ^{14}C-Dichlorobenzidine was bioconcentrated in bluegill sunfish (Lepomis macrochirus) with a factor of 495-507 in whole fish, whereas the bioconcentration factor was 114-170 in edible and 814-856 in non-edible parts of the same fish. The non-edible parts had a higher bioconcentration factor presumably due to higher lipid content [2]. The bioconcentration factors in golden ide (Leuciscus melanotus) and in algae (Chlorella fusca) were 610 and 940, respectively [11].

Soil Adsorption/Mobility: The distribution coefficient of DCB to natural sediments at pH 7 ranged from 26.7 to 128 [1]. The adsorption was initially very rapid [1]. Adsorption at pH 9 was reduced by 30-50% and desorption was low [1]. Attempts to extract the DCB from the sediment revealed that it was very tightly bound [1]. It is strongly adsorbed to Brookston clay loam and Rubicon sand, with the distribution constant being 1100 and 273, respectively [5]. Aromatic amines are known to form covalent bonds to humic materials [17], but no data are available specific to DCB.

Volatilization from Water/Soil: No DCB was lost from soil due to volatilization during persistence studies over 32 and 52 weeks under aerobic and anaerobic conditions, respectively [5]. The rate of evaporation from water is unknown but based on the reported boiling point of unsubstituted benzidine of 402 °C [7], evaporation from water should not be a significant transport process [7].

Water Concentrations: In the USEPA STORET database, DCB was detected in 0.1% of the 863 ambient water samples analyzed at a detection limit of 10 ppb [18].

Effluent Concentrations: Wastewater from metal finishing 10 ppb max; nonferrous metals manufacture 2.0 ppb max, 0.3 ppb avg; paint and ink formulation 10 ppb max; coal mining 3 ppb max [20]. Nationwide urban runoff program (51 catchments in 19 cities - 86 samples) not detected [8]. Analysis of purge wells and seepage water near a waste disposal lagoon receiving

3,3'-Dichlorobenzidine

DCB-manufacture wastes showed levels of DCB ranging from 0.13 to 0.27 mg/L [19]. Water from Sumida River in Tokyo which receives waste effluents from a dye and pigment factory contained DCB, but the conc was not quantified [19]. The STORET database shows that DCB was detected in only 1% of the 1239 effluent samples analyzed at a detection limit of 10 ppb [18].

Sediment/Soil Concentrations: The STORET database indicates that no DCB was detected in 347 sediment samples at a detection limit of 1 mg/kg (dry wt) [18]. DCB was qualitatively identified in sediment/soil/water mixture obtained from the contaminated areas of Love Canal, Niagara Falls, NY [13].

Atmospheric Concentrations:

Food Survey Values:

Plant Concentrations:

Fish/Seafood Concentrations: The STORET database indicates that no DCB was detected in 83 biota samples at a detection limit of 2.5 mg/kg (wet wt) [18].

Animal Concentrations:

Milk Concentrations:

Other Environmental Concentrations:

Probable Routes of Human Exposure: Exposure to DCB is primarily occupational in workers connected with its manufacture, conversion to derived pigments, and in curing polyurethane elastomers [14]. Exposure is most likely to be through inhalation of dust or mist or by dermal adsorption [14,16].

Average Daily Intake:

Occupational Exposures: In 1973, 18 U.S. companies had been confirmed to be using DCB and 166 employees were potentially exposed [14]. Although no survey data is available, NIOSH statistically estimated that approx 1000 people were exposed to

3,3'-Dichlorobenzidine

DCB in 1974 [10]. In pigment manufacturing plants in Japan, the exposure levels are 2 ppb within 10 min of charging reaction vessels, dropping to 0.2 ppb within 20 min [14].

Body Burdens:

REFERENCES

1. Appleton H et al; Pergamon Ser Environ Sci 1: 473-4 (1978)
2. Appleton HT and Sikka HC; Environ Sci Tech 14: 50-4 (1980)
3. Banerjee S et al; Environ Sci Technol 12: 1425-7 (1978)
4. Banerjee S et al; Environ Sci Technol 14: 1227-9 (1980)
5. Boyd SA et al; Environ Toxicol Chem 3: 201-8 (1984)
6. Brown D, Laboureur P; Chemosphere 12: 405-14 (1983)
7. Callahan MA et al; Water-related environmental fate of 129 priority pollutants Vol 2; pp.103-1 to 103-9 USEPA-440/4-79-029b (1979)
8. Cole RH et al; J Water Pollut Control Fed 56: 898-908 (1984)
9. Demirjian YA et al; J Water Pollut Control Fed 59: 32-8 (1987)
10. Fishbein L; Aromatic Amines in The Handbook of Environmental Chemistry: Anthropogenic Substances, Vol 3, Part A, Springer-Verlag (Berlin), Heidelberg, Germany,pp 1-40 (1984)
11. Freitag D et al; Chemosphere 14: 1589-616 (1985)
12. Hansch C, Leo AJ; Medchem Project Issue No 26. Claremont CA: Pomona College (1985)
13. Hauser TR and Bromberg SM; Environ Monit Assess 2: 249-71 (1982)
14. IARC; Some Industrial Chemicals and Dyestuffs 29: 239-56 (1982)
15. Lyman WJ; Estimation of Physical Properties, in Environmental Exposures from Chemicals, Vol I, Neely WB and Blau GE ed, CRC Press, Boca Raton, FL pp 30-2 (1985)
16. NIOSH; Carcinogenicity and Metabolism of Azo Dyes, especially those derived from benzidine (1980)
17. Parris GE; Environ Sci Technol 14: 1099-1106 (1980)
18. Staples CA et al; Environ Toxicol Chem 4: 131-42 (1985)
19. USEPA; Ambient Water Quality Criteria Doc: 3,3'-Dichlorobenzidine p.C-1 to C-4, NTIS PB81-11717, Springfield, VA (1980)
20. USEPA; Treatability Manual; p.I.7.5-1 to I.7.5-5 USEPA-600/2-82-001a (1982)

Diethyl Phthalate

SUBSTANCE IDENTIFICATION

Synonyms:

Structure:

CAS Registry Number: 84-66-2

Molecular Formula: $C_{12}H_{14}O_4$

Wiswesser Line Notation: 2OVR BVO2

CHEMICAL AND PHYSICAL PROPERTIES

Boiling Point: 295 °C

Melting Point: -40.5 °C

Molecular Weight: 222.26

Dissociation Constants:

Log Octanol/Water Partition Coefficient: log Kow = 2.47 [16]

Water Solubility: 1080 mg/L at 25 °C [20]

Vapor Pressure: 1.65 x 10^{-3} mm Hg at 25 °C [20]

Henry's Law Constant: 4.8 x 10^{-7} atm-m³/mole (calculated) [30]

Diethyl Phthalate

ENVIRONMENTAL FATE/EXPOSURE POTENTIAL

Summary: Diethyl phthalate (DEP) may enter the environment in air emissions, aqueous effluent, and solid waste products from manufacturing and processing plants. It is estimated that 0.5% of all DEP produced is lost during its manufacture. DEP may also be emitted in vapor and particulate form during incineration of DEP containing plastics, and this has been estimated to amount to 0.67% of all DEP used. DEP may volatilize from its plastic products or may enter the environment directly during non-plasticizer use. Plastic materials containing DEP in waste disposal sites constitute the major reservoir of this compound in the environment. Volatilization and leaching from these materials are potential sources of transport into air, water, and soil. DEP has been identified in cranberries, baked potatoes, and roasted filberts; however, sufficient evidence is not available to indicate that DEP is a natural product, since it may have been present as a solvent residue. If released to soil, DEP is expected to undergo aerobic biodegradation. Oxidation, chemical hydrolysis, and volatilization from wet soil surfaces are not expected to be significant fate processes. DEP may volatilize from dry soil surfaces. If released to water, DEP is expected to biodegrade (aerobic biodegradation half-life approx 2 days to >2 weeks). Anaerobic biodegradation would be very slow or not occur at all. Volatilization should not be an important removal process in most bodies of water, although it may be important in shallow rivers. Removal by oxidation, chemical hydrolysis, direct photolysis, indirect photolysis, or bioaccumulation in aquatic organisms should not be significant. DEP has accumulated and persisted in the sediments of Chesapeake Bay for over a century. If released to the atmosphere, DEP is expected to exist in vapor form and as adsorbed matter on airborne particulates. DEP vapor is expected to react with photochemically generated hydroxyl radical (estimated half-life 22.2 hours). Physical removal by particulate settling and wash-out in precipitation will also occur. Degradation by direct photolysis is not expected to be significant. The most probable routes of human exposure are inhalation and dermal exposure of workers involved in the manufacture and use of DEP. The most probable routes of exposure to the general population are inhalation, ingestion, and dermal contact due to use of consumer products containing DEP.

Diethyl Phthalate

Natural Sources: DEP has been identified in cranberries [39] and as a volatile component of baked potatoes [5] and roasted filberts [25]. However, these data do not provide sufficient evidence to indicate that DEP is a natural product, since it may have been present as a solvent residue [25,39].

Artificial Sources: DEP may enter the environment in air emissions, aqueous effluent, and solid waste products from manufacturing and plastics processing. It is estimated that 0.5% of the total amount of DEP produced is lost during its manufacture from phthalic anhydride [40]. DEP may also be emitted in vapor and particulate form into the atmosphere during incineration of DEP containing plastics. 0.67% of all DEP used is estimated to vaporize during low temperature incineration [40]. DEP may volatilize from plastic products and it may enter the environment directly due to non-plasticizer usage, e.g., in insecticidal sprays, insect repellents, and perfumes [40]. Disposed plastic materials containing DEP in waste disposal sites constitutes the major reservoir of this compound in the environment [39]. Volatilization and leaching from these materials are potential modes of transport into air, water, and soil.

Terrestrial Fate: If released to soil, DEP is expected to undergo aerobic biodegradation. Oxidation, chemical hydrolysis, and volatilization from wet soil surfaces are not expected to be significant fate processes. DEP may volatilize from dry soil surfaces.

Aquatic Fate: If released to water, DEP is expected to undergo aerobic biodegradation. Under aerobic conditions, biodegradation half-lives ranging from approx 2 days to >2 weeks have been reported and anaerobic biodegradation is reported to occur much more slowly or not at all. Volatilization of DEP should not be an important removal process in most bodies of water although it may be important in shallow rivers. Removal of DEP by oxidation, chemical hydrolysis, direct photolysis, indirect photolysis, or bioaccumulation in aquatic organism should not be significant. Identification of DEP in dated sediment cores from the Chesapeake Bay indicates that this compound has accumulated and persisted in the sediment for over a century [42].

271

Diethyl Phthalate

Atmospheric Fate: If released to the atmosphere, DEP is expected to exist in the vapor form and adsorbed to airborne particulates. DEP vapor is expected to react with photochemically generated hydroxyl radicals with an estimated reaction half-life of 22.2 hours at 25 °C. Physical removal by particulate settling and wash-out in precipitation will also occur. Degradation by direct photolysis is not expected to be significant.

Biodegradation: DEP in a screening test incubated with a mixed microbial population had a degradation half-life (measured as ultimate degradation - CO_2 evolution) of 2.2 days and underwent >99% degradation (measured as primary degradation - loss of parent compound) in 28 days [51]. When incubated in dirty river water taken from the Ogawa River in Japan, 25 ppm DEP had a half-life of approx 3 days and was 100% decomposed in 6 days. When incubated in clean and dirty ocean water taken from Osaka Bay, 25 ppm DEP underwent 14 and 68% degradation, respectively, in 14 days [17]. When 5 and 10 ppm DEP were seeded with domestic wastewater under aerobic conditions at 25 °C, 100% degradation was observed in 7 days [53]. When 1 ppm DEP was incubated in a semi-continuous activated sludge system (SCAS), >94.8% degradation occurred in 24 hours [38]. Bacterial transformation of DEP accounted for all measurable loss of this compound in aufwuch cultures [27]. When DEP was incubated with municipal digester sludge under anaerobic conditions at 35 °C in the dark, degradation as measured by theoretical methane production ranged from 32% after 4 weeks to 0% after more than 8 weeks [19]. When 3-10 ppm DEP was incubated at 25 °C in sterilized and unsterilized soil, 34% and 86% degradation, respectively, was observed after 120 hours [45]. When 3-10 DEP was incubated in sterilized and unsterilized soil sprayed with landfill leachate, 0% and 100% degradation, respectively, was observed in 72 hours [45].

Abiotic Degradation: Chemical hydrolysis of DEP would be a significant process near or above pH 10 [26]. The half-life of DEP resulting from alkaline hydrolysis however is proportional to hydroxyl ion concentration, and aquatic environments are most commonly well below pH 10 [26]. Based on measured second order alkaline hydrolysis rate constants of 2.5×10^{-2} mol-sec^{-1} for DEP at pH greater than or equal to 9 and 30 °C [58] and $2.1 \times$

10^{-2} mol-sec^{-1} for DEP at pH 10-12 and 30 °C [2], hydrolysis half-lives for DEP at pH 7 have been calculated to be 8.8 and 18 years, respectively. DEP is not expected to react with alkyl peroxy radicals or singlet oxygen in water. Using a calculated alkyl peroxy radical reaction rate constant of 1.4 mol-hr^{-1} [31] and an ambient alkyl peroxy radical concentration of 10^{-9} mol/L [34], the alkyl peroxy reaction half-life has been calculated to be 6.5 years. DEP exhibits only slight absorption of UV light wavelengths >290 nm [46], suggesting that this compound has the ability to undergo direct photolysis in air and water, although it is not expected to be an environmentally significant process [2]. Indirect photolysis in water involving interaction of the hydroxyl radical with the aromatic ring, is also considered to be too slow to be of environmental significance [2]. The atmospheric hydroxyl reaction half-life for DEP in the vapor phase is estimated to be 22.2 hours based on an ambient hydroxyl concentration of 8.0 x 10^{+5} molecules/cm^3 and an estimated reaction rate constant of 1.08 x 10^{-11} cm^3/molecule-sec at 25 °C [14].

Bioconcentration: The measured bioconcentration factor (BCF) for DEP is 117 in bluegill sunfish (Lepomis macrochirus) [1,56] and 15-16 in mullet (Mugil cephalus) [47]. Using the water solubility and the recommended value for the log octanol-water partition coefficient, BCF values of 12 and 44, respectively, have been calculated [30]. These BCF values indicate that bioaccumulation in aquatic organisms would not be significant.

Soil Adsorption/Mobility: Based on the water solubility and the recommended value for the log octanol-water partition coefficient, soil adsorption coefficient (Koc) values of 526 and 94, respectively, have been calculated using recommended regression equations [30]. A Koc value of 526 and common occurrence of DEP in bottom sediments indicate that adsorption to sediments may be significant and suggest that DEP would have low mobility in soil [52]. In contrast, a Koc value of 94 suggests that adsorption to sediments may be low and that DEP would be highly mobile in soil [52]. In addition, it was observed that DEP did not absorb significantly to sediments during a study on this compound in an aqueous system [26]. Koc of 69 was measured with Broome County, NY soil [44].

273

Diethyl Phthalate

Volatilization from Water/Soil: The Henry's Law constant suggests that volatilization of DEP would not be significant from most bodies of water, but may be a significant removal process in shallow rivers [30]. There is evidence that phthalate esters are slowly volatilized from plastics into air at high temperature [2], suggesting that DEP may volatilize from dry soil surfaces.

Water Concentrations: SURFACE WATER: Data from the USEPA STORET database: 862 samples, 3% pos., median concentration < 10 ppb DEP [49]. Inner Harbor Navigation Canal of Lake Pontchartrain during 1980, 0.7 ppb DEP detected [33]. Lower Tennessee River water and sediment samples, 11.2 ppb DEP detected [15]. DEP was detected in 6 out of 204 water samples from 14 industrial river basins [11]. Concn (ppt) in Mississippi River, Lake Itasca, MN (river's source), 190; Cario, IL, 84; 20 miles south of Memphis, TN, 350; Carrolton St. water intake in New Orleans, LA, 63 [7].   GROUND WATER: DEP was monitored in New York State public water system wells during 1977, 39 samples, 33% pos, 4.6 ppb max level detected [6]. At a municipal solid waste landfill site in Norman, OK 4.1 ppb DEP detected in ground water [9].   At land application sites in Fort Devens, MA, Boulder, CO, Lubbock, TX, and Phoenix, AZ, levels of DEP ranging from ND to 0.87 ppb were found in ground water [22]. DEP was qualitatively identified in leachate from two low level radioactive waste disposal sites in KY and NY [13]. DRINKING WATER: DEP has been identified in drinking water from the following cities: Cincinnati, 0.1 ppb; Miami, 1.0 ppb; Philadelphia, 0.01 ppb; Seattle, 0.01 ppb; Lawrence, 0.04 ppb; New York City, 0.01 ppb [24]. The max concn of DEP detected in finished drinking water supplies in New Orleans during 1974 was 0.07 ppb [55]. Identified, not quantified, in 1975-76 Philadelphia, PA, finished drinking water from 2 of 3 treatment plants and tap water from 1 hotel [50].   OTHER WATER: Results of the Nationwide Urban Runoff Program as of Aug. 1982: 86 sample (from 3 locations), 4% pos, 2 to 10 ppb DEP detected   [4]. SEAWATER: Mersey Estuary, UK, 17 sites, 47% pos, dissolved fraction, 18% pos, not detected (0)-67 ppt, 58 ppt avg pos; particulate fraction, 35% pos, 14-233 ppb (dry wt) [43].

Effluent Concentrations: Biologically treated bleached kraft mill effluents: 9 samples, 100% pos., concn range 20 to 100 ppb DEP, mean concentration 50 ppb [57]. Data from the STORET database: 1,286 samples, 9.9% pos., median concentration <10 mg/L DEP [49]. 0.06 ppm DEP was found in the effluent from a tire plant which discharged 0.4 million gallons of wastewater per day [23]. Ground water from four U.S. rapid-infiltration sites for municipal primary and secondary effluents, 100% pos, 0.02-0.87 ppb, 0.41 ppb avg of 3 sites, 1 site not quantified [21].

Sediment/Soil Concentrations: Concentrations of DEP in Chesapeake Bay sediment cores were 13 to 49 ppb at 0-20 cm, 14 to 35 ppb at 20-50 cm [42]. In sediments taken from Chester River, MD, DEP concn ranged between 11 to 44 ppb [41]. In sediments taken from Lake Pontchartrain, DEP concentrations ranged between 25 to 65 ppb [32]. In sediments taken from San Luis Pass, TX during 1980, DEP concentrations ranged between <2 to 9 ppb [35]. In surficial bed material taken from Bridgeport, NJ during Jan 1981, trace levels of DEP were detected [18]. STORET database: 322 samples, 10.0% pos, median concn <500 ppb [49].

Atmospheric Concentrations: In air samples obtained in Antwerp, Belgium and rural Bolivia (background level), DEP levels ranged between 2.1 to 5.9 ng/m^3 and 0.51 to 0.80 ng/m^3, respectively [3]. The ratio of DEP dissolved in rainwater to DEP vapor in air has been observed to be 20,000 [28]. DEP was qualitatively identified in rain and snow collected in Europe during 1973 to 1975 [29] and in fly ash collected from a municipal waste incinerator [10].

Food Survey Values: DEP has been identified as a volatile flavor component of Idaho Russet Burbank baked potatoes [5] DEP has been identified as a volatile component of roasted filberts, although it was probably present as a solvent artifact [25]. Identified, not quantified, in volatile flavor components of fried chicken, NJ [54]; raw cassava, Dominican Republic [8]; not detected in processed cassava products, farine (Caribbean), and white gari (Nigeria) [8].

Plant Concentrations:

Fish/Seafood Concentrations: Data from USEPA STORET database indicate that DEP has been detected in aquatic biota: 121 samples, 6% pos., median concn <2.5 ppb (wet basis) [49]. DEP

275

has been found in oysters, 1100 ppb (wet basis), and clams, 340-450 ppb (wet basis), taken from Lake Pontchartrain [33].

Animal Concentrations:

Milk Concentrations:

Other Environmental Concentrations: DEP was detected at a concn of 0.027 ppb in air with cigarette smoke [40].

Probable Routes of Human Exposure: The most probable routes of human exposure are inhalation and dermal exposure of workers involved in the manufacture and use of DEP. The most probable routes of exposure to the general population are inhalation of vapors and contaminated particles, ingestion of contaminated drinking water or shell fish, and dermal contact due to use of consumer products containing DEP such as insecticidal sprays, insect repellents, and perfumes.

Average Daily Intake:

Occupational Exposures: NIOSH (NOHS Survey 1972-1974) has statistically estimated that 889,365 workers are exposed to DEP in the United States [37]. NIOSH (NOES Survey 1981-1983) has statistically estimated that 13,840 workers are exposed to DEP in the United States [36].

Body Burdens: DEP was identified in two human atherosclerotic aortas at levels ranging from 10 to 860 ppb [12]. National Human Adipose Tissue Survey, Fiscal Year 1982, 46 tissue composites, 42% pos, wet tissue concn, not detected (<0.009-0.061)-0.65 ppm, concn in extractable lipid, not detected (<0.009-0.076)-0.97 ppm [48].

REFERENCES

1. Barrows ME et al; pp 379-92 in Dynamics, Exposure and Hazard Assess Toxic Chem Ann Arbor Science Ann Arbor, MI (1980)
2. Callahan MA et al; Water-Related Environ Fate of 129 Priority Pollutants Volume II USEPA 440/4-79-029 (1979)
3. Cautreels W et al; Sci Total Environ 8: 79-88 (1977)

Diethyl Phthalate

4. Cole RH et al; J Water Pollut Control Fed 56: 898-908 (1984)
5. Coleman EC et al; J Agric Food Chem 29: 42-8 (1981)
6. Council Env Quality; The 10 Synthetic Organic Compounds Most Commonly Detected in Public Water System Wells in NY State, 11th Ann Report Wash (1980)
7. DeLeon IR et al; Chemosphere 15: 795-805 (1986)
8. Dougan J et al; J Sci Food Agric 34: 874-84 (1983)
9. Dunlap WJ et al; Organic Pollut Contributed to Groundwater by a Landfill, USEPA 600/9-76-004 (1976)
10. Eiceman GA et al; Anal Chem 53: 955-9 (1981)
11. Ewing BB et al; Monitoring to Detect Previously Unrecognized Pollutants in Surface Waters Appendix Organic Anal Data USEPA 560/6-77-015 (1977)
12. Ferrario JB et al; Arch Environ Contam Toxicol 14: 529-34 (1985)
13. Francis AJ et al; Nuclear Tech 50: 158-63 (1980)
14. GEMS; Graphical Exposure Modeling System. FAP. Fate Atmos Pollut (1986)
15. Goodley PC, Gordon M; Kentucky Acad of Sci 37: 11-5 (1976)
16. Hansch C, Leo AJ; Medchem Project Issue No 26. Claremont CA: Pomona College (1985)
17. Hattori Y et al; Pollut Control Cent Osaka Prefect, Mizu Shori Gijutsu 16: 951-4 (1975)
18. Hochreiter JJ Jr; Chemical Quality Reconnaissance of the Water and Surficial Bed Material in the Delaware River Estuary and Adjacent New Jersey Tributaries, 1980-81 USGS/WRI/NTIS 82-36 (1982)
19. Horowitz A et al; Dev Ind Microbiol 23: 435-44 (1982)
20. Howard PH et al; Environ Toxicol Chem 4: 653-61 (1985)
21. Hutchins SR, Ward CH; J Hydrol 67: 223-33 (1984)
22. Hutchins SR et al; Env Tox Chem 2: 195-216 (1983)
23. Jungclaus GA; Anal Chem 48: 1894-6 (1976)
24. Keith LH et al; pp 329-73 in Ident Anal Organic Pollut Water Keith LH ed Ann Arbor Press Ann Arbor, MI (1976)
25. Kinlin TE et al; J Agric Food Chem 20: 1021 (1972)
26. Lewis DL et al; Environ Tox Chem 3: 223-31 (1984)
27. Lewis DL, Hohn HW Appl Environ Microbiol 42: 698-703 (1981)
28. Ligocki MP et al; Atmos Environ 19: 1609-17 (1985)
29. Lunde G et al; Organic Micropollutants in Precipitation in Norway SNSF p 17 Project Fr - 9/76 (1976)
30. Lyman WJ et al; Handbook of chemical property estimation methods. Environmental behavior of organic compounds McGraw-Hill NY (1982)
31. Mabey WR et al; Aquatic Fate Process Data For Organic Priority Pollutants USEPA 440/4-81-014 (1981)
32. McFall JA et al; Chemosphere 14: 1561-9 (1985)
33. McFall JA et al; Chemosphere 14: 1253-65 (1985)
34. Mill T et al; Science 207: 886 (1980)
35. Murray HE et al; Chemosphere 10: 1327-34 (1981)
36. NIOSH; The National Occupational Exposure Survey (NOES) (1983)
37. NIOSH; The National Occupational Hazard Survey (NOHS) (1974)
38. O'Grady DP et al; Appl Environ Microbiol 42: 698-703 (1981)
39. Peakall DB; Res Rev 54: 1-41 (1975)
40. Perwack J et al; An Exposure and Risk Assessment for Phthalate Esters

Diethyl Phthalate

USEPA 440/4-81-020 (1981)
41. Peterson JC, Freeman DH; Int J Env Anal Chem 18: 237-52 (1984)
42. Peterson JC, Freeman DH; Environ Sci Tech 16: 464-9 (1982)
43. Preston MR, Al-Omran LA; Mar Pollut Bull 17: 548-53 (1986)
44. Russell DJ, McDuffie B; Chemosphere 15: 1003-21 (1986)
45. Russell DJ et al; J Environ Sci Health A 20: 927-41 (1985)
46. Sadtler; Sadtler Standard Spectra UV no. 150 (1988)
47. Shimada T et al; Kenkyu Hokuku- Kanagawa- Ken Kogai Senta 5: 45-8 (1983)
48. Stanley JS; Broad Scan Analysis of the FY82 Nat'l Human Adipose Tissue Survey Specimens pp.36,90-1 USEPA-560/5-86-037 (1986)
49. Staples CA et al; Environ Toxicol Chem 4: 131-42 (1985)
50. Suffet IH et al; Water Res 14: 853-67 (1980)
51. Sugatt RH et al; Appl Environ Microbiol 47: 601-6 (1984)
52. Swann RL et al; Res Rev 85: 17-28 (1983)
53. Tabak HH et al; J Water Pollut Contr Fed 53: 1503-18 (1981)
54. Tang J et al; J Agric Food Chem 31: 1287-92 (1983)
55. USEPA; New Orleans Area Water Supply Study Draft Analytical Report by the Lower Mississippi River Facility, Slidell, LA Dallas TX (1974)
56. Veith GD et al; pp 116-29 in Aquatic Toxicology Easton JG et al eds Amer Soc Test Mater (1980)
57. Voss RH; Env Sci Tech 18: 938-46 (1984)
58. Wolfe NL et al; Chemosphere 9: 403-8 (1980)

Di(2-ethylhexyl) Phthalate

SUBSTANCE IDENTIFICATION

Synonyms:

Structure:

CAS Registry Number: 117-81-7

Molecular Formula: $C_{24}H_{38}O_4$

Wiswesser Line Notation: 4Y2&1OVR BVO1Y4&2

CHEMICAL AND PHYSICAL PROPERTIES

Boiling Point: 230 °C at 5 mm Hg

Melting Point: -50 °C

Molecular Weight: 390.54

Dissociation Constants:

Log Octanol/Water Partition Coefficient: log Kow = 5.11 [21]

Water Solubility: 0.3 mg/L at 25 °C [21]

Vapor Pressure: 6.45 x 10^{-6} mm Hg at 25 °C [21]

Henry's Law Constant: 1.1 x 10^{-5} atm-m^3/mole (calculated) [30]

Di(2-ethylhexyl) Phthalate

ENVIRONMENTAL FATE/EXPOSURE POTENTIAL

Summary: Di(2-ethylhexyl) phthalate (DEHP) is used as a plasticizer for polyvinyl chloride (PVC) and other polymers in large quantities and is likely to be released to air and water during production and waste disposal of these plastic products. DEHP in water will biodegrade (half-life 2-3 weeks), adsorb to sediments, and bioconcentrate in aquatic organisms. Atmospheric DEHP will be carried long distances and be removed by rain. Human exposure will occur in occupational settings and from air, consumption of drinking water, food (especially fish, etc., where bioconcentration can occur), and food wrapped in PVC, as well as during blood transfusions from PVC blood bags.

Natural Sources: DEHP has been suggested as a possible natural product in animals and plants.

Artificial Sources: DEHP is used in large quantities, primarily as a plasticizer for polyvinyl chloride and other polymeric materials [24]. Disposal of these products (incineration, landfill, etc) will result in the release of DEHP into the environment. DEHP has been detected in the effluent of numerous industrial plants.

Terrestrial Fate: DEHP released to soil will neither evaporate nor leach into ground water. Limited data is available to suggest that it may biodegrade in soil under aerobic conditions following acclimation.

Aquatic Fate: DEHP released to water systems will biodegrade fairly rapidly (half-life 2-3 weeks) following a period of acclimation. It will also strongly adsorb to sediments (log Koc 4 to 5) and bioconcentrate in aquatic organisms. Evaporation and hydrolysis are not significant aquatic processes.

Atmospheric Fate: DEHP released to air will be carried for long distances in the troposphere and has been detected in air over the Atlantic and Pacific Oceans [1]. Wash out by rain appears to be a significant removal process [1]. It is unknown whether direct photolysis or photooxidation are important atmospheric processes.

Di(2-ethylhexyl) Phthalate

Biodegradation: DEHP biodegrades rapidly under aerobic conditions following acclimation in screening biodegradation tests [41,44,51,54] in river or lake water [20,44,50], in water/sediment systems [26,60] and in activated sludge [44]. Acclimation is very important in unacclimated systems no biodegradation occurs [54]. River die-away tests have reported half-lives of 2 - 3 weeks [20,44,60]. Half-lives of 0.8 days have been reported for DEHP in activated sludge [44]. In oligotrophic lake samples, no degradation occurs [50]. Under anaerobic conditions in water/sediment mixtures, no biodegradation occurs [26,46]. Limited data suggests some biodegradation in soil [22,31].

Abiotic Degradation: DEHP does not hydrolyze rapidly (half-life of 2000 yr) at pH 7 [7]. The photolysis half-life in water is estimated to be 143 days [59]. Some photodegradation (1.6% to CO_2 in 17 hr) has been noted in a photomineralization screening test [15]. Whether photooxidation of vapor or particulate adsorbed DEHP occurs is unknown; detectable concn of DEHP have been observed in atmospheric samples at remote sites.

Bioconcentration: DEHP does have a tendency to bioconcentrate in aquatic organisms; the experimental BCF values range from a log of 2 to 4 in fish and invertebrates [2,5,32,35,45,49,57]. In fathead minnows, the log BCF was 2.93 [57]; in bluegill sunfish, it was 2.06 [2].

Soil Adsorption/Mobility: DEHP has a strong tendency to adsorb to soil and sediments. Calculated log Koc values of 4 to 5 have been reported [28,59]. Experimental evidence demonstrates strong partitioning to clays and sediments (log K = 5; approx organic content = 1%) [52]. Koc was 87,420 in Broome County, NY soil [43].

Volatilization from Water/Soil: DEHP has a very low vapor pressure and Henry's Law constant, and therefore, it should not evaporate from soil or water. A calculated half-life of evaporation from water of 15 yr has been reported [4]. Evaporation rates from the leaves of Sinapis alba in a closed terrestrial simulation chamber of <0.8 and <0.5 ng/sq cm-hr (below detection limits) during times intervals of 0-1 and 8-15 days after application of 2.78 ug/sq cm onto the leaves, respectively [29].

Di(2-ethylhexyl) Phthalate

Water Concentrations: DRINKING WATER: U.S. Cities - 0.04 to 30 ppb, 4 pos out of 10 [27]; 170 ppt ground water well in NY [6]. GROUND WATER: 2.4 ppb landfill ground water [13]; ground water receiving wastewater applications to land 0.11 to .40 ppb [22]. SURFACE WATER: U.S. Industrialized river basins 132 of 203 samples pos [14] 1 to 80 ppb [14,47,48]. Concn (ppt) in Mississippi River, Lake Itasca, MN (river's source), not detected; Cairo, IL, 620; and 20 miles south of Memphis, TN, 720 [12]. SEAWATER: Gulf of Mexico avg 112 ppt [16] Galveston Bay, TX avg 600 ppt, 54 to 3000 ppt [36]. Mersey Estuary, UK, 17 sites, 47% pos, dissolved fraction, 18% pos, not detected (0)-67 ppt, 58 ppt avg pos; particulate fraction, 35% pos, 14-233 ppb (dry weight) [40]. RAINWATER: North Pacific avg 55 ppt, 5 to 213 ppt [1]. Urban College Station, TX, Feb-May 1979, filtered rain, 2.6

Effluent Concentrations: Industries with mean concentrations of treated wastewater exceeding 200 ppb: auto and other laundries - mean 440 ppb; coal mining - mean 940 ppb; aluminum forming - mean <13,000 ppt; foundries - mean 1100 ppb [56]. Mean concentrations in wastewaters from water-based paint plants and ink manufacturing plants were 400 ppb and 12,500 ppb, respectively [24]. DEHP has been detected in diesel exhaust [24]. Ground water from four U.S. rapid-infiltration sites for municipal primary and secondary effluents, 100% pos, 0.11-1.40 ppb, 0.55 ppb avg of 3 sites, 1 site not quantified [23].

Sediment/Soil Concentrations: SEDIMENT: European rivers 1-70 ppm, Mississippi River delta mean - 0.069 ppm; Gulf of Mexico 0.002-0.007 ppm; Lake Superior - 200 ppm [19]; Galveston Bay, TX - avg 0.022 ppm [36]. Chesapeake Bay - 0.022-0.18 ppm [39]. Portland, ME, 9 coastal sites, all pos, 60-7800 ppb [42].

Atmospheric Concentrations: RURAL: United States 0.14 to 1.0 ppb [1], mean remote ocean air 0.07-0.17 ppb [1]. Gulf of Mexico 0.53 to 1.92 ng/m^3, avg 1.16 [18]. URBAN: United States >0.92 ppb [3], Europe >8 ppb [9]. Urban air, Belgium, avg concn (ug/1000 m^3): particulates, 54.1, gas phase, 127 [10].

Di(2-ethylhexyl) Phthalate

Food Survey Values: Because DEHP is used as a plasticizer in polyvinyl chloride used in contact with food, some DEHP in foods may be due to migration from packaging materials [55]. In Japan, concentrations in foodstuffs that were in contact with packaging material varied from 0-68 ppm [55]; 0.05 to 0.4 ppm in egg whites - Japan retail stores [25].

Plant Concentrations:

Fish/Seafood Concentrations: Marine fish and shellfish - 10-100 ppb [16,17]; processed canned tuna and salmon in Canada - 40-160 ppb [58]; fish, Lake Superior - 0.3-0.7 ppb [53].

Animal Concentrations:

Milk Concentrations: 80 mg/L of milk (whole milk basis, one sample), not detected in later samples from same farm [11].

Other Environmental Concentrations: DEHP has been detected in commercial organic solvents [24].

Probable Routes of Human Exposure: Humans are primarily exposed to DEHP through ingestion of contaminated drinking water, fish, and seafood; foods which came in contact with packaging materials that use DEHP as a plasticizer; and inhalation of contaminated ambient air. Humans also receive significant exposures to DEHP during blood transfusions due to migration of DEHP into the blood from the blood bag [24].

Average Daily Intake: AIR INTAKE (assume 1 ppb) - 0.33 ug; WATER INTAKE (assume 0.04-30 ppb) - 0.08-60 ug; FOOD - insufficient data.

Occupational Exposures: In 1980, NIOSH estimated that 625,000 workers were exposed to DEHP [24]. Industries where concentrations were judged to be the highest were plastics and rubber products manufacture, blast-furnace and steel mill operations, nonferrous wire manufacture, and motor vehicle and aircraft manufacture [24]. Concentrations of 0.25-2 ppm in factory air has been detected [24]. NIOSH (NOHS Survey 1972-1974) has statistically estimated that 612,106 workers are exposed to DEHP

283

Di(2-ethylhexyl) Phthalate

in the United States [38]. NIOSH (NOES Survey 1981-1983) has statistically estimated that 147,848 workers are exposed to DEHP in the United States [37].

Body Burdens: Human adipose tissue 0.30 to 1.15 ppm [8,34].

REFERENCES

1.   Atlas E, Giam CS; Science 211: 163-5 (1981)
2.   Barrows ME et al; Dyn Exp Hazard Assess Toxic Chem, Ann Arbor, MI: Ann Arbor Sci p 379-92 (1980)
3.   Bove JL et al; Int J Environ Anal Chem 5: 189 (1978)
4.   Branson DK; Predicting the fate of chemicals in the aquatic environment. ASTM STP657 p 55-70 ASTM, Philadelphia, PA. (1978)
5.   Brown D, Thompson RS; Chemosphere 11: 417-26 (1982)
6.   Burmaster DE; Environ 24: 6-13 (1982)
7.   Callahan MA et al; Water-related environmental fate of 129 priority pollutants - Vol II. EPA-440/4-79-029B p 94-9 (1979)
8.   Campbell DS et al; Bull Environ Contam Toxicol 12: 721-5 (1974)
9.   Cautreels W et al; Sci Total Environ 8: 79 (1977)
10.   Cautreels W, VanCauwenberghe K; Atmos Environ 12: 1133-41 (1978)
11.   Cerbulis J, Ard JS; J Assoc Off Anal Chem 501: 646-50 (1967)
12.   DeLeon IR et al; Chemosphere 15: 795-805 (1986)
13.   Dunlap WJ et al; Organic Pollutants Contributed to Groundwater by a Landfill. EPA-600/9-76-004 p 96-110 (1976)
14.   Ewing BB et al; Monitoring to Detect Previously Unrecognized Pollutants in Surface Waters. EPA-568/6-77-015 and 015a p 75 (1977)
15.   Freitag D et al; Ecotox Environ Safety 6: 60-81 (1982)
16.   Giam CS et al; Mar Pollut Transfer p 375-86 (1976)
17.   Giam CS et al; Anal Chem 47: 2225-9 (1975)
18.   Giam CS et al; Atmos Environ 14: 65-9 (1980)
19.   Giam CS, Atlas EL; Die Naturwissenschaften 67:508 (1980)
20.   Hattori Y et al; Pollut Control Cent Osaka Prefect Mizu Shori Gijutsu 16: 951-4 (1975)
21.   Howard PH et al; Environ Toxicol Chem 4: 653-61 (1985)
22.   Hutchins SR et al; Environ Toxicol Chem 2: 195-216 (1983)
23.   Hutchins SR, Ward CH; J Hydrol 67: 223-33 (1984)
24.   IARC; Some Industrial Chemicals and Dyestuffs 29: 269 (1980)
25.   Ishida M et al; J Agr Food Chem 29: 72 (1981)
26.   Johnson BT, Lulves W; J Fish Res Board Can 32: 333-9 (1975)
27.   Keith LH et al; Ident and Anal of Organ Pollut in Water. Ann Arbor MI: Ann Arbor Press p 329-73 (1976)
28.   Kenaga EE; Ecotox Environ Safety 4: 26-38 (1980)
29.   Loekke H, Bro-Rasmussen F; Chemosphere 10: 1223-35 (1981)
30.   Lyman WJ et al; Handbook of chemical property estimation methods. Environmental behavior of organic compounds McGraw-Hill NY (1982)
31.   Mathur SP et al; J Environ Qual 3: 207-9 (1974)

Di(2-ethylhexyl) Phthalate

32. Mayer FL, Sanders HD; Environ Health Perspect 3: 153-7 (1973)
33. Mazurek MA, Simoneit BRT; Crit Rev Environ Control 16: 69 (1986)
34. Mes J, Campbell DS; Bull Environ Contam Toxicol 16: 53-60 (1976)
35. Metcalf RL et al; Environ Health Perspect 6: 27-34 (1973)
36. Murray HE et al; Bull Environ Contam Toxicol 26: 769-74 (1981)
37. NIOSH; The National Occupational Exposure Survey (NOES) (1983)
38. NIOSH; The National Occupational Hazard Survey (NOHS) (1974)
39. Peterson JC, Freeman DH; Environ Sci Technol 16: 464-9 (1982)
40. Preston MR, Al-Omran LA; Mar Pollut Bull 17: 548-53 (1986)
41. Price KS et al; J Water Pollut Contr Fed 46: 63-77 (1974)
42. Ray LE et al; Chemosphere 12: 1031-8 (1983)
43. Russell DJ, McDuffie B; Chemosphere 15: 1003-21 (1986)
44. Saeger VW, Tucker ES; Appl Environ Microbiol 31: 29-34 (1976)
45. Sanders HD et al; Environ Res 6: 84-90 (1973)
46. Schwartz HE et al; Int J Environ Anal Chem 6: 133-44 (1979)
47. Sheldon LS, Hites RA; Environ Sci Tech 13: 574-9 (1979)
48. Sheldon LS, Hites RA; Environ Sci Tech 12: 1188-94 (1978)
49. Statham CN et al; Science 193: 680-1 (1976)
50. Subba-Rao RV et al; Appl Environ Microbiol 43: 1139-50 (1982)
51. Sugatt RH et al; Appl Environ Microbiol 47: 601-6 (1984)
52. Sullivan KF et al; Environ Sci Tech; 16: 428-32 (1982)
53. Swain WR; J Great Lakes Res 4:398-407 (1978)
54. Tabak HH et al; J Water Pollut Contr Fed 53: 1503-18 (1981)
55. USEPA; Ambient Water Quality Criteria for Phthalate Esters EPA-440/5-80-067 p.C-4 (1980)
56. USEPA; Treatability Manual page I.65-3 EPA-600/2-82-001a (1982)
57. Veith G et al; J Fish Res Board Canada 36: 1040-8 (1979)
58. Williams DT; J Agr Food Chem 21: 1128-9 (1973)
59. Wolfe NL et al; Chemosphere 9: 393-402 (1980)
60. Wolfe NL et al; Environ Sci Technol 14: 1143-4 (1980)

Dimethyl Phthalate

SUBSTANCE IDENTIFICATION

Synonyms:

Structure:

CAS Registry Number: 131-11-3

Molecular Formula: $C_{10}H_{10}O_4$

Wiswesser Line Notation: 1OVR BVO1

CHEMICAL AND PHYSICAL PROPERTIES

Boiling Point: 283.7 °C at 760 mm Hg

Melting Point: 5.5 °C

Molecular Weight: 194.20

Dissociation Constants:

Log Octanol/Water Partition Coefficient: log Kow = 1.56 [11]

Water Solubility: 4,000 mg/L at 25 °C [16]

Vapor Pressure: 1.65 x 10^{-3} mm Hg at 25 °C [16]

Henry's Law Constant: 1.1 x 10^{-7} atm-m³/mole (calculated) [24]

Dimethyl Phthalate

ENVIRONMENTAL FATE/EXPOSURE POTENTIAL

Summary: Dimethyl phthalate (DMP) is released into the environment principally in industrial wastewater from its production and use as a plasticizer and mosquito repellent. Its primary loss mechanism appears to be biodegradation. Half-lives of 8-11 days and 0.2 days has been determined in river water, but no half-lives are available for soil or ground water, although DMP is utilized by soil microorganisms and degrades under anaerobic conditions. Little adsorption to soil or sediment will occur. DMP will not bioconcentrate in fish. If DMP is emitted into the atmosphere, it will be in the vapor phase and adsorbed to particulate matter, and will be subject to rainout and gravitational settling. Photodegradation by hydroxyl radicals will also occur (estimated half-life 23.8 hr) with vapor phase DMP. Humans will be exposed to DMP occupationally and from continued use of mosquito repellent.

Natural Sources:

Artificial Sources: DMP may be released as emissions (most probably in the form of aerosols) or wastewater during its production, storage, transport and use as a plasticizer, especially in cellulose esters and a component in lacquers, hair spray, coating agents, and vinyl plastic products. DMP may be released as leachate from plastic tubing and other containers [37].

Terrestrial Fate: If spilled on land, DMP should weakly sorb to most soils unless the soil has a very high organic content. Therefore, leaching through most soils to ground water is expected. It is readily biodegraded in screening tests and surfaces waters, especially after a short period of acclimation; however, no data could be found for soil samples.

Aquatic Fate: If released into water, DMP will primarily remain in the water column since it weakly sorbs to soil and sediment. Biodegradation will be the principal loss process in freshwater with a half-life of <11 days in river water. Based on concentration measurements between sampling points on the Rhine River, the half-life was determined to be only 0.2 days [42]. The rate of biodegradation in seawater is much longer. Although DMP biodegrades under anaerobic conditions, its fate in ground water is unknown. In situations where biodegradation is less important, other loss processes may be significant such as volatilization (shallow oligotrophic lakes or salt water bays), hydrolysis (alkaline bodies of water), and photolysis (clear surface waters), although data for the latter process are conflicting.

Atmospheric Fate: If released into air, DMP will be in the vapor phase as well as associated with particulate matter and will be subject to rainout and gravitational settling. It will be attacked by photochemically produced hydroxyl radicals, resulting in an estimated half-life of 23.8 hr.

Biodegradation: Microorganisms isolated from soil [38] and natural waters [35] are capable of utilizing DMP. In a river die-away test using Ogawa River water (Japan), 100% degradation occurred in 8-11 days in two samples [12]. Degradation in ocean water was much slower, with 20 and 32% degradation occurring in 2 weeks [12]. DMP was degraded in a shake-flask biodegradation test utilizing a soil/sewage inoculum with a half-life of 1.90 days after a 2.7-day lag [33]. After 28 days, >99% of the DMP had disappeared and 86% mineralization had occurred [33]. It was completely degraded within 7 days in a static-flask screening test with a wastewater inoculum [34]. DMP was also completely degraded when digested anaerobically in undiluted and 10% municipal sludge in 1 and 10 days, respectively, with 82% mineralization occurring in the 10% sludge [30]. In two operating plants, 88 and 58% of the DMP was mineralized by the digested municipal sludge [15]. Essentially 100% removals were reported in waste water treatment plants resulting from biodegradation [3,28,36]. Biodegradation is expected to be the principal loss process in lakes and ponds, with an estimated half-life of 13-27 hr due to biodegradation in a modelling study of simulated ecosystems [4].

Abiotic Degradation: DMP has a strong UV absorption spectrum from band 283 nm to beyond 290 nm [22] and, therefore, it is a candidate for photolysis with sunlight. When DMP in pure water was irradiated with a UV lamp through a Pyrex filter (>290 nm), its half-life was 12.7 hr [21]. This was reduced to 2.8 hr in the presence of nitrogen dioxide [21]. The half-life for direct photolysis in surface waters estimated from unpublished work is 3500 hr [40], in conflict with the above figure. The rate of oxidation by alkoxy radicals in natural waters is negligible [22]. The hydrolysis half-life of DMP is estimated to be 3.2 yr under neutral conditions at 30 °C [4]. At pH 9, the estimated half-lives are 11.6 days and 25 days at 30 deg and 18 °C, respectively [41]. The hydrolysis rate under acidic environmental conditions is estimated to be very low compared to neutral conditions [40]. Photochemically produced hydroxyl radicals will degrade vapor phase DMP by aromatic ring addition and hydrogen abstraction with an estimated half-life of 23.8 hr [10].

Bioconcentration: The mean bioconcentration factors of DMP in brown shrimp and sheepshead minnows was 4.7 and 5.4 after 24 hr [39], which indicates very low bioconcentration. Bluegill sunfish showed a bioconcentration factor of 57 [2], which may be elevated because only ^{14}C was measured in the experiment and metabolites may be included with the parent compound. The depuration half-life was between 1 and 2 days [2].

Soil Adsorption/Mobility: The Koc for DMP has been estimated to be 160 [40] and 44 [20], which indicates low adsorption to soil and sediment.

Volatilization from Water/Soil: From the calculated Henry's Law constant, it is estimated that the half-life for DMP in a river 1 m deep with a 1 m/sec current and a 3 m/sec wind is 46 days, with the rate being controlled by the diffusion through the air layer [24]. Less than 1% of DMP was lost in a sewage treatment plant by air stripping [28]. Due to its low vapor pressure, evaporation from soil and solid surfaces should not be a significant loss process.

Water Concentrations: DRINKING WATER: Three New Orleans drinking water plants 0.13-0.27 ppb [19]. Philadelphia drinking water, but not quantified [32]. Kitakyushu, Japan tap water, but not quantified [1]. Detected in treated drinking water from 6 of 14 sites tested in England; 5 drinking water sites were from rivers and 1 was from a ground water source [8]. Philadelphia, PA, 1975-76, identified, not quantified, in finished drinking water from 1 of 3 treatment plants, not detected in tapwater from 1 hotel [31].

Dimethyl Phthalate

SURFACE WATER: Merrimack River, detected, not quantified [13]. Rhine River at Lobith, The Netherlands 0.3 ppb [42]. Concn (ppt) in Mississippi River, Lake Itasca, MN (river's source), not detected; Cairo, IL, 2; 20 miles south of Memphis, TN, 5; Carrolton St. water intake in New Orleans, LA, not detected [7]. **RAIN:** Norway - detected, not quantified in the water phase of rain and snow [23].   **SEAWATER:** Mersey Estuary, UK, 17 sites, 94% pos, dissolved fraction, 82% pos, 48-973 ppt; particulate fraction, 65% pos, 14.6-140 ppb (dry wt) [29].

Effluent Concentrations: Mean effluent levels >100 ppb occur from foundries (280 ppb), metal finishing (200 ppb), and the organic chemicals manufacturing/plastics (510 ppb) industries [36]. Industries whose maximum effluent levels exceeded 1000 ppb were: metal finishing (1200 ppb), foundries (3200 ppb), and nonferrous metals manufacturing (1300 ppb) [36]. DMP was not detected in urban runoff in the Nationwide Urban Runoff Program, which included 86 samples from 19 cities across the U.S. [6]. Fort Polk, LA (secondary effluent) 0.77 ppb [17]. Lake Superior - effluent from pulp and paper mill - detected as minor component [9]. Ground water from four U.S. rapid - infiltration sites for municipal primary and secondary effluents, 75% pos, 0.01-0.19 ppb, and 0.10 ppb avg [18].

Sediment/Soil Concentrations: Chester River (downstream from plasticizer manufacturer) near where it joins Chesapeake Bay (12 sites), not detected [27].   Raccoon Creek (tributary of Delaware River) at Bridgeport, NJ (surficial bed material), trace [14].

Atmospheric Concentrations: Urban air, Belgium, avg concn (ug/1000 m^3): particulates, 101, gas phase, 353 [5].

Food Survey Values:

Plant Concentrations:

Fish/Seafood Concentrations:

Animal Concentrations:

Milk Concentrations:

Other Environmental Concentrations:

Probable Routes of Human Exposure: Humans will be exposed to DMP principally in occupational settings. Plastic tubing and containers which use DMP as a plasticizer might leach it into food or beverages.

Average Daily Intake:

Occupational Exposures: NIOSH (NOES Survey 1981-1983) has statistically estimated that 34,134 workers are exposed to DMP in the United States [26]. NIOSH (NOHS Survey 1972-1974) has statistically estimated that 53,096 workers are exposed to DMP in the United States [25].

Body Burdens:

REFERENCES

1. Akiyama T et al; J UOEH 2: 285-300 (1980)
2. Barrows ME et al; pp.379-92 in Dyn Exposure Hazard Assess, Ann Arbor, MI (1980)
3. Barth EF, Branch RL; Biodegradation and treatability of specific pollutants. USEPA-600/9-79-034 (1979)
4. Callahan MA et al; Water-related fate of priority pollutants Vol.II pp.94-1 to 94-28 USEPA-440/4-79-029b (1979)
5. Cautreels W, VanCauwenberghe K; Atmos Environ 12: 1133-41 (1978)
6. Cole RH; J Water Pollut Control Fed 56: 898-908 (1984)
7. DeLeon IR et al; Chemosphere 15: 795-805 (1986)
8. Fielding M et al; Environ Tech Lett A 2: 545-50 (1981)
9. Fox ME; J Fish Res Board Canada 34: 798-804 (1977)
10. GEMS; Graphical Exposure Modeling System. Fate of Atmospheric Pollutants (FAP) Data Base. Office of Toxic Substances, USEPA (1987)
11. Hansch C, Leo AJ; Medchem Project Issue No 26. Claremont CA: Pomona College (1985)
12. Hattori Y et al; Pollut Control Cent Osaka Prefect Mizu Shori Gijutsu 16: 951-4 (1975)
13. Hites RA; Environ Health Perspect 3: 17 (1973)
14. Hochreiter JJ Jr; Chemical-quality reconnaissance of the water and surficial bed material in the Delaware River Estuary and adjacent New Jersey tributaries, 1980-81. USGS/WRI/NTIS 82-36 (1982)
15. Horowitz A et al; Dev Ind Microbiol 23: 435-44 (1982)
16. Howard PH et al; Environ Toxicol Chem 4: 653-61 (1985)
17. Hutchins SR et al Environ Toxicol Chem 2: 195-216 (1983)

291

Dimethyl Phthalate

18. Hutchins SR, Ward CH; J Hydrol 67: 223-33 (1984)
19. Keith LH et al; pp.329-73 in Identification and Analysis of Organic Pollutants in Water Keith LH ed; Ann Arbor Press, Ann Arbor, MI (1976)
20. Kenaga EE; Ecotox Environ Safety 4: 26-38 (1980)
21. Kotzias D et al; Naturwissenschaften 69: 444-5 (1982)
22. Leyder F, Boulanger P; Bull Environ Contam Toxicol 30: 152-7 (1983)
23. Lunde G et al; Organic Micropollutants in Precipitation in Norway pp.17 SNSF Project FR-9/76 (1977)
24. Lyman WJ et al; Handbook of chemical property estimation methods. Environmental behavior of organic compounds McGraw-Hill NY (1982)
25. NIOSH; The National Occupational Hazard Survey (NOHS) (1974)
26. NIOSH; The National Occupational Exposure Survey (NOES) (1983)
27. Peterson JC, Freeman DH; Int J Environ Analyt Chem 18: 237-52 (1984)
28. Petrasek AC et al; J Water Pollut Control Fed 55: 1286-96 (1983)
29. Preston MR, Al-Omran LA; Mar Pollut Bull 17: 548-53 (1986)
30. Shelton DR et al; Environ Sci Technol 18: 93-7 (1984)
31. Suffet IH et al; Water Res 14: 853-67 (1980)
32. Suffet IH; pp.375-97 in Identification and Analysis of Organic Pollutants in Water Keith LH ed; Ann Arbor Press, Ann Arbor, MI (1976)
33. Sugatt RH et al; Appl Environ Microbiol 47: 601-6 (1984)
34. Tabak HH et al; J Water Pollut Control Fed 531: 1503-18 (1981)
35. Taylor BF et al; Appl Environ Microbiol 42: 590-5 (1981)
36. USEPA Treatability Manual; pp.I6.1-1 to I6.1-6 USEPA-600/2-82-001a (1981)
37. Verschueren K; Handbook of Environmental Data on Organic Chemicals 2nd ed; Van Nostrand Reinhold, New York pp.561-2 (1983)
38. Williams GR, Dale R; Int Biodeter Bull 19: 37-8 (1983)
39. Wofford HW et al; Ecotox Environ Safety 5: 202-10 (1981)
40. Wolfe NL et al; Chemosphere 9: 393-402 (1980)
41. Wolfe NL et al; Chemosphere 9: 403-8 (1980)
42. Zoeteman BC et al; Chemosphere 9: 231-49 (1980)

2,4-Dimethylphenol

SUBSTANCE IDENTIFICATION

Synonyms: 2,4-Xylenol

Structure:

CAS Registry Number: 105-67-9

Molecular Formula: $C_8H_{10}O$

Wiswesser Line Notation:

CHEMICAL AND PHYSICAL PROPERTIES

Boiling Point: 211.5 °C

Melting Point: 25.4-26 °C

Molecular Weight: 122.16

Dissociation Constants: 10.63 [24]

Log Octanol/Water Partition Coefficient: 2.30 [11]

Water Solubility: 6,200 mg/L at 25 °C [15]

Vapor Pressure: 0.098 mm Hg at 25 °C [15]

Henry's Law Constant: 6.3 x 10^{-7} atm-m³/mole at 8 °C [15]

ENVIRONMENTAL FATE/EXPOSURE POTENTIAL

Summary: 2,4-Dimethylphenol is released to the environment as fugitive emissions and in wastewater as a result of coal tar refining, coal processing, and in its use in chemical/plastics manufacturing, etc. When released in water, it will degrade principally due to biodegradation with a half-life of hours to days. In humic waters, oxidation by alkyl peroxy radicals may also be important. Adsorption to sediment and particulate matter in the water column will only be moderate and bioconcentration in fish should not be significant. If spilled on soil, 2,4-dimethylphenol will probably adsorb moderately to the soil and biodegrade in several days. In the atmosphere, vapor phase 2,4-dimethylphenol will degrade during daylight hours by reaction with photochemically produced hydroxyl radicals (half-life 8 hr). At night it will probably degrade very rapidly by reaction with nitrate radicals. Washout by rain will also be an effective removal process. Human exposure will be primarily from dermal contact with chemicals containing phenolic mixtures and inhalation by occupationally exposed workers.

Natural Sources: Coal [29]. Dimethylphenols have been identified as naturally-occurring constituents of some plants such as tea, tobacco, marijuana, and the Siberian pine (<u>Abies</u> <u>sibirica</u>) [28].

Artificial Sources: 2,4-Dimethylphenol may be emitted as fugitive emissions and in wastewater during coal processing and coal tar refining [29]. It may also be released during its use in the manufacture of plastics and resins, pharmaceuticals, insecticides and fungicides, disinfectants, solvents, etc [29]. It may also be released in asphalt and roadway runoff and in domestic sewage [29]. Gasoline and diesel exhaust and tobacco smoke are additional sources [9,10].

Terrestrial Fate: If spilled on soil, 2,4-dimethylphenol would adsorb moderately to soil and it is reported to biodegrade in soil in 4 days [20].

Aquatic Fate: When released in water, 2,4-dimethylphenol may adsorb moderately to sediment and will be readily biodegradable. The biodegradation rates are comparable to that of phenol in one river but biodegradation rates were not given. The half-life should be less than several days in humic waters due to photooxidation by alkylperoxy radicals. Photolysis may occur in clear surface waters.

Atmospheric Fate: If released in air, vapor phase 2,4-dimethylphenol should degrade by reaction with photochemically produced hydroxyl radicals (half-life 8 hr). At night it would be rapidly attacked by nitrate radicals in urban areas. Scavenging by rain will be an effective removal process, as is reflected by the concentrations in rainwater.

Biodegradation: 95% removal in 5 days was achieved in screening studies using activated sludge [23] and 100% degradation occurred in 7 days with a sewage seed [27]. It readily degraded in St. Lawrence river water, being the most readily degradable compounds phenol other than phenol itself (30 ug/L-hr) [30]. The 98% loss in a wastewater plant is almost entirely attributable to biodegradation [22]. 2,4-Dimethylphenol degraded in 4 days in a hard, carbonaceous woody capital loam at 19 °C [20]. At 5 °C the degradation time increased 3-4 times [20].

Abiotic Degradation: 2,4-Dimethylphenol has an absorption band (maxima at 296 nm) which extends beyond 320 nm and is therefore a candidate for direct photolysis [5]. No photolysis rates either in air on water could be found in the literature. It will react by ring addition with hydroxyl radicals, having an estimated reaction half-life of 8 hr [7]. o-Cresols react very rapidly with nitrate radicals (half-life 2 minutes) [2] and this should be a very important loss process for 2,4-dimethylphenol at night. Peroxy radicals which are found in humic waters react with phenols and the half-lives are less than several days [18,21].

Bioconcentration: The log bioconcentration factor in bluegills as determined in a 28-day experiment in a flow through system was 1.18 [3].

Soil Adsorption/Mobility: No experimental data on the adsorption of 2,4-dimethylphenol to soil is available. However, based on the log octanol/water partition coefficient, a Koc of 425 was estimated [17], indicating a moderate adsorption to soil.

Volatilization from Water/Soil: Because of the Henry's Law constant, volatilization from water would not be a significant transport process [17]. The estimated loss due to vapor stripping in a wastewater treatment plant has been estimated to be <1% [22].

2,4-Dimethylphenol

Water Concentrations: DRINKING WATER: Listed as having been identified in drinking water in the United States [14,25]. SURFACE WATER: Detected in 1% of 804 samples reported in the USEPA STORET database for 1975-82 [26]. Immediately downstream from a waste input into a creek from a former pinetar manufacturing facility - Gainesville, FL in the 10-200 u/L range [19]. GROUND WATER: Detected in ground water at all 4 sites investigated in a sand aquifer at Pensacola, FL underlying a wood-preserving facility - 0-5.65 mg/L [8]. While detected in samples from 6,12,18 and 24 m depth, not detected at 30 m depth [8]. Detected in 10 of 11 wells underlying a former pinetar manufacturing facility in Gainesville, FL 1-9400 ug/L (including 2,5-dimethylphenol) [19]. RAINWATER: The concentration of 2,4- and 2,5-dimethylphenol, combined in rainwater for seven events in Portland, OR ranged from 300 to 1300 ng/L, 820 ng/L avg [15].

Effluent Concentrations: 2,4-Dimethylphenol was found in 6 effluents in an EPA survey (4000 samples) of effluents covering 46 industrial categories [4]. Industries with positive levels of 2,4-dimethylphenol included iron and steel manufacturing, petroleum refining, organics and plastics, rubber processing, organic chemicals, and publicly owned treatment works [4]. Detected in 3.4% of 1321 effluent samples reported in STORET database (1975-82) [26]. Detected in 2 out of 5 effluents of hazardous waste incinerators [13]. Detected at 10 ug/L in urban runoff in Washington, DC [6]. This constituted a 2% frequency of detection in the National Urban Runoff Program which examined 86 runoff samples from 15 U.S. cities [6]. Final effluent of Los Angeles County Municipal Wastewater Treatment Plant, 5 and <10 ppb in July 1978 and Nov 1980, respectively [31]. Of 18 advanced water treatment effluents analyzed, found in effluents at Lake Tahoe (2 ng/L) and Blue Plains, WA (1 and 8.9 ng/L) [16].

Sediment/Soil Concentrations: 0.3% of 310 samples from STORET database reported detectable amounts of 2,4-dimethylphenol in sediment in 1975-82 [26]. Detected in soil sample at the site of a former pinetar manufacturer in Gainesville, FL [19].

Atmospheric Concentrations: The gas phase concentration of 2,4- and 2,5-dimethylphenol combined during 7 rain events in Portland,

OR ranged from 15-70 ng/m³, 33 ng/m³ mean [15]. The amount associated with particulate matter was <5% in every case and generally <1% of the gas phase concentration [15]. Detected at 1 ug/m³ in air outside an oil shale wastewater facility at Logan Wash, CO but not in an undeveloped site in the oil shale region or in Boulder, CO [12]. Although found in gasoline and diesel motor exhaust, not detected in Allegheny Mountain tunnel, PA [10].

Food Survey Values:

Plant Concentrations:

Fish/Seafood Concentrations: Not detected in 99 samples of fish reported in STORET database in 1975-82 [26].

Animal Concentrations:

Milk Concentrations:

Other Environmental Concentrations:

Probable Routes of Human Exposure: The primary routes of occupational exposure to 2,4-dimethylphenol is considered to be dermal due to occupational and the general public's use of commercial products containing complex mixture of phenols [28]. In addition, workers may be exposed to vapors and dermal contact during petroleum, coal and coke processing, and degreasing operations [28].

Average Daily Intake:

Occupational Exposures:

Body Burdens: Urine from 35 varnishing workers in 6 different workplaces contained 2,4-dimethylphenol mean (SD) 3.8(3.0) mg/L [1]. Phenols are metabolites of the benzene and alkylbenzenes in the varnish [1].

2,4-Dimethylphenol

REFERENCES

1. Angerer J, Wulf H; Int Arch Occup Environ Health 56: 307-21 (1985)
2. Atkinson R; Chem Rev 85: 69-201 (1985)
3. Barrows ME et al; pp 379-92 in Dyn Exposure Hazard Assess Toxic Chem Ann Arbor, MI Ann Arbor Science (1980)
4. Bursey JT, Pellizzari ED; Analysis of Industrial Wastewater for Organic Pollut in Consent Degree Survey p 167 USEPA Contract No. 68-03-2867 (1982)
5. Callahan MA et al; Water-Related Environ Fate of 129 Priority Pollut Vol.II USEPA-440/4-79-029B (1979)
6. Cole et al; J Water Pollut Control Fed 56: 898-908 (1984)
7. GEMS; Graphical Exposure Modeling System. FAP. Fate of Atmos Pollut (1986)
8. Goerlitz DF et al; Environ Sci Technol 19: 955-61 (1985)
9. Graedel TE; Chemical Compounds in the Atmosphere Academic Press NY p 257 (1978)
10. Hampton CV et al; Environ Sci Technol 16: 287-98 (1982)
11. Hansch C, Leo AJ; Medchem Project Issue No 26. Claremont CA: Pomona College (1985)
12. Hawthorne SB, Sievers RE; Environ Sci Technol 18: 483-90 (1984)
13. James RH et al; Proc APCA Annu Meet Vol.1, pp 84-18.5 (1984)
14. Kopfler FC et al; Human Exposure to Water Pollutants Adv Environ Sci Technol 8 (Fate Pollut Air Water Environ): 419-33 (1977)
15. Leuenberger C et al; Environ Sci Technol 19: 1053-8 (1985)
16. Lucas SV; GC/MS Analysis of Organics in Drinking Water concentrates and Advanced Waste Treatment Concentrates, Vol.1 p 321 USEPA-600/1-84-020A NTIS PB85-128221 (1984)
17. Lyman WJ et al; Handbook of Chemical Property Estimation Methods. Environmental Behavior of Organic Compounds. McGraw-Hill NY (1982)
18. Mabey WR et al; Aquatic Fate Process Data for Priority Pollut USEPA-440/4-81-014 (1981)
19. McCreary JJ et al; Chemosphere 12: 1619-32 (1983)
20. Medvedev VA, Davidov VD; pp. 245-54 in Decomposition of Toxic and Nontoxic Compounds in Soil. Overcash MR ed Ann Arbor MI: Ann Arbor Sci Publ (1981)
21. Mill T; Environ Toxicol Chem 1: 135-141 (1982)
22. Petrasek AC et al; J Water Pollut Control Fed 55: 1286-96 (1983)
23. Pitter P; Water Res 10: 231-5 (1976)
24. Riddick JA et al; Organic Solvents: Physical Properties and Methods of Purification, 4th Edit. New York: J Wiley & Sons (1986)
25. Shakelford NM, Keith LH; Frequency of Organic Compounds Identified in Water USEPA-600/4-76-062 (1976)
26. Staples CA et al; Environ Toxicol Chem 4: 131-42 (1985)
27. Tabak HH et al; J Water Pollut Contr Fed 53: 1503-18 (1981)
28. USEPA; AWQC for 2,4-dimethylphenol PB81-117558 (1980)
29. Verschueren K; Handbook of Environ Data on Org Chemicals 2nd ed Von Nostrand Reinhold NY p 1195 (1983)
30. Visser SA et al; Arch Environ Contam Toxicol 6: 455-70 (1977)
31. Young DR et al; 4: 871-84 in Water Chlorin: Envir Impact Health Eff (1983)

2,4-Dinitrophenol

SUBSTANCE IDENTIFICATION

Synonyms:

Structure:

CAS Registry Number: 51-28-5

Molecular Formula: $C_6H_4N_2O_5$

Wiswesser Line Notation:

CHEMICAL AND PHYSICAL PROPERTIES

Boiling Point:

Melting Point: 112 °C

Molecular Weight: 184.11

Dissociation Constants: pKa = 3.94 [14]

Log Octanol/Water Partition Coefficient: 1.54 [4]

Water Solubility: 6000 mg/L at 25 °C [12]

Vapor Pressure: 2 x 10^{-5} mm Hg at 25 °C (calculated) [6]

Henry's Law Constant: 8 x 10^{-10} atm-m³/mole (calculated from water solubility and vapor pressure)

2,4-Dinitrophenol

ENVIRONMENTAL FATE/EXPOSURE POTENTIAL

Summary: 2,4-Dinitrophenol may enter the environment in emissions or effluents from manufacturing plants, mines, foundries, and other facilities at which it is used. It may also be released in automobile exhaust gas, during its use as a pesticide, or as a result of disposal of products which contain this compound. It may form as a result of photochemical reaction between benzene and nitrogen monoxide in polluted air. If released to soil, 2,4-dinitrophenol is expected to be highly mobile, although there is a possibility that some of this compound will adsorb to clay minerals. 2,4-Dinitrophenol may inhibit microbial growth of some aerobic microbes, but there are other microorganisms which may degrade this compound in the environment; possible biotransformation mechanisms include the reduction of the nitro group, hydroxylation of the aromatic ring, and displacement of the nitro group; possible biodegradation products include, 2-amino-4-nitrophenol, 4-amino-2-nitrophenol, and nitrite. 2.4-Dinitrophenol is not expected to volatilize significantly from wet or dry soil surfaces. If released to water, 2,4-dinitrophenol is expected to react with alkylperoxy radicals (calculated half-life 58 days) and it has the potential to photolyze due to absorption of UV light wavelengths >290 nm. 2,4-Dinitrophenol is not expected to undergo aerobic biodegradation, bioaccumulate in aquatic organisms, or volatilize significantly. This compound is not expected to adsorb significantly to suspended solids or sediments, although there is a possibility that some of this compound will adsorb to clay minerals. If released to air, 2,4-dinitrophenol is expected to exist in both vapor and particulate form. It may photolyze, it may be physically removed by settling or washout in precipitation, or it may react with photochemically generated hydroxyl radicals (calculated vapor-phase half-life 14 hr). The most probable route of human exposure to 2,4-dinitrophenol is by inhalation or dermal contact of workers involved in the manufacture, handling, and use of this compound.

Natural Sources:

Artificial Sources: 2,4-Dinitrophenol may enter the environment in emissions or effluents from manufacturing plants, mines, foundries, metal plants, petroleum plants, dye manufacturing plants, and other

300

2,4-Dinitrophenol

facilities at which 2,4-dinitrophenol is used [12]. 2,4-Dinitrophenol may also be released to the environment in automobile exhaust gas [10], during its use as a pesticide [19], or as a result of disposal of products which contain this compound. It may form as a result of photochemical reaction of benzene with nitrogen monoxide in highly polluted air [11].

Terrestrial Fate: If released to soil, 2,4-dinitrophenol is expected to be highly mobile, although there is a possibility that some of this compound will adsorb to clay minerals. 2,4-Dinitrophenol may inhibit microbial growth of some aerobic microbes, but there are soil microorganisms which may degrade this compound in the environment. Possible biotransformation mechanisms include the reduction of the nitro group, hydroxylation of the aromatic ring, and displacement of the nitro group; possible biodegradation products include 2-amino-4-nitrophenol, 4-amino-2-nitrophenol, and nitrite. 2,4-Dinitrophenol is not expected to volatilize from soil surfaces.

Aquatic Fate: If released to water, 2,4-dinitrophenol is not expected to adsorb significantly to suspended solids or sediments, although there is a possibility that some of this compound will adsorb to clay minerals. 2,4-Dinitrophenol has the potential to undergo direct photolysis due to absorption of UV light wavelengths >290 nm or it may react with alkylperoxy radicals (calculated half-life 58 days). 2,4-Dinitrophenol is not expected to undergo aerobic biodegradation, bioaccumulate in aquatic organisms, or volatilize significantly.

Atmospheric Fate: If released to the atmosphere, 2,4-dinitrophenol may exist in either vapor or particulate form. It may undergo direct photolysis due to absorption of UV light wavelengths >290 nm, it may be physically removed by settling or washout in precipitation, or it may react in the vapor phase with photochemically generated hydroxyl radicals (calculated vapor-phase half-life 14 hr).

Biodegradation: It is reported that nitrophenols can inhibit aerobic microbial growth by uncoupling the metabolic process of oxidative phosphorylation [1]. Static incubation of 5 and 10 mg/L of 2,4-dinitrophenol seeded with settled domestic wastewater resulted in 60 and 68% degradation, respectively, in 7 days [18]. Oxygen uptake by mixed cultures of phenol adapted microorganisms suggests that 2,4-dinitrophenol was slowly degraded under aerobic conditions [2,17]. Possible biotransformation processes of 2,4-dinitrophenol are: reduction of the nitro group, hydroxylation of the aromatic ring, and displacement of the nitro group by a hydroxy group [12]. A pure culture of the fungus Fusarium oxysporum was found to reduce 2,4-dinitrophenol to 2-amino-4-nitrophenol and 4-amino-2-nitrophenol [12]. Nitrite release has been observed during the metabolism of 2,4-dinitrophenol by pure cultures of Nocardia alba, Arthrobacter and Corynebacterium simplex [12].

Abiotic Degradation: Chemical hydrolysis of 2,4-dinitrophenol is not expected to be environmentally relevant [8]. 2,4-Dinitrophenol in methanol exhibits strong absorption of UV light in the environmentally significant range (wavelength >290 nm) [13], suggesting that 2,4-dinitrophenol has the potential to undergo direct photolysis in water and in the atmosphere. Nitroaromatic compounds are generally reduced photochemically in the presence of suitable hydrogen donors, thus adsorption of 2,4-dinitrophenol to organic material may lead to photoreduction [1]. The half-life for 2,4-dinitrophenol reacting with alkylperoxy radicals in water has been calculated to be 58 days using a calculated reaction rate constant of $5 \times 10^{+5}$ mol/L-hr [8] and an ambient alkylperoxy radical concn of 1×10^{-9} mol/L [9]. In the atmosphere, the half-life for 2,4-dinitrophenol vapor reacting with photochemically generated hydroxyl radicals has been calculated to be 14 hours using a calculated reaction rate constant of 1.7×10^{-11} cm^3/molecule-sec at 25 °C and an ambient hydroxyl radical concn of $8.0 \times 10^{+5}$ molecules/cm^3 [3].

Bioconcentration: Using the water solubility and the octanol/water partition coefficient, the bioconcentration factor (BCF) for 2,4-dinitrophenol has been calculated to be <10 using recommended regression equations [7]. This value suggests that 2,4-dinitrophenol would not bioaccumulate significantly in aquatic organisms.

Soil Adsorption/Mobility: Using the water solubility and the octanol/water partition coefficient, the soil adsorption coefficient (Koc) for 2,4-dinitrophenol has been calculated to be 36 and 164, respectively, using recommended regression equations [7]. These Koc values indicated that 2,4-dinitrophenol should be highly mobile in soil [16] and should not adsorb significantly to suspended solids or sediments in water. There is a possibility that 2,4-dinitrophenol may adsorb to clay minerals, since 2-nitrophenol and 4-nitrophenol appear to form very stable complexes with clay [1].

Volatilization from Water/Soil: The value for Henry's Law constant indicates that volatilization from water and moist soil surfaces is probably not a significant process, and the relatively low vapor pressure of 2,4-nitrophenol indicates that volatilization from dry soil surfaces should not be a significant process.

Water Concentrations: During 1980, 2,4-dinitrophenol was detected in sediment/water/soil samples at Love Canal [5]. 2,4-Dinitrophenol has been monitored at USEPA STORET stations, 812 samples, 0.4% positive [15].

Effluent Concentrations: 2,4-Dinitrophenol has been detected in effluents at USEPA STORET stations, 1311 samples, <2% pos [15].

Sediment/Soil Concentrations: During 1980, 2,4-dinitrophenol was detected in sediment/water/soil samples at Love Canal [5].

Atmospheric Concentrations: During May 1982, 2,4-dinitrophenol was detected in airborne particulate matter collected in Yokohama, Japan where photochemical smog had been observed [10].

Food Survey Values:

Plant Concentrations:

Fish/Seafood Concentrations:

Animal Concentrations:

Milk Concentrations:

Other Environmental Concentrations: 2,4-Dinitrophenol has been detected in automobile exhaust gas [10].

2,4-Dinitrophenol

Probable Routes of Human Exposure: The most probable route of human exposure to 2,4-dinitrophenol is by inhalation or dermal contact of workers involved in the manufacture, handling, and use of this compound.

Average Daily Intake:

Occupational Exposures:

Body Burdens:

REFERENCES

1. Callahan MA et al; Water-Related Environ Fate of 129 Priority Vol.II USEPA-440/4-79-029B (1979)
2. Chambers CW et al; J Water Pollut Control Fed 35: 1517 (1963)
3. GEMS; Graphical Exposure Modeling System. FAP. Fate of Atmos Pollut (1986)
4. Hansch C, Leo AJ; Medchem Project Issue No.26 Pomona College Claremont, CA (1985)
5. Hauser TR, Bomberg SM; Environ Monit Assess 2: 249-72 (1982)
6. Hoyer H, Peperle W; Z Elaktrochem 62: 61-6 (1958)
7. Lyman WJ et al; Handbook of Chemical Property Estimation Methods. Environmental Behavior of Organic Compounds. McGraw-Hill NY (1982)
8. Mabey WR et al; Aquatic Fate Process Data for Org Priority Pollut USEPA-440/4-81-014 (1981)
9. Mill T et al; Science 207: 886 (1980)
10. Nojima K et al; Chem Pharm Bull 31: 1047-51 (1983)
11. Nojima K et al; Chemosphere 4: 77 (1975)
12. Overcash MR et al; Behavior of Org Priority Pollut in the Terrestrial System, Di-n-butyl phthalate ester, Toluene and 2,4-dinitrophenol NTIS PB82-224544 (1982)
13. Sadtler Res Lab; Sadtler Standard UV Spectra No.3234
14. Schwarzenbach RP et al; Environ Sci Technol 22: 83-92 (1988)
15. Staples CA et al; Environ Tox Chem 4: 131-42 (1985)
16. Swann RL et al; Res Rev 85: 17-28 (1983)
17. Tabak HH et al; J Bacteriol 87: 910-9 (1964)
18. Tabak HH et al; J Water Pollut Control Fed 53: 1503-18 (1981)
19. Windholz M et al; Merck Index 10th ed. Rahway NJ (1983)

2,4-Dinitrotoluene

SUBSTANCE IDENTIFICATION

Synonyms: 1-Methyl-2,4-dinitrobenzene

Structure:

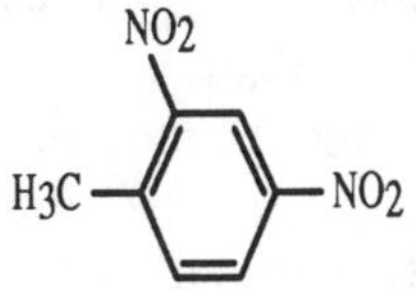

H_3C —benzene ring— NO_2 (with NO_2 at top)

CAS Registry Number: 121-14-2

Molecular Formula: $C_7H_6N_2O_4$

Wiswesser Line Notation: WNR B1 ENW

CHEMICAL AND PHYSICAL PROPERTIES

Boiling Point: 300 °C with slight decomp

Melting Point: 71 °C

Molecular Weight: 182.14

Dissociation Constants:

Log Octanol/Water Partition Coefficient: 1.98 [6]

Water Solubility: 300 mg/L at 22 °C [3]

Vapor Pressure: 1.1 x 10^{-4} mm Hg at 20 °C [26]

Henry's Law Constant: 8.67 x 10^{-7} atm-m^3/mole [25]

ENVIRONMENTAL FATE/EXPOSURE POTENTIAL

Summary: A major use of 2,4-dinitrotoluene (2,4-DNT) is in making 2,4-diaminotoluene. Dinitrotoluenes are also used in organic synthesis, dyes, explosives, and as a propellant additive.

2,4-Dinitrotoluene

2,4-DNT may enter the environment in wastewater from the processes in which it is made and used. In soil, 2,4-DNT will be slightly mobile. Based on aqueous biodegradation tests, 2,4-DNT may biodegrade in both aerobic and anaerobic zones of soil. 2,4-DNT in water will not bioconcentrate significantly and will have a slight tendency to partition to suspended and sediment organic matter. Volatilization of 2,4-DNT from water will not be significant. Photolysis will probably be the most important removal process for 2,4-DNT in water. Photolytic half-lives for 2,4-DNT in river, bay, and pond waters were 2.7, 9.6, and 3.7 hr, respectively. One source gave a theoretical half-life of 11 hr for 2,4-DNT in natural waters, while another estimated the half-life as 1.7 days in the Rhine River. The importance of biodegradation in natural waters is unknown, although a number of conflicting screening test results are available. Anaerobic biodegradation to the amine may be quite rapid. In the atmosphere, 2,4-DNT is estimated to have a half-life of 8 hr. 2,4-DNT has been detected in drinking water, seawater, river water, and in wastewater from 2,4,6-trinitrotoluene production.

Natural Sources:

Artificial Sources: 2,4-DNT is currently produced by a number of companies [31] and the mixture of 2,4- and 2,6-DNT was produced in over 0.5 billion lb/yr in 1982 [33]. A major use of 2,4-DNT is in making 2,4-diaminotoluene [10] for isocyanate production and, therefore, release from its production and use for isocyantes may be a major source of release to the environment. Dinitrotoluenes are also used in organic synthesis, dyes, explosives [9], and as a propellant additive [22], and releases to the environment from these applications may also occur.

Terrestrial Fate: The estimated soil adsorption coefficient (Koc) for 2,4-DNT and the measured sediment adsorption coefficient indicate that it is slightly mobile in soil. Aromatic nitro compounds are not susceptible to hydrolysis [15] and photolysis should not be an important process in soil. No information was found on 2,4-DNT biodegradation in soil; however, based on aqueous biodegradation experiments, some biodegradation may occur in both aerobic and anaerobic zones of soil.

2,4-Dinitrotoluene

Aquatic Fate: The estimated Koc value and the measured sediment adsorption coefficient indicates that 2,4-DNT will have a slight tendency to sorb to sediments, suspended solids, and biota. The volatilization rate constant for 2,4-DNT from a body of water was estimated as 6.6 x $10^{-5}hr^{-1}$ (half-life = 438 days) [27] and an EXAMS II estimate of the half-life of volatilization was 1.1-7.6 years. These estimates indicate that volatilization of 2,4-DNT from water will probably not be a significant transport process. Photolysis is probably the most significant removal process for 2,4-DNT in water. The photolytic half-lives for 2,4-DNT in river, bay, and pond waters were 2.7, 9.6, and 3.7 hr, respectively [27], and the rate will be dependent upon the amount of humic material that is present. Biodegradation data are inconsistent. 2,4-DNT degraded in some screening tests but not in others. In natural water samples, 2,4-DNT does not degrade unless large amounts of supplemental carbon are added. 2,4-DNT degraded fairly rapidly under anaerobic conditions [5] to give the amine derivatives. Aromatic nitro compounds are not susceptible to hydrolysis [15]. One source gave a theoretical half-life for 2,4-DNT in natural waters as 11.0 hr [1], while another estimated the half-life of 2,4-DNT as 1.7 days in the Rhine River [35], the latter being based upon monitored data.

Atmospheric Fate: No information was found in the available literature about the fate of 2,4-DNT in the atmosphere; however, an estimated half-life for 2,4-DNT based upon vapor phase reaction with photochemically produced hydroxyl radical in the atmosphere is 8 hr [4].

Biodegradation: 2,4-DNT, as the sole carbon and energy source, did not degrade in an aerobic batch culture after 14 days of incubation [13]. 2,4-DNT concentration was from 5-25 mg/L and the inoculum was municipal activated sludge [13]. 2,4-DNT degraded rapidly in 2 days in a batch culture containing glucose and inoculated with an industrial seed [2]. The industrial seed contained four bacteria (Actinobacter, Alcaligenes, Flavobacterium and Pseudomonas) and one yeast (Rhodotorula). Municipal seed (conventional activated sludge organisms), however, were inhibited by 2,4-DNT concentrations as low as 10 mg/L [2]. 2,4-DNT biodegradability was studied in a static screening test in which each flask contained 2,4-DNT at either 5 or 10 mg/L, 5 mg/L

307

yeast extract, and settled domestic sewage as the inoculum; percent biodegradation after 7 days in the original culture and the first, second, and third subculture were: (5 mg/L initial concentration) 77, 61, 50, and 27%; (10 mg/L initial concentration) 50, 49, 44, and 23% [32]. The percent of 2,4-DNT biodegraded decreased with each subculture. Degradation of 2,4-DNT in natural waters depends upon whether sediment or yeast extract were added. For example, 0% of 10 ppm were removed from water taken from a pond near Searsville Lake in 42 days, although 100% was removed in 5 days when 500 ppm of yeast extract were added [29]. Of 190 fungi representing 98 genera screened, only 5 organisms were able to transform 2,4-DNT at an initial concentration of 100 mg/L [19]. Under anaerobic conditions with activated sludge, 5 ppm of 2,4-DNT is 100% removed in 14 days with the formation of the amino derivative [13]. Similarly, greater than 90% 2,4-DNT of 10 ppm was removed from Waconda Bay water and sediment in 6 days under anaerobic conditions [27].

Abiotic Degradation: Sunlight photolysis of 1.0 ppm 2,4-DNT in distilled and natural waters gave half-lives of 43 hr in distilled water and 2.7, 9.6, and 3.7 hr in river, bay, and pond waters, respectively [27]. The photolysis quantum yield and photolysis rates in distilled water and distilled water/1% acetonitrile have been measured [17,23]. When adjusted for light intensity over a full year and assuming no light attenuation, the rates varied from 0.23 [17] to 0.72 [23] day^{-1}(half-life of 0.95 to 3 days). The photolysis rate of 2,4-DNT is strongly accelerated (1.3-2.5 times) in natural waters containing humic substances [23,30]. Aromatic nitro compounds are not susceptible to hydrolysis [15] and the concentration of 2,4-DNT did not change after 2 weeks incubation in sterile, natural water [27]. No information was found in the available literature about the fate of 2,4-DNT in the atmosphere; however, an estimated half-life for 2,4-DNT in the atmosphere based upon vapor phase reaction with photochemically produced hydroxyl radical is 8 hours [4].

Bioconcentration: BCFs of 13, 58, and more than 2000 for Daphnia Magna, Lumbriculus variegatus, and the algae, Selanastrum capricornutum exposed to 1 mg/L of 2,4-DNT for 4 days has been reported [14]. In bluegill sunfish, the BCFs were 78 in viscera and 4 in muscle in 4 days [14] and ranged from 11 to

103 in various tissues in 14 days [7]. These values indicate that 2,4-DNT will not bioconcentrate in aquatic organisms.

Soil Adsorption/Mobility: The estimated soil adsorption coefficient (Koc) for 2,4-DNT using a recommended regression equation [15] is 282. A measured sediment (sterilized, 3.3% organic carbon) adsorption coefficient of 12 has been reported [27] with Holston River sediments after 10 days, indicating that sorption kinetics are not fast. These values indicate that 2,4-DNT should be slightly mobile in soil [11] and have little tendency to adsorb to sediment.

Volatilization from Water/Soil: The Henry's Law constant calculated for 2,4-DNT indicates that volatilization from water will probably not be a significant transport mechanism [15]. Experimental results support this conclusion. The volatilization rate constants for 2,4-DNT from distilled water were 0.0028 to 0.0052 hr^{-1} which corresponds to volatilization half-lives of 248 and 133 hr, respectively [24]. The estimated volatilization rate constant for 2,4-DNT from a body of water is 6.6 x 10^{-5} hr^{-1} (half-life = 438 days) [27]. An EXAMS II model of both a pond and Lake Zurich revealed that the volatilization half-life for 2,4-DNT ranged from 1.1-7.6 years, depending on the Henry's Law constant and environment selected.

Water Concentrations: DRINKING WATER: 2,4-Dinitrotoluene was found in drinking water at an unspecified concentration and location [12]. SURFACE WATER: 2,4-Dinitrotoluene was found in Dokai Bay, Japan at concentrations up to 206 ug/L [8]; in Rhine River water (The Netherlands) at 0.3 ug/L [35], and in Waconda Bay, Lake Chickamauga, TN (range of means <0.10-22.1 ppb) [21]. GROUND WATER: 0.002-90.5 ppm, pos 6 of 6 sites in ground water near a nitroaromatic manufacturing facility in Pasadena, TX [16]. Detected, not quantified, in one sample at Hawthorne Naval Ammun Depot, NV [20]

Effluent Concentrations: EFFLUENT: 2,4-DNT was found in condensate wastewater from 2,4,5-trinitrotoluene manufacture at an unspecified concentration [13] and in wastewater from 2,4,6-trinitrotoluene production at an average concentration of 9.7 mg/L [28]. 2,4-DNT has been detected in effluent from coal mining (1 pos of 49, 18 ppb); iron and steel manufacture (1 pos of 5, 530 ppb); aluminum forming (1 pos of 2, 77 ppb); foundries (4 pos of 4, 7-50 ppb, 26 ppb mean); and organic chem manufacture (4 detections, 14,000 ppb mean) [34]

Sediment/Soil Concentrations: 2,4-DNT was detected in one of two soil samples taken near the Buffalo River at the former site of a dye manufacturing plant [18]. 2,4-DNT was detected in sediment from Waconda Bay, Lake Chickamauga, TN at <2.5-7.9 ppm (range of means from 9 sites, 3 replicate samples from each site) [21].

Atmospheric Concentrations:

Food Survey Values:

Plant Concentrations:

Fish/Seafood Concentrations:

Animal Concentrations:

Milk Concentrations:

Other Environmental Concentrations:

Probable Routes of Human Exposure:

Average Daily Intake:

Occupational Exposures: Exposure to dinitrotoluene may occur from its use in the manufacture of toluene diisocyanate for the production of polyurethane plastics, in the production of military and some commercial explosives(used to plasticize cellulose nitrate in explosives, to moderate burning rate of propellants and explosives, in the manufacture of gelatin explosives, as a waterproofing coating for some smokeless powders, and as an intermediate in TNT manufacture), in the manufacture of azo dye intermediates, and in organic synthesis in the preparation of toluidines and dyes.

2,4-Dinitrotoluene

Body Burdens:

REFERENCES

1. Baily HC; ASTM Spec Tech Publ 766: 221-33 (1982)
2. Davis EM et al; p 176-84 in 5th Meet Symp Pap Int Biodeterior (1983)
3. Dunlap KL; p 925-6, 930-1 in Kirk-Othmer Encycl Chem Tech Vol.15 3rd ed John Wiley and Sons NY (1981)
4. GEMS: Graphical Exposure Modeling System. Fate of atmospheric pollutant (FAP). Office of Toxic Substances USEPA (1985)
5. Hallas LE, Alexander M; Appl Environ Microbiol 45: 1234-41 (1983)
6. Hansch C, Leo AJ; Medchem Project CLOGP3 Claremont CA: Pomona College (1985)
7. Hartley WR; Evaluations of selected subacute effects of the nitrotoluene group of munitions compounds on fish and potential use in aquatic toxicity evaluation NTIS AD-A1011829 223 pp (1981)
8. Hashimoto AH et al; Water Res 16: 891-7 (1982)
9. Hawley GG; The Condensed Chemical Dictionary 10th ed Van Nostrand Reinhold NY p 375 (1981)
10. Hoff MC; Kirk-Othmer Encycl Chem Tech 3rd ed 23: 265-66 (1983)
11. Kenaga EE; Ecotox Environ Safety 4: 26-38 (1980)
12. Kool HJ et al; Crit Rev Environ Contam 12: 307-57 (1982)
13. Liu D et al; Appl Environ Microbiol 47: 1295-8 (1984)
14. Liu DHW et al; ASTM Spec Tech Publ 802 (Aquat Toxicol Hazard Assess): 135-150 (1983)
15. Lyman WJ et al; Handbook of Chemical Property Estimation Methods. McGraw-Hill NY (1982)
16. Matson C et al; Feasibility findings for In Situ Biodegradation of Nitro-aromatic Compounds with Specific Gravities Greater than that of Water. In: Proc Conf Southwestern Ground Waster Issues, October 20-22, Tempe, AZ; 256-268 (1986)
17. Mill T, Mabey W; Environ Toxicol Chem 1: 175-216 (1985)
18. Nelson CR, Hites RA; Environ Sci Technol 14: 1147-1150 (1980)
19. Parrish FW; Appl Environ Microbiol 34: 232-33 (1977)
20. Pereira WE et al; Bull Environ Contam Toxicol 21: 554-62 (1979)
21. Putnam HD et al; ASTM Spec Tech Publ 730 (Ecol Assess Effluent Impacts Commun Indig Aquat Org): 220-242 (1981)
22. Sears JK, Touchette NW; Kirk-Othmer Encycl Chem Tech 3rd ed 18: 174 (1982)
23. Simmons MS, Zepp RG; Water Res 20: 899-904 (1986)
24. Smith JH, Bomberger DC Jr.; Chemosphere 10: 281-9 (1981)
25. Smith JH et al; Residue Reviews 85: 73-88 (1983)
26. Spanggord RJ et al; Environmental Fate Studies on Certain Munitions Wastewater Constituents. Final Report, Phase I - Literature Review. DAMD 17-78-C-8081 (1980)
27. Spanggord RJ et al; Environmental Fate Studies on Certain Munitions Wastewater Constituents. Final Report, Phase II - Laboratory Studies U.S. NTIS AD A099256 (1980)
28. Spanggord RJ, Suta BE; Environ Sci Technol 16: 233-6 (1982)

2,4-Dinitrotoluene

29. Spanggord RJ et al; Environmental Fate Studies on Certain Munition Wastewater Constituents. Phase 3, Part 2. Laboratory Studies. NTIS AD-A131 908 pp58 (1981)
30. Spanggord RJ et al; Environmental Fate of Selected Nitroaromatic Compounds in the Aquatic Environment. In: Chemical Industry Institute of Toxicology Series. Toxicity of Nitroaromatic Compounds, Vol XVI, Rickert DE, Ed. Washington, DC: Hemisphere Publishing Corporation 285 pp (1985)
31. SRI International; 1985 Directory of Chemical Producers (1986)
32. Tabak HH et al; J Water Pollut Cont Fed 53: 1503-18 (1981)
33. U.S. International Trade Commission. Synthetic Organic Chemicals U.S. Production and Sales, 1982 USITC Publ. 1422 (1983)
34. U.S. EPA; Treatability Manual EPA-600/8-80-042 (1980)
35. Zoeteman BCJ et al; Chemosphere 9: 231-49 (1980)

2,6-Dinitrotoluene

SUBSTANCE IDENTIFICATION

Synonyms: 1-Methyl-2,6-dinitrobenzene

Structure:

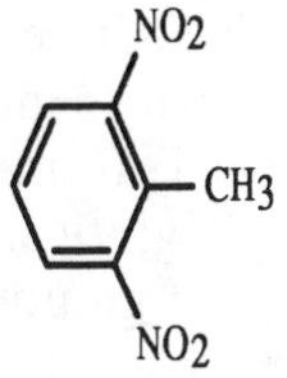

CAS Registry Number: 606-20-2

Molecular Formula: $C_7H_6N_2O_4$

Wiswesser Line Notation: WNR B1 CNW

CHEMICAL AND PHYSICAL PROPERTIES

Boiling Point: 285 °C

Melting Point: 66 °C

Molecular Weight: 182.15

Dissociation Constants:

Log Octanol/Water Partition Coefficient: 1.72 [5]

Water Solubility:

Vapor Pressure: 3.5×10^{-4} mm Hg at 20 °C [16]

Henry's Law Constant: 2.17×10^{-7} atm-m^3/mole [21]

ENVIRONMENTAL FATE/EXPOSURE POTENTIAL

Summary: Recent information indicated that two companies produced 2,6-dinitrotoluene (2,6-DNT) as a mixture with 2,4-DNT. No specific use information was found for 2,6-DNT; however, dinitrotoluenes are used in organic synthesis, dyes, and explosives and probably some of it is consumed in isocyanate production along with 2,4-DNT. 2,6-DNT may enter the environment from its production and the above uses. If released to soil, 2,6-DNT is expected to be slightly mobile (estimated Koc = 204). Information on evaporation from soil, biodegradation in soil, or hydrolysis in soil was not found for 2,6-DNT. If released to water, 2,6-DNT will have a slight tendency to sorb to sediments, suspended solids, and biota. Volatilization of 2,6-DNT from water is relatively slow and hydrolysis is probably not significant since 2,6-DNT does not contain hydrolyzable functional groups. Photolysis is probably the most significant removal mechanism for 2,6-DNT in water. The importance of biodegradation cannot be assessed because biodegradation data are inconsistent. In the atmosphere, 2,6-DNT has an estimated half-life of 8 hr based upon reaction with photochemically-generated hydroxyl radicals. 2,6-DNT has been detected in drinking water, in raw wastewater from a textile plant, in wastewater from 2,4,6-trinitrotoluene production, and in saltwater.

Natural Sources:

Artificial Sources: Recent available public information indicates that two companies produced 2,6-DNT [24]. A mixture of 2,4- and 2,6-DNT was produced in over 0.5 billion lb/yr in 1982 [26]. No specific use information was found for 2,6-DNT; however, dinitrotoluenes are used in organic synthesis, dyes, and explosives [7] and probably some of it is consumed with the 2,4-DNT in the production of isocyanates. This production and these uses may result in some releases to the environment.

Terrestrial Fate: The calculated soil adsorption coefficient (Koc = 204 [12]) for 2,6-DNT indicates that 2,6-DNT is slightly mobile in soil [8]. No information was found about 2,6-DNT evaporation from soil surfaces, photodegradation, biodegradation, or hydrolysis in soil. Aromatic nitro compounds are not susceptible to hydrolysis [12].

Aquatic Fate: 2,6-DNT will have a slight tendency to sorb to sediments, suspended solids, or bioconcentrate in biota. The half-life for 2,6-DNT in river water exposed to sunlight was 12 minutes and was determined to be an indirect photoreaction [28]. The photoreaction will be particularly affected by the humic material concentration. Biodegradation data in natural waters are extremely variable, so the importance of biodegradation is unknown. Volatilization and bioconcentration do not appear to be important processes.

Atmospheric Fate: An estimated atmospheric half-life for vapor phase reaction of 2,6-DNT with photochemically generated hydroxyl radicals is 8 hr [3].

Biodegradation: The biodegradability of 2,6-DNT was tested in a static screening test using flasks that contained either 5 or 10 mg/L 2,6-DNT, 5 mg/L yeast extract, and settled domestic sewage as the inoculum. Subcultures were done every 7 days. The percent biodegradation by the original culture and first, second, and third subcultures at an initial 2,6-DNT concentration of 10 mg/L was: 57, 49, 35, and 13%, respectively [25]. 2,6-DNT at an initial concentration of 10 ug/mL (10 mg/L) did not significantly degrade after 28 days incubation with raw municipal sewage [4]. In another study, 2,6-DNT inhibited municipal sewage at concentrations above 50 mg/L [2]. In the same study, industrial seed degraded 50 mg/L 2,6-DNT to 25 mg/L in 2 days, but after 7 days, 2,6-DNT concentration remained at 25 mg/L. Anaerobic biodegradation of 2,6-DNT was much faster than aerobic degradation and the amine product was formed [4]. In natural water samples (pond near Searsville Lake, 10 ppm; Coyote Creek, 10 ppm, 100 ppm yeast extract added), no degradation of 2,6-DNT was noted in 42 days [22]. When 500 ppm of yeast extract was added to the pond water, 100% degradation occurred in 5 days. The inconsistency of the biodegradation data may be due to different seeds, acclimation procedures, and 2,6-DNT concentrations.

Abiotic Degradation: The half-life for 2,6-DNT in river water exposed to sunlight was 12 minutes and was determined to be an indirect photoreaction [28]. The photolysis quantum yield and photolysis rates in distilled water and distilled water/1% acetonitrile have been measured [14,20]. When adjusted for light intensity over a full year and assuming no light attenuation, the rates varied from 0.67 to 1.0 day^{-1} (half-life of 1.03 to 0.69 days) [14,20]. The photolysis rate of 2,4-DNT is strongly accelerated (11-17 times) in natural waters containing humic substances [14,20]. Aromatic nitro compounds are not susceptible to hydrolysis [12]. A computer estimated atmospheric half-life for vapor phase 2,6-DNT reacting with hydroxyl radical is 8.0 hours [3].

Bioconcentration: A 2,6-DNT bioconcentration factor (BCF) of 5225 was measured [2] for the algal biomass in a model waste stabilization pond. An estimated BCF for 2,6-DNT using an estimated log Kow of 1.72 and recommended repression equations is 12 [12].

Soil Adsorption/Mobility: The calculated soil adsorption coefficient Koc for 2,6-DNT using the estimated log Kow and recommended regression equations is 204 [12]. This indicates that 2,6-DNT is slightly mobile in soil [8].

Volatilization from Water/Soil: The fate of 2,6-DNT, including volatilization, was studied in a model waste stabilization pond in which influent 2,6-DNT concentration was 1 mg/L, temperature was 23 °C, and the detention time in the pond was 12 days. Volatilization accounted for only a 0.3% loss for 2,6-DNT [2]. The Henry's Law constant for 2,6-DNT would suggest that volatilization from water is not an important process. An EXAMS II model of both a pond and Lake Zurich revealed that the volatilization half-life for 2,6-DNT ranged from 6.7-17 years. No information was found about 2,6-DNT volatilization from soil.

Water Concentrations: SEAWATER: 2,6-Dinitrotoluene was found in Dokai Bay, Japan at concentrations ND-14.9 ug/L [6]. SURFACE WATER: 2,6-DNT was found at conc ranging from 1.3 to 38.7 ppb (19.4 ppb mean) at Waconda Bay, Lake Chichamauga, TN [17]. GROUND WATER: 2,6-DNT was detected at a range of ND-76,800 ppb (16,763 ppb mean) in ground water near a nitroaromatic plant in Pasadena, TX [13]. DRINKING WATER: 2,6-Dinitrotoluene was found in drinking water by a number of investigators at unspecified locations at unspecified concentrations [1,9,10,11].

Effluent Concentrations: 2,6-DNT concentration in raw wastewater from a textile plant was 50 mg/m^3 [18] and was an average of 4.3 mg/L in the wastewater from 2,4,6-trinitrotoluene production [23]. It was also detected (5 ppb) in wastewater from a nitrobenzene plant [19]. 2,6-DNT has been detected in effluent from coal mining (1 pos of 49, 30 ppb); iron and steel manufacture (2 pos of 8, 47-140 ppb, 530 ppb mean); nonferrous metals manufacture (16 ppb max conc); foundries (6 pos of 6, 4-50 ppb, 20 ppb mean); organic chemical manufacture (4 detections, 3800 ppb mean); paint and ink formulations (10 ppb max); and textile mills (1 pos of 50, 54 ppb max) [27].

Sediment/Soil Concentrations: 2,6-DNT was detected in one of two soil samples taken near the Buffalo River at the former site of a dye manufacturing plant [15]. 2,4-DNT was detected in sediment from Waconda Bay, Lake Chickamauga, TN at <1.3-17.0 ppm (range of means from 9 sites, 3 replicate samples from each site) [17].

Atmospheric Concentrations:

Food Survey Values:

Plant Concentrations:

Fish/Seafood Concentrations:

Animal Concentrations:

Milk Concentrations:

Other Environmental Concentrations:

Probable Routes of Human Exposure:

Average Daily Intake:

Occupational Exposures:

Body Burdens:

REFERENCES

1. Clark RM et al; Sci Total Environ 53: 153-172 (1986)

2,6-Dinitrotoluene

2. Davis EM et al; Water Res 15: 1125-27 (1981)
3. GEMS; Graphical Exposure Modeling System. Fate of atmospheric pollutants (FAP). Office of Toxic Substances USEPA (1985)
4. Hallas LE, Alexander M; Appl Environ Microbiol 45: 1234-41 (1983)
5. Hansch C, Leo AJ; Medchem Project CLOGP3 Claremont CA: Pomona College (1985)
6. Hashimoto AH et al; Water Res 16: 891-7 (1982)
7. Hawley GG; The Condensed Chem Dict 10th ed Von Nostrand Reinhold, NY p. 375 (1981)
8. Kenaga EE; Ecotox Environ Safety 4: 26-38 (1980)
9. Kool HJ et al; Crit Rev Environ Cont 12: 307-57 (1982)
10. Kopfler FC et al; Adv Environ Sci Technol 8: 419-33 (1977)
11. Kraybill HF; J Environ Sci Health C1: 175-232 (1983)
12. Lyman WJ et al; Handbook of Chemical Property Estimation Methods. McGraw-Hill NY (1982)
13. Matson C et al; In: Proc Conf Southwestern Ground Water Issues, October 20-22, Tempe, AZ; 256-68 (1986)
14. Mill T, Mabey W; Environ Toxicol Chem 1: 175-216 (1985)
15. Nelson CR, Hites RA; Environ Sci Technol 14: 1147-50 (1980)
16. Pella PA; J Chem Thermodyn 9: 301-305 (1977)
17. Putnam HD et al; ASTM Spec Tech Publ 730(Ecol Assess Effluent Impacts Commun Indeg Aquat Org):220-42 (1981)
18. Rawlings GD, Samfield M; in Proc Symp Process Meas Environ Assess. USEPA-600/7-78-168 (1979)
19. Shafer KH; Determination of nitroaromatic compounds and isophorone in industrial and municipal wastewaters EPA 600/4-82-034 NTIS PB82-208398, 37 pp (1982)
20. Simmons MS, Zepp RG; Water Res 20: 899-904 (1986)
21. Society of German Chemists; Dinitrotoluenes: BUA Substance Report 12. The Advisory Board for Environmentally Significant Hazardous Wastes (BUA) of the Society of German Chemists, eds. Weinheim, Fed Rep of Germany: VCH Verlagsgesellschaft p 2 (1987)
22. Spanggord RJ et al; Environmental Fate Studies on Certain Munition Wastewater Constituents. Phase 3, Part 2. Laboratory Studies. NTIS AD-A131 908 pp58 (1981)
23. Spanggord RJ, Suta BE; Environ Sci Technol 16: 233-6 (1982)
24. SRI International; 1985 Directory of Chemical Producers (1986)
25. Tabak HH et al; J Water Pollut Control Fed 53: 1503-18 (1981)
26. U.S. International Trade Commission. Synthetic Chemicals U.S. Production and Sales, 1982 USITC Publ. 1422 (1983)
27. U.S. EPA; Treatability Manual EPA-600/8-80-042 (1980)
28. Zepp RG et al; Fresenius Z Anal Chem 319: 119-25 (1984)

Epichlorohydrin

SUBSTANCE IDENTIFICATION

Synonyms: Chloromethyloxirane; 1-Chloro-2,3-epoxypropane; (Chloromethyl)ethylene oxide

Structure:

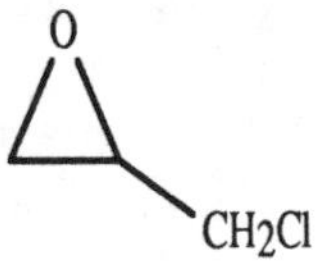

$$\overset{O}{\underset{CH_2Cl}{\triangle}}$$

CAS Registry Number: 106-89-8

Molecular Formula: C_3H_5ClO

Wiswesser Line Notation: T3OTJ B1G

CHEMICAL AND PHYSICAL PROPERTIES

Boiling Point: 116.5 °C at 760 mm Hg

Melting Point: -57.2 °C

Molecular weight: 92.53

Dissociation Constants:

Log Octanol/Water Partition Coefficient: 0.30 [7]

Water Solubility: 65.8 g/L at 20 °C [7]

Vapor Pressure: 16.44 mm Hg at 25 °C [4]

Henry's Law Constant:

ENVIRONMENTAL FATE/EXPOSURE POTENTIAL

Summary: Epichlorohydrin may be released to the atmosphere and in wastewater during its production and use in epoxy resins,

319

Epichlorohydrin

glycerin manufacture, as a chemical intermediate in the manufacture of other chemicals, and other uses. If released into water, it will be lost primarily by evaporation (half-life 29 hr in a typical model river) and hydrolysis (half-life 8.2 days). It should neither adsorb appreciably to sediment nor bioconcentrate in fish. If spilled on land, it will evaporate and leach into the ground water, where it will hydrolyze. Biodegradation and chemical reactions with ions and reactive species may accelerate its loss in soil and water but data from field studies are lacking. In the atmosphere, epichlorohydrin will degrade by reaction with photochemically produced hydroxyl radicals (estimated half-life 4 days). There is a lack of monitoring data for epichlorohydrin in all but occupational settings. Humans will primarily be exposed to epichlorohydrin in occupational settings.

Natural Sources:

Artificial Sources: Estimated total U.S. atmospheric emissions, 1978, 479,000 lb/yr [1]. Emissions from its production (estimated 6.7 x 10^{+4} kg from 3 facilities in 1978) and use in epoxy resins (estimated 1.1 x 10^{+5} kg from 11 facilities in 1978) and as a chemical intermediate (estimated 3.7 x 10^{+4} kg in production of chemicals other than glycerine in 1978); wastewater, and spills [6]. Other uses which may lead to its release include textile treatment, coatings, solvent, surface active agent, stabilizer in insecticide, and elastomer manufacture [6,18].

Terrestrial Fate: If spilled on land, epichlorohydrin will evaporate into the atmosphere and leach into ground water. Although data are lacking, it is probable that biodegradation and chemical degradation will occur in the soil based upon biodegradation screening studies and the rapid hydrolysis rate.

Aquatic Fate: When released into water, epichlorohydrin will be lost by evaporation (half-life 29 hr in a typical model river) and hydrolyze (half-life 8.2 days)(product 1-chloropropan-2,3-diol). In seawater, it will additionally react with chloride ions which will reduce its overall half-life to about 5.3 days, producing 1,3-dichloro-2-propanol as well as the hydrolysis product. Biodegradation is also likely to occur as is reaction with radicals,

Epichlorohydrin

but no estimates for the rate of these processes in natural waters could be found.

Atmospheric Fate: When released to the atmosphere, epichlorohydrin will degrade by reaction with photochemically produced hydroxyl radicals (estimated half-life 4 days; faster under photochemical smog conditions). It is somewhat soluble in water and will therefore be subject to washout by rain.

Biodegradation: In a laboratory biodegradability test using sewage seed, 3% of the theoretical BOD was consumed in 5 days [1]. With acclimated sewage seed, the percent of theoretical BOD consumed increased to 14% [2]. Epichlorohydrin was confirmed to be significantly degraded in another laboratory test that utilizes an inoculum originating from soil, natural waters, and sewage [16]. Although 89% of the chemical oxygen demand was removed in 24 hr in an aerated laboratory system using an acclimated sludge inoculum, most of the loss was due to volatilization [10]. No information could be located on the biodegradability of epichlorohydrin in soil or natural waters.

Abiotic Degradation: Epichlorohydrin hydrolyzes in distilled water to yield 1-chloro-2,3-propanediol (half-life 8.2 days at 20 °C) [9]. Acid catalysis contributes less than 10% to the rate of hydrolysis and base catalysis is not detectable at pH <10 [9]. Anions such as chloride attack the epoxide ring, producing 1,3-dichloro-2-propanol [15]. The half-life in 3% NaCl (approximation for seawater) is calculated to be 5.3 days [15]. Epichlorohydrin reacts with photochemically produced hydroxyl radicals with an estimated atmospheric half-life of 4 days [3]. When irradiated in the presence of 5 ppm nitric oxide to simulate photochemical smog conditions, the half-life was 16 hr [5].

Bioconcentration: Epichlorohydrin would not be expected to bioconcentrate appreciably in aquatic organisms. The log BCF has been estimated to be 0.66 [15].

Soil Adsorption/Mobility: The Koc for epichlorohydrin calculated from its water solubility is 123 [8], which indicates that is not appreciably adsorbed. After a spill of 20,000 gal following a train accident, water in wells closest to the spill were highly contaminated [6].

Volatilization from Water/Soil: The half-life for evaporation of epichlorohydrin from a model river 1 m deep with a 1 m/sec current and 3 m/sec wind is 29 hr, with the gas exchange rate playing a more dominant role than the liquid exchange rate [8]. Epichlorohydrin is relatively volatile and would therefore readily evaporate from near-surface soils and other solid surfaces.

Water Concentrations: SURFACE WATER: Detected, not quantified in unspecified surface water [6]. GROUND WATER: Point Pleasant, WV (1/78) - Closest well to 20,000 gal spill resulting from train accident - 75 ppm [6].

Effluent Concentrations: Detected, not quantified in chemical industry effluent in Louisville, KY [17].

Sediment/Soil Concentrations:

Atmospheric Concentrations:

Food Survey Values:

Plant Concentrations:

Fish/Seafood Concentrations:

Animal Concentrations:

Milk Concentrations:

Other Environmental Concentrations:

Probable Routes of Human Exposure: Humans are most likely to be exposed to epichlorohydrin in occupational settings. The general public may be exposed while using epoxy resins containing epichlorohydrin.

Average Daily Intake:

Epichlorohydrin

Occupational Exposures: NIOSH (NOES Survey 1981-1983) has statistically estimated that 3306 workers are exposed to epichlorohydrin in the United States [12]. NIOSH (NOHS Survey 1972-1974) has statistically estimated that 77,297 workers are exposed to epichlorohydrin in the United States [13]. Hazard survey of electronic component-mold operations where epoxy resins containing bisphenol A and epichlorohydrin - below detection limit of 0.005 ppm [11]. Solvent epichlorohydrin production plant: quality control sampling 4.9-5.5 ppm, filling tanks 3.1-3.9 ppm, during emergency caused by mechanical difficulty 54.6-54.9 ppm [11]. Dow Chemical units employing epichlorohydrin: Epoxy resin unit (1974, 78 samples) range <0.60-13 ppm, 3.17 ppm avg; Glycerine unit 0.01-4.69 ppm; Allyl chloride unit <0.1 ppm [11]. NIOSH estimates that 50,000 workers may be exposed to epichlorohydrin [11]. Time Weighted Avg personal samplers, Epichlorohydrin Manufacturers, plant A: 17 samples from 4 different plant jobs, not detected to 0.4 ppm, 0.07-0.3 range of avgs (highest avgs for chemical operators and tank car loaders), plant B: 17 samples, not detected to 2.1 ppm, range of avgs 0.08-0.3 ppm range of avgs (highest avg for tank car loader); Resin Manufacturers: chemical operators, 39 samples, not detected to 0.8 ppm, 0.04-0.09 range of avgs, operating foremen, 6 samples, not detected to 0.6 ppm [14].

Body Burdens:

REFERENCES

1. Anderson GE; Human Exposure to Atmospheric Concentrations of Selected Chemicals. Volume 1. USEPA Res Triangle Park, NC, Office of Air Quality Planning and Standards p. 20 (1983)
2. Bridie AL et al; Water Res 13: 627-30 (1979)
3. Cupitt LT; Fate of toxic and hazardous materials in the air environment; USEPA-600/3-80-084 (1980)
4. Daubert TE, Danner RP; Data compilation tables of properties of pure compounds. American Institute of Chemical Engineers pp 450 (1985)
5. Dilling WL et al; Environ Sci Technol 10: 351-6 (1976)
6. Keneklis T et al; Health assessment documents for epichlorohydrin. External review draft; p.1.1-3.24 USEPA 600/8-83-032a (1983)
7. Krijgsheld KR, Vandergen A; Chemosphere 15: 881-93 (1986)
8. Lyman WJ et al; Handbook of chemical property estimation methods. Environmental behavior of organic compounds; McGraw Hill New York (1982)
9. Mabey W, Mill T; J Phys Chem Ref Data 7: 383-415 (1978)

Epichlorohydrin

10. Matsui S et al; Prog Water Technol 7: 645-59 (1975)
11. NIOSH; Criteria for a recommended standard. Occupational exposure to epichlorohydrin; p.22-30, 92-94 Sept 1976 (1976)
12. NIOSH; The National Occupational Exposure Survey (NOES) (1983)
13. NIOSH; The National Occupational Hazard Survey (NOHS) (1974)
14. Oser JL; Am Ind Hyg Assoc J 41: 463-8 (1980)
15. Santodonato J et al; Investigation of selected potential environmental contaminants: epichlorohydrin and epibromohydrin; USEPA 560/11-80-006 (1980)
16. Sasaki S; pp.283-98 in Aquatic pollutants : transformation and biological effects; Hutzinger O et al eds; Pergamon Oxford (1978)
17. Shakelford WM, Keith LH; Frequency of organic compounds identified in water; p.117 USEPA-600/4-76-062 (1976)
18. Verscheuren K; Handbook of environmental data on organic chemicals; Van Nostrand Reinhold New York p.611-2 (1983)

Ethylbenzene

SUBSTANCE IDENTIFICATION

Synonyms:

Structure:

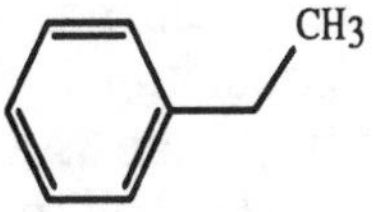

CAS Registry Number: 100-41-4

Molecular Formula: C_8H_{10}

Wiswesser Line Notation: 2R

CHEMICAL AND PHYSICAL PROPERTIES

Boiling Point: 136.2 °C at 760 mm Hg

Melting Point: -94.97 °C

Molecular Weight: 106.16

Dissociation Constants:

Log Octanol/Water Partition Coefficient: 3.15 [27]

Water Solubility: 161 mg/L at 25 °C [75]

Vapor Pressure: 9.53 mm Hg at 25 °C [7]

Henry's Law Constant: 8.44 x 10^{-3} atm-m³/mole [46]

ENVIRONMENTAL FATE/EXPOSURE POTENTIAL

Summary: Ethylbenzene will enter the atmosphere primarily from fugitive emissions and exhaust connected with its use in gasoline. More localized sources will be emissions, wastewater, and spills from its production and industrial use. If ethylbenzene is released to the atmosphere, it will exist predominantly in the vapor phase based on its vapor pressure, where it will photochemically degrade by reaction with hydroxyl radicals (half-life hours to 2 days) and partially return to earth in rain. It will not be subject to direct photolysis. Releases into water will decrease in concentration by evaporation and biodegradation. The time for this decrease and the primary loss processes will depend on the season, and the turbulence and microbial populations in the particular body of water. Representative half-lives are several days to two weeks. Some ethylbenzene may be adsorbed by sediment, but significant bioconcentration in fish is not expected to occur based upon its octanol/water partition coefficient. Ethylbenzene released to soil is only adsorbed moderately by soil and may leach into ground water, where its biodegradation is possible. It will not significantly hydrolyze in water or soil. The primary source of exposure is from the air, especially in areas of high traffic. However, exposure from drinking water is not uncommon.

Natural Sources:

Artificial Sources: Releases to the environment include emissions, wastewater, leaks, and spills connected with its production, and use in the manufacture of styrene and use as a solvent [62]. Other sources that have been mentioned are emissions from petroleum refining, vaporization losses and spills of gasoline and diesel fuel at filling stations and during storage and transit of these fuels, auto emissions, and cigarette smoke [23,49,62,84].

Terrestrial Fate: When released onto soil, part of the ethylbenzene will evaporate into the atmosphere. It has a moderate adsorption in soil but will probably leach into the ground water, especially in soil with a low organic carbon content. While there is no direct data concerning its biodegradability in soil, it is likely that it will biodegrade slowly after acclimation. There is evidence that ethylbenzene slowly biodegrades in ground water. In cases where large concentrations persist in ground water over a year after a spill, it is possible that resident microorganisms were killed by toxic concentrations. It will not hydrolyze in soil or ground water.

Ethylbenzene

Aquatic Fate: When released into water, ethylbenzene will evaporate fairly rapidly into the atmosphere with a half-life ranging from hr to a few weeks. Biodegradation will also be rapid (half-life 2 days) after a population of degrading microorganisms becomes established, which will depend on the particular body of water and the temperature. In one study, this acclimation took 2 days and 2 weeks in summer and spring, respectively. Some ethylbenzene may be adsorbed by sediment, but significant bioconcentration in fish is not expected to occur based upon its octanol/water partition coefficient. It will not significantly photolyze or hydrolyze.

Atmospheric Fate: If ethylbenzene is released to the atmosphere, it will exist predominantly in the vapor phase based on its vapor pressure [20]. It will be removed from the atmosphere principally by reaction with photochemically produced hydroxyl radicals (half-life 0.5 hr to 2 days). Additional quantities will be removed by rain. It will not be expected to directly photolyze.

Biodegradation: After a period of adaptation, ethylbenzene is biodegraded fairly rapidly by sewage or activated sludge inocula [47,71,76,81]. As a component of gas oil, it is completely degraded in ground water in 8 days [34] and seawater in 10 days [83]. In a mesocosm experiment using simulated Narraganset Bay conditions, complete biodegradation occurred in approx 2 days after a 2-week lag in spring and a 2-day lag in summer [85]. Part of the attenuation in concentration from a leaky petroleum storage tank in the chalk aquifer in England has been attributed to biodegradation [78]. No degradation was observed in an anaerobic reactor, even after 110 days acclimation [14] or at low concentrations in a batch reactor in 11 weeks under denitrifying conditions [8]. Percent removal in an anaerobic, continuous-flow, laboratory biofilm column was 7% after a 2-day detention time [9]; 99% removal was observed in a similar aerobic column following a 20-min detention time [9].

Abiotic Degradation: The predominant photochemical reaction of ethylbenzene in the atmosphere is with hydroxyl radicals; the tropospheric half-life for this reaction is 5.5 and 24 hr in the summer and winter, respectively [60,69]. Degradation is somewhat faster under photochemical smog situations [18,86,87].

327

Ethylbenzene

Photooxidation products which have been identified include ethylphenol, benzaldehyde, acetophenone, and m- and p-ethylnitrobenzene [29]. Ethylbenzene does not significantly absorb light above 290 nm in methanol solution [63] and, therefore, direct sunlight photolysis in the gas phase or in surface water will not be expected to be an important removal process. Ethylbenzene is resistant to hydrolysis [45].

Bioconcentration: Experimental data on the bioconcentration of ethylbenzene include a log BCF of 1.19 in goldfish [55] and the low log BCF of 0.67 for clams exposed to the water-soluble fraction of crude oil [54]. Using its octanol/water partition coefficient and a recommended regression equation [45], one can calculate a log BCF in fish of 2.16.

Soil Adsorption/Mobility: Ethylbenzene has a moderate adsorption for soil. The measured Koc for silt loam was 164 [13]. Its presence in bank infiltrated water suggests that there is a good probability of its leaching through soil [59]. Using its octanol/water partition coefficient and using a recommended regression equation [45], one can calculate a log Koc of 2.94.

Volatilization from Water/Soil: Ethylbenzene has a high Henry's Law constant and will evaporate rapidly from water; a half-life for evaporation from water with 1 m/s current, 3 m/s wind, and 1 m depth is 3.1 hr [45]. In a mesocosm experiment using simulated conditions for Narraganset Bay, MA, and seasonal conditions, the loss of ethylbenzene was primarily by evaporation in winter (half-life 13 days) [85]. Since it has a moderately high vapor pressure, it will evaporate fairly rapidly from dry soil.

Water Concentrations: DRINKING WATER: In surveys of representative U.S. municipal water supplies, ethylbenzene has been detected in most cases [4,12,36,37,38,49,62,67]. Values for 3 New Orleans finished drinking waters ranged from 1.6 to 2.3 ppb [36]. Chicago Central Water Works on Lake Michigan measured 4 ppb [38]. It has been found in the water supply for Evansville, IN on the Ohio River [37]. 6 of 10 U.S. cities were found to be positive [4,49]. One U.S. city had 1 of 4 samples pos with a 1 ppb avg, while another reported no positive samples [12]. Tap water from bank infiltrated Rhine River water in The Netherlands measured 30

328

ppb in one study [59]. Zurich, Switzerland tap water - detected, not quantified [62]. 1982 U.S. Ground Water Supply Survey, random samples of finished water supplies using ground water as a source, 466 random samples, 0.6% pos (0.5 ppb detection limit), 0.8 ppb median, 1.1 ppb max [17]. GROUND WATER: A well in Ames, IA measured 15 ppb 50 yr after tar residues were buried at a nearby coal gas plant [62]. Two aquifers near the Hoe Creek underground coal gasification site in Wyoming were sampled 15 month after gasification was complete, giving values of 82-400 ppb [74]. In a U.S. survey, 1970-76, it was detected but not quantified in well waters [67]. In Jackson Township, NJ, drinking water wells measured 2000 ppb [11]. Chalk aquifer in East Anglia, England - 210 m from petroleum storage - 0.15 ppb; 10 m distance - 1110 ppb; and 100-200 m - <250 ppb [78]. SURFACE WATER: Ethylbenzene has been detected but not quantified in a 1970-76 U.S. survey [5,67]. 14 heavily industrialized U.S. river basins, 5 of 204 sites pos - 1-4 ppb; Chicago area and Illinois River Basin, 5 of 31 sites pos - 1-4 ppb [21]. Two representative U.S. cities, city A - 41% of 28 samples pos, 5.0 ppb avg; city B - 40% of 48 samples pos 3.2 ppb avg [12]. Lower Tennessee River near Calvert City, KY reported 4.0 ppb [22]. Lake Michigan, Chicago Sanitary and Ship Channel measured 1-2 ppb [38]. North Sea, max conc 0.02 ppb [82]. River Glatt, Switzerland - detected, not quantified [89]. USEPA STORET database, 1101 data points, 10% pos, <5.0 ppb median [73]. SEAWATER: Gulf of Mexico, unpolluted areas - 0.4 to 5 ppb [64,65], while an area of anthropogenic influence ranged from 5 to 15 ppb [64]. Cape Cod, MA measured a trace to 22 ppb [25,48] with 11 ppb avg [25]. Concn (ppt) Dutch North Sea coastal waters, 108 samples, not detected (<5)-20, avg 4 [3]. RAIN WATER: West Los Angeles, CA - 9 ppt [35]. Concn (ppt) dissolved in rain, Portland, OR, Feb-Apr 1984, 7 rain events, 100% pos, 6.9-72, 34 avg [40].

Effluent Concentrations: Industries with mean raw wastewater concentrations >2000 ppb: gum and wood chemicals (11,000 ppb), pharmaceutical manufacturing (10,000 ppb), paint and ink formulation, and auto and other laundries [81]. Effluents from representative water treatment plants in Southern California were variable: <10 ppb at San Diego City to 130 ppb at Los Angeles Co (both measurements following primary treatment) [88]; <10 ppb detected following secondary treatment [88]. In a U.S. city survey,

17% of 6 samples were positive, 6.0 ppb avg [12]; Lake Michigan, North Side sewage treatment plant - 1 ppb [38]. USEPA STORET database, 1368 data points, 7.4% pos, <3.0 ppb median [73]. MN municipal solid waste landfills, leachates, 6 sites, 100% pos, 12-820 ppb; contaminated ground water (by inorganic indices), 13 sites, 61.5% pos, 1.2-590 ppb; other ground water (apparently not contaminated as indicated by inorganic indices), 7 sites, 14.3% pos, 9.4 ppb [61].

Sediment/Soil Concentrations: Sediments from the lower Tennessee River below Calvert City, KY measured 4.0 ppb [22]. USEPA STORET database, 350 data points, 11% pos, 5.0 ppm median, dry wt) [73].

Atmospheric Concentrations: RURAL/REMOTE: Areas in the continental United States ranged between 0.5 to 2.2 ppb [3,10]. Jones State Forest north of Houston TX ranged from 0.8 to 10.4 ppb [66]. Air intake fan rooms of the Allegheny Mt. tunnel measured 0.07 to 0.16 ppb [26]. Air in England - 11.3 ppt avg [79]; The Netherlands - 0.8 ppb avg; and Belgium 0.01 to 15 ppb [77]. Concn (ppb) at rural site in UK, May-Aug 1983, 204 samples, not detected- 0.70, 0.14 avg [15]; July 1982, 175 samples, not detected- 0.6, 0.12 avg [3]. URBAN/SUBURBAN: Values for major western U.S. cities ranged from 0.1 to 27.7 ppb [3,43,66,68,69], with the avg being 2.68 ppb. Representative centers in New Jersey had a range of 0.17 to 0.33 ppb avg, 107 of 110 samples pos [28]. Ethylbenzene was detected but not quantified in another New Jersey study [41]. It has been detected in 6 USSR cities, including Leningrad as well as New York and Paris [30,31,32]. The Hague, Netherlands - 5 ppb [6]; Sidney Australia - 1.3 ppb [50]; Japan - 0.2 ppb and Frankfurt am Main, Germany - 1 ppb [62]. Zwuch, Switzerland 8.7 ppm [24]. 3 sites in England away from traffic - 16.1 to 18.8 ppt avg; 2 sites with heavy traffic 28.7 to 33.9 ppt avg [79]. 669 samples from the United States had a median concentration of 1.2 ppb [10]. 36 Chicago metropolitan area homes tested - 57% frequency in indoor air and 36% in outdoor air [33]. Gas-phase concn (ng/m³) during 7 rain events, Portland, OR, Feb-Apr 1984, 100% pos, 780-2800, 1300 avg [40]. Concn (ppb), Exhibition Road, London, May-Aug 1983, 267 samples, 100% pos, 0.05-2.17, 0.78 avg [15]; June-July 1982, 256 samples, not detected - 3.3, 0.88 avg [16]. United States

Ethylbenzene

1979-1984, 15 cities, 1-2 weeks of sampling/site, overall range not detected - 31.5 ppb; range of avg, 0.6-4.6 ppb, avg of avg, 1.9 ppb [70]. INDUSTRIAL: Houston, TX; 21 individuals and urban sites reported a range of 2.5 to 154.2 ppb [42]. A natural gas facility in Rio Blanco County, CO measured 3.6 ppb, and a Texaco refinery in Tulsa, OK ranged from 4.7 to 7.9 ppb [3]. England - car park - 115 ppb; motorway - 92 ppb [57], while 6 sites at Gatwich airport ranged from 0.46 to 1.8 ppb, 1.4 ppb avg [80]. The Maastunnel, The Netherlands, measured 6 ppb avg [6]. 181 samples of U.S. source dominated areas - 0.63 ppb [10]. Allegheny Mt. Tunnels 0.5-2.6 ppb with concentrations directly corresponding to the number of vehicles passing through the tunnels [26]. Concn (ppb) at rural motorway in UK, May-Aug 1983, 184 samples, not detected- 1.14, 0.17 avg [15]; Aug 1982, not detected- 0.70, 0.25 avg [16].

Food Survey Values: Detected, not quantified in roasted filbert nuts [62] and mountain Beaufort cheese [19]. Detected in dried legumes: beans, not detected- 11 ppb, 5 ppb avg; split peas, 13 ppb; lentils, 5 ppb [44].

Plant Concentrations:

Fish/Seafood Concentrations: Bottomfish, Commencement Bay and adjacent waterways, Tacoma, WA, 1982, highest avg level, 0.01 ppm [53].

Animal Concentrations:

Milk Concentrations:

Other Environmental Concentrations: Detected in cigarette smoke [49].

Probable Routes of Human Exposure: Human populations are primarily exposed to ethylbenzene from ambient air, particularly in areas of heavy traffic, tunnels, parking lots, and around filling stations since it is a component of gasoline. High levels of exposure may exist near production and manufacturing facilities and in occupational settings where ethylbenzene is used as a solvent. Nonoccupational exposure may result from indoor air

Ethylbenzene

containing cigarette smoke. Ethylbenzene is a contaminant in many drinking water supplies and levels can be quite high for wells near leaky gasoline storage tanks and for many surface supplies.

Average Daily Intake: AIR INTAKE (assume 0.2-2.7 ppb [1,3,43,68,69]) 17-235 ug; WATER INTAKE (assume 0-4 ppb [38]) 0-8 ug; FOOD INTAKE - insufficient data.

Occupational Exposures: NIOSH (NOES Survey 1981-1983) has statistically estimated that 31,890 workers are exposed to ethylbenzene in the United States [51]. NIOSH (NOHS Survey 1972-1974) has statistically estimated that 57,637 workers are exposed to ethylbenzene in the United States [52].

Body Burdens: Detected, not quantified in 8 of 8 samples of mother's milk from 4 U.S. urban areas [56]. Concn in expired air of 54 normal, healthy, urban volunteers, 387 samples, 16.5% pos, 1.8 ng/L expired air [39]. Whole blood samples from 250 subjects, not detected- 59 ppb, 1.0 ppb avg [2]. United States FY82 National Human Adipose Tissue Survey specimens, 46 composites, 96% pos, (>2 ppb, wet tissue concn), 280 ppb max [72].

REFERENCES

1. Altshuller AP et al; Environ Sci Technol 5: 1009-16 (1971)
2. Antoine SR et al; Bull Environ Contam Toxicol 36: 364-71 (1986)
3. Arnts RR, Meeks SA; Biogenic hydrocarbon contribution to the ambient air of selected areas 31 p USEPA 600/3-80-023 (1980)
4. Bedding ND et al; Sci Total Environ 25: 143-67 (1982)
5. Bertsch W et al; J Chromatogr 112: 701-18 (1975)
6. Bos R et al; Sci Total Environ 7: 269-81 (1977)
7. Boublik T et al; The Vapor Pressures of Pure Substances Vol 17 Amsterdam, Netherlands: Elsevier Science Publ (1984)
8. Bouwer EJ, McCarty PL; Appl Environ Microbiol 45: 1295-99 (1983)
9. Bouwer EJ, McCarty PL; Ground Water 22: 433-40 (1984)
10. Brodzinsky R, Singh HB; Volatile organic chemicals in the atmosphere: An assessment of available data 198 p SRI contract 68-02-3452 (1982)
11. Burmaster DE; Environ 24: 6-13, 33-6 (1982)
12. Callahan MA et al; p 55-61 in 8th Natl Conf Munic Sludge Manage Proc (1979)
13. Chiou CT et al; Environ Sci Technol 17: 227-31 (1983)
14. Chou WL et al; Biotechnol Bioeng Symp 8: 391-414 (1979)
15. Clark AI et al; Sci Total Environ 39: 265-79 (1984)
16. Clark AI et al; Environ Pollut (Series B) 7: 141-58 (1984)

Ethylbenzene

17. Cotruvo JA; Sci Total Environ 47: 7-26 (1985)
18. Dilling WL et al; Environ Sci Technol 10: 351-6 (1976)
19. Dumont JP, Adda J; J Agric Food Chem 26: 264-7 (1978)
20. Eisenreich SJ et al; Environ Sci Technol 15: 30-8 (1981)
21. Ewing BB et al; Monitoring to detect previously unrecognized pollutants in surface waters 75 p USEPA 560/6-77-015 (appendix USEPA 560/6-77-015a) (1977)
22. Goodley PG, Gordon M; Kentucky Acad Sci 37: 11-5 (1976)
23. Graedel TE; Chemical compounds in the atmosphere p.110 New York, NY Academic Press (1978)
24. Grob K, Grob G; J Chromatogr 62: 1-13 (1971)
25. Gschwend PM et al; Environ Sci Technol 16: 31-8 (1982)
26. Hampton CV et al; Environ Sci Technol 17: 699-708 (1983)
27. Hansch C, Leo AJ; Medchem Project Issue No 26. Claremont CA: Pomona College (1985)
28. Harkov R et al; J Air Pollut Control Assoc 33: 1177-83 (1983)
29. Hoshino M et al; Kokuritsu Kogai Kenkyusho Kenkyu Hokoku 5: 43-59 (1978)
30. Ioffe BV et al; Dokl Akad Nauk SSSR 243: 1186-9 (1978)
31. Ioffe BV et al; J Chromatogr 142: 787-95 (1977)
32. Ioffe BV et al; Environ Sci Technol 13: 864-8 (1979)
33. Jarke FH et al; ASHRAE Trans 87: 153-66 (1981)
34. Kappeler T, Wuhrmann K; Water Res 12: 327-33 (1978)
35. Kawamura K, Kaplan IR; Environ Sci Technol 17: 497-501 (1983)
36. Keith, LH et al; p 329-73 in Identification and analysis of organic pollutants in water. Keith LH ed (1976)
37. Kleopfer RD, Fairless BJ; Environ Sci Technol 6: 1036-7 (1972)
38. Konasewich D et al; Status report on organic and heavy metal contaminants in the Lakes Erie, Michigan, Huron and Superior basins. Great Lakes Quality Review Board (1978)
39. Krotoszynski BK et al; J Anal Toxicol 3: 225-34 (1979)
40. Ligocki MP et al; Atmos Environ 19: 1609-17 (1985)
41. Lioy P et al; J Water Pollut Control Fed 33: 649-57 (1983)
42. Lonneman WA et al; Hydrocarbons in Houston air 44 p USEPA 600/3-79-018 (1979)
43. Lonneman WA et al; Environ Sci Technol 2: 1017-20 (1968)
44. Lovegren NV et al; J Agric Food Chem 27: 851-3 (1979)
45. Lyman WJ et al; Handbook of chemical property estimation methods. Environmental behavior of organic compounds New York, NY McGraw Hill Co (1982)
46. Mackay D et al; Environ Sci Technol 13: 333-6 (1979)
47. Malaney GW, McKinney RE; Water Sewage Works 113: 302-9 (1966)
48. Mantoura RFC; Environ Sci Technol 16: 38-45 (1982)
49. NAS; The Alkylbenzenes USEPA contract 68-01-4655 (1980)
50. Nelson PF, Quigley SM; Atmos Environ 17: 659-62 (1983)
51. NIOSH; The National Occupational Exposure Survey (NOES) (1983)
52. NIOSH; The National Occupational Hazard Survey (NOHS) (1974)
53. Nicola RM; J Environ Health 49: 342-7 (1987)
54. Nunes P, Benville PE Jr; Bull Environ Contam Toxicol 21: 719-24 (1979)
55. Ogata M et al; Bull Environ Contam Toxicol 33: 561-7 (1984)

Ethylbenzene

56. Pellizzari ED et al; Bull Environ Contam Toxicol 28:322-8 (1982)
57. Perry R, Twibell JD; Atmos Environ 7: 329-37 (1973)
58. Piet GJ, Morra CF; p 608-20 in Oxidation techniques in drinking water treatment Kuehn IW, Sontheimer H, eds USEPA 570/9-79-020 (1979)
59. Piet GJ, Morra CF; p 31-42 in Artificial Groundwater Recharge; Huisman L, Olsthorn TN, eds (1983)
60. Ravishankara AR et al; Int J Chem Kinet 10: 783-804 (1978)
61. Sabel GV, Clark TP; Waste Manag Res 2: 119-30 (1984)
62. Santodonato J et al; Investigation of selected potential environmental contaminants: styrene, ethylbenzene and related compounds USEPA 560/11-80-018 (1980)
63. Sadtler Standard Spectra; UV No. 97
64. Sauer TC Jr; Org Geochem 3: 91-101 (1981)
65. Sauer TC Jr et al; Mar Chem 7: 1-16 (1978)
66. Seila RL; Non-urban hydrocarbon concentrations in ambient air north of Houston, TX 38 p USEPA 500/3-79-010 (1979)
67. Shackelford WM, Keith, LH; Frequency of organic compounds in surface waters USEPA 600/4-76-062 (1976)
68. Singh HB et al; Atmospheric measurements of selected toxic organic chemicals USEPA 600/3-80-072 (1980)
69. Singh HB et al; Atmos Environ 15: 601-12 (1981)
70. Singh HB et al; Atmos Environ 19: 1911-9 (1985)
71. Slave T et al; Rev Chim 25: 666-70 (1974)
72. Stanley JS; Broad Scan Analysis of the FY82 National Human Adipose Tissue Survey Specimens Vol. I Executive Summary p 5 USEPA-560/5-86-035 (1986)
73. Staples CA et al; Environ Toxicol Chem 4: 131-42 (1985)
74. Stuermer DH et al; Environ Sci Technol 16: 582-7 (1982)
75. Sutton C, Calder JA; J Chem Eng Data 20: 320-2 (1975)
76. Tabak HH et al; J Water Pollut Control Fed 53: 1503-18 (1981)
77. Termonia M; pp 356-61 in Comm Eur Comm EUR7624 Phys Chem Behav Atmos Pollut (1982)
78. Tester DH, Harker RJ; Water Pollut Control 80: 614-31 (1981)
79. Thornburn S, Colenutt BA; Int J Environ Stud 13: 265-71 (1979)
80. Tsani-Bazaca E et al; Chemosphere 11: 11-23 (1982)
81. USEPA; Treatability Manual p 1.9.8-1 to 1.9.8-5 USEPA 600/2-82-001a (1981)
82. Van de Meent D et al; Wat Sci Tech 18: 73-81 (1986)
83. Van der Linden AC; Dev Biodegrad Hydrocarbons 1: 165-200 (1978)
84. Verschueren K; Handbook of Environmental Data on Organic Chemicals 2nd ed p. 628-9, New York, NY Van Nostrand Reinhold Co, Inc. (1983)
85. Wakeham SG et al; Environ Sci Technol 17: 611-7 (1983)
86. Washida N et al; Bull Chem Soc Japan 51: 2215-21 (1978)
87. Yanagihara S et al; 4th Int Clean Air Congr Proc p 472-7 (1977)
88. Young DR; 1978 Ann Rep Southern Calif Coastal Water Res Proj p 103-12 (1978)
89. Zuercher F, Giger W; Vom Wasser 47: 37-55 (1976)

Ethylene Oxide

SUBSTANCE IDENTIFICATION

Synonyms: 1,2-Epoxyethane; Oxirane

Structure:

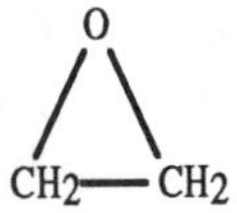

CAS Registry Number: 75-21-8

Molecular Formula: C_2H_4O

Wiswesser Line Notation: T3OTJ

CHEMICAL AND PHYSICAL PROPERTIES

Boiling Point: 10.7 °C at 760 mm Hg

Melting Point: -112.5 °C

Molecular Weight: 44.06

Dissociation Constants:

Log Octanol/Water Partition Coefficient: log Kow= -0.30 [11]

Water Solubility: 8,700 mole/L H_2O at 20 °C and 760 mm Hg [20]

Vapor Pressure: 1094 mm Hg at 20 °C [5]

Henry's Law Constant: 1.2×10^{-4} atm-m^3/mole [7]

ENVIRONMENTAL FATE/EXPOSURE POTENTIAL

Summary: Ethylene oxide will enter the atmosphere primarily in association with its production and use as a chemical intermediate as well as its relatively minor use as a sterilant and fumigant.

Ethylene Oxide

Although the use of ethylene oxide as a sterilant is minor compared to other uses, release from sterilizing operations may account for a large percentage of the total environmental burden, depending on the control technology used. From its industrial use, some ethylene oxide will be discharged into water. Once in the atmosphere it will degrade by reaction with hydroxyl-radicals (estimated half-life 120 days), water vapor (estimated half-life 16 days at 50% relative humidity and room temperature), or possibly return to earth in rain. Releases into water will be removed by evaporation, hydrolysis, and, to a lesser extent, biodegradation. The half-life for its removal from the aquatic environment probably will range from hours to 2 weeks. Ethylene oxide will not adsorb strongly to soil or bioconcentrate in fish, although its presence in some food items may result from its use as a fumigant and sterilant. Major human exposure will be from occupational atmospheres.

Natural Sources: Ethylene oxide is a product of combustion of hydrocarbon fuels [2,9], and may be produced to some extent during the combustion of naturally-occurring hydrocarbons.

Artificial Sources: Ethylene oxide is present in vent gases and fugitive emissions from production and use as a chemical intermediate in the manufacture of ethylene glycol, ethoxylates, glycol ethers, and ethanolamines [14]. Aqueous effluents are associate, with its production and use as a chemical intermediate. Fugitive emissions also result from its use as a fumigant and sterilant of food, cosmetics, and hospital supplies [3]; auto and diesel exhaust - combustion product of hydrocarbon fuels [3,9]; tobacco smoke [2].

Terrestrial Fate: When released on land, ethylene oxide will tend to volatilize rapidly. It is miscible in water (although solubility is pressure dependant) and poorly adsorbed to soil so leaching into the ground water may occur, particularly when ethylene oxide is released in water solution and when ground water is near the soil surface. Where ground water is deep, ethylene oxide would be expected to hydrolyze before reaching the saturated zone. Although experimental data are lacking concerning the rate of ethylene oxide hydrolysis in soil, hydrolysis in soil is probable.

Ethylene Oxide

Aquatic Fate: When released into water, ethylene oxide will be lost primarily by three processes: volatilization, hydrolysis, and biodegradation, in that order of importance. Volatilization will depend on wind and mixing conditions and is expected to occur in hours to days. The half-life for hydrolysis is 9-14 days; hydrolysis products are biodegradable. Because of the lack of data, it is difficult to estimate the rate of biodegradation; the available data suggest that the biodegradation rate is slower than the volatilization or hydrolysis rates. Ethylene oxide has a very low octanol/water partition coefficient and hence is not expected to adsorb to sediments. If ethylene oxide migrates to ground water, degradation by hydrolysis will continue.

Atmospheric Fate: Ethylene oxide will degrade in the atmosphere primarily by reaction with hydroxyl radicals and water vapor. Based on limited experimental data, the estimated half-life of ethylene oxide from reaction with hydroxyl radicals in the atmosphere will be approx 120 days. The reaction half-life of ethylene oxide at room temperature with water vapor at relative humidities above 50% is about 16 days. Since ethylene oxide is miscible in water, it is expected to wash out in rain to some extent, but its high vapor pressure and rapid volatilization rate may limit this.

Biodegradation: Based on limited data, ethylene oxide appears to biodegrade at a reasonable rate after a period of acclimation. In a dilution bottle test, 3-5% degradation was reported after 5 days and 52% degradation after 20 days [4,7]. Since ethylene oxide hydrolyzes to ethylene glycol, which is readily biodegraded, the biodegradability measurements are uncertain [7]. In a river die-away test, the rate of degradation was not significantly different than for hydrolysis [7].

Abiotic Degradation: Ethylene oxide hydrolyzes slowly in fresh water to give ethylene glycol and saltwater to give ethylene glycol and ethylene chlorohydrin [7]. The half-life for this reaction is 12-14 days for pH's between 5-7 in fresh water [3,7,16] and 9-11 days in salt water [7]. The ratio of chlorohydrin to glycol formed was found to be 0.11 and 0.23 in 1% and 3% sodium chloride solutions, respectively [7]. The hydrolysis rate is increased considerably in acidic or basic solutions [3]. There are two measurements of the rate of reaction between hydroxyl radicals and ethylene oxide, $(5.3 \pm 1) \times 10^{-14}$ and $(8.1 \pm 1.6) \times 10^{-14}$ cm^3/molecule-sec [1]. Based on an average hydroxyl radical reaction rate of 6.7×10^{-14} cm^3/molecule-sec and a hydroxyl radical concentration of $1 \times 10^{+6}$ molecules/cm^3, ethylene oxide is predicted to have a half-life of about 120 days in the atmosphere. At a relative humidity of 50% and higher and at room temperature, the upper limit of the loss rate for gas phase ethylene oxide from hydrolysis was reported to be 3×10^{-5} min^{-1} [22], yielding a half-life of approx 16 days.

Bioconcentration: Although no studies of bioconcentration for ethylene oxide were found in the literature, it is not expected to bioconcentrate due to its low octanol/water partition coefficient [10].

Soil Adsorption/Mobility: No data could be found concerning the adsorption of ethylene oxide to soil. Ethylene oxide is not expected to absorb due to its low octanol/water partition coefficient [10]. Based on a regression analysis with the partition coefficient [15], a Koc of 16 may be calculated.

Volatilization from Water/Soil: The half-life for evaporation of ethylene oxide from water is 1 hr with no wind and 0.8 hr with a 5 m/sec wind as determined in a laboratory experiment [7]. Although no data on the volatilization of ethylene oxide from soil could be found, a study of the dissipation of ethylene oxide from fumigated commodities gave a half-lives of 4 hr to 17.5 days [3].

Water Concentrations:

Effluent Concentrations: Detected, not quantified, in 1 effluent sample in Brandenburg, KY in Feb 1974, a production facility [3,21].

Sediment/Soil Concentrations:

Atmospheric Concentrations: Air concentrations of 2.4-17 ppb were estimated to be present near large industrial sources based on the Industrial Source Complex - Short Term model. Other sources are expected to contribute to the overall environmental burden to a lesser extent [22]. These concentrations are lower than the current detection limits for ethylene oxide [22].

Food Survey Values: 1970-76 FDA Monitoring Program - detected, not quantified in 1 out of 2372 samples of eggs in 1975 [8].

Plant Concentrations:

Fish/Seafood Concentrations: 1970-1976 FDA Food Monitoring Program - detected, not quantified in 1 out of 3262 samples of fish (1975); not found in 443 samples of shellfish for this period [8].

Animal Concentrations:

Milk Concentrations:

Other Environmental Concentrations:

Probable Routes of Human Exposure: Ethylene oxide enters the environment during its manufacture and captive use in the manufacture of ethylene glycol, ethoxylates, glycol ethers, and ethanolamines primarily in the vent stream and in liquid effluent [5,6]. Small amounts (about 2%) of ethylene oxide are used as a fumigant for spices, tobacco, furs, bedding, etc., a food and cosmetic sterilant, and in hospital sterilization [3,5,6], but this has the largest worker exposure [19]. All the ethylene oxide from the above uses enters the atmosphere [3,5].

Average Daily Intake: The lack of monitoring data for ethylene oxide in air and water probably indicates that it is absent or present in only very low concentrations in air and water. AIR INTAKE: Approx 7 million people in the Los Angeles basin are exposed to a population-weighted annual mean concentration of 0.05 ppb ethylene oxide [22]; WATER INTAKE: (assume 0 ppb, suggestive due to lack of data) 0 ug; FOOD INTAKE: insufficient data.

Occupational Exposures: OSHA estimates that approx 80,000 and 144,000 workers are directly and indirectly exposed to ethylene oxide in ethylene oxide production, chemical synthesis by ethoxylation, health care facilities (sterilization), medical products (sterilization), and miscellaneous manufacturers (e.g., spice sterilization) [19]. The number of workers exposed in the various industries are: production and synthesis 3676; sterilization - health care facilities 62,370 (25,000 indirectly); sterilization - medical products manufacture 14,000 (116,900 indirectly); sterilization - spice manufacturers 160 [19]. Typical exposures are usually high during short periods in which sterilizer doors are opened, typically 5-10 ppm for 20 minutes [19]. Some typical survey results are: medical products manufactures 0.1-2.0 ppm 8 hr TWA; hospital sterilizer chamber operators 2.5 ppm TWA; 121 use sites in Southern California <5 ppm (TWA) in 114/121 sites; 2 hospitals 3-6 ppm and <5 ppm resp; survey of 27 hospitals TWA exposures less than or equal to 1, <4 and >10 ppm in 9/27, 16/27, and 5/27, respectively [19]. Union Carbide production plant in Texas City, 5-33 ppm and 7.25 and 10.25 ppm avg in 2 control rooms and 0-56 ppm; 11.6 ppm avg throughout plant [13]. In-depth survey of 2 Union Carbide production facilities in West Virginia 2 of 48 and 4 of 41 samples positive, TWA exposure of pos samples 1.5-82 ppm [17,18]. Production and maintenance workers in the 1960's avg exposure levels 0.6-60 ppm [12].

Body Burdens:

REFERENCES

1. Atkinson R; Chem Rev 85:69-201 (1985)
2. Binder H, Lindner W; Fachliche Mitt Oesterr Tabakregie 13:215-220 (1972)
3. Bogyo DA et al; Investigations of Selected Potential Environmental Contaminants: Epoxides. EPA-560/11-80-005 p 70-90 (1980)
4. Bridie AL et al; Water Res 13:627-30 (1979)
5. Cawse et al; Kirk-Othmer Handbook of Chemical Technology 3rd edition 9: 432-71 (1980)
6. Chemical and Engineering News p 12 Aug 22 (1983)
7. Conway RA et al; Environ Sci Technol 17:107-12 (1983)
8. Duggan RE et al; Pesticide residue levels in foods in the U.S. from July 1 to June 30, 1976. Food and Drug Admin. page 10-18 (1983)
9. Graedel TE; Chemical Compounds in the Atmosphere Academic Press, NY p 272 (1978)
10. Hansch C, Leo AJ; Substituent constants for correlation analysis in chemistry and biology. New York, NY: John Wiley and Sons 339 p (1979)
11. Hansch C, Leo AJ; Medchem Project Issue No 26. Claremont CA: Pomona College (1985)

Ethylene Oxide

12. Hogstedt C et al; Brit J Ind Med 36:276-80 (1979)
13. Joyner RE; Arch Environ Health 8:700-10 (1964)
14. Kalcevik V, Lawson JF; Organic Chemical Manufacturing Volume 9: Selected Processes. Ethylene Oxide EPA-450/3-80-028d (2) Bogyo DA et al; Investigations of Selected Potential Environmental Contaminants. Epoxides EPA-560/11-80-005 p 60-69 (1980)
15. Kenaga EE, Going CAI; Aquatic toxicology, proceedings of the third annual symp on Aquatic Toxicology, ASTM, Philadelphia, PA (1980)
16. Mabey W, Mill T; J Phys Chem Ref Data 7:383-415 (1978)
17. Oser JL; In-depth Industrial Hygiene Report of Ethylene Oxide Exposure at Union Carbide Corp., South Charleston, WV NIOSH IWS-67.10 25 p (1979)
18. Oser JL; In-depth Industrial Hygiene Report of Ethylene Oxide Exposure at Union Carbide Corp., WV NIOSH IWS-67.17B 47 p (1978)
19. OSHA; Occupational Exposure to Ethylene Oxide: Proposed Rule 48 FR 17283-17319. 4,12 (1983)
20. Seidell, A; Solubilities of Organic Compounds D. Van Nostrand Co., Inc., NY (1941)
21. Shalkelford WM, Keith LH; Frequency of Organic Compounds Identified in Water. EPA-600/4-76-062 p 129 (1976)
22. State of California; Report to the Air Resources Board on Ethylene Oxide (1987)

Formaldehyde

SUBSTANCE IDENTIFICATION

Synonyms:

Structure:

CAS Registry Number: 50-00-0

Molecular Formula: CH_2O

Wiswesser Line Notation: VHH

CHEMICAL AND PHYSICAL PROPERTIES

Boiling Point: -19.5 °C

Melting Point: -92 °C

Molecular Weight: 30.03

Dissociation Constants:

Log Octanol/Water Partition Coefficient: 0.35 (calculated) [16]

Water Solubility: Very soluble, up to 55% [35]

Vapor Pressure: 3883 mm Hg at 25 °C [10]

Henry's Law Constant: 3.27×10^{-7} atm-m^3/mole [12]

ENVIRONMENTAL FATE/EXPOSURE POTENTIAL

Summary: Formaldehyde is produced in large quantities (5.7 billion lb in 1983) primarily for use in the manufacture of resins and as a chemical intermediate. Much of this use is captive and not released into the environment. Most of the formaldehyde entering the environment is produced directly or indirectly in combustion processes. The indirect production is derived from the atmospheric photochemical oxidation by sunlight of hydrocarbons or other formaldehyde precursors that have been released from combustion processes. This tremendous input of formaldehyde is removed by direct photolysis and oxidation by photochemically produced hydroxyl radicals (half-life a few hours). Additional quantities are removed by dry deposition, rain, or by dissolving in the ocean and other surface waters. In the aqueous compartment, biodegradation takes place in a few days. Human exposure to formaldehyde is from ambient air in heavy traffic, particularly during photochemical smog episodes, occupational atmospheres where resins are used, or where formaldehyde is used as a fumigant, disinfectant, embalming fluid, etc. Homes, particularly energy efficient ones, can have high levels of formaldehyde from stoves and the emission of the gas from insulation, furniture, resin-coated rugs, and other fabrics.

Natural Sources: Forest fire, animal wastes, microbial products of biological systems, and plant volatiles [17,27] are natural sources of formaldehyde. Formaldehyde can also be formed in seawater by photochemical processes [36]. Production of the chemical displays diurnal variation with maximal concentrations occurring in the late afternoon [36]. Calculations of sea-air exchange indicates that this process is probably a minor source of formaldehyde in the sea [36].

Artificial Sources: Exhaust from diesel and gasoline powered motor vehicles; major products of hydrocarbon photooxidation (major source of hydrocarbons is vehicular exhaust); vent gas from formaldehyde production; wastewater from its production and use in the manufacture of urea-formaldehyde, phenol-formaldehyde, and melamine resins and as a chemical intermediate; wastewater from the use of these resins, combustion sources such as power plants, incinerators, refineries, wood stoves, cigarettes, etc.; emissions from its use as a fumigant, soil disinfectant, embalming fluid, and in leather tanning; emissions from resins in particleboard, plywood, foam insulation, and emissions from resin-treated fabrics and paper [17,27,28,37,50]. Estimates of emission factor for many of these industries are available [48]. Emission rates from samples of particleboard, paneling, and fiberboard varied over two orders of magnitude and averaged 0.30, 0.17, and 1.5 mg/sq m-hr, respectively [34]. Emissions from urea-formaldehyde resins may continue after free formaldehyde is removed due to slow hydrolysis from contact with moisture [48].

Terrestrial Fate: When released on soil, aqueous solutions containing formaldehyde will leach into the soil. While formaldehyde is biodegradable under both aerobic and anaerobic conditions, its fate in soil is unknown.

Aquatic Fate: When released into water, formaldehyde will biodegrade to low levels in a few days. Little adsorption to sediment would be expected to occur. In nutrient-enriched seawater, there is a long lag period (approx 40 hr) prior to measurable loss of added formaldehyde by presumably biological processes [36]. Its fate in ground water is unknown.

Atmospheric Fate: Formaldehyde is released to the atmosphere in large amounts and formed in the atmosphere by the photooxidation of hydrocarbons. This input is counterbalanced by several important removal paths. It both photolyzes and reacts rapidly with reactive free radicals, principally hydroxyl radicals, which are formed in the sunlight-irradiated atmosphere. The half-life in the sunlit troposphere is a few hours. Reaction with nitrate radicals, insignificant during the day, may be an important removal mechanism at night [37]. Because of its high solubility there will be efficient transfer into rain and surface water, which may be an important sink [37]. One model predicts dry deposition and wet removal half-lives of 19 and 50 hr, respectively [32]. Although formaldehyde is found in remote areas, it is probably not transported there, but rather a result of the local generation of formaldehyde from longer-lived precursors which have been transported there [37].

Biodegradation: Formaldehyde in aqueous effluent is degraded by activated sludge and sewage in 48-72 hr [21,23,27,50]. In a die-away test using water from a stagnant lake, degradation was complete in 30 hr under aerobic conditions and 48 hr under anaerobic conditions [27].

Formaldehyde

Abiotic Degradation: Solutions containing formaldehyde are unstable, both oxidizing slowly to form formic acid and polymerizing [26]. In the presence of air and moisture, polymerization readily takes place in concentrated solutions at room temperatures to form paraformaldehyde, a solid mixture of linear polyoxymethylene glycols containing 90-99% formaldehyde [48]. In dilute aqueous solution, formaldehyde exists almost exclusively as the hydrated gem-diol $(CH_2(OH)_2)$ [12]. Formaldehyde absorbs UV radiation at wavelengths of 360 nm and longer [20], so is capable of photolyzing in sunlight. The measured half-life for photolysis as measured in simulated sunlight is 6.0 hr [47]. There are two photolytic channels; one producing H_2 and CO and the H and HCO [32]. When the rates of these reactions are combined with estimates of actinic irradiance, one predicts that the half-life of formaldehyde due to photolysis in the lower atmosphere is 1.6 hr at a solar zenith angle of 40 deg [6]. The hydrate does not have a chromophore that is capable of adsorbing sunlight and photolytically decomposing [7]. Based on its rate of reaction with photochemically produced hydroxyl radicals, formaldehyde will have a half-life of approx 19 hours in clean air and about half that long in polluted air [2,13,20]. One investigator reports the lifetime of formaldehyde in the sunlit atmosphere (due to photolysis and reaction with hydroxyl radicals) as 4 hours [32]. The hydroxy radical initiated oxidation of formaldehyde also occurs in cloud droplets to form formic acid, a component of acid rain [7]. When formaldehyde is irradiated in a reactor, however, the half-life is 50 min and 35 min in the absence and presence of NO_2, respectively [5,27]. The primary products formed are formic acid, HCl and CO [47]. Formaldehyde reacts with the NO_3 radical by H-atom abstraction with a half-life of 12 days, assuming an average NO_3 radical concentration of $2 \times 10^{+9}$ per cm^3 [1]. In water, formaldehyde is hydrated and the hydrate does not have a chromophore that is capable of adsorbing sunlight and photolytically decomposing [7].

Bioconcentration: Experiments performed on a variety of fish and shrimp show no bioconcentration of formaldehyde [24,43].

Soil Adsorption/Mobility: No information concerning the adsorption of formaldehyde to soil could be found in the literature. Formaldehyde gas adsorbs somewhat to clay mineral at high concentrations of the gas, which is important to its use as a soil fumigant [11]. Its low log octanol/water partition coefficient, 0.35 [16], indicates that soil adsorption will be low.

Volatilization from Water/Soil: Based upon the Henry's Law constant, volatilization should not be significant [33]. Since formaldehyde is generally available as aqueous solutions, its volatilization from spills on soil will also be low.

Water Concentrations: DRINKING WATER: Not detected in National Organics Reconnaissance Survey of Suspected Carcinogens in Drinking Water [49]. SURFACE WATER: 14 heavily industrialized river basins in United States - 1/204 sites pos, 12 ppb [14]. Detected only in hypolimnion of stagnant lake in Japan [27]. SEAWATER: Not detected in surface waters [27]. RAINWATER: Mainz and Deuselbach, Germany and Ireland 0.111-0.174 ppm [29]; Eniwetek Atoll (Central Pacific) 6.2-11.3 ppb [51]; 5 sites in CA - 1/6 samples pos, 0.06 ppm [18]. ICE FOG: Fairbanks, AK - 0.50-1.16 ppm [18]. MIST: 2 sites in CA 0.25-0.56 ppm [18]. FOG: 4 sites in California 0-2.3 ppm [18].

Effluent Concentrations: Detected in 3 effluent streams, two from chemical plants and one from a sewage treatment plant [42]. Effluent from urea and melamine production contained 4% formaldehyde and from phenolic resin production 0.1% formaldehyde [25]. Effluent of plywood industry which uses phenol and urea-formaldehyde resin glue contains formaldehyde [25].

Sediment/Soil Concentrations:

Atmospheric Concentrations: RURAL/REMOTE: Clean marine air <0.03-4 ppb, avg generally <0.5 ppb [15,31,38,40,41,51]. URBAN/SUBURBAN: Various sites in United States (749 samples) 26% pos, 25% of samples >2.7 ppb, max 27 ppb [4]; 6 cities in United States 11.3-20.6 ppb avg, 4 ppb max [44,45]; 4 cities in NJ 3.8-6.6 ppb daily median, 14-20 ppb 1 hr max [8]; and 2 cities in Southern California during photochemical smog episodes 2-48 ppb [19]. Concentrations are higher when there is more traffic and when there is more photochemical activity; concns tend to increase with traffic, peaking in the late morning and again late in the afternoon [8]. Julich, a moderately polluted area of West Germany (174 measurements from Aug 1979 to Aug 1980) - 0.11-10 ppb,

Formaldehyde

1.28 ppb mean [30]. Formaldehyde concn decrease as one goes up several hundred feet in altitude [32]. INDOOR AIR: Studies conducted in Denmark, Sweden, West Germany, and the United States frequently found indoor formaldehyde levels in excess of 0.12 ppm, and in several cases >3.0 ppm [9]. In an energy efficient research house, formaldehyde levels were 65 ppb without furniture, 182 ppb with furniture, 212 ppb occupied day, and 114 ppb occupied night [9]; mobil homes in 3 states which have brought complaints has formaldehyde concn that ranged up to 3.0 ppm, with mean levels of 0.1-0.88 ppm [37]. Noncomplaint mobile homes in Wisconsin: <3 yr old - 0.54 ppm mean, >3 yr old - 0.19 ppm mean [30].

Food Survey Values: Formaldehyde may be present naturally in food or as a result of contamination. Determination of formaldehyde concn in fruits and vegetables by two different methods produced the following results: tomato, 5.7 and 7.3 ppm; apple, 17.3 and 22.3 ppm; cabbage, 4.7 and 5.3 ppm; spinach, 3.3 and 7.3 ppm; green onion, 13.3 and 26.3 ppm; carrot, 6.7 and 10.0 ppm; and white radish, 3.7 and 4.4 ppm [25].

Plant Concentrations:

Fish/Seafood Concentrations:

Animal Concentrations:

Milk Concentrations:

Other Environmental Concentrations: Concentrations of formaldehyde in cigarette smoke have been reported as 45.2 to 73.1 ug/cigarette [1]. Free formaldehyde is emitted from formaldehyde resins used in durable-press cotton when they are heat-cured and stored. In the U.S., concn in 112 fabric samples ranged from 1 to 3517 ppm; 18 samples had a free formaldehyde content greater than 750 ppm [25].

Probable Routes of Human Exposure: Humans are exposed to formaldehyde from a variety of sources. The major source of atmospheric discharge is from combustion processes, specifically from auto emissions and also from the photooxidation of hydrocarbons in auto emissions [27,37]. Additional exposure to formaldehyde emissions comes from its use as an embalming fluid in anatomy labs, morgues, etc., and its use as a fumigant and sterilant [27]. Resin treated fabric, rugs, paper, etc., and materials such as particleboard and plywood which use resin adhesives and foam insulation release formaldehyde which may build up in homes and occupational atmospheres [27,37]. Contact with industrial wastewater, especially from production facilities and lumber-related operations where formaldehyde is used in adhesives, has resulted in the Pacific Northwest, Northeast, parts of Texas, and lumber areas of the Southeast [27]. The estimated daily intake of formaldehyde among exposed Finnish workers is 3000 ug, whereas heavily exposed workers (particleboard and glue production, foundry work) is 10,000 ug [22].

Average Daily Intake: AIR INTAKE:(assume 2-20 ppb) 50-500 ug; in energy efficient houses (assume 212 ppb day, 114 ppb night) 4500 ug. The estimated daily exposure of the Finnish population to formaldehyde from community air is 100 ug and from the home environment, 1000 ug [22]. WATER INTAKE: (assume 0 ppb) 0 ug. FOOD: insufficient data. TOBACCO: 50 ug [22].

Occupational Exposures: NIOSH (NOES 1981-1983) has statistically estimated that 551,795 workers are exposed to formaldehyde in the United States [39]. In a 12-week study of exposure in a gross anatomy lab of a medical school, 44% of breathing room samples and 11% of ambient air samples were >1.0 ppm, the ceiling recommended by ACGIH; half the breathing zone samples were between 0.6-1.0 ppm and the range was 0.3-2.63 ppm [46]. A 1976 report estimates that 8000 U.S. workers were potentially exposed to formaldehyde during its production [25]. A more recent estimate of the number of exposed workers in industries producing and using formaldehyde and its derivatives ranges from 1.4 to 1.75 million [25]. Concentrations of formaldehyde in occupational areas dating from the 1960's and early 1970's are: textile plant 0-2.7 ppm, 0.68 ppm avg; garment factory 0.9-2.7 ppm; clothing store 0.9-3.3 ppm; laminating plant 0.04-10 ppm; funeral homes 0.09-5.26 ppm, 0.25-1.39 ppm avg; resin manufacture and paper production 16-30 ppm; paper conditioning 0.9-1.6 ppm; wood processing 31.2 ppm max [25].

Formaldehyde

Concns in occupational settings dating from the late 1970's are: textile plants 0.1-0.5 ppm, 0.2 ppm avg; shoe factory 0.9-2.7 ppm, 1.9 ppm avg; particleboard plant 0.1-4.9 ppm, 1.15 ppm avg; plywood plant 0.1-1.2 ppm, 0.35 ppm avg; wooden furniture manufacturing plant 0.1-5.4 ppm, 1.35 ppm avg; adhesive plants 0.8-3.5 ppm, 1.75 ppm avg; foundries 0.05-2.0 ppm, 0.6 ppm avg; construction sites 0.5-7.0 ppm, 2.8 ppm avg; hospitals and clinics 0.05-3.5 ppm, 0.7 ppm avg [25]. More recent survey results for occupational environments include: fertilizer production 0.2-1.9 ppm; dyestuffs <0.1-5.8 ppm; textile manufacture <0.1-1.4 ppm; resins (foundry) <0.1-5.5 ppm; bronze foundry 0.12-0.8 ppm; iron foundry <0.02-18.3 ppm; treated paper 0.14-0.99 ppm; hospital autopsy room 2.2-7.9 ppm; plywood industry 1.0-2.5 ppm; urea-formaldehyde foam applicators <0.08-2.4 ppm [3].

Body Burdens:

REFERENCES

1. Atkinson R et al; J Phys Chem 88: 1210-5 (1984)
2. Atkinson R, Pitts JN Jr; J Chem Phys 68: 3581-4 (1978)
3. Bernstein RS et al; Am Ind Hyg Assoc J; 45: 778-85 (1984)
4. Brodzinsky R, Singh HB; Volatile Organic Chemicals in the Atmosphere: An Assessment of Available Data p. 119 SRI 68-02-3452 (1982)
5. Bufalini JJ et al; J Environ Sci Health A14: 135-41 (1979)
6. Calvert JG et al; Science 175: 751-2 (1972)
7. Chameides WL, Davis DD; Nature 304: 427-9 (1983)
8. Cleveland WS et al; Atmos Environ 11: 357-60 (1977)
9. Council on Environmental Quality; Environmental Quality-1980 p. 185-7 (1980)
10. Daubert TE, Danner RP; Data Compilation Tables of Properties of Pure Compounds NY Amer Inst Chem Eng (1985)
11. De SK, Chandra K; Sci Cult 44: 462-4 (1978)
12. Dong S, Dasgupta PK; Environ Sci Technol 20: 637-40 (1986)
13. Edney E et al; Atmospheric Chemistry of Several Toxic Compounds USEPA 600/53-82-092 (1983)
14. Ewing BB et al; Monitoring to Detect Previously Unrecognized Pollutants in Surface Waters p 75 USEPA 560/6-77-015, appendix USEPA 560/6-77-015a (1977)
15. Fushimi K, Miyake Y; J Geophys Res 85: 7533-6 (1980)
16. GEMS; Graphical Exposure Modeling System. CLOGP. USEPA (1987)
17. Graedel TE; Chemical Compounds in the atmosphere. New York, NY Academic Press p. 161 (1978)
18. Grosjean D, Wright B; Atmos Environ 17: 2093-6 (1983)
19. Grosjean D; Environ Sci Technol 16: 254-62 (1982)

Formaldehyde

20. Hampson RF; Chemical Kinetic and Photochemical Data Sheets for Atmospheric Reactions FAA-EE-80-17 (1980)
21. Hatfield R; Ind Eng Chem 49: 192-6 (1957)
22. Hemminki K, Vainio H; Human Exposure to Potentially Carcinogenic Compounds. IARC Sci Publ 59: 37-45 (1984)
23. Heukelekian H, Rand MC; J Water Pollut Control Assoc 29: 1040-53 (1955)
24. Hose JE, Lightner DV; Aquaculture 21: 197-201 (1980)
25. IARC; Monograph. Some Industrial Chemicals and Dyestuffs 29: 345-89 (1982)
26. Kirk Othmer Encycl Chem Tech 3rd edition 11: 231-58 (1980)
27. Kitchens JF et al; Investigation of Selected Potential Environmental Contaminants: Formaldehyde USEPA 560/2-76-009 (1976)
28. Kleindienst TE et al; Environ Sci Technol 20: 493-501 (1986)
29. Klippel W, Warneck P; Geophys Res Lett 5: 177-9 (1978)
30. Lowe DC, Schmidt U; J Geophys Res 88: 10,844-58 (1983)
31. Lowe DC et al; Environ Sci Technol 15: 819-23 (1981)
32. Lowe DC et al; Geophys Res Lett 7: 825-8 (1980)
33. Lyman WJ et al; Handbook of Chem Property Estimation Methods. NY: McGraw-Hill (1982)
34. Matthews TG et al; Environ Internat 12: 301-9 (1986)
35. Merck Index; 10th ed Rahway NJ: Merck & Co (1983)
36. Mopper K, Stahovec WL; Marine Chem 19: 305-21 (1986)
37. National Research Council; Formaldehyde and Other Aldehydes USEPA 600/6-82-002 (1982)
38. Neitzert V, Seiler W; Geophys Res Lett 8: 79-82 (1981)
39. NIOSH; National Occupational Exposure Survey (1985)
40. Platt U et al; J Geophys Res 84: 6329-35 (1979)
41. Platt U, Perner D; J Geophys Res 85: 7453-8 (1980)
42. Shakelford WM, Keith LH; Frequency of Organic Compounds Identified in Water USEPA 600/4-76-062 (1976)
43. Sills JB, Allen JL; Prog Fish Cult 4: 67-8 (1979)
44. Singh HB et al; Environ Sci Technol 16: 872-80 (1982)
45. Singh HB et al; Atmospheric Measurements of Selected Hazards Organic Chemicals USEPA 600/13-80-072 (1981)
46. Skisak, CM; Amer Ind Hyg Assoc J 44: 948-50 (1983)
47. Su F et al; J Phys Chem 83: 3185-91 (1979)
48. USEPA; Locating and Estimating Air Emissions From Sources of Formaldehyde. USEPA-450/4-84-007E (1984)
49. USEPA; Preliminary Assessment of Suspected Carcinogens in Drinking Water. Office of Toxic Substances (1975)
50. Verschueren K; Handbook of Environmental Data on Organic Chemicals. 2nd ed New York, NY Van Nostrand Reinhold p. 678-9 (1983)
51. Zafiriou OC et al; Geophys Res Lett 7: 341-4 (1980)

Hexachlorobenzene

SUBSTANCE IDENTIFICATION

Synonyms:

Structure:

CAS Registry Number: 118-74-1

Molecular Formula: C_6Cl_6

Wiswesser Line Notation: GR BG CG DG EG FG

CHEMICAL AND PHYSICAL PROPERTIES

Boiling Point: 323-326 °C

Melting Point: 231 °C

Molecular Weight: 284.80

Dissociation Constants:

Log Octanol/Water Partition Coefficient: 5.31 [32]

Water Solubility: 0.0062 ppm at 25 °C [23]

Vapor Pressure: 0.000019 mm Hg at 25 °C [23]

Henry's Law Constant: 0.0013 atm-m³/mole at 23 °C [79]

ENVIRONMENTAL FATE/EXPOSURE POTENTIAL

Summary: Hexachlorobenzene (HCB) is formed as a waste product in the production of several chlorinated hydrocarbons and

351

is a contaminant in some pesticides. It may enter the environment in air emissions and wastewater in connection with the above and in flue gases and fly ash from waste incineration. Non-point source dispersal of hexachlorobenzene results from its presence as a contaminant in pesticides. HCB is a very persistent environmental chemical due to its chemical stability and resistance to biodegradation. If released to the atmosphere, HCB will exist primarily in the vapor phase and degradation will be extremely slow (estimated half-life with hydroxyl radicals is 2 years). Long range global transport is possible. Physical removal from the atmosphere can occur via washout by rainfall and dry deposition. If released to water, HCB will significantly partition from the water column to sediment and suspended matter. Volatilization from the water column is rapid; however, the strong adsorption to sediment can result in long periods of persistence. If released to soil, HCB will be strongly adsorbed and not generally susceptible to leaching. Hexachlorobenzene will bioconcentrate in fish and enter into the food chain (has been detected in food during market basket surveys). Human exposure will be from ambient air, contaminated drinking water, and food, as well as contact with contaminated soil or occupational atmospheres.

Natural Sources: Not known to occur as a natural product [34].

Artificial Sources: HCB is produced as a by-product or waste material in the production of tetrachloroethylene, trichloroethylene, carbon tetrachloride, chlorine, dimethyl tetrachloroterephthalate, vinyl chloride, atrazine, propazine, simazine, pentachloronitrobenzene, and mirex [34,93]. It has been detected in treated wastewater from nonferrous metal manufacturing [92]. It is a contaminant in several pesticides including dimethyl tetrachlorophthalate and pentachloronitrobenzene [34]. HCB is emitted to the atmosphere in flue gases and fly ash generated at waste incineration facilities [19,69,70,90].

Terrestrial Fate: HCB released to soil is likely to remain there for extended periods of time due to its strong adsorption to soil (a half-life of 1530 days has been reported). Little biodegradation will occur and transport to ground water is expected to be slow, depending upon the organic carbon content of the soil; some evaporation from surface soil to air may occur, the extent of which is dependant upon the organic content of the soil [29].

Aquatic Fate: HCB released to water will evaporate rapidly (half-life of ca. 8 hr has been measured in the laboratory), adsorb to sediments, and bioconcentrate in fish and other aquatic organisms. Hydrolysis and biodegradation will not be significant processes in water.

Atmospheric Fate: HCB released to the atmosphere can exist in both the vapor phase and adsorbed phase; however, monitoring studies have demonstrated that the vapor phase should strongly dominate [9]. Degradation of HCB in the atmosphere appears to be extremely slow (estimated half-life with hydroxyl radicals is 2 years). Long range global transport is possible and has been observed [71]. Physical removal of HCB from the atmosphere to aquatic and soil environments is possible via washout by rainfall and by dry deposition.

Biodegradation: No significant biodegradation was noted in screening biodegradation tests [46,81,89], in activated sludge [45,55], or in soil [6,30,35]. A half-life in soil of 1530 days has been reported [8] and the major loss mechanism is volatilization [6,35]. Measurement of CO_2 evolution from suspended soil cultures over a 14-day incubation period were as follows: 0.4% under aerobic conditions, 0.2% under anaerobic conditions [82].

Abiotic Degradation: Although limited data are available, HCB appears to be stable to hydrolysis, photolysis, and oxidation [14]. Screening photodegradation tests indicate no degradation in water [33] and very little degradation on silica gel [24]. The half-life for the vapor-phase reaction of hexachlorobenzene with sunlight produced hydroxyl radicals in an average ambient atmosphere can be estimated to be about 2 years [2].

Bioconcentration: HCB bioconcentrates extensively in a number of fish and invertebrates [28,44,47,50,64,75,76,94,97]. Log BCF in trout, 3.7-4.3 [64,75,94]; sunfish, 3.1-4.3 [50,94]; and fathead minnow, 4.2-4.5 [47,94]. Similar high BCF values (log BCF 2-3) have been measured in aquatic microcosms [35,54].

Hexachlorobenzene

Soil Adsorption/Mobility: HCB adsorbs strongly to soil and sediment; measured log Koc ranged from 4 to 5 [31,60]. No movement on soil was observed with thin layer chromatography [31] and very little (<0.016%) appeared in leachate from a soil column [4]. However, HCB was not eliminated during bank infiltration of river water, suggesting that it may be transported in low organic carbon soils [84].

Volatilization from Water/Soil: The Henry's Law constant indicates that HCB will rapidly evaporate from water; a calculated value for the half-life for evaporation from a 1 m column of water is approx 8 hr [14]. Evaporation of hexachlorobenzene from both water and soil have been observed [42].

Water Concentrations: DRINKING WATER: 3 cities - Canada 0.06-0.2 ppt, mean 0.1 ppt [74]. SURFACE WATER: Niagara Falls dumpsite-water and sediment draining into Niagara R 8-30 ppm [22]; Great Lakes 0.02-0.1 ppt [74]; Lake Erie 4 of 5 sites pos, 0-0.04 ppt [43]; U.S. industrialized river avg [52]. SEAWATER: Mediterranean Sea coastal water Italy 16% pos 0.002-0.01 ppb [52]. RAINWATER: Great Lakes 1-4 ppt [3], North Pacific - 0.03 ppt [3], Lake Superior 2.8 ppt [43]. SEAWATER: Southern North Sea: 0.002-0.02 ng/L in solution and 0.01-6.0 ng/g in suspended particles [11]. Mediterranean Sea near Egypt (1982-3) 0.1-12.6 ng/L [21].

Effluent Concentrations: Wastewater effluent - nonferrous metals manufacturing 26 samples, 2 pos 220 ppb max [92]. Wastewater from four Canadian plants - 1-2 ppt, 1.5 ppt mean [74]. Geismar, LA - pond and ditch water on an industrial site - 170-75,000 ppb [50]. Hexachlorobenzene has been detected in fly ash and effluent gases released from municipal refuse incinerators and other combustion facilities; levels in flue gas ranged from 9.5 ng/m^3 to 11 ug/m^3 [19,69,70,90].

Sediment/Soil Concentrations: SOIL: 37 states, 0.7% pos, 0.01 ppm avg, 0-0.44 ppm [15]. Rome, Italy - 40 ppb [52]. Soil from Love Canal area, Niagara Falls, NY, 55 ug/g soil [85]. SEDIMENT: Germany and Libya, rivers and lakes - 0-15 ppb [13]. Mississippi River 0-900 ppb [51]. Lake Superior - 13 sites all pos, 0.2 ppb avg; Lake Huron - 42 sites all pos, 2 ppb avg; Lake

Hexachlorobenzene

Erie - 5 sites all pos, 3 ppb avg; Lake Ontario - 11 sites all pos, 97 ppb avg; Lake Ontario (Niagara Basin) - various depths, 0.5-460 ppb [74]. Oslo Fjord - 2-227 ppb [10]; Gulf of Mexico - 0.49 ppb [62]. Portland, MA - 9 coastal samples all pos, 0.05-0.37 ppb [78].

Atmospheric Concentrations: Great Lakes 0.008-0.024 ppt [20]; South Carolina, Florida and Texas 0.001-0.016 ppt; North Atlantic 0.04 ppt; Eniwetak Atoll 0.008 ppt [3,93]. Near chlorinated solvent or pesticide manufacturing plants 0.006-2 ppb [56,93]. Portland, OR in Feb 1984 0.05-0.11 ng/m^3 [53]. South Pacific Ocean near New Zealand (1983) and Samoa (1981) 0.055-0.061 ng/m^3 [27]. Arctic station in Norway (1983) 0.1-0.15 ng/m^3 [71].

Food Survey Values: FDA total diet samples - percent pos samples (daily intake, ug): 1971 - 1.7 (0.04), 1972 - 1.0 (0.03), 1973 - 2.8 (0.4), 1974 - 4.7 (0.07), 1975 - 8.3 (0.3), 1976 - 7.9 (0.1) [17]; 1976-77 U.S. adult total diet 300 samples, 33 pos, 29 trace, 0.001-0.002 ppb [40]; 1976-77 U.S. infant total diet 117 samples, 13 pos, 9 trace, 0.01-0.006 ppm [39]; 1976-77 U.S. toddler total diet 132 samples, 27 pos, 12 trace, 0.001-0.006 ppm [39]; 1977-78 U.S. adult total diet 240 samples, 49 pos, 40 trace, 0.001-0.002 [77]; U.S. FDA avg daily intake 1973 - 0.40 ug/day, 1974 - 0.073 ug/day [39]. U.S. FDA Total Diet Study for fiscal years 1981-1982: positive detections in 46 of 327 food composites (concn range of trace-0.005 ppm) collected from markets in 27 cities [26]. U.S.D.A. positive detections in 251 of 6138 U.S. meat and poultry samples in 1984 [12].

Plant Concentrations: Hexachlorobenzene levels of 1-5 ng/g (dry wt) were detected in fallen leaves and lichens from various trees (pine, oak, ivy, chestnut, juniper, and strawberry) in Italy [25].

Fish/Seafood Concentrations: Detected in freshwater fish (0.001-0.34 ppm) [38,41,43,65,72,74,75,87,91], marine fish (0.001-0.6 ppm) [5,37,62,96], and seafood (0.001-0.350 ppm) [10,18,63]. United States - 7 species of fish, 51 samples, 0-0.34 ppm [38]. U.S. domestic fish, 1969-1972, 2901 samples, 5.5% pos, 0.002 ppm avg; imported fish, 361 samples, 11.2% pos, 0.003 ppm avg [17]. U.S. domestic shellfish, 291 samples, 0.7% pos, 0.001 ppm avg; imported shellfish, 152 samples, 2.1% pos, 0.0003 ppm avg [17]. Survey of major watersheds near Great Lakes, 1979 - 48 samples, <5-150 ppb, 23% >10 ppb; 33% <5 ppb [49]. U.S. - 24.3% positive detections in 315 fish collected nationwide at 107 stations with levels of 0.12 mg/kg (wet wt) and 1.2 mg/kg (lipid wt) [83].

Animal Concentrations: Maine - gull eggs, 0-0.25 ppm [88]. Germany - fox, 0.26 ppm avg; wild boar, 0.31 ppm avg; deer, 0.03 ppm avg [48]. U.S. - 16 states, herons, 0.05-1.3 ppm [73]. Washington - hawk and kestrel eggs, 0.08-5.2 ppm [7].

Milk Concentrations: U.S. - 1970-76, 4638 samples, 3.3 pos, 0.001 ppm avg [17]. Yugoslavia milk - 1.31 ng/g [36]. West Germany - monitoring of cow's milk for a 10-yr period (1974-84) showed a decrease from 0.16 to 0.02 mg/kg milk fat [66].

Other Environmental Concentrations:

Probable Routes of Human Exposure: Humans will be exposed to the persistent hexachlorobenzene in a variety of ways. Occupational exposure will occur at plants manufacturing chlorinated solvents and some pesticides. Farmers will be exposed in handling and applying pesticides in which it is a contaminant, or handling contaminated soil. General population exposure will occur in people living near these industrial plants (ppb in air) and consuming contaminated drinking water; however, major general population exposure will occur through consumption of food and milk, both of which frequently have detectable levels of HCB.

Average Daily Intake: AIR: (assume 0.001-0.016 ppt) 0.20-4 ng. WATER: (assume 0.06-0.2 ppt) 0.12-0.4 ng. FOOD: 0.03-0.3 ug [17].

Occupational Exposures: Detected in workplace air during production of pentachlorophenol [34]. NIOSH has estimated that 1038 workers are potentially exposed to hexachlorobenzene based on statistical estimates derived from a survey conducted between 1981-1983 in the United States [67]. NIOSH has estimated that 4399 workers are potentially exposed to hexachlorobenzene based on statistical estimates derived from a survey conducted between 1972-1974 in the United States [68].

Body Burdens: BLOOD: chlorinated solvents factory workers, 14-233 ppb, vegetable spraymen, 0-310 ppb [34]. ADIPOSE TISSUE: U.S., 6115 samples. 0.05 ppm (avg) (U.S. National Human Adipose Tissue Survey) [80]. Canada, 99 samples, 10-667 ppb [57]. Japan, 0.06-0.21 ppm [58,59]. Rome, Italy, 28 samples, 0.49 ppm, avg [52]. United Kingdom, 0.02-3.2 ppm [1]. HUMAN MILK: Canada, 0-5.13 ppm [16]. Norway, 133 women, 2.1 ppb (fat basis) [86]. Finland, 50 samples, 0.7-6 ppb [95]. SERUM: pos detections in 3.3% of 2269 U.S. persons (aged 12-74 yr) (at levels of 1-17 ppb) monitored between 1976 and 1980 [61].

REFERENCES

1. Abbott DC et al; Br Med J 283: 1425-8 (1981)
2. Atkinson RA; Chem Rev 85: 69-201 (1985)
3. Atlas E, Giam CS; Science 211: 163-5 (1981)
4. Ausmus BS et al; Environ Pollut 20: 103-11 (1979)
5. Ballschmiter K, Zell M; Int J Environ Anal Chem 8: 15-35 (1980)
6. Beall ML Jr; J Environ Qual 5: 367-9 (1976)
7. Bechard M; Bull Environ Contam Toxicol 26: 248-53 (1981)
8. Beck J, Hansen KE; Pestic Sci 5: 41-8 (1974)
9. Bidleman TF et al; Environ Sci Technol 20: 1038-43 (1986)
10. Bjerk JE, Brevik EM; Arch Environ Contam Toxicol 9: 743-50 (1980)
11. Boon JP, Duinker JC; Environ Monitor Assess 7: 189-208 (1986)
12. Brown EA et al; p. 99-108 in Hexachlorobenzene: Proceedings of an International Symposium, IARC Publ. No. 77 (1986)
13. Buchert H et al; Chemosphere 10: 945-6 (1981)
14. Callahan MA et al; Water-related environmental fate of 129 priority pollutants p 77-1,2 USEPA-440/4-79-029b (1979)
15. Carey AE et al; Pest Monit J 12: 209-29 (1979)
16. Currie RA et al; Pest Monit J 13: 52-5 (1979)
17. Duggan RE; Pesticide Residue Levels in Foods in U.S. FDA and AOAC (1983)
18. Eder G et al; Netherlands J Sea Res 15: 78-87 (1981)
19. Eiceman GA et al; Anal Chem 53: 955-9 (1981)
20. Eisenreich SJ et al; Environ Sci Technol 15: 30-8 (1981)
21. El-Dib MA, Badaway MI; Bull Environ Contam Toxicol 34: 216-27 (1985)
22. Elder VA et al; Environ Sci Technol 15: 1237-43 (1981)
23. Farmer WJ et al; pp.83-6 in EPA/EQS/IT Natl Conference on Disposal of Residues on Land, Proc (1976)
24. Freitag D et al; Ecotox Environ Safety 6: 60-81 (1982)
25. Gaggi C et al; Chemosphere 14: 1673-86 (1985)
26. Gartell MJ et al; J Assoc Off Anal Chem 69: 146-61 (1986)
27. Giam CS, Atlas E; Amer Chem Soc 25: 5-7 (1985)
28. Giam CS et al; Bull Environ Contam Toxicol 25: 891-7 (1980)
29. Gile JD, Gillett JW; J Agr Food Chem 27: 159-64 (1979)

Hexachlorobenzene

30. Griffin RA, Chou SFJ; p 60 USEPA-600/2-81-191 (1981)
31. Griffin RA, Chou SFJ; Water Sci Technol 13: 1153-63 (1981)
32. Hansch C, Leo AJ; Medchem Project Issue No 26. Claremont CA: Pomona College (1985)
33. Hustert K et al; Chemosphere 10: 995-8 (1981)
34. IARC; Monograph. Some Halogenated Hydrocarbon 20:155-178 (1979)
35. Isensee AR et al; J Agr Food Chem 24: 1210-4 (1976)
36. Jan J; Mitt Geb Lebensmittelunters Hyg 74: 420-5 (1983)
37. Jan J, Malnersic S; Bull Environ Cont Toxicol 24: 824-7 (1980)
38. Johnson JL et al; Bull Environ Contam Toxicol 11: 393-8 (1974)
39. Johnson RD et al; J Assoc Off Anal Chem 67: 145-54 (1984)
40. Johnson RD et al; J Assoc Off Anal Chem 67: 154-66 (1984)
41. Kaiser KLE; Can J Fish Aquat Sci 39: 571-9 (1982)
42. Kilzer L et al; Chemosphere 8: 751-61 (1979)
43. Konasewich D et al; Great Lakes Water Qual Board Status Report p. 373 (1978)
44. Konemann H, Vanleeuwen K; Chemosphere 9: 3-19 (1980)
45. Korte F et al; Chemosphere 1: 79-102 (1978)
46. Korte F, Klein W; Ecotox Environ Safety 6: 311-27 (1982)
47. Kosian P et al; The precision of the ASTM bioconcentration test 24 p USEPA-600/3-81-022 (1981)
48. Koss G, Manz D; Bull Environ Contam Toxicol 15: 189-91 (1979)
49. Kuehl DW et al; Environ Int 9: 293-9 (1983)
50. Laseter JL et al; NTIS PB-252651 Gov Rep Announce Index 76: 66 (1976)
51. Laska AL et al; Bull Environ Contam Toxicol 15: 535-42 (1976)
52. Leoni V, Darca SU; Sci Total Environ 5: 253-72 (1976)
53. Ligocki MP et al; Atmos Environ 19: 1609-17 (1985)
54. Lu PY, Metcalf RL; Environ Health Perspect 10: 269-84 (1975)
55. Malaney GW, McKinney RE; Water Sewage Works 113: 302-9 (1966)
56. Mann JB et al; Environ Sci Technol 8: 584-5 (1974)
57. Mes J et al; Bull Environ Contam Toxicol 28: 97-104 (1982)
58. Mori Y et al; Bull Environ Contam Toxicol 30: 74-9 (1983)
59. Morita M et al; Environ Pollut 9: 175-9 (1975)
60. Muller-Wegener U; Z Fur Pflanzenernahrung und Biodenkunde 144: 456-62 (1981)
61. Murphy R, Harvey C; Environ Health Persp 60: 115-20 (1985)
62. Murray HE et al; Chemosphere 10: 1327-34 (1981)
63. Murray HE et al; Bull Environ Contam Toxicol 25: 663-7 (1980)
64. Neely WB et al; Environ Sci Tech 8: 1113-5 (1974)
65. Niimi AJ; Bull Environ Contam Toxicol 23: 20-4 (1979)
66. Nijhuis H, Heeschen W; p. 133-7 in Hexachlorobenzene: Proceedings of an International Symposium, IARC Publ No. 77 (1986)
67. NIOSH; National Occupational Exposure Survey (NOES) (1983)
68. NIOSH; National Occupational Hazard Survey (NOHS) (1974)
69. Oberg T, Bergstrom JGT; Chemosphere 14: 1081-6 (1985)
70. Oehme M et al; Chemosphere 16: 143-53 (1987)
71. Oehme M, Ottar B; Geophys Res Lett 11: 1133-6 (1984)
72. Ohio R Valley Water Sanit Comm; ORVWSC (1978)
73. Ohlendorf HM et al; Pest Monit J 14: 125-35 (1981)
74. Oliver BG, Nicol KD; Environ Sci Technol 16: 532-6 (1982)

Hexachlorobenzene

75. Oliver BG, Niimi AJ; Environ Sci Technol 17: 287-91 (1983)
76. Parrish PR et al; Chronic toxicity of chlordane, trifluralin, and pentachlorophenol to sheepshead minnows (Cyprinodon variegatus) 67 p USEPA NTIS PB-272101 (1978)
77. Podrebarac DS; J Assoc Off Anal Chem 67: 176-85 (1984)
78. Ray LE et al; Chemosphere 12: 1031-8 (1983)
79. Rippen G, Frank R; pp.45-52 in Hexachlorobenzene: Proceedings of an International Symposium. Morris CR, Cabral JRP eds. Lyon, France: IARC (1986)
80. Robinson ED et al; p. 182-83 in Hexachlorobenzene: Proceedings of an International Symposium, IARC Publ. No. 77 (1986)
81. Rott B et al; Chemosphere 11: 531-8 (1982)
82. Scheunert I et al; Chemosphere 16: 1031-41 (1987)
83. Schmitt CJ et al; Arch Environ Contam Toxicol 14: 225-60 (1985)
84. Schwarzenbach RP et al; Environ Sci Tech 17: 472-9 (1983)
85. Silkworth JB et al; Fundam Appl Toxicol 4: 231-9 (1984)
86. Skaare JV; Acta Pharmacol Toxicol 49: 384-9 (1981)
87. Swain WR; J Great Lakes Res 4: 398-407 (1978)
88. Szaro RC et al; Bull Environ Contam Toxicol 22: 394-9 (1979)
89. Tabak HH et al; J Water Pollut Contr Fed 53: 1503-18 (1981)
90. Tiernan TO et al; Environ Health Persp 59: 145-58 (1985)
91. Tsui PTP, McCarty PJ; Int J Environ Anal Chem 10: 277-85 (1981)
92. USEPA; Treatability Manual - p I.9.7-3 USEPA 600/2-82-001a (1981)
93. USEPA; Health Assessment Document for Chlorinated Benzenes. Environmental Criteria and Assessment Office, Cincinnati, OH USEPA 600/8-84-015a (1984)
94. Veith GD et al; J Fish Res Board Can 36: 1040-8 (1979)
95. Wiskstrom H et al; Bull Environ Contam Toxicol 31: 251-6 (1983)
96. Young DR et al; pp.471-86 in Water Chlorination: Environmental Impact and Health Effects; Jolley RL et al eds (1980)
97. Zitko V, Hutzinger D; Bull Environ Contam Toxicol 16: 665-73 (1976)

Hexachloro-1,3-butadiene

Synonyms:

Structure:

CAS Registry Number: 87-68-3

Molecular Formula: C_4Cl_6

Wiswesser Line Notation: GYGUYGYGUYGG

CHEMICAL AND PHYSICAL PROPERTIES

Boiling Point: 215 °C

Melting Point: -21 °C

Molecular Weight: 260.76

Dissociation Constants:

Log Octanol/Water Partition Coefficient: 4.90 [5]

Water Solubility: 2.55 mg/L at 20 °C [5]

Vapor Pressure: 0.15 mm Hg at 25 °C [30]

Henry's Law Constant: 1.03×10^{-2} atm-m^3/mol [36]

ENVIRONMENTAL FATE/EXPOSURE POTENTIAL

Summary: Recent information on hexachlorobutadiene (HCBD) production was not available; however, total U.S. production of HCBD was reported as 8.0 million lb/yr in 1975 and losses of

360

Hexachloro-1,3-butadiene

HCBD to the environment estimated at 0.1 million lb/yr. HCBD is used as a solvent for elastomers, heat transfer liquid, transformer and hydraulic fluids, and as a wash liquor for removing C_4 and higher hydrocarbons. When released to soil surfaces, HCBD is expected to rapidly evaporate. In aerobic zones of soil, HCBD may biodegrade. HCBD will probably adsorb strongly to soils (estimated Koc = 5181), so leaching will probably not be rapid except in sandy soils. Estimated half-lives for HCBD are 3-30 days in river water and 30-300 days in lake and ground waters. Because of its relatively high log Kow value (4.90), HCBD will sorb to sediments, suspended sediments, and biota. Because of the high Henry's Law constant volatilization from unfrozen water will be rapid. It has a long half-life in the atmosphere, with estimates ranging from months to over a year. Therefore, considerable dispersion is expected; HCBD is found in remote areas of the globe far removed from emission sources. HCBD has been detected in a number of U.S. surface and drinking waters. For example, Mississippi River water near Baton Rouge contained 1.9 ppb of HCBD and New Orleans drinking water from the Carrollton water plant contained from 0.04 to 0.70 ppb. HCBD has been measured in the ambient air of Houston, TX and Niagara Falls, NY at 117 ng/m³ and 0.39 ug/m³, respectively. HCBD has been detected in fish from the Mississippi, Ashtabula (OH) and Wabash (IN) Rivers and from Lakes Ontario, Erie, Huron, and Superior. The highest exposure to HCBD will probably be in occupational settings while the primary exposure of the general public to HCBD will probably be from drinking water.

Natural Sources: Hexachlorobutadiene is not known to occur as a natural product [10].

Artificial Sources: Based on a model of global mixing in conjunction with monitoring measurements at a remote site, the atmospheric burden of HCDB in the northern and southern hemisphere has recently been estimated to be 3.2 and 1.3 million kg/yr, respectively [6]. In 1975, total U.S. production of hexachlorobutadiene (HCBD) was reported as 8.0 million pounds (3.6 million kg)/yr and HCBD losses to the environment as 0.1 million pounds (0.05 million kg)/year [4]. HCBD is used as a solvent for elastomers, heat transfer liquid, transformer and hydraulic fluid, and as a wash liquor for removing C_4 and higher hydrocarbons [9]. In 1975, the largest U.S. use of HCBD was for recovery of "sniff" (chlorine-containing) gas in chlorine production plants [10]. Li et al [20], in a survey of six industries, found that higher concn of HCBD were associated with the production of perchloroethylene and trichloroethylene than with plants producing chlorine and triazine herbicides. Many of the above uses can result in environmental releases. HCBD is also released during refuse combustion and is found in fly ash [12].

Terrestrial Fate: When hexachlorobutadiene (HCBD) is released to soil surfaces, evaporation is expected to be a significant transport mechanism. HCBD may biodegrade in aerobic zones of soil based on aerobic biodegradation studies in water [38]. In anaerobic batch tests in water, HCBD did not biodegrade [11]; thus, HCBD may not biodegrade in anaerobic zones of soil. No information was found about HCBD adsorption to soil, but a calculated Koc of 5181 suggests that HCBD will adsorb strongly to organic material in soils and is not expected to rapidly migrate. However, migration will be more rapid in sandy soils. Dune infiltration studies in The Netherlands indicate that HCBD is mobile in sandy soil. Dune infiltration and effluent HCBD concentrations (respectively) were 0.02 and 0.01 ug/L in 1976, and 0.01 and 0.01 ug/L in 1977 [33]. Average residence time was 100 days. These results also suggest that degradation of HCBD under these conditions is very slow. The maximum level of HCBD in vineyard soil 8 months after its last application as a fumigant at 250 kg/ha was 4.36 ppm in the 75-100 cm layer, and 7.3 ppm in the 50-75 cm layer [41]. After 32 months, the levels in the 50-75 cm and 75-100 cm layers were 0.65 and 2.99 ppm, respectively.

Aquatic Fate: Hexachlorobutadiene (HCBD) may biodegrade in natural waters, since 100% HCBD degradation occurred in 7 days in an aerobic batch culture incubated at 25 °C and inoculated with settled domestic sewage [38]. No information was found on HCBD biodegradation rates in natural waters. Estimated half-lives for HCBD degradation are 3-30 days in river water and 30-300 days in lake and ground waters based upon monitoring data [43]. No information was found on hydrolysis or photolysis. Because of its relatively high log Kow value, HCBD will sorb to sediments, suspended sediments, and biota. The mean bioconcentration factor (BCF) for rainbow trout exposed to 0.10 ng/L and 3.4 ng/L of

Hexachloro-1,3-butadiene

HCBD was 5800 and 17,000, respectively [28], which would suggest extensive bioconcentration. The high Henry's Law constant indicates that HCBD will rapidly volatilize from water [22].

Atmospheric Fate: In the atmosphere, hexachlorobutadiene should degrade primarily by addition of photochemically derived hydroxyl radicals to its double bonds. By analogy with tetrachloroethylene one would anticipate a half-life of 2 months. Using a mass balance approach a half-life estimate of 1.6 years was obtained for the northern hemisphere. Therefore HCBD will be long lived in the atmosphere and considerable dispersion would be expected.

Biodegradation: Hexachloro-1,3-butadiene (HCBD) did not biodegrade in an anaerobic batch culture incubated for 48 hours at 37 °C [11]. However, under aerobic batch conditions with domestic wastewater as the inoculum, 100% of the HCBD was removed after 7 days incubation at 25 °C [38].

Abiotic Degradation: Hexachlorobutadiene (HCBD) will react with photochemically produced hydroxyl radicals and ozone in the atmosphere by addition to the double bond. Estimation methods are unreliable for predicting atmospheric half-lives of completely chlorinated alkenes and the best estimate of atmospheric lifetime would be obtained from experimental values for tetrachloroethylene. For this compound the OH radical and ozone addition rates are 0.167×10^{-12} and $<2.00 \times 10^{-23}$ cm^3/molecule-sec, respectively [1,2]. The half-life of tetrachloroethene with ambient concentrations of hydroxyl radicals is 60 days whereas that for ozone is insignificant. The half-life of HCBD should be expected to be approx the same. Another estimate of tropospheric half-life obtained from monitoring data at remote sites is 1.6 yr in the northern hemisphere and 0.6 yr in the southern hemisphere [6]. In the laboratory, HCBD was adsorbed on silica gel, placed in a pure oxygen atmosphere, and exposed to an artificial light source through quartz or Pyrex glass. After a 3-day exposure through the quartz filter, 10-50% of the theoretical HCl and/or Cl_2 and 50-90% of the CO_2 were formed; after 6 days exposure through the Pyrex filter (>290 nm wavelength), 50-90% of the theoretical HCl and/or CO_2 and 50-90% of the CO_2 were formed [8]. No information was found on hydrolysis, but based on structure, hydrolysis should not be an important process.

Bioconcentration: The mean bioconcentration factor (BCF) for rainbow trout exposed to 0.10 ng/L and 3.4 ng/L of HCBD was 5800 and 17,000, respectively [28]. HCBD concn in Mississippi River mosquito fish at Baton Rouge, LA averaged 827.3 ppb HCBD, while the river water contained 1.9 ppb HCBD [19]. The mean bioconcentration factor for HCBD between oligocheate worms and sediment in Lake Ontario near the Niagara River was 0.43 [26]. The concn of the chemical in the sediment pore water was the main factor affecting bioconcentration.

Soil Adsorption/Mobility: The calculated Koc value for HCBD [22] using the water solubility is 5181. Thus, HCBD will adsorb strongly to organic material in soils and is not expected to rapidly migrate. However, migration will be more rapid in sandy soils. Despite its strong adsorption to soil, HCBD has been found in bank-filtered Rhine water in which the retention time in soil is >1 yr [32].

Volatilization from Water/Soil: The Henry's Law constant indicates that volatilization from water will be rapid and that the liquid phase will control the volatilization rate [22]. Thus, volatilization of HCBD will be more rapid from a turbulent stream than from a lake. Because of its high log Kow and log Koc, HCBD volatilization from water may be decreased due to adsorption to sediments, suspended sediments, and biota. With its vapor pressure of 0.15 mm Hg at 25 °C [30], HCBD will probably evaporate from surfaces. However, the high Koc value for HCBD should make the evaporation rate from soil less than from other surfaces. One day and 1 month after a spring application of 250 kg hexachlorobutadiene (HCBD)/ha to a vineyard, the air contained 0.08 and 0.003 mg/m³ of HCBD, respectively. After an application of 150 kg/ha, corresponding values were 0.06 and 0.001 mg/ha. HCBD vaporization from light soils was more rapid than from heavy soils. Raising temperatures from 13-18 °C hastened HCDB vaporization 5-fold under static conditions. Its application on granulated superphosphate decreased vaporization in comparison with liquid application.

Water Concentrations: DRINKING WATER: Hexachlorobutadiene (HCBD) concentration ranged from 0.04 to 0.70 ug/L in New

Orleans drinking water from the Carrollton water plant [14]. The highest HCBD concentration found in tap water from houses near Love Canal (Niagara Falls, NY) was 170 ng/L [3]. SURFACE WATER: Hexachlorobutadiene (HCBD) was found at concentrations of 10 and 23 ug/L in water sampled beyond the boundries of two chemical plants [20]. HCBD concentration in an open waste treatment pond was 244 ug/L [20]. Mississippi River water at Baton Rouge, LA contained 1.9 ppb of HCBD [19]. HCBD was found in the Ashtabula River (Ohio) at an unspecified concentration [18]. HCBD has also been detected in Lake Erie water [15]. Grab samples of water from Lake Ontario, the Niagara River, and the Detroit River contained mean HCBD concns of 0.01, 0.2, and 0.1 ppt, respectively [27]. Between 1982 and 1985, the average concentration of HCBD in the Rhine River ranged from 0.01 to 0.1 ppb [29]. The concentration of HCBD in the North Sea off the coast of The Netherlands during 1983-1984 ranged from <0.02 to 1.30 ppt, median 0.23 ppt [39]. Since the concentration in the Rhine River was 15 ppt, it was concluded that the river was a source of this contaminant [39].

Effluent Concentrations: Effluent from the Diamond Shamrock Corp in Deerpark, TX contained 2 ug/L HCBD [20]. HCBD was found to be discharged from the Niagara Falls Sewage treatment plant [35]. No levels were given.

Sediment/Soil Concentrations: HCBD concentration in soil near a chemical plant was 0.11 ug/g [20]. HCBD levels of 0.15 and 0.34 ug/g were found at the boundaries of the two other plants [20]. HCBD was found in soil near the Mississippi River at 433.0 ppb (dry basis) [19]. Sediment from Lake Ontario near the Niagara River contained 7.3-11 ppb HCBD [13,26]. HCBD has also been detected in Lake Michigan sediment [15] and from suspended matter of the Hylebos Waterway in Tacoma, WA [34]. HCBD was found in Ohio River sediment near Louisville at 2 and 17 ng/g [20]. Sediments in southern Lake Huron, Lake St. Clair, Western Lake Huron, Central Lake Erie, and Eastern Lake Erie contained mean HCBD concns of 0.08, 7.3, 1.6, 0.2, and 0.2 ppb (dry wt), respectively [25].

Atmospheric Concentrations: The maximum hexachlorobutadiene (HCBD) concentration was 460 ug/m^3 in the air at a plant that produced perchloroethylene, carbon tetrachloride, and chlorine [20]. Maximum HCBD air concentration off plant property was 0.22 ug/m^3 [20]. At another plant, HCBD concentration was 10 ug/m^3 at the plant boundary [20]. HCBD was found at 0.39 ug/m^3 in the ambient air of Niagara Falls, NY and at 0.41 ug/m^3 in the ambient air of a household basement in Niagara Falls, NY [31]. In Houston, HCBD ambient air concentration was 117 ng/m^3 [37]. The mean concentration (standard deviation) of HCBD in remote sites in the northern and southern hemisphere selected for being influenced by long range transport only is 0.17 (0.05) and 0.07 (0.03) ppt, respectively [6].

Food Survey Values: HCBD concentrations (ppm) in United Kingdom foodstuffs were: butter (2.0); vegetable cooking oil (0.2); ale (0.2); tomatoes (0.8); and black grapes (3.7) [23]. HCBD was also found in the following foodstuffs and feeds (country not specified) (ppm): evaporated milk (4.0); egg yolk (42); vegetable-oil margarine (33); chicken grain feed (39); and chicken laying rations (20) [16]. In a study of foods collected within a 25-mile (40-km) radius of factories producing tetrachloroethylene or trichloroethylene, no HCBD residues were found in any of 15 egg samples or 20 samples of a variety of vegetables [42]. Of 20 milk samples collected, only one contained HCBD residues (1.32 mg/kg on a fat basis); in a later follow-up sample, no residues were detected [42].

Plant Concentrations: Five species of marine algae in the Mersey Estuary in England were found to contain hexachlorobutadiene residues at levels ranging from 0 to 8.9 ppb [30]. Grapes contained 0.006 mg HCDB/kg (fresh wt) 4 months after fumigation [21].

Fish/Seafood Concentrations: Mosquito fish taken from the Mississippi River near Baton Rouge, LA contained an avg of 827.3 ppb HCBD [19]. Crayfish from a ditch near Baton Rouge, LA contained 70.1 ppb HCBD [19]. HCBD was found, but not quantified, in fish from the Ashtabula River, OH [18] and the Wabash River, IN [40]. However a study of 9 Great Lakes' harbors and tributaries in Ohio and Wisconsin, including the Astabula River, between 1980 and 1981 found no HCBD in composite fish samples [7]. Mean HCBD concentration in Lake Ontario rainbow trout was 0.2 ng/g [28]. HCBD has also been detected in fish from Lakes Erie, Huron, and Superior [15]. A study of the distribution of HCBD in the St. Clair and Detroit Rivers using clams (E. complanatus) suspended in cages, found that none of the sites in the Detroit River and 55% of the sites in the St. Clair River contained levels of HCBD above the detection limit, 1 ppb, wet wt [17]. The highest mean concentration was 83 ppb and results indicated that sources of pollution were present in the Sarnia area [17].

Animal Concentrations: Oligochaete worms from Lake Ontario sediment near the Niagara River 2.0-8.6 ppb dry weight [26]. In a British study, 14 species of invertebrates were found to contain HCBD residues at levels ranging from 0 to 7 ppb (wet wt) [30]. Eggs or organs of 8 species of sea, and freshwater birds, 0-9.9 ppb [30]. Grey seal and common shrew, 0.4-3.6 and 0 ppb, respectively [30].

Milk Concentrations: HCBD was found in fresh milk at a concentration of 0.08 ug/kg [23].

Other Environmental Concentrations:

Probable Routes of Human Exposure: The highest exposure to HCBD will probably be in occupational settings, while the primary exposure of the general public to HCBD will probably be from contaminated drinking water. A smaller population will be exposed to HCBD by ingesting fish and other edible aquatic organisms.

Average Daily Intake: AIR: insufficient data; WATER: (assume 0.04-0.7 ppb, data from [14]): 0.08-1.4 ug; FOOD: insufficient data.

Occupational Exposures: The maximum HCBD concentration at nine manufacturing plants was 460 ug/m³ [20].

Body Burdens: Average HCBD concentration in Canadian adipose tissue (human) was 0.004 ug/g [24]. HCBD was found at a concentration of 13.7 ug/kg (wet tissue) in the liver of a 75-year-old male and 1.8 ug/kg in the body fat of a 48-year-old male [23].

REFERENCES

1. Atkinson R, Carter WPL; Chem Rev 84:437-70 (1984)
2. Atkinson R; Int J Chem Kinet 19:799-828 (1987)
3. Barkley J et al; Biomed Mass Spectra 7:139-47 (1980)
4. Brown SL et al; Research Program on Hazard Priority Ranking of Manufactured Chemicals (Chemicals 1-20) p.9-A-1 NTIS PB-263161 (1975)
5. Chiou CT; Environ Sci Technol 19: 57-62 (1985)
6. Class T, Ballschmiter K; Freznius Z Anal Chem 327: 198-204 (1987)
7. DeVault DS; Arch Environ Contam Toxicol 14:587-94 (1985)
8. Gaeb S et al; Nature 270: 331-3 (1977)
9. Hawley GG; The Condensed Chemical Dictionary 10 ed New York Nan Nostrand Reinhold Co (1981)
10. IARC; IARC Monograph on the Evaluation of the Carcinogenic Risk of Chemicals to Human. Some Halogenated Hydrocarbons. 20:181-193 (1979)
11. Johnson LD, Young JC; J Water Pollut Control Fed 55: 1441-9 (1983)
12. Junk GA, Ford CS; Chemosphere 9: 187-230 (1980)
13. Kaminsky R et al; J Great Lakes Res 9: 183-9 (1983)
14. Keith LH et al; in: Identification and Analysis of Organic Pollutants in Water Keith LH ed Ann Arbor Press p.329-73 Ann Arbor, MI (1976)
15. Konasewich D et al; Status Report on Organic and Heavy Metal Contaminants in the Lakes Erie, Michigan, Huron and Superior Basins. Great Lake Water Qual Board (1978)
16. Kotzias D et al; Chemosphere 4: 247-50 (1975)
17. Krauss PB, Hamdy YS; J. Great Lakes Res 11:247-663 (1985)
18. Kuehl DW et al; J Assoc Off Anal Chem 63: 1238-44 (1980)
19. Laska AL et al; Bull Environ Contam Toxicol 15: 535-42 (1976)
20. Li RT et al; Sampling and Analysis of Selected Toxic Substances Task IB Hexachlorobutadiene USEPA-560/6-76/015 pp.152 (1976)
21. Litvinov PI, Gorenshtein RS; Zashch Rast (Moscow) 8: 33-4 (1982)
22. Lyman WJ et al; Handbook of Chemical Property Estimation Methods. Environmental Behavior of Organic Compounds McGraw-Hill NY (1982)
23. McConnell G et al; Endeavour 34: 13-8 (1975)
24. Mes J et al; Bull Environ Contam Toxicol 28: 97-104 (1982)
25. Oliver BG, Bourbonniere RA; J Great Lakes Res 11: 366-72 (1985)
26. Oliver BG; Environ Sci Technol 21: 785-90 (1987)
27. Oliver BG, Nicol KD; Intern J Anal Environ Chem 25: 275-85 (1986)
28. Oliver BG, Niimi AJ; Environ Sci Technol 17: 287-91 (1983)
29. Pagga U; Z Wasser Abwasser Forsch 20: 101-7 (1987)
30. Pearson CR, McConnell G; Proc Roy Soc London Ser B 189: 305-32 (1975)
31. Pellizzari ED; Environ Sci Technol 16: 781-5 (1982)
32. Piet GJ et al; pp. 69-80 in Hydrocarbon Halogenated Hydrocarbons in the Aquatic Environment. Afghan, BK and Mackay, D. ed, NY: Plenum Press (1980)
33. Piet GJ, Zoeteman BCJ; J Amer Water Works Assoc 72: 400-4 (1980)
34. Riley RG et al; Quantitation of Pollutants in Suspended Matter and Water from Puget Sound p.110 NOAA-80061003 (NTIS PB80-203524) (1980)

Hexachloro-1,3-butadiene

35. Rohmann SO et al; Tracing a River's Toxic Pollution. NY: Inform, Inc (1985)
36. Shen TT; Environ Management 6: 297-305 (1982)
37. Singh HB et al; Environ Sci Technol 16: 872-80 (1982)
38. Tabak HH et al; J Water Pollut Control Fed 53: 1503-18 (1981)
39. Van de Meent D et al; Wat Sci Technol. 18:73-81 (1986)
40. Veith GD et al; Pest Monit J 13: 1-11 (1979)
41. Vorobeva TN; Khim Selsk Khoz 18: 39-40 (1980)
42. Yip G; J Assoc Off Anal Chem 59: 559-61 (1976)
43. Zoeteman BCJ et al; Chemosphere 9: 231-49 (1980)

Hexachloroethane

SUBSTANCE IDENTIFICATION

Synonyms:

Structure:

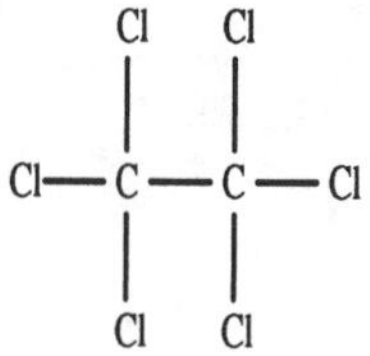

CAS Registry Number: 67-72-1

Molecular Formula: C_2Cl_6

Wiswesser Line Notation:

CHEMICAL AND PHYSICAL PROPERTIES

Boiling Point: 186 °C at 777 mm Hg

Melting Point: 186.8-187.4 °C

Molecular Weight: 236.74

Dissociation Constants:

Log Octanol/Water Partition Coefficient: 3.82 [15]

Water Solubility: 50 mg/L at 22.3 °C [2]

Vapor Pressure: 0.21 mm Hg at 20 °C [2]

Henry's Law Constant: 2.8 x 10^{-3} atm-m³/mole at 20 °C [25]

ENVIRONMENTAL FATE/EXPOSURE POTENTIAL

Summary: Potential sources of hexachloroethane release to the environment include: formation during combustion and incineration of chlorinated wastes (e.g., polyvinyl chloride (PVC)); release to air due to volatility; and inefficient solvent recovery and recirculation (hexachloroethane is an impurity in some chlorinated solvents), and formation of very small amounts during chlorination of sewage effluent prior to discharge. Hexachloroethane is also reported to be produced in very small quantities from chlorination of raw water during drinking water treatment. If released to unadapted soil, this compound may persist for greater than 2 years and could potentially contaminate ground water since it is not strongly adsorbed to soil. Hexachloroethane should volatilize slowly from dry soil surfaces. If released to water, volatilization appears to be the dominant removal mechanism (half-life 15 hours from a model river). Moderate to slight adsorption to suspended solids and sediments may occur. Biodegradation, photolysis, oxidation by reaction with singlet oxygen, alkylperoxy radicals or hydroxyl radicals, and chemical hydrolysis are not expected to be important fate processes in water. If released to air, hexachloroethane should exist almost entirely in the vapor phase. This compound is not expected to degrade in the troposphere. It should diffuse slowly into the stratosphere (half-life approx 30 years), where it is predicted to photodegrade. As a result of its persistence in the troposphere, long range transport is expected to occur. The most probable route of human exposure to hexachloroethane is inhalation of contaminated occupational or ambient air. Some segments of the general population may also be exposed by ingestion of contaminated drinking water.

Natural Sources: Hexachloroethane is not known to occur as a natural product.

Artificial Sources: Significant commercial quantities of hexachloroethane apparently are not produced intentionally in the United States [20]. However, it is formed in minor amounts in many industrial chlorination processes designed to produce lower chlorination products [2]. Potential sources of hexachloroethane release to the environment include: formation during combustion and incineration of chlorinated wastes (PVC); release to air due to volatility, and inefficient solvent recovery and recirculation; and formation of very small amounts during chlorination of sewage effluent prior to discharges [1,7]. Hexachloroethane is also reported to be produced in very small quantities from chlorination of raw water during drinking water treatment [1].

Hexachloroethane

Terrestrial Fate: If released to soil, hexachloroethane is expected to have medium to low mobility. Chemical hydrolysis is not expected to be an important fate process. This compound may volatilize slowly from dry soil surfaces. It has been reported that hexachloroethane may persist in unadapted soil for greater than 2 years [1]. Hexachloroethane injected into a sandy aquifer decreased to below detectable levels in 330 days [31].

Aquatic Fate: If released to water, volatilization appears to be the dominant removal mechanism (half-life 15 hours from a model river). One biodegradation screening study has shown that a mixed microbial population was capable of degrading hexachloroethane. However, biodegradation is not expected to occur at a rate which would make this an important fate process in natural water systems. Moderate to slight adsorption to suspended solids and sediments may occur. Photolysis, oxidation by reaction with singlet oxygen, hydroxyl radicals, or alkyl peroxy radicals, and chemical hydrolysis are not expected to be important.

Atmospheric Fate: Based on its vapor pressure, hexachloroethane is expected to exist almost entirely in the vapor phase in the atmosphere [10]. Hexachloroethane is persistent in the troposphere and as a result, long-range transport should occur. This compound is expected to diffuse slowly into the stratosphere (half-life approx 30 yr [5]), where photodegradation may be an important fate process.

Biodegradation: Conflicting data are available on the microbial breakdown of hexachloroethane. Hexachloroethane is reported to be resistant to aerobic biological treatment and inhibitory to anaerobic biological reactions [1,18]. Results of one biodegradation screening study indicated that <30% degradation occurred after 2-weeks incubation of 100 ppm hexachloroethane in 30 ppm activated sludge under aerobic conditions (Japanese MITI Test) [19,33]. In contrast, 100% loss was observed when 5 and 10 mg/L hexachloroethane underwent a 7-day static incubation in the dark at 25 °C under aerobic conditions using settled domestic wastewater as inoculum (0% loss attributed to volatility) [38].

Hexachloroethane

Abiotic Degradation: Hexachloroethane would not be susceptible to chemical hydrolysis under environmental conditions [24,40]. Oxidation by reaction with single oxygen, hydroxyl, or alkylperoxy radicals in water would probably not be an important fate process [24,40]. Hexachloroethane is not expected to photodegrade in water or the troposphere, although photolysis may be an important degradation mechanism in the stratosphere [24,40]. Hexachloroethane is not expected to react with photochemically generated hydroxyl radicals or ozone in the atmosphere [13].

Bioconcentration: Bioconcentration factors for hexachloroethane have been measured to be 139 in bluegill sunfish [3] and 510 in rainbow trout [28]. The half-life for hexachloroethane in tissue of bluegill sunfish has been found to be <1 day [3]. These data indicate that hexachloroethane will not bioaccumulate significantly in aquatic organisms. Monitoring data supports this conclusion.

Soil Adsorption/Mobility: A soil adsorption factor (Koc) of 173 has been estimated using a molecular topology and quantitative structure-activity relationship [32]. A Koc of 960 has been estimated using a linear regression equation based on the water solubility [23]. These Koc values suggest that hexachloroethane would have medium to low mobility in soil and adsorb moderately to slightly to suspended solids and sediments in water [37].

Volatilization from Water/Soil: The half-life of hexachloroethane volatilizing from a model river 1 m deep, flowing 1 m/sec with a wind speed of 3 m/sec has been estimated to be 15 hours based on the Henry's Law constant [23]. The volatilization half-life of a dilute solution of hexachloroethane in a beaker 6.5 cm deep, stirred 200 rpm in still air has been determined to be 40.7 minutes [9]. The vapor pressure suggests that this compound should volatilize slowly from dry soil surfaces.

Water Concentrations: SURFACE WATER: USEPA STORET Database - 882 whole water samples, 0.1% pos, median concn <10 ug/L [35]. Fall 1980 in Lake Ontario - 0.2 ng/L [28]. Detected in Ganaraska River and detected in Cuyahoga River [14]. 1982 U.S. EPA Nationwide Urban Runoff Program (NURP) - 86 samples urban runoff from 15 cities, 0% pos. [8]. Identified in 1/204 water samples collected from 14 heavily industrialized river basins across

U.S., detection limit 1 ug/L [11]. DRINKING WATER: Detected in 4/14 U.S. drinking water supplies sampled between July 1977 and June 1979, raw water supply - surface water [12]. Identified in finished drinking water from Philadelphia, PA, Cincinnati, OH, Miami, FL, New Orleans, LA, Jefferson City, MO and Evansville, IN [1,22,34,36,41]. Highest concentration found in finished drinking water during 1975 U.S. EPA National Organics Reconnaissance Survey (NORS) - 4.4 ug/L [41]. Not detected in finished drinking water from 10 Canadian water treatment plants [29]. GROUND WATER: Identified in leachate from the Occidental Chemical Co S-Area landfill in Niagara Falls, NY [39]. During 1978, detected in ground water contaminated by leachate from a pesticide waste dump in Hardeman County, TN: concn range trace-4.6 ug/L; median concn 0.26 ug/L [6].

Effluent Concentrations: USEPA STORET database - 1253 effluent sample 0.2% pos, median concn <10 ug/L [35]. Concn in treated wastewater from coal mining <0.4 ug/L [40]. Identified in wastewater from paper mills, concn <1 ug/L [20].

Sediment/Soil Concentrations: USEPA STORET database - 356 sediment samples, 0% pos [35]. Detected in sediment/soil/water samples taken from Love Canal in Niagara Falls, NY [16].

Atmospheric Concentrations: Mean background level in northern hemisphere, 0.5 ppt [7]. Above tradewinds and in areas with descending air, mean background level 0.2 ppt [7]. Concentration in air samples over Atlantic Ocean during 1982-85 ranged from 0.2 to 0.7 ppt [7]. Concentration in air samples obtained during rain events in Portland, OR during Feb to April 1982 ranged between 0.28 to 0.41 ppt [21]. Not detected in rainwater collected in Portland, OR during 1982 [30]. During 1976-78 in US: rural/remote areas - 6 samples, 100% pos, mean concn 3.2 ppt, median concn 5.5 ppt; urban/suburban areas - 76 samples, 0% pos.; source dominated areas - 42 samples, 0% pos [4]. Identified but not quantified in emissions from a hazardous waste incineration test burn [17].

Food Survey Values:

Plant Concentrations:

Hexachloroethane

Fish/Seafood Concentrations: USEPA STORET database - 116 samples of biota, 0% pos [35]. During spring 1981, concentration in Lake Ontario rainbow trout ranged between 0.1 and 0.6 ng/g [28].

Animal Concentrations:

Milk Concentrations:

Other Environmental Concentrations:

Probable Routes of Human Exposure: The most probable route of human exposure to hexachloroethane is expected to be inhalation of contaminated air. Some segments of the general population may be exposed by ingestion of contaminated drinking water.

Average Daily Intake:

Occupational Exposures: NIOSH has estimated that 8335 workers are potentially exposed to hexachloroethane based on statistical estimates derived from a survey conducted during 1981-1983 in the United States [27]. NIOSH has estimated that 1489 workers are potentially exposed to hexachloroethane based on statistical estimates derived from a survey conducted during 1972-1974 in the United States [26].

Body Burdens:

REFERENCES

1. Abrams EF et al; Identification of Organic Compound in Effluents from Industrial Sources USEPA-560/3-75-002 (1975)
2. Archer WL; Kirk-Othmer Encycl Chem Tech 3rd ed NY: Wiley 5: 722-42 (1979)
3. Barrows ME et al; pp. 379-92 in Dyn Exp Hazard Assess Toxic Chem Ann Arbor, MI: Ann Arbor Sci (1980)
4. Brodzinsky R, Singh HB; Volatile Organic Chemicals in the Atmosphere. Menlo Park, CA; SRI Inter p. 198 (1982)
5. Callahan MA et al; Water-Related Fate of 129 Priority Pollutants Vol. 2 USEPA-400/4-79-029b (1979)
6. Clarke CS et al; Arch Environ Health 37: 9-18 (1982)

Hexachloroethane

7. Class T, Ballschmitier K; Chemosphere 15: 413-27 (1986)
8. Cole RH et al; J Water Pollut Control Fed 56: 898-908 (1984)
9. Dilling WL; Environ Sci Technol 11: 405-9 (1977)
10. Eisenreich SJ et al; Environ Sci Tech 15: 30-8 (1981)
11. Ewing BB et al; Monitoring to Detect Previously Unrecognized Pollutants in Surface Waters p.75 USEPA-560/6-77-015 (1977)
12. Fielding M et al; Organic Micropollutants in Drinking Water. Medmenham, UK Water Res Ctr p. 49 (1981)
13. GEMS; Graphical Exposure Modeling System. FAP. Fate of Atmospheric Pollutants (1987)
14. Great Lakes Water Quality Board; an Inventory of Chemical Substances Identified in the Great Lakes Ecosystem Vol 1 (1983)
15. Hansch C, Leo AJ; Medchem Project Issue No 26. Claremont CA: Pomona College (1985)
16. Hauser TR, Bromberg SM; Env Monit Assess 2: 249-72 (1982)
17. James RH et al; in Proc APCA Ann Mtg 77 Vol 1 p. 25 (1984)
18. Johnson LD, Young JC; J Water Pollut Control Fed 55: 1441-9 (1983)
19. Kawasaki M; Ecotox Environ Safety 4: 444-54 (1980)
20. Konietzko H; Hazard Asses Chem Curr Dev 3: 401-48 (1984)
21. Ligocki MP et al; Atmos Environ 19: 1609-17 (1985)
22. Lucas SV; GC/MS Analysis of Organics in Drinking Water Concentrates and Advanced Waste Treatment Concentrates Vol. 2 p. 397 USEPA-600/1-84-020b (1984)
23. Lyman WJ et al; Handbook of Chemical Property Estimation Methods NY: McGraw-Hill p. 4-9 (1982)
24. Mabey WR et al; Aquatic Fate Process Data for Organic Priority Pollutants p. 434 USEPA-400/4-81-014 (1981)
25. Munz C, Roberts PV; Res Technol 79: 62-9 (1987)
26. NIOSH; National Occupational Hazard Survey (NOES) (1974)
27. NIOSH; National Occupational Exposure Survey (NOES) Sept 20 (1985)
28. Oliver BG, Niimi AJ; Environ Sci Technol 17: 287-91 (1983)
29. Otson R et al; Water Res 20: 775 (1986)
30. Pankow JF et al; Environ Sci Technol 18: 310-8 (1984)
31. Roberts PV et al; Water Resources 22: 2047-58 (1986)
32. Sabljic A; J Agric Food Chem 32: 243-6 (1984)
33. Sasaki S; in Aquatic Pollutants: Transformation and Biological Effects. Hutzinger O et al eds. Oxford Pergamon Press pp. 283-91 (1978)
34. Shackelford NM, Keith LH; Frequency of Organic Compounds Identified in Water. USEPA-600/4-76-062 (1976)
35. Staples CA et al; Environ Toxicol Chem 4: 131-42 (1985)
36. Suffet IH et al; Water Res 14: 853-67 (1980)
37. Swann RL et al; Res Rev 85: 17-28 (1983)
38. Tabak HH et al; in Test Protocols for Environmental Fate and Movements of Toxicants. AOAC 94th Mtg pp. 267-388 (1981)
39. Talian SF et al; in Proc AWWA Water Qual Technol Conf Harrisburg, PA: Gannett Flemin Water Resources Eng Inc pp. 525-42 (1986)
40. USEPA; Treatability Manual Vol 1 USEPA-600/8-80-042 (1980)
41. USEPA; Preliminary Assessment of Suspected Carcinogens in Drinking Water Interim Report (1975)

Maleic Acid

Synonyms: Cis-1,2-Ethylenedicarboxylic acid; (Z)-2-Butenedioic acid

Structure:

CAS Registry Number: 110-16-7

Molecular Formula: $C_4H_4O_4$

Wiswesser Line Notation: QV1U1VQ -C

CHEMICAL AND PHYSICAL PROPERTIES

Boiling Point: 135 °C (decomp)

Melting Point: 130-131 °C

Molecular Weight: 116.07

Dissociation Constants: pKa-1: 1.83; pKa-2: 6.07 [18]

Log Octanol/Water Partition Coefficient: -0.50 [7]

Water Solubility: 788 g/L water at 25 °C [6]

Vapor Pressure: 0.975 mm Hg at 25 °C [16]

Henry's Law Constant: 0.74 x 10^{-13} [10] (estimated by the bond contribution method)

Maleic Acid

ENVIRONMENTAL FATE/EXPOSURE POTENTIAL

Summary: Maleic acid may be released into wastewater during its production and use in the manufacture of polymer products. Maleic acid is also released into the atmosphere from motor exhaust and is a constituent of aerosols in urban air. If released on land, maleic acid will leach into the ground and probably biodegrade. If released into water, maleic acid will also probably biodegrade. Adsorption to sediment, bioconcentration in aquatic organisms, and volatilization should not be significant. It will be primarily associated with aerosols in the atmosphere and be subject to gravitational settling and degradation by reaction with ozone and photochemically produced hydroxyl radicals (vapor phase half-life 1.1 hr). The general population is exposed to maleic acid in areas with heavy traffic, since it is found in aerosols from auto exhaust. Occupational exposure would be via dermal contact and inhalation of aerosols containing maleic acid.

Natural Sources:

Artificial Sources: Maleic acid may be released into wastewater during its production and use in the manufacture of alkyd and other resins [12], organic synthesis, dyeing and finishing of textiles, as a preservative for fats and oils [8], and as a by-product of phthalic anhydride production [1]. Motor exhaust is an important primary source in urban areas [11].

Terrestrial Fate: If released on land, maleic acid will leach into the ground and probably biodegrade. While maleic acid is readily biodegraded in screening tests, no degradability data were found for soil systems.

Aquatic Fate: If released into water, maleic acid will probably biodegrade based on the results of screening studies. No biodegradability studies were found in environmental waters. Adsorption to sediment and volatilization should not be significant.

Atmospheric Fate: Due to its polar nature, maleic acid released into the atmosphere will be primarily associated with aerosols. It will be subject to gravitational settling and also degrade in the

378

vapor phase by reaction with ozone and photochemically produced hydroxyl radicals (half-life 1.1 hr).

Biodegradation: Maleic acid is readily degradable in biodegradability screening tests [2,4,9,17]. The results of several of these tests are: 77.4 and 91.9% of theoretical BOD after 5 and 20 days, respectively [17]; 21% of theoretical BOD after 5 days [2]; and 46% of theoretical BOD after 5 days [9]. In another test that utilized an activated sludge inoculum to simulate maleic acid's fate in a municipal sewage treatment plant, 26.3% mineralization occurred in 5 days and an additional 41% of the chemical was degraded to metabolites that were associated with the sludge [4]. While the results of these screening studies suggest that biodegradation will occur in environmental waters and soil, no studies performed in these media were available.

Abiotic Degradation: Maleic acid's pKa's suggest that it will exist largely in the dissociated form in the environment and form salts with cations. The vapor should react with ozone and photochemically produced hydroxyl radicals in the atmosphere by addition to the double bond with a resulting estimated vapor phase half-life of 1.1 hr [5]. When adsorbed on silica gel and irradiated with light >290 nm for 17 hr, 17% of the maleic acid is mineralized [3].

Bioconcentration: The BCF of maleic acid in fish (golden ide) was <10 after 3 days of exposure, while that in algae (Chlorella fusca) was 11 after 24 hr [3]. The BCF calculated from the water solubility using a recommended regression equation is 3 [13].

Soil Adsorption/Mobility: Maleic acid is extremely soluble in water and therefore should not adsorb appreciably to soil [13].

Volatilization from Water/Soil: Volatilization from water should not be a significant transport process, since the Henry's Law constant is so low [13].

Water Concentrations:

Effluent Concentrations: Motor exhaust of two automobiles contained 2.66 and 25.5 ug/m^3 of maleic acid [11].

Sediment/Soil Concentrations: Soil samples taken at the University of California, Los Angeles campus contained ND and 270 ppb of maleic acid, whereas bog sediment samples from the Sierra Nevada foothills contained 930 ppb [11].

Atmospheric Concentrations: The concn of maleic acid in aerosol samples from West Los Angeles and Los Angeles were 9-204 and 64-95 ng/m³, respectively [11]. Dust samples from these areas contained 6.4 and 2.4 ppm of maleic acid, respectively [11].

Food Survey Values:

Plant Concentrations:

Fish/Seafood Concentrations:

Animal Concentrations:

Milk Concentrations:

Other Environmental Concentrations:

Probable Routes of Human Exposure: The general population is exposed to maleic acid in aerosols from auto exhaust. Occupational exposure would be via dermal contact and inhalation of aerosols containing maleic acid.

Average Daily Intake:

Occupational Exposures: NIOSH has estimated that 17,037 workers are potentially exposed to maleic acid based on statistical estimates derived from a survey conducted in 1972-1974 in the United States [15]. NIOSH has estimated that 14,286 workers are potentially exposed to maleic acid based on statistical estimates derived from a survey conducted in 1981-1983 in the United States [14].

Body Burdens:

Maleic Acid

REFERENCES

1. Bemis AG et al; In: Kirk-Othmer Encyclopedia Chemical Technology 3rd ed. 17: 732-77 (1982)
2. Dore M, et al; Trib Cebedeau 28: 3-11 (1975)
3. Freitag D, et al; Ecotox Environ Safety 6: 60-81 (1982)
4. Freitag D, et al; Chemosphere 14: 1589-616 (1985)
5. GEMS; Graphical Exposure Modeling System. FAP. Fate of Atmos Pollut (1986)
6. Geyer H, et al; Chemosphere 10: 1307-13 (1981)
7. Hansch C, Leo AJ; Medchem Project Issue No 26. Claremont CA: Pomona College (1985)
8. Hawley GG; Condensed Chem Dictionary 10th ed Von Nostrand Reinhold NY pp 638 (1981)
9. Heukelekian H, Rand MC; J Water Pollut Contr Assoc 29: 1040-53 (1955)
10. Hine J, Mookerjee PK; J Org Chem 40: 292-8 (1975)
11. Kawamura K, Kaplan IR; Environ Sci Technol 21: 105-10 (1987)
12. Kuney JH; Chemcyclopedia American Chemical Soc Washington DC p. 87 (1987)
13. Lyman WJ et al; Handbook of Chem Property Estimation Methods. Environ Behavior of Organic Compounds. McGraw-Hill NY (1982)
14. NIOSH; National Occupational Exposure Survey (1985)
15. NIOSH; National Occupational Health Survey (1975)
16. Robinson WD, Mount RA; Kirk-Othmer Encycl Chem Technol 14: 770-93 3rd ed (1981)
17. Wagner R; I Monovalent Alcohols Vom Wasser 42: 271-305 (1974)
18. Weast RC; Handbook of Chemistry and Physics (1972)

381

Maleic Anhydride

SUBSTANCE IDENTIFICATION

Synonyms: 2,5-Furandione; cis-Butenedioic anhydride; Dihydro-2,5-dioxofuran

Structure:

CAS Registry Number: 108-31-6

Molecular Formula: $C_4H_2O_3$

Wiswesser Line Notation: T5VOVJ

CHEMICAL AND PHYSICAL PROPERTIES

Boiling Point: 202.0 °C

Melting Point: 52.8 °C

Molecular Weight: 98.06

Dissociation Constants:

Log Octanol/Water Partition Coefficient:

Water Solubility: decomposes [13]

Vapor Pressure: 0.41 mm Hg at 25 °C [3]

Henry's Law Constant:

ENVIRONMENTAL FATE/EXPOSURE POTENTIAL

Summary: Maleic anhydride is an important industrial chemical used as a chemical intermediate and in the production of resins and coatings. Since it hydrolyzes rapidly in water to the acid, maleic anhydride would not appear in wastewater and would quickly degrade to maleic acid if spilled in water. If emitted to the atmosphere from fugitive emissions or vent gases, it will degrade in a few hours by reaction with ozone or photochemically produced hydroxyl radicals. If spilled on land its fate is unknown, but it will probably hydrolyze. Exposure to maleic anhydride would be primarily limited to occupational settings.

Natural Sources:

Artificial Sources: Maleic anhydride may be spilled or emitted into the atmosphere during its manufacture, transport, or use in the manufacture of alkyd and polyester resins, surface coatings, agricultural chemicals, copolymers, malic and fumaric acids, and as an oil additive. Because of its ease of hydrolysis to the acid, it should not appear in wastewater.

Terrestrial Fate: Maleic anhydride's fate in soil is unknown. It will likely hydrolyze in moist soils and volatilize from dry soil surfaces.

Aquatic Fate: Maleic anhydride released into water will hydrolyze rapidly (half-life 0.37 min) to maleic acid.

Atmospheric Fate: Maleic anhydride released into the atmosphere will degrade by reaction with ozone and photochemically produced hydroxyl radical (estimated half-life 1.7 hr).

Biodegradation: Maleic anhydride has been characterized as biodegradable during biological sewage treatment [11]. In one report, 99% removal was achieved in 4 hr by activated sludge [5]. Another investigator obtained 68.9 and 74.0% theoretical BOD in fresh and salt water, respectively [10]. Others report 40-60% theoretical BOD in 5 days with sewage inoculum [4,12]. In view of the rapid hydrolysis of maleic anhydride, the reported biodegradabilities probably relate to maleic acid rather than the anhydride.

Abiotic Degradation: Maleic anhydride hydrolyzes rapidly in water at room temperature to give maleic acid (half-life 0.37 min at 25 °C in initially neutral solution) [1]. In the vapor phase, hydrolysis is completed in 21 hr at 96% relative humidity, while no hydrolysis occurs at 50% relative humidity [8]. It readily reacts with ozone and hydroxyl radicals in the atmosphere (estimated atmospheric half-life 1.7 hr [2]). Although maleic anhydride absorbs UV radiation above 290 nm [9], no information could be found concerning its photolysis.

Bioconcentration:

Soil Adsorption/Mobility:

Volatilization from Water/Soil: Maleic anhydride should not volatilize appreciably from water or moist soil due to its rapid hydrolysis. Volatilization from dry soil or surfaces should be slow due to its relatively low vapor pressure.

Water Concentrations:

Effluent Concentrations:

Sediment/Soil Concentrations:

Atmospheric Concentrations:

Food Survey Values:

Plant Concentrations:

Fish/Seafood Concentrations:

Animal Concentrations:

Milk Concentrations:

Other Environmental Concentrations:

Probable Routes of Human Exposure: Exposure to maleic anhydride would primarily be occupational from contact with spills, fugitive emissions, or vent gases.

Maleic Anhydride

Average Daily Intake:

Occupational Exposures: NIOSH (NOHS Survey, 1972-75) has statistically estimated that 71,397 workers are exposed to maleic anhydride in the United States [6]. NIOSH (NOES Survey, 1981-83) has statistically estimated that 21,989 workers are exposed to maleic anhydride in the United States [7].

Body Burdens:

REFERENCES

1. Bunton CA et al; J Chem Soc 1963: 2918-26 (1963)
2. Cupitt LT; Fate of Toxic and Hazardous Materials in the Air Environment; USEPA-600/3-80-084 (1980)
3. Daubert TE, Danner RP; Data Compilation Tables of Properties of Pure Compounds. American Institute of Chemical Engineers (1985)
4. Heukelekian H, Rand MC; J Water Pollut Control Assoc 29: 1040-53 (1955)
5. Matsui S et al; Prog Water Technol 7: 645-59 (1975)
6. NIOSH; National Occupational Health Survey (1975)
7. NIOSH; National Occupational Exposure Survey (1985)
8. Rosenfeld JM, Murphy CB; Talenta 14: 91 (1967)
9. Sadtler Index 163 UV
10. Takemoto S et al; Suichitsu Odaku Kenkyu 4: 80-90 (1981)
11. Thom NS, Agg AR; Proc R Soc London B 189: 347-57 (1975)
12. Verschueren K; Handbook of Environmental Data on Organic Chemicals; 2nd ed Van Nostrand Reinhold New York p.803 (1983)
13. Weast RC; Handbook of Chemistry and Physics 67th ed (1986)

Methyl Bromide

SUBSTANCE IDENTIFICATION

Synonyms: Bromomethane

Structure:

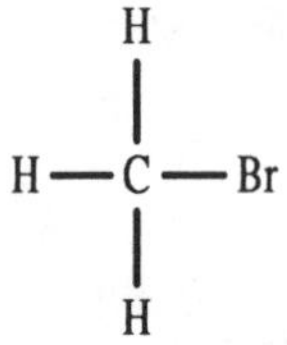

$$H - \underset{\displaystyle H}{\overset{\displaystyle H}{\underset{\displaystyle |}{\overset{\displaystyle |}{C}}}} - Br$$

CAS Registry Number: 74-83-9

Molecular Formula: CH$_3$Br

Wiswesser Line Notation: E1

CHEMICAL AND PHYSICAL PROPERTIES

Boiling Point: 3.6 °C

Melting Point: -93.66 °C

Molecular Weight: 94.95

Dissociation Constants:

Log Octanol/Water Partition Coefficient: 1.19 [14]

Water Solubility: 17,500 mg/L at 20 °C [24]

Vapor Pressure: 1633.0 mm Hg at 25 °C [28]

Henry's Law Constant: 6.24 x 10^{-3} atm m^3/mol [17]

ENVIRONMENTAL FATE/EXPOSURE POTENTIAL

Summary: The primary source of methyl bromide in the environment is the oceans. Release to the environment also results from the use of methyl bromide as a soil and space fumigant and

386

its occurrence in auto exhaust. Methyl bromide released to soil is expected to be primarily lost by volatilization. Methyl bromide may also leach into ground water due to its weak adsorption to soil. Hydrolysis of methyl bromide to methanol and bromide ions and biodegradation may also occur in soil. Release of methyl bromide to water is expected to be lost primarily by volatilization. Hydrolysis to methanol and bromide ions will occur with a half-life of 20-26.7 days. Bioconcentration is not expected to be significant. In the atmosphere, methyl bromide will react with photochemically generated hydroxyl radicals with half-lives ranging from 0.29 years to 1.6 years. Direct photolysis is not expected to be important in the troposphere, but is expected to be the predominant fate of methyl bromide in the stratosphere. Methyl bromide is a major contaminant of air and a minor contaminant of surface, drinking, and ground water. Human exposure to methyl bromide is expected to result primarily from inhalation of contaminated ambient or occupational air.

Natural Sources: The bulk of the methyl bromide detected in the environment is believed to be released from oceans [34,38].

Artificial Sources: Methyl bromide is used as a soil fumigant, for the disinfection of potatoes, tomatoes and other crops, in organic syntheses, as an extraction solvent for vegetable oils [16], in ionization chambers, for degreasing wool, for extracting oils from nuts and seeds, and as a fumigant in mills, warehouses, vaults, ships, and freight cars [24]. As a soil fumigant, methyl bromide is used for the control of fungi and nematodes [42]. It is also used to fumigate seeds [42]. The use pattern of methyl bromide is soil fumigant (65%), space fumigant (15%), chemical processes (10%), and exports (10%) [5]. Methyl bromide is released from turbines [10] and in auto exhaust, possibly from the catalytic decomposition of ethylene dibromide [33].

Terrestrial Fate: The primary fate of methyl bromide in soil is expected to be volatilization. Due to its weak adsorption to soil, methyl bromide is expected to also leach into ground water. Experiments in glasshouses in which the soil was fumigated with methyl bromide and irrigated after venting, showed that methyl bromide appeared in the drainage water shortly after irrigation commenced, reached a maximum concentration after 10 hr, and then decreased by 50%/day [45]. Degradation is by hydrolysis, which is soil catalyzed and is much faster in peat soil than in loam, and in loam than in sand [3]. Biodegradation may occur, but experimental data in soil are lacking.

Aquatic Fate: The primary fate of methyl bromide in water is expected to be volatilization (half-life 3 hr in a model river). Hydrolysis of methyl bromide to methanol and bromide ion will also occur with half-lives of 20-26.7 days (calculated from hydrolysis rate constants of 4.09 x 10^{-7} sec^{-1} [23] and 3 x 10^{-7} sec^{-1} [4]). Bioconcentration in fish is not expected to be significant.

Atmospheric Fate: In the atmosphere, methyl bromide will react with hydroxyl radicals with half-lives ranging from 0.29 yr at 25 °C and at 2 x 10^{+6} hydroxyl radicals/cm^3 to 1.6 yr at -8 °C and at 5 x 10^{+5} hydroxyl radicals/cm^3 [7]. Direct photolysis is not expected to be important in the troposphere, since methyl bromide does not absorb light of >290 nm [29]. The estimated residence time of methyl bromide in the atmosphere is 289 days based on the reaction of methyl bromide with hydroxyl radicals with a 24-hr average concentration of 1 x 10^{+6} molecules [33]. Molecules with tropospheric lifetimes of the same order of magnitude as methyl bromide may eventually diffuse in the stratosphere [1], so methyl bromide may also be expected to do so. In the stratosphere, into which shorter wavelength light can penetrate, photolysis is expected to be the predominant removal mechanism [29].

Biodegradation: Methylotrophic bacteria are capable of oxidizing methyl bromide [13,40]. Methyl bromide was oxidized to formaldehyde by a soluble methane monooxygenase isolated from Methylbacterium sp. CRL-26 [26]. Cell-free extracts and whole cell suspensions of Methylococcus capsulatus oxidized methyl bromide to formaldehyde [40,41], but Methylococcus capsulatus was unable to grow on methyl bromide [40].

Abiotic Degradation: The hydrolysis rate constant of methyl bromide at 25 °C and pH 7 is 4.09 x 10^{-7} sec^{-1} which translates into a half-life of 20 days [23]. Another hydrolysis rate constant for methyl bromide is 3 x 10^{-7} sec^{-1} at 25 °C [4], which translates into a half-life of 26.7 days. The reaction is both neutral and base catalyzed and gives methanol and bromide ions [23]. The hydrolysis of gaseous methyl bromide in soil water during its passage through soil is highly dependent on the soil type, increasing as the organic content of the soil increases [3]. The reaction is first order in the concentration of methyl bromide in the soil atmosphere [3]. A proposed mechanism for this behavior is that the methyl group is transferred to carboxyl or N- and S-containing groups in the soil organic matter [3]. The half-lives for the reaction of methyl bromide with hydroxyl radicals in the gas phase were determined from experimental data obtained at various hydroxyl radical concentrations and temperatures. The results were as follows: at $5 \times 10^{+5}$ molecules/cm^3 - 1.6 yr (-8 °C), 1.1 yr (25 °C); at $1 \times 10^{+6}$ molecules/cm^3 - 0.79 yr (-8 °C), 0.57 yr (25 °C), and at $2 \times 10^{+6}$ molecules/cm^3 0.40 yr (-8 °C), 0.29 yr (25 °C) [7]. The estimated residence time of methyl bromide in the atmosphere is 289 days based on the reaction of methyl bromide with hydroxyl radicals with a 24 hr average concentration of $1 \times 10^{+6}$ molecule/cm^3 [33]. Due to the lack of absorbance of methyl bromide at <290 nm [29], direct photolysis will not be important in the troposphere. Upward diffusion is believed to be the dominant loss mechanism of methyl bromide from the troposphere [29]. Molecules with tropospheric lifetimes of the same order as methyl bromide eventually diffuse into the stratosphere [1], so methyl bromide may also be expected to do so. In the stratosphere, into which shorter wavelength light can penetrate, photolysis is expected to be the predominant removal mechanism [29].

Bioconcentration: Using a measured log octanol/water partition coefficient of 1.19 [14], a bioconcentration factor (BCF) of 4.7 was estimated [22]. A BCF of this magnitude indicates that bioconcentration will not be significant in fish.

Soil Adsorption/Mobility: The Koc values for gaseous methyl bromide were 172, 174, and 164 for Naaldwijk loamy sand, Aalsmeer loam, and Boskoop peaty clay, respectively [6]. Dry soils have a much greater adsorptivity than do moist soils [3]. Using a measured water solubility of 9.5×10^{-3} mol/L [44], a log Koc of 2.1 was estimated [22]. A Koc of this magnitude indicates that methyl bromide will not adsorb strongly to the soil and may, therefore, leach into ground water.

Methyl Bromide

Volatilization from Water/Soil: WATER: A volatilization half-life of 72 min for methyl bromide was obtained by an experiment in which 100 ppm of methyl bromide was placed in a 39.4 cm aeration cylinder with air bubbled through it for 4 hr [25]. The mass transfer coefficient of methyl bromide is 22.56 cm/hr at 25 °C [25], and this value was used to calculate a volatilization half-life of 3.1 hr for methyl bromide from 1 m of water [22]. From the Henry's Law constant, one can estimate the volatilization half-lives of methyl bromide from streams, rivers and lakes. The wind velocity was assumed to be 3 m/sec. The current velocities of the streams, rivers and lakes were assumed to be 2, 1, and 0.01 m/s, respectively. The depths of the streams and rivers were assumed to be 1 m and that of the lakes 50 m. The estimated volatilization half-lives for the streams, rivers and lakes were 1.6 hr, 3 hr and 3419 hr (142 days), respectively [22]. SOIL: Using a soil screening model, the half-lives for the volatilization of methyl bromide from 1 and 10 cm of soil were estimated to be 0.2 and 0.5 days, respectively [19].

Water Concentrations: DRINKING WATER: Methyl bromide has been detected but not quantified in drinking water from Miami, FL [20], and three unidentified U.S. drinking waters [31]. Methyl bromide was not detected in a screening of 1174 community wells and 617 private wells in Wisconsin [21]. GROUND WATER: Methyl bromide was detected but not quantified in New Jersey ground water [11]. SURFACE WATER: Methyl bromide was detected in 1.4% of 941 water samples listed in USEPA STORET database [39]. The median methyl bromide concentration was <10.0 ug/L [39].

Effluent Concentrations: Methyl bromide was detected in 1.4% of 1,317 samples of effluent in the STORET database and the median methyl bromide concentration was <10.0 ug/L [39]. In a comprehensive survey of wastewater from 4000 industrial and publicly owned treatment works (POTWs) sponsored by the Effluent Guidelines Division of the U.S. EPA, methyl bromide was identified in discharges of the following industrial category (frequency of occurrence, median concn in ppb): organics and plastics (1, 32.6); plastics and synthetics (1, 21.1) [30].

Sediment/Soil Concentrations:

Atmospheric Concentrations: Northern California, shore - 116-205 ng/m^3 avg, 1975; Stanford Hills, CA - 60 ng/m^3 avg, Nov 1975; Poing Reyes, CA - 360 ng/m^3 avg, Dec 1975 [38]. Houston, TX - 388 ng/m^3 avg, May 1980; St. Louis, MO - 314 ng/m^3 avg, May-June 1980; Denver, CO - 481 ng/m^3 avg, June 1980; Riverside, CA - 1 ng/m^3 avg, July 1980; Staten Island, NY - 326 ng/m^3 avg, Mar-Apr 1981; Pittsburgh, PA - 159 ng/m^3 avg, May 1981; Chicago, IL - 182 ng/m^3 avg, May 1981 [37]. Los Angeles, CA - 419 ng/m^3 avg, May 1976, Palm Springs, CA - 93.1 ng/m^3 avg, May 1976; Yosemite, CA - 19.4 ng/m^3 May 1976 [36]. Los Angeles, CA - 947 ng/m^3 mean, May 1979; Phoenix, AZ - 260 ng/m^3 mean, Apr-May 1979; Oakland, CA - 210 ng/m^3 mean, Jun-Jul 1979 [35]. Southern California coast - ND-2170 ng/m^3 monthly mean, Nov-Dec 1983 [32]. Rural Northwest U.S. - 19 ng/m^3, Dec 1974-Feb 1975 [12]. Washington State (rural) - 1.9-3.5 ng/m^3, 1976 [15]. Barrow, AL - 35-57 ng/m^3 monthly average [27]. Norway - 56 ng/m^3, spring 1983 [18]. Stratosphere, 44 deg N - 0.388-3.88 ng/m^3, Sept 1980 [9]. Artic - 42.7 ng/m^3, Mar-Apr 1983 [2].

Food Survey Values: Methyl bromide was found infrequently in leaf and stem vegetables [8]. Reported residues in fumigated wheat, flour, raisins, corn, sorghum, cottonseed meal, rice, and peanut meal were reduced to less than 1 mg/kg within a few days [43]. No residual bromomethane was found in asparagus, avocados, peppers, or tomatoes after 2-hr fumigation at 320 mg methyl bromide/m^3 air [43]. Only trace amounts were present in wheat flour and other products fumigated at 370 mg/m^3 methyl bromide after nine days of aeration [43].

Plant Concentrations:

Fish/Seafood Concentrations:

Animal Concentrations:

Milk Concentrations:

Other Environmental Concentrations:

Methyl Bromide

Probable Routes of Human Exposure: Inhalation, percutaneous absorption, ingestion, skin and eye contact.

Average Daily Intake: AIR INTAKE: (Assume ND-2.17 ng/m^3) 0-43.4 ng. WATER INTAKE: Insufficient data. FOOD INTAKE: Insufficient data.

Occupational Exposures:

Body Burdens:

REFERENCES

1. Altshuller AP; Comments on the Lifetimes of Organic Molecules in Air USEPA-600/9-80-003 (1980)
2. Berg WW et al; Geophys Res Lett 11: 429-32 (1984)
3. Brown BD, Rolstron DE; Soil Sci 130: 68-75 (1980)
4. Castro CE, Belser NO; J Agric Food Chem 29: 1005-8 (1981)
5. Chemical Marketing Reporter; Chemical profile - Methyl bromide. Feb 18 (1985)
6. Daelmans A, Sienbering H; Meded Fac Landbouweet, Rijksunic Gent 42: 1729-38 (1979)
7. Dilling WL; pp. 154-97 in Environmental Risk Analysis for Chemicals Conway RA ed, Von Nostrand Reinhold C New York, NY (1982)
8. Duggan RE et al; Pesticide Residue Levels in Food in the U.S. From July 1, 1969 to June 30, 1976, FDA and AOAC (1983)
9. Fabian P et al; Nature 294: 733-5 (1981)
10. Graedel TE; Chemical Compounds in the Atmosphere. Academic Press, New York p. 325 (1978)
11. Greenburg M et al; Environ Sci Technol 16: 14-9 (1982)
12. Grimsrud EP, Rasmussen RA; Atmos Environ 9: 1014-7 (1975)
13. Haber CL et al; Science 221: 1147-53 (1983)
14. Hansch C, Leo AJ; MEDCHEM Project Claremont CA: Pomona College (1985)
15. Harsch DE, Rasmussen RA; Anal Lett 10: 1041-7 (1977)
16. Hawley GG; Condensed Chem Dictionary 10th ed Von Nostrand Reinhold NY p. 670 (1981)
17. Hine J, Mookerjee PK; J Org Chem 40: 292-98 (1975)
18. Hov O et al; Geophys Res Lett 11: 425-8 (1984)
19. Jury WA et al; J Environ Qual 13: 573-9 (1984)
20. Kool HJ et al Crit Rev Env Control 12: 307-57 (1982)
21. Krill RM, Sonzogni WC; J Amer Water Works Assoc 78: 70-5 (1986)
22. Lyman WJ et al; Handbook of Chem Property Estimation Methods Environ Behavior of Organic Compounds NY: McGraw-Hill (1982)
23. Mabey W, Mill T; J Phys Chem Ref Data 7: 383-415 (1978)
24. Merck Index; An Encyclopedia of Chemicals, Drugs and Biologicals 10th ed p. 865 (1983)

Methyl Bromide

25. Neily WB; Predicting the Flux of Organics Across the Air/water Interface. Control Hazard Materials Spills Proc Nat Conf 3rd pp. 197-200 (1976)
26. Patel RN et al; Appl Environ Microbiol 44: 1130-37 (1982)
27. Rasmussen RA, Khalil MA; Geophys Res Lett 11: 433-6 (1984)
28. Riddick JA et al; Organic Solvents New York: Wiley Interscience (1986)
29. Robbins DE; Geophys Res Lett 3: 213-6 (1976)
30. Shackelford WM et al; Analyt Chim Acta 146: 15-27 (1983)
31. Shackelford WM, Keith LH; Frequency of Organic Compounds Identified in Drinking Water USEPA-600/4-76-062 (1976)
32. Shikiya JG et al; Proc APCA Ann Mtg 77: 1-21 (1984)
33. Singh HB et al; Atmospheric Measurements of Selected Toxic Organic Chemicals USEPA-600/3-80-072 (1980)
34. Singh HB et al; J Geophys Res 88: 3684-90 (1983)
35. Singh HB et al; Atmos Environ 15: 601-12 (1981)
36. Singh HB et al; Atmospheric Distributions, Sources and Sinks of Selected Halocarbon, Hydrocarbons, SF_6 and N_2O USEPA-600/S3-81-032 (1979)
37. Singh HB et al; Environ Sci Technol 16: 872-80 (1982)
38. Singh HB et al; J Air Pollut Cont Assoc 27: 332-6 (1977)
39. Staples CA et al; Environ Technol Chem 4: 131-42 (1985)
40. Stirling DI, Dalton H; J Gen Microbiol 116: 227-83 (1980)
41. Stirling DI, Dalton H; FEMS Microbiol Lett 5: 315-8 (1979)
42. The Pesticide Manual; 7th ed Worthing CR ed p. 372 (1983)
43. USEPA; Ambient Water Quality Criteria Doc: Halomethanes EPA 440/5-80-051 p.C-9,10 (1980)
44. Verschueren K; Handbook of Environ Data on Organic Chemicals 2nd ed Von Nostrand Reinhold, NY p. 835 (1983)
45. Wegmann RCC et al; Water Air, Soil Pollut 16: 2-11 (1981)

Methyl Chloride

SUBSTANCE IDENTIFICATION

Synonyms: Chloromethane

Structure:

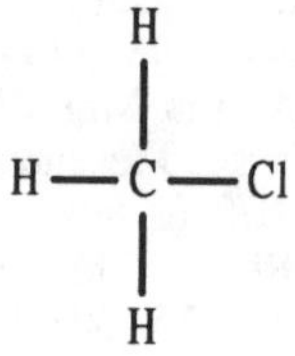

$$H - \underset{\underset{H}{|}}{\overset{\overset{H}{|}}{C}} - Cl$$

CAS Registry Number: 74-87-3

Molecular Formula: CH_3Cl

Wiswesser Line Notation: G1

CHEMICAL AND PHYSICAL PROPERTIES

Boiling Point: -23.7 °C

Melting Point: -97 °C

Molecular Weight: 50.49

Dissociation Constants:

Log Octanol/Water Partition Coefficient: 0.91 [19]

Water Solubility: 0.648% wt. at 30 °C [31]

Vapor Pressure: 4309.7 mm Hg at 30 °C [31]

Henry's Law Constant: 2.4×10^{-2} atm-m³/mol [24]

ENVIRONMENTAL FATE/EXPOSURE POTENTIAL

Summary: Methyl chloride is produced naturally in the oceans by mechanisms which are not entirely understood. One source is believed to be the reaction of biologically produced methyl iodide with chloride ions. Other significant natural sources include forest and brush fires and volcanoes. Although the atmospheric budget of methyl chloride can be accounted for by volatilization from the oceanic reservoir, it is apparent that man-made sources arising from its production and use in the manufacture of silicones and other chemicals, and as a solvent and propellant can make a significant impact on the local atmospheric concn of methyl chloride. If released into water, methyl chloride will be rapidly lost by volatilization (half-life in a model river 2.4 hr). It will also be rapidly lost from soil by volatilization, although there is a potential for it to leach into ground water where it may very slowly biodegrade and hydrolyze (half-life may exceed a yr). Once in the atmosphere, it will disperse and will be lost primarily by upward dispersion. Above the tropopause, reaction with hydroxyl radicals aid in the removal of methyl chloride and above 30 km, photodissociation, diffusion, and reaction with hydroxyl radicals make roughly equal contributions to its removal. Humans are exposed to methyl chloride by inhalation of ambient air.

Natural Sources: Volcanoes, plant volatiles, and forest fires are natural sources of methyl chloride [13]. Methyl chloride is produced in seawater by the reaction of methyl iodide with chloride ions [43]. The methyl iodide is produced photosynthetically by several marine organisms [43]. It is claimed that atmospheric methyl chloride results largely from these natural sources [8], although others point to the fact that methyl chloride and methyl iodide concentrations are uncorrelated and suggest that there is an independent oceanic source of methyl chloride [35]. For the eastern Pacific, the mean ocean air flux of 13 x 10^{-7} g/cm^2-yr when extrapolated to global waters provide an adequate source to explain the atmospheric reservoir of methyl chloride [35]. Methyl chloride is released from brush and forest fires [39]. The estimated 0.6 g of methyl chloride per kg vegetation burned is believed to make a significant impact on the global burden of methyl chloride [39]. Plant volatiles from cedar and cypress have also been indicated as sources of methyl chloride [20].

Artificial Sources: Methyl chloride may be emitted as fugitive emissions and in wastewater during its production and use in the manufacture of silicones, agrichemicals, methyl cellulose, quarternary amines, butyl rubber, and tetraethyl lead [2,6]. It is released in tobacco smoke and turbine exhaust [13] as well as in its use as a solvent, propellant, and in the manufacture of fumigants [42]. Just as forest fires contribute to natural sources of methyl chloride, so does wood burning, field burning, and backyard burning [10]. It is formed in the chlorination of drinking water and sewage effluent [1] and is found in the effluent of some publically owned treatment works [40].

Terrestrial Fate: Methyl chloride has a very high vapor pressure and if released on land will be rapidly lost by volatilization. It may also leach into ground water, where it should very slowly biodegrade and hydrolyze.

Aquatic Fate: If methyl chloride is released into water, it will be lost primarily by volatilization (half-life 2.4 hr in a model river).

Atmospheric Fate: The dominant loss mechanism for methyl chloride in the troposphere is upward diffusion, although washout by rain may also be important. From the tropopause to about 30 km, both upward diffusion and reaction with hydroxyl radicals will be of approx equal importance, and above 30 km in the stratosphere diffusion, reaction with hydroxyl radicals and photodissociation will have approx equal weight [32]. The surface half-life resulting from upward diffusion is 80 days [32].

Biodegradation: Field and laboratory tests demonstrate that several halogenated aliphatics may biodegrade slowly under anaerobic conditions, but not under aerobic conditions [22,26]. Chlorinated methanes released 50-70% of bound Cl when incubated anaerobically for 4-5 days with arable soil or sewage sludge [18].

Abiotic Degradation: Aliphatic halides hydrolyze in water by neutral and base catalyzed reactions to give the corresponding alcohol. The half-life for the hydrolysis of methyl chloride extrapolated from data obtained at higher temperatures is 0.93 yr at pH 7 and 25 °C [25]. The rate is independent of pH below pH 10 [25]. Calculated values of the half-life of methyl chloride in water are 88, 14, and 2.5 yr at 0, 10, and 20 °C, respectively [43]. Methyl chloride reacts with photochemically produced hydroxyl radicals via H-atom abstraction with a half-life of 1 yr [3]. In the stratosphere, photodissociation will occur at a rate approximate equal to its reaction with hydroxyl radicals [32].

Bioconcentration: Methyl chloride has a very low log octanol/water partition coefficient, indicating that it would not have a significant tendency to bioconcentrate in aquatic organisms.

Soil Adsorption/Mobility: Methyl chloride has a very low log octanol/water partition coefficient, indicating that it would not have a significant tendency to adsorb to soil [24].

Volatilization from Water/Soil: The volatilization half-life for 1 ppm methyl chloride from a stirred beaker 6.5 cm deep is 27.6 min [9], which converts to a half-life at a 1 m depth of 7.1 hr. From its Henry's Law constant, one can estimate that the volatilization half-life for methyl chloride from a model river with a 1 m/sec current, 3 m/sec wind speed, and 1 m depth is 2.4 hr [24]. For a chemical with such a high Henry's Law constant, the volatilization is controlled by the rate of movement throughout the liquid phase [24]. The evaporation from soil will be rapid due to methyl chloride's very high vapor pressure and low adsorption to soil.

Water Concentrations: DRINKING WATER: Treated water from 30 Canadian potable water treatment facilities - 2 samples pos, mean - 5 ppb [29]. Drinking water well in Maine reported in a Council on Environmental Quality survey of contaminated drinking water from ground water sources - 44 ppb [5]. Highest reported concn of methyl chloride in surface water derived drinking water - 12 ppb [5]. Identified, not quantified in drinking water in New Orleans, Cincinnati, Miami, Philadelphia, and Ottumwa, IA of the 10 cities surveyed [1]. GROUND WATER: Detected, not quantified in 11 of 20 ground waters underlying municipal solid waste landfills in MN [33]. SURFACE WATER: 895 stations in the USEPA STORET database 1.4% pos, median <10 ppb [38]. Raw water from 30 Canadian potable water treatment facilities - 1 sample pos, mean <5 ppb [29]. Detected in the Niagara River and the open water of Lake Ontario [14]. SEAWATER: Pacific Ocean 26.8 ppt at surface, 3.3 ppt at 300 m depth [37]. Eastern Pacific surface water (latitude 29 deg N to -29 deg S) 6.3-42 ppt, 11.5 ppt mean, 200-300% supersaturation [35]. Point Reyes, CA (near shore) - 1200 ppt [36].

Effluent Concentrations: 1298 stations in the USEPA STORET database 3.5% pos, median <10 ppb [38]. Detected in 1 of 5 leachates from municipal waste landfills in WI, 170 ppb and detected; not quantified in 4 of 6 leachates from municipal landfills in MN [33]. Methyl chloride has been detected in treated wastewater from the following industries (industry (mean concn)): pharmaceutical manufacturing (2000 ppb); organic chemical manufacturing/plastics (0.1 ppb); timber products processing (140 ppb); and raw wastewater from metal finishing (610 ppb) [41]. Ratios of methyl chloride (ppmv) to carbon dioxide (1 x 10^{-6} ppmv) in wood smoke ranged from 0.66 to 2.63 [11]. In a comprehensive survey of wastewater from 4000 industrial and publicly owned treatment works (POTWs) sponsored by the Effluent Guidelines Div of the U.S. EPA, methyl chloride was identified in discharges of the following industrial categories (frequency of occurrence, median concn in ppb): nonferrous metals (1, 21.6); paint and ink (2, 4128.7); printing and publishing (1, 6.0); organics and plastics (1, 156.7); pharmaceuticals (1, 2558.3); organic chemicals (3, 49.0) [34]. The highest effluent concn was 4194 ppb in the paint and ink industry [34].

Sediment/Soil Concentrations: 345 stations in the USEPA STORET database 0.3% pos, median < 5 ppb [38].

Atmospheric Concentrations: RURAL/REMOTE: United States (191 samples at 4 sites) 590-1300 ppt, 1300 ppt median [4]. Southwest Washington 530 ppt [16]. Eight background locations on earth (1980, time series over seasons) 564-687 ppt; concn highest in spring and lowest in fall and highest in the tropics; however, there is no significant difference between hemispheres [21]. Two rural coastal sites (192 samples) near San Francisco 953 ppt [36]. Mean concn of methyl chloride over the eastern Pacific between 40 deg N and 32 deg S latitudes 630 ppt; contrary to other results, the concn was independent of latitude [35]. The concn of methyl chloride decreases with altitude, declining to 50 ppt at 29 km [12]. The island of Terschelling, The Netherlands (least populated area of country) 700 ppt, mean [17]. From the coast to the forest in Guyana - 630-730 ppb v/v [15]. URBAN/SUBURBAN: United

States (389 samples, 12 sites) 570-5700 ppt, 1000 ppt median [4]. Delft, The Netherlands (densely populated area of the country) 3000 ppt mean, 7000 ppt max [17]. A suburban site in Hillsboro, OR had peak methyl chloride concn in Dec and May of 680 and 700 ppt, respectively, that has been attributed to wood burning and backyard burning [10]. INDOOR AIR: Methyl chloride concn are elevated due to biomass combustion. In rural Nepal, where stoves are used for cooking and heating, methyl chloride levels in one house were 6950 ppt [7].

Food Survey Values:

Plant Concentrations:

Fish/Seafood Concentrations: 84 samples in the USEPA STORET database 1% pos, median <50 ppb [38].

Animal Concentrations:

Milk Concentrations:

Other Environmental Concentrations:

Probable Routes of Human Exposure: Humans are exposed to methyl chloride by inhalation of ambient air.

Average Daily Intake: AIR INTAKE: (assume concn of 1000 ppt) 42 ug; WATER INTAKE: (assume concn of 0 ppt) 0 ug; FOOD INTAKE: insufficient data.

Occupational Exposures: NIOSH (NOHS Survey 1972-1974) has statistically estimated that 40,545 workers are exposed to methyl chloride in the United States [27]. NIOSH (NOES Survey 1981-1983) has statistically estimated that 8853 workers are exposed to methyl chloride in the United States [28].

Body Burdens: Mother's milk from 4 urban areas of the United States - 2 of 8 samples positive [30]. Methyl chloride was detected in expired air from a sample of 62 nonsmoking individuals [23].

Methyl Chloride

REFERENCES

1. Abrams EF et al; Identification of Organic Cmpds in Effluents from Industrial Sources USEPA-560/3-75-002 (1975)
2. Ahlstrom RC, Steele JM; Kirk-Othmer Encycl Chem Tech 3rd ed 5: 677-85 (1979)
3. Atkinson R; Chem Rev 85: 69-201 (1985)
4. Brodzinsky R, Singh HB; Volatile Organic Chemicals in the Atmos, An Assess of Available Data, Menlo Park, CA pp.198 No 68-02-3452 (1982)
5. Burmaster DE; Environ 24: 6-13, 33-36 (1982)
6. Chemical Marketing Reporter; March 3, 1986 (1986)
7. Davidson CI et al; Environ Sci Technol 20: 561-7 (1986)
8. Derwent RG, Eggleton AEG; Atmos Environ 12: 1261-9 (1978)
9. Dilling WL; Environ Sci Technol (1977)
10. Edgerton SA et al; J Air Pollut Control Assoc 34: 661-4 (1984)
11. Edgerton SA et al; Environ Sci Technol 20: 803-807 (1986)
12. Fabian P, Goemer D; Fresenius' Z Anal Chem 319: 890-7 (1984)
13. Graedel TE; Chemical Compounds in the Atmosphere, Academic Press New York, NY pp.324 (1978)
14. Great Lakes Water Quality Board; Report to the Great Lakes Water Quality Board, Windsor Ontario, Canada 1: 195 (1982)
15. Gregory G et al; J Geophys Res 91: 8603-12 (1986)
16. Grimsrud EP, Rasmussen RA; Atmos Environ 9: 1014-7 (1975)
17. Guicherit R, Schulting FL; Sci Total Environ 43: 193-219 (1985)
18. Haider K; pp 200-4 in Comm Eur communities, Rep Eur 1980 Eur 6388, Environ Res Programme (1980)
19. Hansch C, Leo AJ; Medchem Project Issue No 26. Claremont CA: Pomona College (1985)
20. Isidorov VA et al; Atmos Environ 19: 1-8 (1985)
21. Khalil MAK, Rasmussen RA; Chemosphere 10: 1019-23 (1981)
22. Kobayashi H, Rittmann BE; Environ Sci Technol 16: 170A-82A (1982)
23. Krotoszynski BK, O'Neill HJ; J Environ Sci Health A17: 855-83 (1982)
24. Lyman WJ et al; Handbook of Chem Property Estimation Methods. McGraw-Hill NY (1982)
25. Mabey W, Mill T; J Phys Chem Ref Data 7: 383-415 (1978)
26. Newsom JM; Groundwater Monit Rev 5: 28-36 (1985)
27. NIOSH; National Occupational Health Survey (1975)
28. NIOSH; National Occupational Exposure Survey (1985)
29. Otson R et al; J Assoc Offic Analyst Chem 65: 1370-4 (1982)
30. Pellizzari ED et al; Bull Environ Contam Toxicol 28: 322-8 (1982)
31. Riddick JA et al; Organic Solvents: Physical Properties and Methods of Purification, 4th Edit. New York: J Wiley & Sons (1986)
32. Robbins DE; Geophys Res Lett 3: 213-6 (1976)
33. Sabel GV, Clark TP; Waste Manag Res 2: 119-30 (1984)
34. Shackelford WM et al; Analyt Chim Acta 146: 15-27 (1983)
35. Singh HB et al; J Geophys Res 88: 3684-90 (1983)
36. Singh HB et al; J Air Pollut Control Assoc 27: 332-6 (1977)
37. Singh HB et al; Atmos Distributions, Sources and Sinks of Selected Halocarbon, Hydrocarbons, SF_6 + N_2O, pp.134 USEPA-600/3-79-107 (1979)
38. Staples, CA et al; Environ Toxicol Chem 4: 131-42 (1985)

Methyl Chloride

39. Tassios S, Packham DR; J Air Pollution Control Association 33: 41-2 (1985)
40. USEPA; Fate of Priority Pollut in Publically Owned Treatment Works Final Report Vol 1 USEPA-440/1-82-303 (1982)
41. USEPA; Treatability Manual pp. I.12.1-I.12.5 USEPA-600/282-001a (1981)
42. Verschueren K; Handbook of Environmental Data on Organic Chemicals 2nd ed Van Nostrand Reinhold NY (1983)
43. Zafiriou OC; J Mar Res 33: 75-81 (1975)

Methyl Methacrylate

SUBSTANCE IDENTIFICATION

Synonyms: Methyl 2-methyl-2-propenoate

Structure:

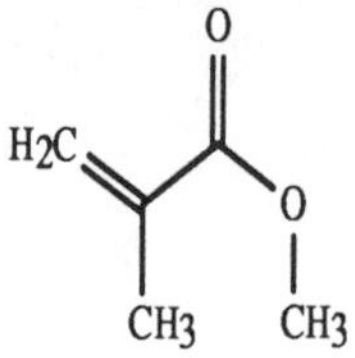

CAS Registry Number: 80-62-6

Molecular Formula: $C_5H_8O_2$

Wiswesser Line Notation: 1UY1&VO1

CHEMICAL AND PHYSICAL PROPERTIES

Boiling Point: 100-101 °C

Melting Point: -48 °C

Molecular Weight: 100.13

Dissociation Constants:

Log Octanol/Water Partition Coefficient: 1.38 [6]

Water Solubility: 15.6 g/L at 20 °C [19]

Vapor Pressure: 38.4 mm Hg at 25 °C [2]

Henry's Law Constant: 3.24×10^{-4} atm-m³/mol (calc from the water solubility and vapor pressure)

ENVIRONMENTAL FATE/EXPOSURE POTENTIAL

Summary: Methyl methacrylate may enter the atmosphere or be released into wastewater or on land during its production, use in the manufacture of resins and plastics, transport, or storage. If released into water, it will principally be lost by volatilization (half-life 6.3 hr from a model river). If spilled on land, it will volatilize and leach into the ground water where its fate is unknown. If emitted into the atmosphere, it will photodegrade (half-life 2.7 hr in urban areas and >3 hr in rural areas). It would not be expected to bioconcentrate in fish. Human exposure will be primarily in the workplace. It is possible that the monomer can migrate into food from polymethyl methacrylate wrappers resulting in exposure to the population at large from food.

Natural Sources: Methyl methacrylate is not known to occur naturally [8].

Artificial Sources: Methyl methacrylate may be emitted or released in wastewater during its production and use in the manufacture of resins and plastics [8]. An estimated 75% of the U.S. production is used captively in the production of homopolymers or copolymers with acrylic esters; such captive use would be expected to have limited release [8]. Sources of release include spills and emissions during storage and transport [8]. In 1974, total emissions from production, end-product manufacture and bulk storage were estimated to be 1.7, 1.7, and 0.2 million kg, respectively [8].

Terrestrial Fate: When spilled on soil, methyl methacrylate would be expected to both volatilize and leach into the ground water. Some biodegradation would be expected to occur, especially where acclimated microorganisms exist.

Aquatic Fate: When released into water, methyl methacrylate will primarily be lost through volatilization (half-life 6.3 hr for a typical river). Hydrolysis, which is base catalyzed, is not significant at neutral and acid pH. At pH 9, the half-life is 14 days. Some biodegradation or degradation due to reaction with photolytically produced radicals may occur, but no estimates for rates of these reactions are available. No appreciable adsorption to sediment or particulate matter will occur.

Methyl Methacrylate

Atmospheric Fate: Methyl methacrylate released to the atmosphere will degrade by reaction with reactive atmospheric species with typical half-lives of 2.7 hr in urban and >3 hr in rural areas.

Biodegradation: Methyl methacrylate is confirmed to be significantly degraded in the biodegradability test of the Japanese Ministry of International Trade and Industry (MITI), which uses a mixed inoculum of soil, surface water, and sewage [21]. It was reported to be completely degraded by activated sludge in approx 20 hr [22]. In a standard biodegradability test using sewage seed, 42% of the theoretical BOD was consumed in 19 days, including a 3- to 4-day lag period [18]. With acclimated seed, 66% of the theoretical BOD was consumed in 22 days [18].

Abiotic Degradation: Methyl methacrylate is readily polymerized by light and heat [7]. Therefore, an inhibitor, generally 10-15 ppm of monomethyl ester of hydroquinone, is frequently added to the chemical [11]. Methyl methacrylate was moderately reactive in a smog chamber [10]. With methyl methacrylate and nitrogen oxides concentration ratios typical of urban areas, the half-life for photodegradation was 2.7 hr; the half-life exceeded 3 hr when the concentration ratios were typical of rural areas [10]. Simple esters are fairly resistant to hydrolysis [14]. While the hydrolysis is acid and base catalyzed, little data are available for the hydrolysis in weakly acidic solutions [14]. The basic rate constant is 200 l/mole-hr which translates into a hydrolysis half-life of 3.9 yr at pH 7 and 14 days at pH 9 [4]. The adsorption maximum for methyl methacrylate is 231 nm [8], so it should not adsorb radiation >290 nm and photolyze. However, free radicals formed in natural waters by the action of light might react with methyl methacrylate. Environmentally pertinent data in this area is lacking. Irradiation >300 nm resulted in 0.43% polymerization in 1.2 hr of methyl methacrylate in dioxane solution [17].

Bioconcentration: From the octanol/water partition coefficient, one can calculate a log BCF of 0.55 [13]. Thus no bioconcentration of methyl methacrylate would be expected to occur in fish.

Soil Adsorption/Mobility: No data on the adsorption of methyl methacrylate could be found. Using the octanol/water partition coefficient, one can calculate a Koc of 87 [13], which indicates little adsorption to soil or sediment should occur.

Volatilization from Water/Soil: No experimental data on the rate of evaporation of methyl methacrylate from water could be found. From the Henry's Law constant, one can estimate a half-life of 6.3 hr for evaporation from a body of water 1 m deep with a 1 m/sec current and 3 m/sec wind [13]. Due to its high vapor pressure and low adsorption to soil, evaporation from soil would be relatively rapid.

Water Concentrations: DRINKING WATER: Finished water <1.0 ppb [8]. Detected in commercial deionized charcoal-filtered water [3]. It was suggested that this could have originated from the plastics used in the preparation or storage of the ion-exchange resin or charcoal [3]. SURFACE WATER: 14 heavily industrialized river basins in U.S. (204 sites) - 1 site positive, 10 ppb in Chicago area [5]; Lake Michigan (91 sites) - 1 site positive 10 ppb [12].

Effluent Concentrations: Concentration of methyl methacrylate in exhaust stacks from a plant where acrylic resin based paints were dried 5-20 ppm [8].

Sediment/Soil Concentrations:

Atmospheric Concentrations:

Food Survey Values: No data on methyl methacrylate levels in food could be located. However, polymethyl methacrylate is used for food wrap and methyl methacrylate monomer can migrate from the plastic into ethanolic solutions at room temperature or into water at elevated temperatures [9] and, therefore, methyl methacrylate may migrate into food from packaging material.

Plant Concentrations:

Fish/Seafood Concentrations:

Animal Concentrations:

Milk Concentrations:

Other Environmental Concentrations: Residual methyl methacrylate monomer has been detected in commercial polystyrene plastics at 36 ppm [8]. Residual methyl methacrylate in 5 commercial acrylic bone cements has been reported to have migrated into prepared tissue medium and concentrations as high as 0.7-5.1 wt% were detected in fatty components of bone marrow [8].

Probable Routes of Human Exposure: Exposure to methyl methacrylate will be primarily occupational.

Average Daily Intake:

Occupational Exposures: Five plants manufacturing polymethyl methacrylate sheet, 4-88 ppm 8 hr TWA exposure [8]. Polystyrene production plant TWA concentration in breathing zone and in workplace area is 66 and 169 ppb, respectively, and the maximum concentration levels were 378 and 3300 ppb [20]. Detected at concentrations below the hygienic threshold limit in dental work rooms and in operating rooms during total hip replacement surgery [1,23]. Atmospheres above surfaces freshly painted with commercial acrylic latexes in the USSR 1075 ppb [8]. NIOSH (NOHS Survey, 1972-75) has statistically estimated that 90,205 workers are exposed to methyl methacrylate in the United States [15]. NIOSH (NOES Survey, 1981-83) has statistically estimated that 88,701 workers are exposed to methyl methacrylate in the United States [16].

Body Burdens:

REFERENCES

1. Brune D, Beltesbrekke H; Scan J Dent Res 89: 113-16 (1981)
2. Daubert TE, Danner RP; Data Compilation Tables of Properties of Pure Compounds. American Institute of Chemical Engineers. (1985)
3. Dowty BJ et al; Environ Sci Technol 9: 762-5 (1975)
4. Ellington JJ et al; Measurement of hydrolysis rate constants for evaluation of hazardous waste land disposal: Volume 2. Data on 54 chemicals. EPA/600/S3-87/019 (1987)
5. Ewing BB et al; Monitoring to detect previously unrecognized pollutants in surface waters. USEPA-560/6-77-015, UESPA-560/6-77-015a (1977)

Methyl Methacrylate

6.   Hansch C, Leo AJ; Medchem Project Issue no 20; Pomona College, Claremont, CA (1982)
7.   Hawley GG; Condensed Chemical Dictionary 10th ed. Van Nostrand Reinhold (1981)
8.   IARC; Some Monomers, Plastics and Synthetic Elastomers, and Acrolein; 19: 187-211 (1979)
9.   Inoue T et al; Bull Nat. Inst Hyg Sci (Tokyo) 0(99); 144-147 (1981)
10. Joshi SB et al; Atmos Environ 16: 1301-10 (1982)
11. Kirk-Othmer Encycl Chem Tech; 3rd ed; Wiley; 15: 346-76 (1981)
12. Konasewich D et al; Status report on organic and heavy metal contaminants in the lakes Erie, Michigan, Huron, and Superior Basins; Great Lakes Quality Review Board (1978)
13. Lyman WJ et al; Handbook of chemical property estimation methods. Environmental behavior of organic compounds; McGraw-Hill New York (1982)
14. Mabey W, Mill T; J Phys Chem Ref Data 7: 383-415 (1978)
15. NIOSH; National Occupational Health Survey (1975)
16. NIOSH; National Occupational Exposure Survey (1985)
17. Otsu T et al; Polymer 20: 55-8 (1979)
18. Pahren HR, Bloodgood DE; Water Pollut Control Fed J 33: 233-8 (1961)
19. Riddick JA et al; Organic Solvents New York: Wiley Interscience (1986)
20. Samimi B, Falbo L; Amer Indust Hyg Assoc J 43: 858-62 (1982)
21. Sasaki S; pp.283-98 in Aquatic pollutants: transformation and biological effects. Hutzinger O et al eds; Pergamon Press (1978)
22. Slave T et al; Rev Chim 25: 666-70 (1974)
23. Vedel P et al; Ugeokr Laeg 143: 2734-5 (1981)

Naphthalene

SUBSTANCE IDENTIFICATION

Synonyms:

Structure:

CAS Registry Number: 91-20-3

Molecular Formula: $C_{10}H_8$

Wiswesser Line Notation: L66J

CHEMICAL AND PHYSICAL PROPERTIES

Boiling Point: 217.9 °C at 760 mm Hg

Melting Point: 80.2 °C

Molecular Weight: 128.16

Dissociation Constants:

Log Octanol/Water Partition Coefficient: 3.30 [35]

Water Solubility: 31.7 mg/L at 25 °C [77]

Vapor Pressure: 0.082 mm Hg at 25 °C [77]

Henry's Law Constant: 4.83 x 10^{-4} atm-m³/mol [66]; 5.53 x 10^{-4} atm m³/mol [92]

ENVIRONMENTAL FATE/EXPOSURE POTENTIAL

Summary: Naphthalene enters the atmosphere primarily from fugitive emissions and exhaust connected with production and use of fuel oil and gasoline. In addition, there are discharges on land

408

Naphthalene

and into water from spills during the storage, transport, and disposal of fuel oil, coal tar, etc. Once in the atmosphere, naphthalene rapidly degrades due to its reaction with hydroxyl radicals (half-life <1 day). Releases into water are lost due to volatilization, photolysis, adsorption, and biodegradation. The principal loss processes will depend on local conditions, but half-lives can be expected to range from a couple of days to a few months. When adsorbed to sediment biodegradation occurs much more rapidly than in the overlying water column. When spilled on land, naphthalene is adsorbed moderately to soil and undergoes biodegradation. However, in some cases it will appear in the ground water where biodegradation still may occur if conditions are aerobic. The equilibrium bioconcentration factor is low since metabolism readily proceeds in aquatic organisms. The primary source of exposure is from air, especially in areas of heavy traffic or where fumes from evaporating gasoline or fuel oil exists, or in the vicinity of petroleum refineries and coal coking operations.

Natural Sources: It is a component of crude oil; since naphthalene is a product of natural uncontrolled combustion, forest fires, etc., may be a source of naphthalene [70].

Artificial Sources: Emissions may occur during its production from petroleum refining and coal tar distillation [111]; during its use as a chemical intermediate, e.g., phthalic anhydride manufacture [31,54]; from vehicular emissions [31]; combustion processes including refuse combustion [25,31]; tobacco smoke [31,105]; coal tar pitch fumes [111]; oil spills.

Terrestrial Fate: The sorption of naphthalene to soil will be low to moderate depending on the organic carbon content of the soil. Its passage through sandy soil will be rapid. It will undergo biodegradation, which may be rapid when the soil has been contaminated with other polycyclic aromatic hydrocarbons (PAHs) (half-life a few hours to days) but slow otherwise (half-life > 80 days). Evaporation of naphthalene from the top soil layer will be important, but the importance of the process will gradually decrease as the soil depth increases.

Aquatic Fate: Volatilization, photolysis, sorption, and biodegradation are the important loss mechanisms for naphthalene discharged into water. In a particular water, one or several processes may predominate depending on the nature of water. For example, the contributions of different processes in the removal of naphthalene from a shallow, rapidly flowing uncontaminated stream were estimated as follows: photolysis, <1%; microbial degradation, 4.4%; sediment sorption, 15.5%; and volatilization, 80% [40]. In deeper and slower moving contaminated water, biodegradation may be the most important process, with a half-life of 1-9 days [38,114]. Removal through sediment sorption may become important in water where the concentration of suspended solid is relatively high and water movement is slow, as in lakes and reservoirs. The photolysis half-life in near-surface water is about 3 days, but is about 550 days at a depth of 5 m [118]. In the Rhine River the overall half-life has been estimated to be 2.3 days based upon monitoring data [120].

Atmospheric Fate: The most important process for the removal of naphthalene from the atmosphere is its reaction with photochemically produced hydroxyl radicals and the half-life of this reaction is <1 day. In polluted urban air, reaction with NO_3 radicals may be an additional sink for nighttime loss.

Biodegradation: WATER: Data concerning the biodegradability of naphthalene both in standard biodegradability tests and in natural systems suggest that naphthalene degrades after a relatively short period of acclimation and that degradation can be rapid in oil polluted water, slow in unpolluted water, and that the rate of degradation increases with the concentration of naphthalene [109]. Bacteria can utilize naphthalene only when it is in the dissolved state [101]. In die-away tests, reported half-lives include about 3 days in water with high PAH levels [40]; 7, 24, 63, and 1700 days in an oil polluted estuarine stream, clean estuarine stream, coastal waters, and in the Gulf Stream, respectively [63]; 3-9 days in water near a coal-coking wastewater discharge [38,40]. In water from the Alaskan Continental Shelf, degradation rates avg 0.5%/wk; however, when nutrient levels are lower, as in late spring-early summer (after algae blooms), the degradation rate is reduced [80]. In a mesocosm experiment using Narragansett Bay seawater, the half-life in late summer was 0.8 days, principally due to biodegradation [114]. Biodegradation half-life 43 days in microbe-supplemented filtered Lake Superior harbor water and 39 days with nutrient and microbe-supplemented harbor water [107,108]. SEDIMENT: Degradation rates in sediment are much higher than in water, being 8- to 20-fold higher than in the water column above the sediment [39]. Half-lives in sediment include 4.9 hr and >88 days in oil contaminated and uncontaminated sediment, respectively [39], 9 days in sediment near a coal coking discharge [38], and 3, 5, and >2000 hours in sediments with high, medium, and low PAH levels, respectively [40]. When incubated in a slurry with sediment from an uncontaminated pond, the mineralization rate increases, reaching a peak after 6-12 days corresponding to a half-life of 78 days [82]. Biodegradation half-life ranged from 2.4 weeks in sediments chronically exposed to petroleum hydrocarbons to 4.4 weeks in sediments from a pristine environment [37]. SOIL: The overall half-life of naphthalene in a solid waste site was estimated to be 3.6 months [121]. In typical soils, the half-life is expected to be lower due to faster biodegradation. ANAEROBIC: No degradation under anaerobic conditions was observed in 6 and 11 weeks in a lab reactor with seed from a well near a source of contamination [24], or with sewage seed [10], respectively. GROUND WATER: Complete degradation occurred in 8 days in gas-oil contaminated ground water which was circulated through sand that had been inoculated with ground water under aerobic conditions [51]. Biodegradation occurred in ground water contaminated with creosote [102].

Abiotic Degradation: Naphthalene absorbs light in the solar wavelength region 288-330 nm and will photolyze in water [117,118]. The presence of algae in the water can increase the rate of photolysis of naphthalene by a factor of 1.3 to 2.7 [119]. Photolysis should also occur in air, but no experimental data could be found. The half-life in surface waters is estimated to be 71 hours [40,118] and 550 days in a 5 m deep inland water [118]. Reaction with oxidizing species in natural waters as well as hydrolysis will not be significant [117]. Naphthalene in air reacts with photochemically generated hydroxy radicals with a half-life of <1 day [4,6,55]. The loss of naphthalene due to reactions with photochemically generated N_2O_3 and O_3 in the atmosphere are negligible [4,6]. In polluted urban air, reaction with NO_3 radicals may be an additional sink for nighttime loss [5].

Naphthalene

Bioconcentration: Naphthalene bioconcentrates to a moderate amount in fish and aquatic invertebrates (log BCF 1.6-3.0) [23,28,36,63,81,94,110,]. However, for most invertebrates, depuration is rapid when the organism is placed in water free of the pollutant [23,98] and naphthalene is also readily metabolized in fish [15]. Some marine organisms that have no detectable aryl hydroxylase enzyme systems, e.g., phytoplankton, certain zooplankton, mussels (<u>Mytilus</u> <u>edulis</u>), scallops (<u>Placopecten</u> sp), and snails (<u>Litternia</u> <u>littorea</u>) tend to accumulate polynuclear aromatic hydrocarbons [67].

Soil Adsorption/Mobility: Naphthalene is adsorbed moderately by soil and sediment. 17 soils and sediment had a mean Koc of 871 [52] and soils from Switzerland had a Koc of 812 [86]. A mean Koc of 2400 was measured for 4 silt loams and a sandy loam soil [12] and a Koc of 843 was estimated for a sediment [113]. Measured Koc's ranged from 400 and 1000 for soils of varying organic carbon content [93] and 4100 for natural estuarine colloids [115]. Although it adsorbs to aquifer material [24], in simulations of ground water transport systems and rapid infiltration sites, and in field studies, naphthalene frequently appears in the effluent [29,42,73,78,79,85] and the sorption is weak in sandy soils. A half-life of 65 hr due to sediment adsorption in a flowing river of 1 m depth and 0.5 m/sec has been predicted [40]. In a variety of surface waters, only 0.1-.8% of the naphthalene was sorbed to particulate matter [40].

Volatilization from Water/Soil: The laboratory determined half-life for the evaporation of naphthalene from water 1 m deep with a 1 m/sec current velocity and a 3 m/sec wind speed is 4.1-5 hr [65,92]. In the case of naphthalene, the rate of volatilization is much more sensitive to the current velocity and a 10-fold decrease in current to 0.1 m/sec will increase the half-life to 32 hours, whereas 10-fold decrease in wind speed to 0.3 m/sec will increase the half-life to 11 hr [92]. The rate of evaporation of naphthalene in jet fuel from water relative to the oxygen reaeration rate ranged from 0.2 to 0.5. When combined with typical reaeration rates for natural bodies of water [69], these values give a half-life for evaporation of 50 and 200 hr in a river and lake, respectively [91]. Estimated volatilization half-lives from a soil containing 1.25% organic carbon were 1.1 day for 1 cm soil depth and 14.0 days for

Naphthalene

10 cm soil depth [49]. In moisture-saturated soil, as in the case of flooded soil, volatilization may not be important [11].

Water Concentrations: DRINKING WATER: Naphthalene concentrations measured as follows: Washington DC tap water - 1 ppb [83]. 3 New Orleans area drinking water plants sampled - detected but not quantified [53]. 12 Great Lake municipalities drinking water supplies - 0.9-1271 ppb, with levels being generally higher in winter [116]. Cincinnati, OH, Feb 1980 - 5 ppt [19]. Drinking waters in unspecified U.S. cities - up to 1.4 ppb [105]. 2 representative tap waters in U.S. cities - not detected, but other sources indicated averages of 7.8 ppb and 23.0 ppb [16]. Bank-filtered tap water from Rhine River in The Netherlands - 100 ppt [74]. Kitakyushu area, Japan - 2.2 ppb [2]. Zurich Switzerland, tap water - 8 ppt [32,58]. Ottawa, Ontario - January 1978 - 4.8 ppt; February 1978 - 6.8 ppt [8]. 4 of 5 Nordic tap water samples at conc range 1.2-8.8 ppt [60]. GROUND WATER: Naphthalene was detected as follows: Hoe Creek, NY, underground coal gasification site, 2 aquifers sampled 15 months after gasification completed - 380-1800 ppb [97]. Samples from East Anglica, England chalk aquifer at distances 10, 100-120, and 210 m from gasoline storage - 150, 30, and 0.1 ppb, respectively [100]. 3 rapid infiltration sites at Fort Polk, LA - 0.03 to 0.22 ppb [42]. Zurich, Switzerland - not detected [58]. One of 5 landfill leachates in Denmark - 1-10 mg/L [84]. SURFACE WATER: Lake Michigan - a trace detected at 5 of 9 sites [57]. Delaware River ranged from a trace to 0.9 ppb [87,88]. Ohio River between Wheeling and Evansville (5 samples) and 3 tributaries - detected at a detection limit of 0.1 ppb [71]. Charles River, Boston - detected at a detection limit of 0.1 ppb [41]. Lower Tennessee River, Calvert City, KY - 30.4 ppb (water/sediment mixture) [30]. Unspecified U.S. river near industrial sites - 6 to 10 ppb [47]. Natural waters - up to 2 ppb [105]. Mississippi River during summer 1984 - 4-34 ppb [21]. Lake Zurich, Switzerland - surface water, 8 ppt; water at 30 m depth, 52 ppt [32,58]. Kitakyusku area, Japan - detected, not quantified in river water [2]. River Glatt, Switzerland - detected, not quantified [122]. SEAWATER: Naphthalene measured as follows: Cape Cod, MA - Vineyard Sound - 0.5-35 ppt range, 12 ppt avg, and results displayed a strong seasonal pattern, highest concentrations noted in winter, which suggests a source from heating fuels [33]. Chemotaxis

Naphthalene

Dock, Vineyard Sound, MA, Dec 78 to Mar 79 - 0-27 ppt, with low levels reported in Dec and Jan; high level reported in Feb, correlating with a late heavy snowfall, indicating runoff or atmospheric inputs [68]. Dohkai Bay, Japan - area polluted by domestic and industrial waste as well as airborne particulates - detected, not quantified [90]. Kitakyusku area, Japan - detected, not quantified [2].

Effluent Concentrations: Industrial effluents - up to 3200 ppb; discharges from sewage treatment plants - up to 22 ppb [105]. Water sample from a stream running through an oil tank farm, Knoxville TN - 8 ppb [18], tire manufacturing plant wastewaters - 100 ppb [18,46]. Spent chlorination liquor from bleaching of sulfite pulp - 0.8-2.0 g/ton pulp [17]. Bekkelaget Sewage treatment plant, Oslo, Norway, secondary sewage water effluent - 88 ppt (dry period, Nov 1979), 303 ppt (dry period, spring 1980), 1504 ppt (after rainfall, summer 1980) [60]. Gas phase emission rates, diesel trucks - 7.4 mg/km (filtered), 9.2 mg/km (nonfiltered); gasoline-powered vehicles - 8.6 mg/km (filtered), 8.1 mg/km (unfiltered) [34]. 2 representative U.S. cities, sewage treatment plant influent, city A - 13 ppb avg, city B - 14.8 ppb avg; city B effluent - not detected [16]. Industries with mean treated wastewater concentrations range of >200 ppb-<920 ppb - paint and ink formulation, electrical/electronic components, auto and other laundries, and iron and steel manufacturing [106]. Maxey Flats, KY and West Valley, NY - trench leachates - 0.12 to 1.7 ppm [26].

Sediment/Soil Concentrations: Detected in only 1 sediment sample from an industrial location on an unspecified U.S. river [47]. Royal Botanical Gardens, Hamilton, Ontario - 2.0 ppb in pond sediment [50]. Lower Tennessee River, Calvert, KY - 30.4 ppb water and sediment [30]. Kitakyusku area, Japan - detected in sediment but not quantified [2]. Soil near Al-reduction plant, 48.3 ppb; unpolluted soil, 46.2 ppb; soil under a marsh, 57.7 ppb [112]. Dohkai Bay, Japan, area polluted by domestic and industrial waste and airborne particulates - detected in sediment, not quantified [90]. Saudajord, Norway, suggested sources - ferro alloy smelter, 6 sites, station 1 closest to smelter - 483.8 ppb (0-2 cm), 685.9 ppb (2-4 cm), 278.7 ppb (4-6 cm), 328.3 ppb (6-8 cm); station 2 2479.5 ppb (0-2 cm); station 3 48.3 ppb (0-2 cm); station 4 10.9 ppb (4-6 cm); not detected stations 5 and 6 (farthest away) [9].

South Texas coast, samples taken following the blowout of an exploratory oil well - detected at trace amount in 3 of 3 samples [7]. Cascoe Bay, ME, detected in 1 of 30 samples at 113 ppb [62]. Windsor Cove, Buzzards Bay, MA following an oil spill - 9.2 ppm (Oct 1974), 0.63 ppm (May 1975), and 0.11 ppm (June 1977) [99]. Wild Harbor, Buzzards Bay, MA - detected, not quantified immediately following Sept 1969 oil spill, not detected from 1971-1976 [99]. Various fjords in Norway situated near industrial and urban sites, 0-5 cm samples: 41.5-2870 ppb; North Sea 500 m from oil field - 31.6 ppb; North Sea, 10 km from oil field - 4.32 ppb, and Framvaren, a permanent anoxic fjord with no potential local pollution but high PAH values - 292 ppb (0-10 cm) and 272 ppb (14-20 cm) [95]. Surficial sediments of Boston Harbor, MA near areas of stormwater runoff and sewage outfalls - <10-43,628 ng/g dry wt [89]. March Point, Strait of Juan de Fuca and Northern Puget Sound, unpolluted area, baseline study - not detected [14].

Atmospheric Concentrations: RURAL: Narragansett Bay, RI coastal area - 3.18 pg/m³ (particulates >1.0 um), 49.10 pg/m³ (very fine particulates <1.0 um) [59]. URBAN/SUBURBAN: 11 U.S. cities 180 ppt median, 11-480 ppt range [13]. Kingston, RI - 31.1 pg/m³ (particulates >1.0 um), 27.90 pg/m³ (very fine particulates <1.0 um) [59]. USSR industrial cities and Leningrad - detected, not quantified [43,44]. Providence, RI, industrialized urban - 248.0 pg/m³ (particulates >1.0 um), 100.70 pg/m³ (very fine particulates < 1.0 um) [59]. Lillestrom and Oslo Norway - detected, not quantified [103]. Air over residential areas near Al-reduction plant - 11.3-117 ng/m³ [104,112] Three large South African cities - detected, not quantified [64]. Paris, France - 730-2100 ppt; Zurich Switzerland - 320 ppt [76]. Torrance, CA during a high pollution episode - 2.9-3.3 ug/m³ [3]. Chicago area homes - 43% frequency in indoor air, 21% frequency of occurrence in outdoor air [45]. Northern Italy indoor air, 11 ug/m³ (mean), 70 ug/m³ (max); outdoor air, 2 ug/m³ (mean), 11 ug/m³ (max) [20]. SOURCE DOMINATED: U.S. source dominated areas; 95 samples 400 ppt median, 16000 ppt max. [13]. Allegheny Mt Tunnel, Pennsylvania Tpk. - 3.1 to 10.0 ug/m³ (592-1910 ppt) (filtered), 3.5 to 10.1 ug/m³ (nonfiltered), low values correspond to low traffic volume [34]. Air near hazardous sites - 0.1-0.88 ppb (mean), 5.2 (max); near a landfill - 0.08 ppb (mean), 0.31 ppb (max) [61].

Naphthalene

Gaseous effluents from a coal-fired power plant under near-ideal conditions - 0.01-1.8 ug/m³ [48].

Food Survey Values:

Plant Concentrations: Southern Norway area, various species marine algae - not detected to 2109 ppb [56].

Fish/Seafood Concentrations: Pike from Detroit River, and carp and pike from Hamilton Harbor - detected, not quantified; trout from Lake Superior - detected, not quantified [57]. Cepangopaludina chinensis, Royal Botanical Gardens, Hamilton Ontario - <0.01 ppb [50]. Polychaetes 4.2-5.5 ppm, clam 0.43 ppm [18]. Mussels, Saudafjord, Norway suggested source - ferro alloy smelter, 4 stations - not detected [9]. Mussels sampled near the Bekkelaget sewage treatment plant, Oslo, Norway - not detected [60]. Four of 7 mussels from Finnish Archipelago Sea - 5-41 ppm wet wt; not detected in muscle of herring, pike-perch, and burbot caught from same location [75]. Southern Norway Coast, mussels, detected in 7 of 9 samples at trace-516 ppb; various invertebrates not detected-241 ppb, although compound not separable from methyl naphthalene [56]. Mussels and oysters from more than 100 U.S. east, west, gulf coast sites as analyzed by 3 different laboratories: Woods Hole Lab - 2.8 ppb mean, USEPA Natl Res Lab, Narragansett - 4.8 ppb mean, Univ New Orleans Lab - 96 ppb mean [27]. Several dried Nigerian freshwater fish species: traditionally smoked - 1.75-7.88 ppb, traditionally solar dried - 0.96-7.38 ppb, oven dried - 0.19-4.42 ppb [1]. March Point mussels, Strait of Juan de Fuca and Northern Puget Sound, unpolluted area baseline study: detected in 3 of 6 samples at 3.3-13 ppb [14].

Animal Concentrations:

Milk Concentrations: Mother's milk from 4 U.S. urban areas - detected but not quantified in 6 of 8 samples [72].

Other Environmental Concentrations:

Probable Routes of Human Exposure:- People are commonly exposed to naphthalene from inhalation of ambient air, particularly in areas with heavy traffic and at gasoline filling stations. Spill on the hand during filling may cause dermal exposure. Another source of inhalation exposure is from tobacco smoke. Although data is scanty, moderate ingestion exposure may occur from some supplies of drinking water and consumption of contaminated foods.

Average Daily Intake: AIR INTAKE (URBAN/SUBURBAN): (assume an air conc of 0.95 ug/m^3 and an inhalation rate of 20 m^3/day) 19 ug; WATER INTAKE: (assume water concn range of 0.001-2 ppb and an ingestion of 2 liters/day) 0.002-4 ug; FOOD INTAKE: insufficient data.

Occupational Exposures: Exposure of up to 220 ppm (vapor) and 4.4 ug/m^3 (particulates) are possible in industrial situations [105]. Naphthalene exposed workers include those who make beta naphthol, celluloid, dye chemicals, fungicide, hydronaphthalene, lampblack, phthalic anhydride, and smokeless powder as well as those who work with/in coal tar, moth repellants, tanneries, textile chemicals, and aluminum reduction plants [105]. Air levels of naphthalene in an aluminum reduction plant - 0.72-311.3 ug/m^3 (0.1-59.5 ppb) (vapor), 0.090-4.00 ug/m^3 (particulate); coke oven 11.35-1,120 ug/m^3 (2-214 ppb) (vapor), 0-4.40 ug/m^3 (particulate) [105]. Air concn in different work areas of Silicon Carbide plant - 1.3-58.0 ug/m^3 [22].

Body Burdens: A National Human Adipose Tissue Survey (NHATS) by EPA for fiscal year 1982 detected naphthalene in wet adipose tissue with a frequency of 40% at concn range of <9-63 ppb [96].

REFERENCES

1. Afolabi OA et al; J Agric Food Chem 31: 1083-90 (1983)
2. Akiyama T et al; J UOEH 2: 285-300 (1980)
3. Arey J et al; Atmos Environ 21: 1437-45 (1987)
4. Atkinson R et al; Environ Sci Technol 18: 110-3 (1984)
5. Atkinson R et al; J Phys Chem 88: 1210-5 (1984)
6. Atkinson R et al; Environ Sci Technol 21: 1014-22 (1987)
7. Bedinger CA Jr, Nulton CP; Bull Environ Contam Toxicol 28: 166-71 (1982)
8. Benoit FM et al; Int J Environ Anal Chem 6: 227-87 (1979)
9. Bjoerseth A et al; Sci Total Environ 13: 71-86 (1979)
10. Bouwer EJ, McCarty PL; Appl Environ Microbiol 45: 1295-9 (1983)

Naphthalene

11. Bouwer EJ et al; Water Res 18: 463-72 (1984)
12. Briggs GG; J Agric Food Chem 29: 1050-9 (1981)
13. Brodzinsky R, Singh HB; Volatile Organics in the Atmosphere: an assessment of available data; pp.198 SRI 68-02-3452 (1982)
14. Brown DW et al; Investigation of Petroleum in the Marine Environs of the Strait of Juan de Fuca and Northern Puget Sound; p.33 USEPA 600/7-79-164 (1979)
15. Callahan MA et al; Water-related Environmental Fate of 129 Priority Pollutants; pp.95-1 to 95-20 USEPA-440/4-79-029b (1979)
16. Callahan MA et al; 8th Natl Conf Munic Sludge Manag Proc; p.55 (1979)
17. Carlberg GE et al; Sci Total Environ 48: 157-67 (1986)
18. Carlson RM et al; Implications to the Aquatic Environment of Polynuclear Aromatic Hydrocarbons Liberated from Northern Great Plains Coal; pp.156 USEPA 600/3-79-093 (1979)
19. Coleman WE et al; Arch Environ Contam Toxicol 13: 171-8 (1984)
20. DeBaortoli M et al; Environ Int 12: 343-50 (1986)
21. DeLeon IR et al; Chemosphere 15: 795-805 (1986)
22. Dufresne A et al; Am Ind Hyg Assoc J 48: 160-6 (1987)
23. Eastmond DA et al; Arch Environ Contam Toxicol 13: 105-11 (1984)
24. Ehrlich GG et al; Ground Water 20: 703-10 (1982)
25. Eklund G et al; Chemosphere 16: 161-166 (1987)
26. Francis AJ et al; Nuclear Tech 50: 158-63 (1980)
27. Galloway WB et al; Environ Toxicol Chem 2: 395-410 (1983)
28. Geyer H et al; Chemosphere 11: 1121-34 (1982)
29. Goerlitz DF; Bull Environ Contam Toxicol 32: 37-44 (1984)
30. Goodley PC, Gordon M; Kentucky Acad Sci 37: 11-5 (1976)
31. Graedel TE; Chemical Compounds in the Atmosphere; Academic Press New York NY p.124 (1978)
32. Grob K, Grob G; J Chromatogr 90: 303-13 (1974)
33. Gschwend PM et al; Environ Sci Technol 16: 31-8 (1982)
34. Hampton CV et al; Environ Sci Technol 17: 699-708 (1983)
35. Hansch C, Leo AJ; Medchem Project Issue No 26. Claremont CA: Pomona College (1985)
36. Hawker DW, Connell DW; Ecotox Environ Safety 11: 184-97 (1986)
37. Heitkamp MA et al; Appl Environ Microbiol 53: 129-36 (1987)
38. Herbes SE; Appl Environ Microbiol 41: 20-8 (1981)
39. Herbes SE, Schwall LR; Appl Environ Microbiol 35: 306-16 (1978)
40. Herbes SE et al; pp.113-28 in The Scientific Basis of Toxicity Assessment; Witschi H ed; Elseveir/North Holland Biomed Press (1980)
41. Hites RA, Biemann K; Science 178: 158-60 (1972)
42. Hutchins SR et al; Environ Toxicol Chem 2: 195-216 (1983)
43. Ioffe BV et al; Environ Sci Technol 13: 864-8 (1979)
44. Ioffe BV et al; J Chromatogr 142: 787-95 (1977)
45. Jarke FH et al; ASHRAE Trans 87: 153-66 (1981)
46. Jungclaus GA et al; Anal Chem 48: 1894-6 (1976)
47. Jungclaus GA et al; Environ Sci Technol 12: 88-96 (1978)
48. Junk GA et al; ACS Symp Ser 319 (Fossil Fuels Util.): 109-23 (1986)
49. Jury WA et al; J Environ Qual 13: 573-9 (1984)

Naphthalene

50. Kalas L et al; pp.567-76 in Hydrocarbons and Halogenated Hydrocarbons in the Aquatic Environment; Afghan BK, Mackay D eds; New York NY Plenum Press (1980)
51. Kappeler T, Wuhrmann K; Water Res 12: 327-33 (1978)
52. Karickhoff SW; Chemosphere 10: 833-46 (1981)
53. Keith LH et al; pp.329-73 in Identification and Analysis of Organic Pollutants in Water; Keith LH ed; Ann Arbor MI Ann Arbor Press (1976)
54. Kirk-Othmer Encycl Chem Tech; 3rd ed 15:698 (1978)
55. Kloepffer W et al; Chem -Zig 110: 57-61 (1986)
56. Knutzen J, Sortland B; Water Res 16: 421-8 (1982)
57. Konasewich D et al; States Report on Organic and Heavy Metal Contaminants in the Lake Erie, Michigan, Huron, and Superior Basins. Great Lakes Quality Board; pp. 273 (1978)
58. Korte F, Klein W; Ecotox Environ Safety 6: 311-27 (1982)
59. Krstulovic AM et al; Am Lab 9: 11-8 (1977)
60. Kveseth K et al; Chemosphere 11: 623-39 (1982)
61. LaRegina J et al; Environ Prog 5: 18-27 (1986)
62. Larsen PF et al; Bull Environ Contam Toxicol 30: 530-5 (1983)
63. Lee RF; 1977 Oil Spill Conf, Amer Petrol Inst pp 611-6 (1977)
64. Louw CW et al; Atmos Environ 11: 703-17 (1977)
65. Lyman WJ et al; Handbook of Chemical Property Estimation Methods Environmental behavior of organic chemicals; McGraw Hill New York NY p.960 (1982)
66. Mackay D et al; Environ Sci Technol 13: 333-336 (1979)
67. Malins DC; Ann NY Acad Sci 298: 482-96 (1977)
68. Mantoura RFC et al; Environ Sci Technol 16: 39-45 (1982)
69. Mill T et al; Laboratory Protocols for Evaluating the Fate of Organic Chemicals in Air and Water; p.255 USEPA-600/3-82-022 (1982)
70. NAS; Biological Effects of Atmospheric Pollutants: Particulate Polycyclic Organic Matter. pp 361 Wasington, D.C. National Academy of Sciences (1972)
71. Ohio R Valley Water Sanit Comm; Assessment of Water Qaulity Condition Ohio R Mainstreams 1978-9 Cincinnati OH (1980)
72. Pellizzari ED et al; Bull Environ Contam Toxicol 28: 322-8 (1982)
73. Piet GJ et al; Int Symp Quality of Groundwater Studies in Environ Sci 17: 557-64 (1981)
74. Piet GJ and Morra CF; pp.31-42 in Artificial Groundwater Recharge (Water Res Eng Ser); Huisman L, Olsthorn TN eds; Pitman Pub (1983)
75. Raino K et al; Bull Environ Contam Toxicol 37: 337-43 (1986)
76. Raymond A, Guiochon G; Environ Sci Technol 8: 143-8 (1974)
77. Riddick JA et al; Organic Solvents: Physical Properties and Methods of Purification, 4th Edit. New York: J Wiley & Sons (1986)
78. Rittmann BE et al; Ground Water 18: 236-43 (1980)
79. Roberts PV et al; J Water Pollut Control Fed 52: 161-71 (1980)
80. Roubal G, Atlas RM; Appl Environ Microbiol 35: 897-905 (1978)
81. Roubal WT et al; Arch Environ Contam Toxicol 7: 237-44 (1978)
82. Saylor GS, Sherrill TW; Bacterial Degradation of Coal Conversion Byproducts (polycyclic aromatic hydrocarbons) in Aquatic Environments; Knoxville TN pp.80 NTIS Report No PB83-187161, Springfield, VA (1981)
83. Scheiman MA et al; Biomed Mass Spectrom 4: 209-11 (1974)

Naphthalene

84. Schultz B, Kjeldsen P; Water Res 20: 965-70 (1986)
85. Schwarzenbach RP et al; Environ Sci Technol 17: 472-9 (1983)
86. Schwarzenbach RP, Westall J; Environ Sci Technol 15: 1360-7 (1981)
87. Sheldon LS, Hites RA; Environ Sci Technol 13: 574-9 (1979)
88. Sheldon LS, Hites RA; Environ Sci Technol 12: 1188-94 (1978)
89. Shiaris MP, Jambard-Sweet D; Mar Pollut Bull 17: 469-72 (1986)
90. Shinohara R et al; Environ Int 4: 163-74 (1980)
91. Smith JH, Harper JC; 12th Conf on Environ Toxicol; pp.336-53 (1982)
92. Southworth GR; Bull Environ Contam Toxicol 21: 507-14 (1979)
93. Southworth GR, Keller JL; Water Air Soil Poll 28: 239-48 (1986)
94. Southworth GR et al; Water Res 12: 973-7 (1978)
95. Sporstal S et al; Environ Sci Technol 17: 282-6 (1983)
96. Stanley JS; Broad Scan Analysis of the FY82 National Human Adipose Tissue Survey Specimens. Vol III. Semivolatile Organic Compounds. EPA-560/5-860-037, Washington, D.C., USEPA, pp 148 (1986)
97. Stuermer DH et al; Environ Sci Technol 16: 582-7 (1982)
98. Tarshis IB; Arch Environ Contam Toxicol 10: 79-86 (1981)
99. Teal JM et al; J Fish Res Board Canada 35: 510-20 (1978)
100. Tester DJ, Harker RJ; Water Pollut Control 80: 614-31 (1981)
101. Thomas JM et al; Appl Environ Microbiol 52: 290-6 (1986)
102. Thomas JM et al; Environ Toxicol Chem 6: 607-14 (1987)
103. Thrane KE, Mikalsen A; Atmos Environ 15: 909-18 (1981)
104. Thrane KE; Atmos Environ 21: 617-28 (1987)
105. U.S. EPA; Ambient Water Quality Criteria: Naphthalene, NTIS PB 81-117707, Springfield, VA (1980)
106. USEPA; Treatability Manual; p.I.10.15-1 to 15-5 USEPA 600/2-82-001a (1981)
107. Vaishnav DD, Babeu L; Bull Environ Contam Toxicol 39: 237-44 (1987)
108. Vaishnav DD and Babeu L; J Great Lakes Res 12: 184-92 (1986)
109. Van der Linden AC; Dev Biodegrad Hydrocarbons 1: 165-200 (1978)
110. Veith GD et al; J Fish Res Board Canada 36: 1040-8 (1979)
111. Verschueren K; Handbook of Environmental Data on Organic Chemicals; 2nd ed Van Nostrand Reinhold Co New York NY p.892 (1983)
112. Vogt NB et al; Environ Sci Technol 21: 35-44 (1987)
113. Vowles PD, Mantoura RFC; Chemosphere 16: 109-16 (1987)
114. Wakeham SG et al; Environ Sci Technol 17: 611-7 (1983)
115. Wijayaratne RD, Means JC; Mar Environ Res 11: 77-89 (1984)
116. Williams DT et al; Chemosphere 11: 263-76 (1982)
117. Zadelis D and Simmons MS; pp 1279-90 in Polynuclear Aromatic Hydrocarbons, 7th Meeting, 1982, Cooke M and Dennis AJ ed; Battelle Press, Columbus, OH (1983)
118. Zepp RG, Schlotzhauer PF; pp 141-58 in Polynuclear Aromatic Hydrocarbons; Jones PW and Leber P ed; Ann Arbor Press, Ann Arbor, MI (1979)
119. Zepp RG, Scholzhauer PF; Environ Sci Technol 17: 462-8 (1983)
120. Zoeteman BCJ et al; Chemosphere 9: 231-49 (1980)
121. Zoeteman BCJ et al; Sci Total Environ 21: 187-202 (1981)
122. Zuercher F, Giger W; Vom Wasser 47: 37-55 (1976)

Nitrobenzene

SUBSTANCE IDENTIFICATION

Synonyms:

Structure:

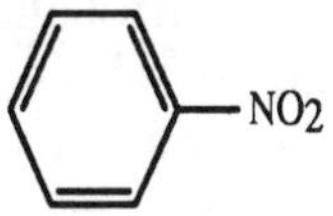

CAS Registry Number: 98-95-3

Molecular Formula: $C_6H_5NO_2$

Wiswesser Line Notation: WNR

CHEMICAL AND PHYSICAL PROPERTIES

Boiling Point: 210.8 °C

Melting Point: 5.7 °C

Molecular Weight: 123.11

Dissociation Constants:

Log Octanol/Water Partition Coefficient: 1.85 [24]

Water Solubility: 1900 mg/L at 20 °C [44]

Vapor Pressure: 0.15 mm Hg at 20 °C [60]

Henry's Law Constant: 2.44 x 10^{-5} atm-m^3/mole [34]

ENVIRONMENTAL FATE/EXPOSURE POTENTIAL

Summary: Nitrobenzene is produced in large quantities and may be released to the environment in emissions and wastewater during its production and use. Since 98% of nitrobenzene is used captively to produce aniline in five regions of the country, industrial releases will be fairly localized. Nitrobenzene is also produced by the photochemical reaction of benzene with oxides of nitrogen. This is a general source since benzene is found in petroleum products. It is difficult to estimate the ambient concentrations of nitrobenzene from this source because available air monitoring data are for areas of the country near production facilities. If released on land, nitrobenzene would leach into the ground and probably biodegrade within a few months. If released in wastewater, it should biodegrade (two experimental values for half-lives are 1 and 3.8 days). Some volatilization would be expected, but adsorption to sediment and bioconcentration in aquatic organism should not be significant. In the atmosphere, nitrobenzene will degrade primarily by photolysis (38% degradation in 5 hr). Human exposure will be primarily occupational via inhalation of the vapor or dermal contact with the vapor or liquid.

Natural Sources:

Artificial Sources: Nitrobenzene is produced in large quantities and may be released to the environment in emissions and wastewater during its production and use, 98% of which is used for producing aniline [11,18]. This production is focused geographically in five regions of the county [18]. Production losses are estimated to be 8.3 million lb yearly, with principal losses occurring in the reactor and acid concentration vents as well as the spent acid [18]. In one plant the loss of nitrobenzene in waste water totaled 0.09% and at another, 2.0% of production [18]. Another study estimated the emissions of nitrobenzene as 13 million lbs/yr as of 1978 [4]. Other uses of nitrobenzene, principally as a solvent for the production of cellulose ethers and in Freidel Craft alkylations [18], should also contribute to releases as fugitive emissions, in wastewater, and from spills. Nitrobenzene may also form in the atmosphere from the photochemical reaction of benzene with oxides of nitrogen [58].

Terrestrial Fate: Nitrobenzene is moderately adsorbed to soil and should leach into the ground if released on land and probably biodegrade within a few months. Nitrobenzene was completely removed from Rhine River water during soil filtration (bank or dune infiltration) [66]. Generally bank filtration takes 1-12 months and dune filtration 2-3 months [66], and the soil microorganism would be well acclimated. In another study, only 60% removal was obtained during infiltration through dunes consisting of fine-grained sand mixed with clay and lens-shaped peat layers [42].

Aquatic Fate: The half-life of nitrobenzene in the Rhine River in The Netherlands was estimated to be 1 day by measuring the concn reduction between sampling points [66]. In model waste stabilization ponds that were continuously fed with a synthetic waste feedstock and detained for 12 days, 89.5% of the added nitrobenzene was degraded, 4.9% volatilized, 2.3% adsorbed to sediment, 2.3% lost in effluent, and 1% remained in the water column [15]. The biodegradation half-life in the pond was 3.8 days.

Atmospheric Fate: Nitrobenzene will degrade in the atmosphere primarily by photolysis (38% degradation in 5 hr in laboratory tests). The rate of reaction with photochemically produced hydroxy radicals and ozone is relatively low. Results of modeling studies and field experiments suggest that wet deposition will have little effect on nitrobenzene loss in plumes within kilometers of a source [14].

Biodegradation: Results of screening tests on nitrobenzene are conflicting, with results ranging from rapid degradation to no degradation. These results include: 100% degradation in 7 days with a sewage inoculum [55]; 98% removal after 5 days with activated sludge [43]; 99.6 and 20% degradation after 6 days using municipal and industrial sewage seeds, respectively [16]; 100% degradation in 10 days, including a 6-day lag [23]; 0.4% mineralization in 5 days [21]; no BOD removed after 5 days using a sewage seed [8,27]; degradation resistant according to results of the MITI test, the biodegradation test of the Japanese Ministry of International Trade and Industry [28]; no degradation in 10 days with an activated sludge inoculum [56]; and degrades in >64 days using a soil inoculum [3]. While many factors affect biodegradation such as concn of the test compound, inoculum, acclimation of the seed, environmental factors, and incubation time, no simple explanation for these disparate results is apparent. Several investigators have reported, however, that nitrobenzene is toxic to microorganism at higher concentrations [33,36]. An overall removal of 97.8% was obtained in a completely mixed, continuous-flow activated sludge system with a 2- to 6-hr retention time after 1 month [29]. All losses were ascribed to biodegradation [29]. In a model aquatic ecosystem, little degradation occurred over the 2-day course of the experiment, since most of the radioactivity was recovered as the parent compound [32]. Complete removal of nitrobenzene was obtained when Rhine River water underwent bank infiltration [66]. The investigators ascribed the removal to microbial processes [66]. In a 45-day soil column transport experiment in which a solution of nitrobenzene and other pollutants was passed through a column packed with Lincoln fine sand, 20-40% of the chemical was degraded [61]. Under anaerobic conditions using a sewage inoculum, 50% degradation occurred in 14 days including a 8-day lag [23]. Aniline was detected in the reactor as soon as degradation began [23]. In an anaerobic reactor with a 2- to 10-hr hydraulic retention time and an inoculum maintained on acetate, 81% utilization was obtained after 110 days of operation [12].

Abiotic Degradation: Nitrobenzene absorbs UV light to about 400 nm [35] and can therefore undergo photolysis. In organic solvents containing abstractable H-atoms, it undergoes photoreduction when irradiated with UV light [6,45]. The presence of oxygen does not appear to affect the reaction rates [6,45]. In one study using a petroleum solvent and light >290 nm, 26% degradation occurred in 5 hr [6]. Azobenzene and aniline were the main products formed [6]. In near-surface pure water at 40 °C N latitude, the average annual photolytic half-life is estimated to be 133 days [49]. This long half-life is due to low quantum yield and sunlight adsorption rate [49,62]. The photolysis of humic substances in natural water gives rise to hydrated electrons that can reduce organic compounds [65]. The calculated half-life of nitrobenzene in a eutrophic Swiss lake due to this reaction is 22 days [65]. In the Aucilla River in northern Florida, humic substances accelerated the photolysis rate by a factor of 1.4 [49]. Nitrate ions in water can promote the photochemical oxidation of trace organic substances through the production of hydroxyl radicals [65]. In a clear, shallow body of water rich in nitrate (14 mg nitrate-N/L), the half-life of nitrobenzene exposed to midsummer, midday sunlight is 11 hr [17,65]. Hydrolysis is not environmentally significant [35]. When nitrobenzene was photolyzed in air, 38% degradation occurred in 5 hr, producing 2- and 4-nitrophenol [41]. When absorbed on silica gel and irradiated with light >290 nm, 6.7% mineralization occurred in 17 hr [30]. Nitrobenzene reacts with photochemically produced hydroxyl radicals in the vapor phase resulting in a half-life of 125 days in the clean troposphere and 62 days in a moderately polluted atmosphere [5]. Under atmospheric conditions, the reaction with ozone will be of negligible importance [5]. In smog chamber experiments, nitrobenzene has low reactivity in terms of ozone production [48,51].

Bioconcentration: The log BCF of nitrobenzene in golden orfe (Leuciscus idus melanotus) was <1.0 in a 3-day static test [21]. In a 28 day test using fathead minnows, the log BCF was 1.18 [59]. Another investigator obtained a log BCF of 0.78 in fish (P. reticulata) [10] and the bioconcentration test of the Japanese Ministry of International Trade and Industry reported a log BCF of <1 [28]. In green algae (Chlorella fusca), a log BCF of 1.38 was obtained [21]. No biomagnification of nitrobenzene was observed in an aquatic ecosystem containing algae, Daphnia magna, mosquito larvae, snails, and mosquito fish [32].

Soil Adsorption/Mobility: The leachability of nitrobenzene was studied in three typical Norwegian soils, one which was sandy with a low organic content and two organic soils [46]. The resulting Koc and retardation factor for the sandy soil was 30.6 and 1.27, while for the two organic soils the Koc values were 42.8 and 69.6 and the retardation factors 3.36 and 5.52 [46]. Koc values for two Danish subsoils were 170 and 370 [31]. When a mixture of pollutants including nitrobenzene in springwater was added to a column of Lincoln fine sand over a 45-day period, the retardation factor of nitrobenzene was 1.9 [61]. The Koc calculated from this experiment was 200 [61]. For clay, 15-31 mg of nitrobenzene were retained by 1 g of sodium montmorillonite in batch experiments [63].

Volatilization from Water/Soil: Using the Henry's Law constant, one can estimate that the volatilization half-life of nitrobenzene in a model river 1 m deep with a 1 m/sec current and a 3 m/sec wind is 45 hr [34]. Volatilization is primarily controlled by transport through the gas phase [34].

Nitrobenzene

Water Concentrations: DRINKING WATER: Nitrobenzene was detected but not quantified in finished water from the Carrollton Water Plant in LA [57] and in drinking water in Cincinnati [1]. In a survey of 14 treated drinking water supplies of varied sources in England, nitrobenzene was detected in one supply which came from an upland reservoir [20]. SURFACE WATER: Of the 836 stations reporting nitrobenzene in ambient water in the USEPA STORET database, 0.4% contained detectable levels of the chemical [53]. No nitrobenzene was reported in the Buffalo and Cuyahoga Rivers in the Lake Erie basin or the St. Joseph River in the Lake Michigan basin [22]. Average and maximum levels of nitrobenzene were 1.70 and 13.8 ppb in the Waal River and <0.1 and 0.3 ppb in the Maas River, both in The Netherlands [37]. Rhine River water in The Netherlands contained 0.5 ppb [66]. A 2-week composite sample taken from the Rhine River near Dusseldorf in 1984 contained a mean nitrobenzene concn of 0.42 ppb [50]. Japanese river water and seawater contained 0.16-0.99 ppb of nitrobenzene [54] and it was detected but not quantified in seawater in the Kitakyushi area of Japan [2].

Effluent Concentrations: In a comprehensive survey of wastewater from 4000 industrial and publicly owned treatment works (POTWs) sponsored by the Effluent Guidelines Division of the U.S. EPA, nitrobenzene was identified in discharges of the following industrial category (frequency of occurrence, median concn in ppb): leather tanning (1, 3.7); petroleum refining (1, 7.7); nonferrous metals (1, 47.7); organics and plastics (13, 3876.7); inorganic chemicals (3, 1995.3); pulp and paper (1, 124.3); auto and other laundries (1, 40.4); pesticides manufactures (1, 16.3); explosives (8, 51.7); and organic chemicals (36, 43.7) [47]. The highest effluent concn was 100,245 ppb in the organics and plastics industry [47]. Of the 1245 stations reporting nitrobenzene in industrial effluents in the STORET database, 1.8% contained detectable levels of the chemical [53]. Nitrobenzene was detected in the final effluent of 3 publicly owned treatment works (POTWs) and an oil refinery [19]. Two samplings of final effluent of the Los Angeles County Municipal Wastewater Treatment Plant contained 20 and <100 ppb of nitrobenzene [64]. It was not detected in 28 samples of industrial effluents and polluted fjords in Norway [52]. In the National Urban Runoff Program in which samples of runoff were

426

collected from 19 cities (51 catchments) in the U.S., no nitrobenzene was found [13].

Sediment/Soil Concentrations: Nitrobenzene was detected at a concn of 8 ppm in soil along the Buffalo River in Buffalo, NY but not detected in three samples of bottom sediment from the river [38]. None of the 349 stations reporting nitrobenzene in sediment in the USEPA STORET database contained detectable levels of the chemical [53].

Atmospheric Concentrations: URBAN/SUBURBAN: U.S. (595 sites/samples) 170 ppt mean, 2800 ppt max., 75% of samples were <92 ppt, 0 ppt median [9]. As part of the 1981 Airborne Toxic Element and Organic Substances program, sites in 3 NJ cities that are influenced by industry were monitored continuously for 6 weeks. The results for nitrobenzene were (city, number of samples, geometric mean concn, % pos): Newark, 37, 0.07 ppb, 81%; Elizabeth, 36, 0.10 ppb, 86%; and Camden, 37, 0.07 ppb, 86% [26]. SOURCE AREAS: U.S. (14 sites/samples) - 2 ppt median, 25 ppt max [9]. Trace levels of nitrobenzene were found in 2 of 13 air samples from the Lipari and BFI landfills in New Jersey [7]. Mean air concn of nitrobenzene at 5 abandoned hazardous landfill sites in New Jersey ranged from 0.01 to 1.32 ppb, with a max concn of 3.46 ppb [25]. No nitrobenzene was found in the air above an active sanitary landfill [25].

Food Survey Values:

Plant Concentrations:

Fish/Seafood Concentrations: None of the 122 stations reporting nitrobenzene in fish in the STORET database contained detectable levels of the chemical [53].

Animal Concentrations:

Milk Concentrations:

Other Environmental Concentrations:

Probable Routes of Human Exposure: Exposure to nitrobenzene is primarily occupational via cutaneous absorption of the liquid or vapor or inhalation of the vapor [18]. The general population may be exposed to nitrobenzene vapor in urban areas.

Average Daily Intake:

Occupational Exposures: Based on statistical estimates derived from the NIOSH National Occupational Hazard Survey (NOHS) conducted in 1972-74, 13,546 workers are potentially exposed to nitrobenzene [39]. According to statistical estimates derived from the NIOSH National Occupational Exposure Survey (NOES) conducted in 1981-83, 4335 workers are potentially exposed to nitrobenzene [40]. The total U.S. dosage (exposed persons times average atmospheric concn to which they are exposed) is estimated to be 1,114,000 ug/m^3 [4]. In a nitrobenzene factory in England, workers received a maximum dose of 70 mg/day by vapor adsorption through lungs and skin [18]. At a dyestuff factory the maximum dose is 80 mg/day, and most workers absorbed <35 mg/day [18]. The estimated nitrobenzene concn at ground level 500 m downwind from an emission source of the largest production facility has been estimated to be 1.56 mg/m^3 [18].

Body Burdens: Workers' exposure to nitrobenzene can be monitored by measuring 4-nitrophenol in the urine [18]. Levels reach a maximum 4 hr after exposure and may be detected up to 100 hr later [18].

REFERENCES

1. Abrams EF et al; Springfield,VA: Versar Inc (1975)
2. Akiyama T et al; J UOEH 2: 285-300 (1980)
3. Alexander M, Lustigman BK; J Agric Food Chem 14: 410-3 (1966)
4. Anderson GE; Human Exposure to Atmospheric Concentrations of Selected Chemicals Volume 2 (NITS PB83-265249) USEPA, Res Triangle Park, NC Off Air Qual Planning And Standards p. 737 (1983)
5. Atkinson R et al; Environ Sci Technol 21: 64-72 (1987)
6. Barltrop JA, Bunce NJ; J Chem Soc Section C 1968: 1467-74 (1968)
7. Bozzelli JW et al; Analysis of selected toxic and carcinogenic substances in ambient air in New Jersey State Of New Jersey, Department Of Environmental Protection (1980)
8. Branson DR; Predicting The Fate Of Chemicals In The Aquatic Environment From Laboratory Data ASTM STP 657 Phila PA: American Society For Testing And Materials pp. 55-70 (1978)

Nitrobenzene

9. Brodzinsky R, Singh HB; Volatile Organic Chemicals In The Atmosphere: An Assessment of Available Data Menlo Park,CA: Atmospheric Science Center SRI International Contract 68-02-3452 198 pp. (1982)
10. Canton JH et al; Regul Toxicol Pharmacol 5: 123-31 (1985)
11. Chemical Profiles; Chemical Marketing Reporter August 3 (1987)
12. Chou WL et al; Bioeng Symp 8: 391-414 (1979)
13. Cole RH et al; J Water Pollut Control Fed 56: 898-908 (1984)
14. Dana MT et al; Hazardous Air Pollutants EPA-600/3-84-113 (NTIS PB 85 138-626) Richland,WA: USEPA pp.106 (1985)
15. Davis EM et al; Partitioning Of Selected Organic Pollutants In Aquatic Ecosystems pp.176-84 in Biodeterioration 5 Oxley TA & Barry S eds (1983)
16. Davis EM et al; Water Res 15: 1125-7 (1981)
17. Dorfman LM, Adams GE; Reactivity of The Hydroxyl Radical In Aqueous Solution NSRD-NBS-46 (NTIS COM-73-50623) Washington, DC: NBS (1973)
18. Dorigan J, Hushon J; Air Pollution Assessment Of Nitrobenzene NTIS PB 257-776 McClean, VA: MITRE Corp pp. 96 (1976)
19. Ellis DD et al; Arch Environ Contam Toxicol 11: 373-82 (1982)
20. Fielding M et al; Organic Micropollutants In Drinking Water TR-159 Medmenham, Eng Water Res Cent 49 pp. (1981)
21. Freitag D et al; Ecotox Environ Safety 6: 60-81 (1982)
22. Great Lakes Water Quality Board; An Inventory Of Chemical Substances Identified In The Great Lakes Ecosystem Volume 1 - Summary Report To The Great Lakes Water Quality Board Windsor Ontario, Canada pp.195 (1983)
23. Hallas LE, Alexander M; Appl Environ Microbiol 45: 1234-41 (1983)
24. Hansch C, Leo AJ; Medchem Project Issue No 26. Claremont CA: Pomona College (1985)
25. Harkov R et al; J Environ Sci Health 20: 491-501 (1985)
26. Harkov R et al; J Air Pollut Control Assoc 33: 1177-83 (1983)
27. Heukelekian H, Rand MC; J Water Pollut Contr Assoc 29: 1040-53 (1955)
28. Kawasaki M; Ecotoxic Environ Safety 4: 444-54 (1980)
29. Kincannon DF et al; J Water Pollut Control Fed 55: 157-63 (1983)
30. Kotzias D et al; Chemosphere 5: 301-4 (1979)
31. Loekke H; Water Air Soil Pollut 22: 373-87 (1984)
32. Lu PY, Metcalf RL; Environ Health Perspect 10: 269-84 (1975)
33. Lutin PA et al; Purdue Univ Eng Bull, Ext Series 118: 131-45 (1965)
34. Lyman WJ et al; pp.15-1 to 15-34 in Handbook of Chem Property Estimation Methods McGraw-Hill NY (1982)
35. Mabey WR et al; Aquatic Fate Process Data For Organic Priority Pollutants USEPA-440/4-81-014 (1981)
36. Marion CV, Malaney GW; pp. 297-308 Proc 18th Ind Waste Conf, Eng Bull Purdue Univ, Eng Ext Ser (1964)
37. Meijers AP, Vanderleer RC; Water Res 10: 597-604 (1976)
38. Nelson CR, Hites RA; Environ Sci Technol 14: 1147-9 (1980)
39. NIOSH; National Occupational Health Survey (1975)
40. NIOSH; National Occupational Exposure Survey (1985)
41. Nojima K, Kanno S; Chemosphere 6: 371-6 (1977)
42. Piet GJ et al; Proc Internat Symp The Netherlands Mar 23-27, 1981 Van Duijvenboden W et al eds. Studies in Environ Sci 17: 557-64 (1981)

Nitrobenzene

43. Pitter P; Water Res 10: 231-5 (1976)
44. Riddick JA et al; Organic Solvents: Physical Properties and Methods of Purification, 4th Edit. New York: J Wiley & Sons (1986)
45. Sandus O, Slagg N; Reactions Of Aromatic Nitrocompounds I NTIS AD 753 923 Dover, NJ: Picatinny Arsenal; Tech Rep 4385 (1972)
46. Seip HM et al; Sci Total Environ 50: 87-101 (1986)
47. Shackleford WM et al; Analyt Chim Acta 146: 15-27 (1983)
48. Sickles JE II et al; IN: Proc 73rd Annual Meet Air Pollut Cont Assoc Paper 80-501 (1980)
49. Simmons MS, Zepp RG; Water Res 20: 899-904 (1986)
50. Sontheimer H et al; Sci Tot Env 47: 27-44 (1985)
51. Spicer CW et al; Atmospheric Reaction Products From Hazardous Air Pollutant Degradation EPA-600/3-85-028 Research Triangle Park, NC: USEPA p. 4 (1985)
52. Sporstoel S et al; Int J Environ Anal Chem 21: 129-38 (1985)
53. Staples CA et al; Environ Toxicol Chem 4: 131-42 (1985)
54. Sugiyama H et al; Eisei Kagaku 24: 11-8 (1978)
55. Tabak HH et al; J Water Pollut Contr Fed 53: 1503-18 (1981)
56. Urano K, Kato Z; J Hazardous Materials 13: 147-59 (1986)
57. USEPA; Industrial pollution of the lower Mississippi River in Louisiana. Technical Report. Dallas, TX: USEPA (1972)
58. USEPA; Ambient Water Quality Criteria for Nitrobenzene (PB81-117723) (1980)
59. Veith GD et al; J Fish Res Board Can 36: 1040-8 (1979)
60. Verschueren K; Handbook of Environmental Data on Organic Chemicals. New York, NY: Van Nostrand Reinhold Co (1977)
61. Wilson JT et al; J Environ Qual 10: 501-6 (1981)
62. Wojtczak J, Maciejewski A; Poznan Tow Przyj Nauk, Pr Kom Mat-Przyr, Pr Chem 13: 117-27 (1972)
63. Wolfe TA et al; J Water Pollut Control Fed 58: 68-76 (1986)
64. Young DR et al; pp. 871-84 in Water Chlorination II (1983)
65. Zepp RG et al; Environ Sci Technol 21: 443-50 (1987)
66. Zoeteman BCJ et al; Chemosphere 9: 231-49 (1980)

2-Nitrophenol

SUBSTANCE IDENTIFICATION

Synonyms: o-Nitrophenol

Structure:

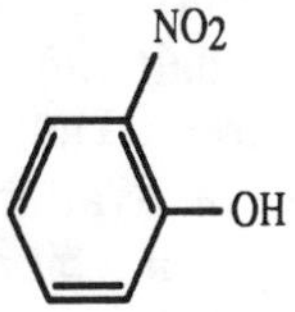

NO$_2$

OH

CAS Registry Number: 88-75-5

Molecular Formula: C$_6$H$_5$NO$_3$

Wiswesser Line Notation: WNR BQ

CHEMICAL AND PHYSICAL PROPERTIES

Boiling Point: 216 °C

Melting Point: 44-45 °C

Molecular Weight: 139.11

Dissociation Constants: pKa = 7.230 [27]

Log Octanol/Water Partition Coefficient: 1.79 [15]

Water Solubility: 1,060 mg/L at 20 °C [26]; 2,500 mg/L at 25 °C [7]

Vapor Pressure: 0.20 mm Hg at 25 °C [26]

Henry's Law Constant: 3.5 x 10^{-6} atm-m^3/mole [17]

ENVIRONMENTAL FATE/EXPOSURE POTENTIAL

Summary: 2-Nitrophenol may be released to the environment in wastewater and as fugitive emissions during its production and use

431

as a chemical intermediate. It is also found in the atmosphere originating mostly from secondary photochemical reactions in the air and partly from emissions of vehicular exhaust gas. When released on soil, appreciable quantities should volatilize and it will slowly biodegrade (10% mineralization in 30 days). It would not be adsorbed strongly to soil and, therefore, may leach into the ground water. If released into water, it will be lost by a combination of volatilization (half-life 12 days in a model river), photolysis, and biodegradation. Half-lives for photolysis and biodegradation in water are not available, but acclimation will be very important in the case of biodegradation. Adsorption to sediment should be minor under most circumstances and little or no bioconcentration should take place. If released in air, it will be partially associated with particulate matter and be removed by gravitational settling and washout by rain, as well as by photolysis and vapor-phase reaction with photochemically produced hydroxyl radicals (estimated half-life 14 hr). Human exposure will be primarily from ambient air via inhalation.

Natural Sources:

Artificial Sources: 2-Nitrophenol may be released to the environment in fugitive emissions or in wastewater during its production and use as a chemical intermediate [20]. It is a photooxidation product of nitrobenzene [12] and is a product of the photooxidation of aromatic hydrocarbons such as benzene, toluene, and phenanthrene with nitric oxide in air [21]. It is emitted in vehicular exhaust gas from both gasoline and diesel engines [21].

Terrestrial Fate: When radiolabelled 2-nitrophenol was incubated in a laboratory terrestrial ecosystem for 30 days, 44.8% of the ^{14}C resided in soil and plants and 9.9% had been transformed to $^{14}CO_2$ [9]. 45.3% of the radioactivity found in air was not CO_2 [9]. According to a modeling calculation on this system, 37.2% of the 2-nitrophenol should volatilize, which suggests that most of the ^{14}C in the vapor phase that is not $^{14}CO_2$ is parent compound, rather than a volatile metabolite. 2-Nitrophenol would not be expected to adsorb appreciably to soil. 4-Nitrophenol complexes with montmorillonite clay, which suggests that 2-nitrophenol might do likewise; however, experimental data are lacking. Based on screening studies, 2-nitrophenol should biodegrade more slowly than the other nitrophenols; 10% mineralization in 30 days in a terrestrial ecosystem [9]. Its half-life in flooded soil is 10 days. While its low adsorption to soil suggests that 2-nitrophenol may leach into ground water, it has not been reported in ground water in monitoring studies.

Aquatic Fate: If released into water, one would not expect 2-nitrophenol to sorb appreciably to sediment because of its low Koc. Losses may result from biodegradation, volatilization, and photolysis. No studies were available that would indicate what the half-life for biodegradation is in natural waters. Screening studies indicate that 2-nitrophenol biodegrades more slowly than 4-nitrophenol (half-life in freshwater is 1-8 days) with acclimation being extremely important. Biodegradation is moderately fast under anaerobic conditions, which implies that degradation may be important in anaerobic sediment. Volatilization (half-life 12 days in a model river) and photolysis may be important in some environmental waters.

Atmospheric Fate: 2-Nitrophenol released to the atmosphere may be associated with suspended particulate matter or aerosols as well as exist in the vapor phase. It will be scavenged by rain, be subject to gravitational settling, and undergo photolysis. The photolysis half-life will depend on the amount of sunlight, pH, and dissolved nitrate in the aerosol water and the nature of the particulate matter on which the 2-nitrophenol is adsorbed; however, data are lacking on which to base an estimated photolysis half-life. The estimated half-life for the reaction of vapor phase 2-nitrophenol with photochemically produced hydroxyl radicals is 14 hr.

Biodegradation: The results of the biodegradability screening studies on 2-nitrophenol are conflicting, with results ranging from no degradation to rapid degradation using soil, sewage, or activated sludge inocula. The majority of the results are that biodegradation is slow [1,11,14,22,24,32,33]. In 7 screening tests, 2-nitrophenol degraded much more slowly than 3- and 4-nitrophenol [11]. In one screening study that used a soil inoculum, degradation took >64 days, much longer than for the other nitrophenol isomers [1]; however, another investigator obtained degradation in 7-14 days [14]. Conflicting results may be due to differences in concentrations of the test chemical, inocula, toxicity at higher concentrations, or insufficient acclimation. No die-away test results were found for 2-nitrophenol in water under aerobic conditions. In a terrestrial ecosystem, 10% mineralization was obtained in 10 days [9]. Under anaerobic conditions, 2-nitrophenol was mineralized in 1 week when incubated with digester sludge [3]. In flooded (anaerobic) alluvial and Pokkali acid sulfate soils, 63 and 50% degradation occurred in 10 days [31]. Nitrite accumulated only in the alluvial soil, suggesting that differences may exist in the degradative pathways in the two soils [31].

Abiotic Degradation: Based upon the acid dissociation constant, 2-nitrophenol is expected to exist to an appreciable extent as the anion in environmental waters. It has several absorption bands >290 nm and a maximum absorption of 346 nm in neutral solutions; the absorption >400 nm increases markedly in basic solutions [23,35]. No information could be found in the literature pertaining to the photolysis of 2-nitrophenol in water; however, it is probable that it will undergo some direct photolysis as does 4-nitrophenol (half-life hr to 2 weeks in clear surface waters). As with 4-nitrophenol, it should also react with hydroxyl radicals that are photochemically produced in waters containing nitrate and nitrite ions. 20% of the 2-nitrophenol was degraded in vapor-phase photolysis test of unspecified duration and 11% mineralization occurred in 3 days when 2-nitrophenol adsorbed on silica gel was exposed to light >290 nm [25]. While these tests are designed to predict the photolytic behavior of chemicals in the vapor and adsorbed on siliceous material, no rates can be inferred from the results. Vapor phase 2-nitrophenol will react with photochemically produced hydroxyl radicals by H-atom abstraction and ring addition with an estimated half-life of 14 hr [10]. 2-Nitrophenol is resistant to hydrolysis [19].

Bioconcentration: From the log octanol/water partition coefficient for 2-nitrophenol, one can estimate a BCF of 14 [15], indicating little tendency for bioconcentration in fish.

Soil Adsorption/Mobility: The Koc estimated from its water solubility is approx 65 [18], which suggests only a slight potential for adsorption by organic matter in soil. The Koc for 2-nitrophenol in Brookston clay loam was 114 [2]. 4-Nitrophenol is known to complex with montmorillonite clay, which suggests that 2-nitrophenol may do likewise. The observation that 2-nitrophenol is an effective flocculating agent for clays and soils in aqueous suspension gives support to the idea that complexation with certain clays is also characteristic of 2-nitrophenol [5].

Volatilization from Water/Soil: Using the Henry's Law constant for 2-nitrophenol, the volatilization half-life can be estimated to be 296 hr from a model river 1 m deep with a 1 m/sec current and a 3 m/sec wind [18], which is in rough agreement with another calculated value (390 hr) for an unspecified body of water 1 m deep [25]. In a terrestrial laboratory ecosystem, 45.3 and 37.2% of radiolabelled 2-nitrophenol was found experimentally and predicted to be in the air compartment, respectively, indicating that volatilization from soil is an important loss mechanism [9].

Water Concentrations: SURFACE WATER: Ambient water concentrations reported in the USEPA STORET database, 1980-1982 (811 samples) 0.0% pos [30]. Not detected in the Lake Erie or Lake Michigan basin [13]. 2-Nitrophenol has been detected, not quantified, in river water [28]. RAIN/SNOW: Portland, OR (7 rain events): 26-130 ng/L, 59 ng/L mean dissolved in rain [17].

Effluent Concentrations: Effluent concentrations reported in the USEPA STORET database, 1980-1982 (1318 samples) 1.8% pos [30]. Norway - 30 samples representing polluted fjord areas as well as effluent from municipal treatment plants, refineries, petrochemical, and metallurgic industries: 11% pos at a 1 ppb limit of quantitation [29]. Confirmed in effluents from the photographic and electronics industries [4]. Detected in effluent from the Sauget wastewater treatment plant which receives wastes from heavy chemical manufacturing suggestive of azo dye wastes [8]. Detected in the treated effluents of the following industries (industry (concn)): iron and steel manufacture (21 ppb max); foundries (20 ppb mean, 40 ppb max); pharmaceuticals (10 ppb max); organic chemicals manufacturing/plastics (130 ppb max); rubber processing (4.9 ppb max); and textile mills (4.1 ppb max) [34]. Additionally, the raw wastewater of the following industries not listed above contained 4-nitrophenol (industry (concn)): coal mining (17 ppb max); electrical/electronic components (<75 ppb mean, 320 ppb max); metal finishing (72 ppb mean, 320 ppb max); photographic equipment/supplies (19 ppb mean, 32 ppb max); and petroleum refining (1400 ppb) [34]. National Urban Runoff Program in which 19 cities and metropolitan councils across the United States (51 catchments) were sampled - not detected [6].

Sediment/Soil Concentrations: Sediment concentrations reported in the USEPA STORET database, 1980-1982 (311 samples) 0% pos [30]. Detected in water/soil/sediment samples from the Love Canal [16].

Atmospheric Concentrations: URBAN/SUBURBAN: Yokahama, Japan where photochemical smog has frequently been observed ND-3.9 ppm in suspended particulates in five 24-hr samples [21]. Portland, OR: 11-39 ng/m^3, 24 ng/m^3, mean during 7 rain events [17].

Food Survey Values:

Plant Concentrations:

Fish/Seafood Concentrations: Biota concn reported in the USEPA STORET database, 1980-1982 (104 samples) 0% pos [30].

Animal Concentrations:

Milk Concentrations:

Other Environmental Concentrations: Exhaust gas from gasoline and diesel engines contained 3.1 and 6.4 ppb of 2-nitrophenol [21].

Probable Routes of Human Exposure: Human exposure will be primarily from ambient air especially in urban areas via inhalation.

Average Daily Intake: AIR INTAKE: (assume air concn 24 ng/m^3) 0.48 ug; WATER INTAKE: insufficient data; FOOD INTAKE: insufficient data.

Occupational Exposures:

Body Burdens:

2-Nitrophenol

REFERENCES

1. Alexander M, Lustigman BK; J Agric Food Chem 14: 410-3 (1966)
2. Boyd SA; Soil Science 134: 337-43 (1982)
3. Boyd SA et al; Appl Environ Microbiol 46: 50-4 (1983)
4. Bursey JT, Pellizzari ED; Analysis of Industrial Wastewater for Organic Pollut in Consent Degree pp 167 Contract No. 68-03-2867 USEPA Environ Res Lab (1982)
5. Callahan MA et al; Water-Related Environ Fate of 129 Priority Pollut Vol.II USEPA 440/4-79-029B (1979)
6. Cole RH et al; J Water Pollut Control Fed 56: 898-908 (1984)
7. Duff JC; J Chem Soc 1929: 2789-96 (1929)
8. Ellis DD et al; Arch Environ Contam Toxicol 11: 373-82 (1982)
9. Figge K et al; Regul Toxicol Pharmacol 3: 199-215 (1983)
10. GEMS; Graphical Exposure Modeling System. FAP. Fate of Atmos Pollut (1986)
11. Gerike P, Fischer WK; Ecotox Environ Safety 3: 159-73 (1979)
12. Graedel TE; Chem Compounds in the Atmos, Academic Press NY pp 297 (1986)
13. Great Lakes Water Quality Board; An Inventory of Chem Substances Identified in the Great Lakes Ecosystem 1: 195 (1983)
14. Haller HD; J Water Pollut Control Fed 50: 2771-7 (1978)
15. Hansch C, Leo AJ; Medchem Project Issue No 26. Claremont CA: Pomona College (1985)
16. Hauser TR, Bromberg SM; Environ Monit Assess 2: 249-72 (1982)
17. Leuenberger C et al; Environ Sci Technol 19: 1053-8 (1985)
18. Lyman WJ et al; Handbook of Chemical Property Estimation Methods. Environmental Behavior of Organic Compounds. McGraw-Hill NY (1982)
19. Mabey WR et al; Aquatic Fate Process Data For Organic Priority Pollut pp 434 USEPA 440/4-81-014 (1981)
20. Merck Index; An Encyclopedia of Chemicals, Drugs and Biologicals 10th ed p 6465 (1983)
21. Nojima K et al; Chem Pharm Bull 31: 1047-51 (1983)
22. Pitter P; Water Res 10: 231-5 (1976)
23. Sadtler; Sadtler Standard Spectra Philadelphia, PA
24. Sasaki S; pp 283-98 in Aquatic Pollut Transformation and Bio Effects, Hutzinger O, et al eds. Pergamon (1978)
25. Schmidt-Bleek F et al; Chemosphere 11: 383-415 (1982)
26. Schwarzenbach RP et al; Environ Sci Technol 22: 83-92 (1988)
27. Serjeant EP, Dempsey B; Ionisation Constants of Organic Acids in Aqueous Solution, IUPAC Chem Data Series pp 989 (1979)
28. Shackelford WM, Keith LH; Frequency of Organ Compounds Identified in Water USEPA 600/4-76-062 (1976)
29. Sporstoel S et al; Int J Environ Anal Chem 21: 129-38 (1985)
30. Staples CA et al; Environ Toxicol Chem 4: 131-42 (1985)
31. Sudhakar-Barik, Sethunathan N; J Environ Qual 7: 349-52 (1978)
32. Tabak HH et al; J Water Pollut Contr Fed 53: 1503-18 (1981)
33. Urushigawa Y et al; Kogai Shigen Kenkyusho Iho 12: 37-46 (1983)

2-Nitrophenol

34. USEPA; Treatability Manual 1: I.8.6-1 to I.8.6-5 USEPA 600/8-80-042 (1980)
35. Weast RC; Handbook of Chem and Physics 53rd ed. Cleveland, OH (1972)

3-Nitrophenol

SUBSTANCE IDENTIFICATION

Synonyms: m-Nitrophenol

Structure:

CAS Registry Number: 554-84-7

Molecular Formula: $C_6H_5NO_3$

Wiswesser Line Notation: WNR CQ

CHEMICAL AND PHYSICAL PROPERTIES

Boiling Point: 194 °C at 70 mm Hg

Melting Point: 97 °C

Molecular Weight: 139.11

Dissociation Constants: pKa = 8.360 [13]

Log Octanol/Water Partition Coefficient: 2.00 [9]

Water Solubility: 13500 mg/L at 25 °C [11]

Vapor Pressure: 0.75 mm Hg at 20 °C [18]

Henry's Law Constant: 1.02×10^{-5} atm-m³/mole (calculated from the water solubility and vapor pressure)

ENVIRONMENTAL FATE/EXPOSURE POTENTIAL

Summary: 3-Nitrophenol may be released to the environment in wastewater and as fugitive emissions during its production and use

439

as a laboratory reagent. When released on soil, significant quantities may volatilize and biodegradation should be important. It should not be adsorbed strongly by the soil and, therefore, may leach into the ground water. If released into water, biodegradation may occur and will be particularly rapid if the resident microorganisms are well acclimated. Volatilization (half-life 4.2 days in a model river) and photolysis may be important in situations where microbial populations are low. Adsorption to sediment should be minor under most circumstances and little or no bioconcentration should take place. If released in air, it will be partially associated with particulate matter and be removed by gravitational settling, washout by rain, as well as removed by photolysis and vapor-phase reaction with photochemically produced hydroxyl radicals (estimated half-life 14 hr). Human exposure will be primarily from ambient air via inhalation.

Natural Sources:

Artificial Sources: 3-Nitrophenol is produced in only limited quantities [16] and no commercial use for the chemical could be found. Therefore, it is probably used primarily as a laboratory reagent and may be released to the environment in fugitive emissions or in wastewater during its production and use.

Terrestrial Fate: If released on soil, 3-nitrophenol would not be expected to adsorb appreciably to the soil. It should biodegrade, especially where acclimated populations of microorganisms are available. The experimental half-life in flooded soil (anaerobic conditions) is 6.3 days.

Aquatic Fate: If released into water, one would not expect 3-nitrophenol to sorb appreciably to sediment because of its low Koc. Both in the water and sediment, biodegradation should be an important loss mechanism. However, volatilization (half-life 4.2 day in a model river) and photolysis should be important in some environmental systems.

Atmospheric Fate: 3-Nitrophenol released to the atmosphere should be associated with suspended particulate matter or aerosols as well as exist in the vapor phase. It is fairly soluble in water and should be scavenged by rain. It should also be subject to gravitational settling and photolysis. The photolysis half-life will depend on the amount of sunlight, pH, and dissolved nitrate/nitrite in the aerosol water and the nature of the particulate matter on which the 3-nitrophenol is adsorbed, but no half-life estimate can be made with the data on hand.

Biodegradation: 3-Nitrophenol has been shown to be biodegradable from results of screening tests using sewage, activated sludge, and soil inocula [1,6,8,12,15,19]. The importance of acclimation is shown, for example, in the fact that 90% degradation was obtained with activated sludge in 16 days including a 10-day lag, while with acclimation only 6 days were required for 90% degradation [19]. It was established that no recalcitrant metabolites were formed in the course of biodegradation [7]. No die-away test results were found for 3-nitrophenol in water or soil under aerobic conditions. In several screening studies employing soil, wastewater, and sludge inocula, the biodegradation of the three nitrophenol isomers were compared. In all cases, 3-nitrophenol was the most readily degradable isomer [1,6,8]. Under anaerobic conditions, 3-nitrophenol was mineralized in 1 week when incubated with digester sludge [4]. In flooded alluvial and Pokkali acid sulfate soils, 72 and 67% degradation occurred in 10 days [14]. This corresponds to a half-life of 6.3 days or less. Nitrite accumulated only in the alluvial soil, suggesting that differences may exist in the degradative pathways in the two soils [14].

Abiotic Degradation: From the acid dissociation constant, 3-nitrophenol will exist to an appreciable extent as the anion in environmental waters. It has a maximum absorption at 328 nm in water [17] and is therefore a candidate for direct photolysis. No rates for photolysis could be found for 3-nitrophenol. However, susceptibility of nitrophenols to photolyze is seen in the results for 4-nitrophenol (half-life hr to 2 weeks in clear surface waters). As with 4-nitrophenol, it should also react with hydroxyl radicals that are photochemically produced in waters containing nitrate and nitrite ions. Vapor phase 3-nitrophenol will react with photochemically produced hydroxyl radicals by H-atom abstraction and ring addition with a half-life of 14 hr [5]. 3-Nitrophenol should be resistant to hydrolysis [10].

3-Nitrophenol

Bioconcentration: From the log octanol/water partition coefficient for 3-nitrophenol, one can estimate a BCF of 19 [10], indicating little tendency for bioconcentration in fish.

Soil Adsorption/Mobility: The Koc estimated from its water solubility is 23 [10], which suggests only a slight potential for adsorption by organic matter in soil. The Koc for 3-nitrophenol in Brookston clay loam was 53 [3]. 4-nitrophenol complexes with montmorillonite clay, which suggests that 3-nitrophenol might do likewise; however, experimental data are lacking.

Volatilization from Water/Soil: The Henry's Law constant for 3-nitrophenol can be used to estimate the volatilization half-life of 101 hr from a model river 1 m deep with a 1 m/sec current and a 3 m/sec wind [10]. Since it has a moderately low vapor pressure and low adsorption to soil, it should volatilize slowly from soil.

Water Concentrations: Abandoned creosote facility in Conroe, TX ND-43 ppb in one of the more polluted shallow wells [2].

Effluent Concentrations:

Sediment/Soil Concentrations:

Atmospheric Concentrations:

Food Survey Values:

Plant Concentrations:

Fish/Seafood Concentrations:

Animal Concentrations:

Milk Concentrations:

Other Environmental Concentrations:

Probable Routes of Human Exposure:

Average Daily Intake:

442

Occupational Exposures:

Body Burdens:

REFERENCES

1. Alexander M, Lustigman BK; J Agric Food Chem 14: 410-3 (1966)
2. Bedient PB et al; Ground Water 22: 318-29 (1984)
3. Boyd SA; Soil Science 134: 337-43 (1982)
4. Boyd SA et al; Appl Environ Microbiol 46: 50-4 (1983)
5. GEMS; Graphical Exposure Modeling System. FAP. Fate of Atmos Pollut (1986)
6. Gerike P, Fischer WK; Ecotox Environ Safety 3: 159-73 (1979)
7. Gerike P et al; Chemosphere 13: 121-41 (1984)
8. Haller HD; J Water Pollut Control Fed 50: 2771-7 (1978)
9. Hansch C, Leo AJ; Medchem Project Issue No 26. Claremont CA: Pomona College (1985)
10. Lyman WJ et al; Handbook of Chemical Property Estimation Methods. Environmental Behavior of Organic Compounds. McGraw-Hill NY (1982)
11. Matsuguma HJ; Kirk-Othmer Encycl Chem Tech 2nd ed., 13: 888-94 (1963)
12. Pitter P; Water Res 10: 231-5 (1976)
13. Serjeant EP, Dempsey B; Ionisation Constants of Organic Acids in Aqueous Solution, IUPAC Chemical Data Series No.23 pp 989 New York, NY (1979)
14. Sudhakar-Barik, Sethunathan N; J Environ Qual 7: 349-52 (1978)
15. Urushigawa Y et al; Kogai Shigen Kenkyusho Iho 12: 37-46 (1983)
16. USEPA; TSCA Production File (1986)
17. Weast RC; Handbook of Chem and Physics, 53rd ed. Cleveland, OH (1972)
18. Weber RC et al; Vapor Pressure Distribution of Selected Organic Chemicals pp 39 USEPA 600/2-81-021 (1981)
19. Zahn R, Wellens H; Z Wasser Abwasser Forsch 13: 1-7 (1980)

4-Nitrophenol

SUBSTANCE IDENTIFICATION

Synonyms: p-Nitrophenol

Structure:

CAS Registry Number: 100-02-7

Molecular Formula: $C_6H_5NO_3$

Wiswesser Line Notation: WNR DQ

CHEMICAL AND PHYSICAL PROPERTIES

Boiling Point: 279 °C (decomposes)

Melting Point: 113-114 °C

Molecular Weight: 139.11

Dissociation Constants: 7.156 [46]

Log Octanol/Water Partition Coefficient: 1.91 [20]

Water Solubility: 11,300 mg/L at 20 °C [44]; 25,000 mg/L at 25 °C [11]

Vapor Pressure: 0.001 mm Hg at 25 °C [44]

Henry's Law Constant: 3.31×10^{-8} atm-m³/mole [44]

ENVIRONMENTAL FATE/EXPOSURE POTENTIAL

Summary: 4-Nitrophenol will be released to the environment in wastewater and as fugitive emissions during its production and use as a chemical intermediate primarily in the manufacture of parathion, methyl parathion, and N-acetyl-p-aminophenol. It is also found in suspended particulate matter in the atmosphere, originating mostly from secondary photochemical reactions in the air and partly from emissions of vehicular exhaust gas. When released on land, it will biodegrade with a half-life of approx a day. If released in water, it will also primarily biodegrade with a half-life of approx 1-8 days. Half-lives will be markedly shortened when the microorganisms are well acclimated. Half-lives will be much longer in marine waters. Photolysis may also be important in clear surface waters (half-lives 2-14 days), being somewhat faster when the water is acidic or contains nitrate or nitrite ions. Adsorption to sediment should be minor under most circumstances and little or no bioconcentration should take place. If 4-nitrophenol is released into the atmosphere, it will be predominately adsorbed to particulate matter, be in aerosols, or in the vapor phase. In the vapor phase, 4-nitrophenol will react with photochemically produced hydroxyl radicals with a half-life of 14 hr. In particulate matter and in aerosols, it will be subject to gravitational settling and photolyze. Human exposure will be primarily from ambient air via inhalation. Agricultural workers using parathion may be exposed to 4-nitrophenol dermally or via inhalation, since it is both a degradation product and impurity in that pesticide.

Natural Sources:

Artificial Sources: 4-Nitrophenol may be released to the environment in fugitive emissions or in wastewater during its production and use as a chemical intermediate in the manufacture of methyl and ethyl parathion, N-acetyl-p-aminophenol, and dyestuffs, as well as a leather treatment agent [8]. It is a photooxidation product of nitrobenzene in air [17] and aromatic hydrocarbons such as benzene, toluene, and phenanthrene with nitric oxide in air [35]. It is emitted in vehicular exhaust from both gasoline and diesel engines [35]. 4-Nitrophenol is also a degradation product of parathion [24] and an impurity in the parathion formulation Thiophos and, therefore, will be released during the application of the insecticide [2].

Terrestrial Fate: If released on soil, 4-nitrophenol would not be expected to adsorb appreciably to soil except for some clays to which it will strongly bind. It will rapidly biodegrade, having a half-life of approx 1 day in agricultural topsoil and 10 days in flooded soil. Should 4-nitrophenol leach into the subsoil, its biodegradation half-life is much longer, 40 days in one report under aerobic conditions and longer still under anaerobic conditions. However, no reports could be found concerning its detection in ground water. When radiolabelled 4-nitrophenol was incubated in a laboratory terrestrial ecosystem, the label was found largely in the top 5 cm of soil after 26 days [16]. After 30 days in another terrestrial ecosystem, 79.1% of the ^{14}C resided in soil and plants and 19.1% had been transformed to $^{14}CO_2$ [12].

Aquatic Fate: If released into water, one would not expect 4-nitrophenol to sorb appreciably to sediment because of its low Koc, unless the sediment contains appreciable amounts of montmorillonite or other clay to which 4-nitrophenol forms chemical bonds. Both in the water and sediment, biodegradation will be the predominant loss mechanism. Half-lives which have been reported are 1-8 days. Half-lives decrease markedly when the microbial populations are acclimated and the 4-nitrophenol may then completely degrade in under a day. Degradation is much slower in marine systems than in freshwater ones, with half-lives of 1-3 yr being reported in water from sites in Pensacola Bay, FL. The presence of sediment in the water markedly increased the degradation rate, as the half-lives in ecocores from the same sites ranged 13-20 days. No degradation rates have been reported for freshwater sediment. In estuarine sediment suspensions, 7 and 19% mineralization occurred in 10 days under anaerobic and aerobic conditions, respectively. The mean half-life predicted for 4-nitrophenol by a non-steady-state equilibrium model is 7.7 days [60]. The model predicts that 94.6% of the 4-nitrophenol will partition into the water and 4.44% into the sediment [60]. Photolysis may also be important in clear surface waters (half-lives 2-14 days), being somewhat faster when the water is acidic or contains nitrate or nitrite ions.

Atmospheric Fate: 4-Nitrophenol released to the atmosphere will most likely be in suspended particulate matter or aerosols. It will be subject to gravitational settling and photolysis. The photolysis half-life will depend on the amount of sunlight, pH and dissolved nitrate in the aerosol water, and the nature of the particulate matter on which the 4-nitrophenol is adsorbed. Half-lives may range from hours to a week or more.

Biodegradation: SCREENING TESTS: 4-Nitrophenol is a benchmark chemical for biodegradability test and therefore there are numerous results on its behavior in screening tests. The results of the biodegradability screening studies are conflicting, ranging from no degradation to rapid degradation using soil, sewage, activated sludge, sediment, and freshwater inocula. The results are roughly divided between 4-nitrophenol biodegrading moderately slowly and rapidly; acclimation is generally important [1,10,15,19,25,33,36,38,39,42,43,49,56], while some of the conflicting results may be due to differences in concentrations of the test chemical, inocula, toxicity at higher concentrations [25], or insufficient acclimation. Many results were obtained in interlaboratory comparisons in which different laboratories report 0 and 100% degradation for the same test [36]. The importance of acclimation is illustrated in results of 100% degradation in 15 and 3 days without and with acclimation, respectively, the latter also at much higher concentration levels [10]. At very low concentration levels (ppt range), mineralization is achieved without a lag period [39]. 4-Nitrophenol biodegrades rapidly in simulated biological treatment plants, but only after adequate acclimation [10,15,33,59]. SOIL: When 2 ppm of 4-nitrophenol was incubated in agricultural topsoil at 10 °C, the half-life was 0.7-1.2 and 14 days under aerobic and anaerobic conditions, respectively [30]. In subsoil the half-life was 40 days under aerobic conditions when the soil moisture was at 79% field capacity [30]. Under anaerobic conditions, no degradation was observed after 90 days, but 97% of the 4-nitrophenol had disappeared in 201 days [30]. As the concn of 4-nitrophenol in Mardin silt loam soil was increased from 1 ppb to 50 ppm, the lag period increased and time for 10% mineralization increased from 5 hr to 40 hr [45]. The biodegradation of 4-nitrophenol in agricultural soil is greatly increased by pretreatment of the soil with 4-nitrophenol, and decreased by the presence of some fertilizers, fungicides, and herbicides [29]. The experiments which showed this involved monitoring the rate of mineralization of ring-labeled parathion in which 4-nitrophenol is an intermediate [29]. The addition of captafol, maneb, and benomyl reduced the mineralization after 3 weeks from 6.4-41% of controls, while PCNB had no effect [29].

Pretreatment of the soil with 4-nitrophenol increased the amount mineralized after 1 day from 2 to 34% [29]. Addition of 100 ppm-N ammonium sulfate and potassium nitrate reduced the 3 weeks mineralization 83 and 35%, respectively, although ammonium nitrate did not have a similar effect [29]. ANAEROBIC STUDIES: In anaerobic screening tests using digester sludge inocula, 4-nitrophenol completely disappeared in 1 week [5] and was 75-100% mineralized in 56 days [47]. When 4-nitrophenol was incubated with an estuarine sediment suspension from the Louisiana Gulf Coast, only 7% mineralization occurred under anaerobic conditions, while 19% mineralization took place in 10 days under aerobic conditions but then showed no further mineralization over the next 40 days [48]. In flooded alluvial and Pokkali acid sulfate soils, 51 and 70% degradation occurred in 10 days, respectively [55]. WATER/SEDIMENT TESTS: In die-away tests in which a chemical is added to a water sample from a site, 4-nitrophenol degrades rapidly with acclimation and half-lives decrease with increasing concns of 4-nitrophenol. Typical half-lives are: approx 1.5 days in the Vistula River in Warsaw, Poland for 5-20 ppm of chemical [10], 7.6 days, mean in 5 ponds [37]; 50% mineralization in 300 hr at 1 ppm increasing to 700 hr at 0.5 ppb including a 250-hr lag in a eutrophic lake [23]; and 58% mineralization in 14 days in an oligotropic lake for 1 ppb of chemical [39]. When 4-nitrophenol was incubated in freshwater/sediment eco-cores from 5 sites along the Escabia River in FL, 10-63% mineralization occurred in 160 hr including a 40-120-hr lag [51]. However, when the eco-cores were preexposed, there was no longer a lag and 12-58% mineralization occurred in 80 hr [51]. Similarly in eco-cores from a freshwater pond, 4-nitrophenol disappeared within 80 hours in previously treated cores and within 15 hr when treated a third time [49]. At the time of disappearance approx 40-50% of the nitrophenol had mineralized [49]. Mineralization was much slower in eco-cores from a salt marsh than from a river, 9-21% in 250 hr versus 46-49% in 100 hr in the freshwater system [50]. The half-life was 348-1225 days in salt water from three sites in Pensacola Bay [58], although the half-lives were decreased to 9-291 days by the addition of sediment from the site into the test flasks and to 13-20 days in eco-cores from the same sites [58]. When initially dosed, 4-nitrophenol disappeared slowly from a laboratory microcosm (100% disappearance in 7.5 days) and a test pond; however, on redose 10 days after initial treatment, it disappeared within a day [49]. Nine months after the study, the microbial community in the pond had apparently reverted, as an acclimation period was again required before the 4-nitrophenol degraded [49].

Abiotic Degradation: The acid dissociation constant of 4-nitrophenol suggests that it will exist to an appreciable extent as the anion in environmental waters. It has several absorption bands >290 nm and a maximum absorption of 316 nm in neutral solutions and the absorption >400 nm increases markedly in basic solutions [28,32]. When 4-nitrophenol in water was exposed to sunlight, the half-life was 5.7, 6.7, and 13.7 days at pH 5, 7, and 9, respectively [22]. Half-lives in water were calculated to range from 18 to 22 hr in Chicago under a noontime midsummer sun [28]. The results of quantum yield determinations by 13 laboratories predicts a yearly avg surface half-life of 4-nitrophenol ranging from 3.1 hr at pH 4 to 5 hr at pH 9 exposed to midday sun at 40 to 50 deg N latitude [27]. Photolysis half-lives in a typical pond may be 100 to 300 times longer [27]. Photodecomposition in deionized water was completed within 1-2 mo yielding 4-nitrocatechol, hydroquinone, and a dark polymer [34]. 39% of the 4-nitrophenol adsorbed on silica gel was mineralized to CO_2 in 17 hr when exposed to 290 nm radiation [13]. The presence of substances like nitrite and nitrate ions in the water result in the formation of hydroxyl radicals under illumination, which can react with 4-nitrophenol [26] to form 1,4-hydroquinone and 1,4-benzoquinone [54]. In laboratory experiments, the half-life which was 16 hr for direct photolysis was reduced to 1.2 and 3.5 hr in the presence of nitrite and nitrate ions [26]. Vapor phase 4-nitrophenol will react with photochemically produced hydroxyl radicals by H-atom abstraction and ring addition with a half-life of 14 hr [14]. 4-Nitrophenol is resistant to hydrolysis [32].

Bioconcentration: The BCF of 4-nitrophenol in fathead minnow is 79 [6] and in golden orfe 58 [13], as determined in 28- and 3-day tests, respectively. Little or no bioconcentration in fish is reported in bioconcentration tests of the Japanese Ministry of Industry and Trade (MITI) [41]. Bioconcentration in green algae (Chlorella fusca) is only 11 [13].

4-Nitrophenol

Soil Adsorption/Mobility: The Koc estimated from its water solubility (25 g/L [11]) is 21 [31] which suggests only a slight potential for adsorption by organic matter in soil. The Koc for 4-nitrophenol in Brookston clay loam was 55 [4] and the soil/water distribution coefficient ranged from 0.6 to 1.6 in 5 clay soils with organic carbon concentrations ranging from 0.3 to 0.9% [3]. 4-Nitrophenol forms strong complexes with montmorillonite clays [40]. 50% of radiolabelled 4-nitrophenol was adsorbed from a water column onto lake sediment in 5 days and little of the sorbed 4-nitrophenol was subsequently released [48]. The experimental procedure did not say what the composition of the sediment was other than that it contained 5.1% organic matter [48]. A possible explanation for these contradictory results is that the sediment was largely montmorillonite or other clay to which 4-nitrophenol formed chemical bonds. When 4-nitrophenol was applied to a synthetic soil medium consisting of illite clay, sea sand, and a commercial mix in a laboratory model ecosystem, it remained in the top 5 cm of soil [16].

Volatilization from Water/Soil: The Henry's Law constant for 4-nitrophenol would suggest that the chemical is essentially non-volatile [31].

Water Concentrations: SURFACE WATER: Ambient water concentrations reported in the USEPA STORET database, 1980-82 (807 samples) 0.0% pos [53]. Not detected in the Lake Erie or Lake Michigan basin [18].

Effluent Concentrations: Effluent concentrations reported in the USEPA STORET database, 1980-1982 (1322 samples) 3.3% pos [53]. Norway - 30 samples representing polluted fjord areas as well as effluent from municipal treatment plants, refineries, petrochemical, and metallurgic industries: 0% pos at a 5 ppb limit of quantitation [52]. Detected in the treated effluents of the following industries (industry (concn)): electrical/electronic components (<22 ppb mean, 35 ppb max); organic chemicals manufacturing/plastics (190 ppb max); petroleum refining (<10 ppb max); and textile mills (<10 ppb max) [57]. Additionally, the raw wastewater of the following industries not listed above contained 4-nitrophenol (industry (concn)): auto and other laundries (14 ppb mean); aluminum forming (18 ppb max); metal finishing (10 ppb max); and photographic equipment/supplies (57 ppb max) [57]. National Urban Runoff Program in which 19 cities and metropolitan councils across the United States (51 catchments) were sampled - detected in Long Island, NY, Washington, DC, Little Rock, AK, and Eugene, OR in the range 1-19 ppb, 9% frequency of detection [9]. Not detected in the effluent from 4 Southern Californian municipal wastewater treatment plants [61].

Sediment/Soil Concentrations: Sediment concentrations reported in the USEPA STORET database, 1980-82 (301 samples) 2% pos [53]. Detected in water/soil/sediment samples from the Love Canal [21].

Atmospheric Concentrations: Yokahama, Japan where photochemical smog has frequently been observed 5.1-42 ppm in suspended particulates in 24 hr samples [35].

Food Survey Values: Lettuce sprayed with parathion (0.5 lb/acre) contained 0.061 ppm 4-nitrophenol at harvest on the trimmed head [2].

Plant Concentrations: Lettuce sprayed with parathion (0.5 lb/acre) contained 0.071 and 0.045 ppm 4-nitrophenol after 0 and 7 days, respectively [2].

Fish/Seafood Concentrations: Biota concn reported in the USEPA STORET database, 1980-82 (101 samples) 0% pos [53].

Animal Concentrations:

Milk Concentrations:

Other Environmental Concentrations: Exhaust gas from gasoline and diesel engines contained trace amounts and 2.5 ppb of 4-nitrophenol, respectively [35].

Probable Routes of Human Exposure: Human exposure will be primarily from ambient air, especially in urban areas via inhalation. Agricultural workers using parathion may be exposed to 4-nitrophenol dermally or via inhalation since it is both a degradation product of and impurity in that pesticide.

4-Nitrophenol

Average Daily Intake:

Occupational Exposures:

Body Burdens: Health and Nutrition Examination Survey II (6990 samples collected from the general U.S. population, 1976-1980) 2.4% pos in urine [7]. People had been exposed to methyl and ethyl parathion [7].

REFERENCES

1. Alexander M, Lustigman BK; J Agric Food Chem 14: 410-3 (1966)
2. Archer TE; J Agric Food Chem 23: 858-60 (1975)
3. Artiola-Fortuny J, Fuller WH; Soil Sci 133: 218-27 (1982)
4. Boyd SA; Soil Science 134: 337-43 (1982)
5. Boyd SA et al; Appl Environ Microbiol 46: 50-4 (1983)
6. Call DJ et al; Arch Environ Contam Toxicol 9: 699-714 (1980)
7. Carey AE, Kutz FW; Environ Assess 5: 155-63 (1985)
8. Chemical Marketing Reporter; Chemical Profiles Aug 6, (1984)
9. Cole RH et al; J Water Pollut Control Fed 56: 898-908 (1984)
10. Dojlido JR; Investigations of Biodeg and Toxicity of Organic Compound pp 118 USEPA 600/2-79-263 (1979)
11. Duff JC; J Chem Soc 1929: 2789-96 (1929)
12. Figge K et al; Regul Toxicol Pharmacol 3: 199-215 (1983)
13. Freitag D et al; Ecotox Environ Safety 6: 60-81 (1982)
14. GEMS; Graphical Exposure Modeling System. FAP. Fate of Atmos Pollut (1986)
15. Gerike P, Fischer WK; Ecotox Environ Safety 3: 159-73 (1979)
16. Gile JD, Gillett JW; J Agric Food Chem 2: 616-21 (1981)
17. Graedel TE; Chem Compounds in the Atmos, Academic Press NY p 297 (1978)
18. Great Lakes Water Quality Board; An Inventory of Chemical Substances Identified in the Great Lakes Ecosystem 1: 195 (1983)
19. Haller HD; J Water Pollut Control Fed 50: 2771-7 (1978)
20. Hansch C, Leo AJ; Medchem Project Issue No 26. Claremont CA: Pomona College (1985)
21. Hauser TR, Bromberg SM; Environ Monit Assess 2: 249-72 (1982)
22. Hustert K et al; Chemosphere 10: 995-8 (1981)
23. Jones SH, Alexander M; Appl Environ Microbiol 51: 891-7 (1986)
24. Kirk-Othmer; Encycl Chem Tech 3rd ed New York Wiley 13: 438 (1981)
25. Kool HJ; Chemosphere 13: 751-61 (1984)
26. Kotzias D et al; Naturwissenschaften 69: 444-5 (1982)
27. Lamaire J et al; Chemosphere 14: 53-77 (1985)
28. Lamaire J et al; Chemosphere 11: 119-64 (1982)
29. Lichtenstein E; Fate of Persistence of (14) C Residues in Different Soils, Internatl Atomic Energy Agency, Vienna IAEA-R-2032-F, pp 1-20 NTIS DE83704120 (1982)

4-Nitrophenol

30. Loekke H; Environ Pollut Ser A 38: 171-81 (1985)
31. Lyman WJ et al; Handbook of Chemical Property Estimation Methods. Environmental Behavior of Organic Compounds. McGraw-Hill NY (1982)
32. Mabey WR et al; Aquatic Fate Process Data for Organic Priority Pollut pp 434 USEPA 440/4-81-014 (1981)
33. Means JL, Anderson SJ; Water Air Soil Poll 16: 301-15 (1981)
34. Nakagawa M, Crosby DG; J Agric Food Chem 22: 849-53 (1974)
35. Nojima K et al; Chem Pharm Bull 31: 1047-51 (1983)
36. Organ Econ Coop Devel; OECD Chemical Testing Programme, Expert Group Degradation/Accumulation 1: 141 (1979)
37. Paris DF et al; Appl Environ Microbiol 45: 1153-5 (1983)
38. Rott B et al; Chemosphere 11: 531-8 (1982)
39. Rubin HE et al; Appl Environ Microbiol 43: 1133-8 (1982)
40. Saltzman S, Yariv S; Soil Sci Soc Amer Proc 39: 474-9 (1975)
41. Sasaki S; pp 283-98 in Aquatic Pollut, Transformation and Biological Effects, Hutzinger O, Letyoeld LH, and Zoeteman BCJ, Eds (1978)
42. Schefer W, Waelchli O; Z Wasser Abwasser Forsch 13: 205-9 (1980)
43. Schmidt-Bleek F et al; Chemosphere 11: 383-415 (1982)
44. Schwarzenbach RP et al; Environ Sci Technol 22: 83-92 (1988)
45. Scow KM et al; Appl Environ Microbiol 51: 1028-35 (1986)
46. Serjeant EP, Dempsey B; Ionization Constants of Organic Acids in Aqueous Solution, IUPAC Chemical Data Series pp 989 No.23 (1979)
47. Shelton DR, Tiedje JM; Appl Environ Microbiol 47: 850-7 (1984)
48. Siragusa GR, Deluane RD; Environ Toxicol Chem 5: 175-8 (1986)
49. Spain JC et al; Appl Environ Microbiol 48: 944-50 (1984)
50. Spain JC et al; Appl Environ Microbiol 40: 726-34 (1980)
51. Spain JC, Van Veld PA; Appl Environ Microbiol 45: 428-35 (1983)
52. Sporstoel S et al; Int J Environ Anal Chem 21: 129-38 (1985)
53. Staples CE et al; Environ Toxicol Chem 4: 131-42 (1985)
54. Suarez C et al; Tetrahedron Lett 8: 575-8 (1970)
55. Sudhakar-Barik, Sethunathan N; J Environ Qual 7: 349-52 (1978)
56. Urushigawa Y et al; Kogai Shigen Kenkyusho Iho 12: 37-46 (1983)
57. USEPA; Treatability Manual 1: I.8.7-1 to I.8.7-4 USEPA-600/8-80-042 (1980)
58. Van Veld PA, Spain JC; Chemosphere 12: 1291-1305 (1983)
59. Wilderer P; AICHE Symp Ser 77: 205-13 (1981)
60. Yoshida K et al; Ecotox Environ Safety 7: 179-90 (1983)
61. Young DR; Priority Pollut in Municipal Wastewaters, Annu Rep South Calif Coastal Water Res Proj pp 103-12 (1978)

2-Nitrotoluene

SUBSTANCE IDENTIFICATION

Synonyms: o-Nitrotoluene; 1-Methyl-2-nitrobenzene

Structure:

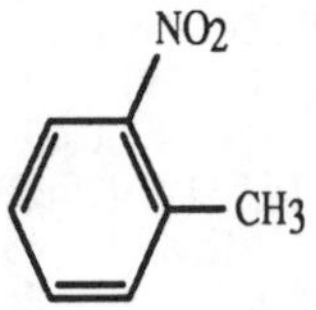

CAS Registry Number: 88-72-2

Molecular Formula: $C_7H_7NO_2$

Wiswesser Line Notation: WNR B1

CHEMICAL AND PHYSICAL PROPERTIES

Boiling Point: 222 °C

Melting Point: -10 °C

Molecular Weight: 137.13

Dissociation Constants:

Log Octanol/Water Partition Coefficient: 2.30 [5]

Water Solubility: 652 mg/L at 30 °C [24]

Vapor Pressure: 0.1 mm Hg at 20 °C [24]

Henry's Law Constant: 5.5 x 10^{-5} atm-m³/mole at 25 °C [6]

ENVIRONMENTAL FATE/EXPOSURE POTENTIAL

Summary: The major sources of release of 2-nitrotoluene to the environment appears to be production and use facilities and plants which produce this compound as a by-product. This would include

454

manufacturers of dinitrotoluene, trinitrotoluene, intermediates for rubber and agricultural chemicals and various azo and sulfur dye intermediates such as 2-toluidine, 2-nitrobenzaldehyde, and 2-nitro-4-chlorotoluene. If released to soil, 2-nitrotoluene should be resistant to oxidation and chemical hydrolysis. One study on 2-nitrotoluene under aerobic conditions in a mixed culture of soil microorganisms in aqueous mineral salts media resulted in persistence of >64 days, suggesting that biodegradation may not be important in soil. 2-Nitrotoluene is expected to be moderately to highly mobile in soil and volatilize slowly from dry soil surfaces. If released to water, 2-nitrotoluene would be susceptible to direct photolysis, indirect photolysis (half-life <1 hr in river water containing a high concn of humic substances), volatilization (estimated half-life 21 hr in water 1 m deep flowing 1 m/sec with a wind speed of 3 m/sec), and possibly aerobic biodegradation provided suitable acclimation has taken place. Oxidation, chemical hydrolysis, adsorption to suspended solids and sediments, and bioaccumulation in aquatic organisms are not expected to be significant fate processes. Based on monitoring data, the half-life of 2-nitrotoluene in a river 4 to 5 m deep has been estimated to be 3.2 days. If released to the atmosphere, 2-nitrotoluene is expected to exist entirely in the vapor phase. The dominant removal mechanisms would be direct photolysis (half-life <5 hr) and reaction with photochemically generated hydroxyl radicals (estimated half-life 8 hr). 2-Methyl-6-nitrophenol and 2-methyl-4-nitrophenol are photoproducts of 2-nitrotoluene. The most probable routes of human exposure to 2-nitrotoluene are inhalation and dermal contact of workers involved in the production and use of this compound, dinitrotoluenes, and trinitrotoluene.

Natural Sources:

Artificial Sources: In general, the major sources of release of nitroaromatic compounds to the environment appears to be production and use facilities and plants which produce these compounds as by-products [7]. 2-Nitrotoluene may be released to the environment by manufacturers of dinitrotoluene, trinitrotoluene, intermediates for rubber and agricultural chemicals, and various azo and sulfur dye intermediates such as 2-toluidine, 2-nitrobenzaldehyde, 2-nitro-4-chlorotoluene, 2-nitro-6-chlorotoluene,

2-amino-4-chlorotoluene, and 2-amino-6-chlorotoluene [3]. 2-Nitrotoluene may also enter the environment from the disposal of waste products which contain this compound.

Terrestrial Fate: If released to soil, 2-nitrotoluene should be resistant to oxidation and chemical hydrolysis. One study on 2-nitrotoluene under aerobic conditions in a mixed culture of soil microorganisms in aqueous mineral salts media resulted in persistence of >64 days, suggesting that biodegradation may not be significant. 2-Nitrotoluene is predicted to be moderately to highly mobile in soil and volatilize slowly from dry soil surfaces.

Aquatic Fate: If released to water, 2-nitrotoluene would be susceptible to direct photolysis, indirect photolysis (half-life <1 hr in river water containing a high concn of humic substances), volatilization (estimated half-life 21 hr in water 1 m deep flowing 1 m/sec with a wind speed of 3 m/sec), and possibly aerobic biodegradation provided suitable acclimation has taken place. Oxidation, chemical hydrolysis, adsorption to suspended solids and sediments, and bioaccumulation in aquatic organisms are not expected to be significant fate processes. Based on monitoring data, the half-life of 2-nitrotoluene in a river 4 to 5 m deep has been estimated to be 3.2 days [26].

Atmospheric Fate: If released to the atmosphere, 2-nitrotoluene is expected to exist entirely in the vapor phase. The dominant removal mechanisms would be reaction with photochemically generated hydroxyl radicals (estimated half-life 8 hr) and direct photolysis. 2-Methyl-6-nitrophenol and 2-methyl-4-nitrophenol are photoproducts of 2-nitrotoluene.

Biodegradation: 100 ppm 2-nitrotoluene inoculated with 30 ppm activated sludge under aerobic conditions at 25 °C was <30% degraded after 2 weeks [8,20]. 2-Nitrotoluene (200 mg/L COD) inoculated with adapted activated sludge under aerobic conditions at 20 °C underwent 98% degradation in 5 days as measured by COD removal [18]. 2-Nitrotoluene should be degraded by biological sewage treatment provided suitable acclimation can be achieved [23]. 10 ug/L 2-nitrotoluene inoculated with a mixed culture of soil microorganisms under aerobic conditions in aqueous mineral salts media persisted >64 days as measured by UV absorbency [1]. In general, anaerobic biodegradation of nitroaromatic compounds results in the reduction of the nitro group to an amino group [12].

Abiotic Degradation: Chemical hydrolysis and oxidation of 2-nitrotoluene are not expected to be important removal processes since this compound contains no functional groups which are susceptible to these types of reactions [11]. Absorption of UV light in the environmentally significant range (>290 nm) by 2-nitrotoluene in cyclohexane [19] indicates that the potential exists for photolysis in water and air. Irradiation (at >300 nm) of 2-nitrotoluene vapor in air for 5 hours resulted in 79% loss of the 2-nitrotoluene initially present and formation of 2-methyl-6-nitrophenol (6.1% yield) and 2-methyl-4-nitrophenol (7.5% yield) [15]. 2-Nitrotoluene in triethylamine underwent 100% conversion when irradiated (wavelengths >290 nm) for 3 hours, forming aniline (32%), azoxybenzene (5%), azobenzene (2.5%), and 2-hydroxyazo compound (10%) [2]. The midday half-life of 2-nitrotoluene in Aucilla River water due to indirect photolysis has been calculated to be 45 minutes using an experimentally determined reaction rate constant of 0.92 hr^{-1} [25] (Aucilla River water contains a high concn of humic substances, which appears to act as a photosensitizer [25]). The half-life for 2-nitrotoluene vapor reacting with photochemically generated hydroxyl radicals in the atmosphere has been estimated to be 8.0 hr based on a reaction rate constant of 3.0 x 10^{-11} cm^3/molecules-sec at 25 °C and ambient hydroxyl radical concn of 8.0 x 10^{-5} molecules/cm^3 [4].

Bioconcentration: The bioconcentration factor (BCF) for 2-nitrotoluene has been experimentally determined to be less than 100 in carp (Carprinus carpio) [8,20]. BCF values of 33 and 16 have been calculated based on the log octanol/water partition coefficient and the measured water solubility [11]. These BCF values suggest that 2-nitrotoluene will not bioaccumulate significantly in aquatic organisms.

Soil Adsorption/Mobility: The soil adsorption coefficient for 2-nitrotoluene has been calculated to be 425 and 124 based on the log octanol/water partition coefficient and the measured water solubility, respectively [11]. These Koc values suggest that

2-Nitrotoluene

2-nitrotoluene would be moderately to highly mobile in soil and would adsorb slightly to suspended solids and sediments in water [22].

Volatilization from Water/Soil: The volatilization half-life of 2-nitrotoluene from water 1 m deep, flowing 1 m/sec with a wind speed of 3 m/sec has been calculated to be 21 hr based on the measured Henry's Law constant [11]. The vapor pressure for 2-nitrotoluene suggests that this compound will volatilize slowly from dry soil surfaces.

Water Concentrations: SURFACE WATER: During 1974, 2- and 4-nitrotoluene were detected in the Waal River (Netherlands), avg concn 4.5 ug/L, max concn 18.1 ug/L, and in the Maas River (Netherlands), max concn 0.3 ug/L [13]. 2-Nitrotoluene has been detected in Rhine River water at a concn of 10 ug/L [17]. DRINKING WATER: 2-Nitrotoluene has been qualitatively identified in German drinking water [9].

Effluent Concentrations: 2-Nitrotoluene has been detected in the effluent from a plant manufacturing trinitrotoluene in Radford, VA, 0.32-16 mg/L detected [7]. 2-Nitrotoluene has been detected in the wastewater resulting from the production and purification of 2,4,6-trinitrotoluene, 0.02-0.14 mg/L detected, 38% samples pos. [21]. 2-Nitrotoluene has also been detected in raw effluent from a plant manufacturing dinitrotoluene, 7.8 mg/L detected, and in a waste treatment lagoon of a paper mill [7].

Sediment/Soil Concentrations:

Atmospheric Concentrations: 2-Nitrotoluene has been detected in the ambient air at the Du Pont plant in Deepwater, NJ at a concn of 47 ng/m^3 [16].

Food Survey Values:

Plant Concentrations:

Fish/Seafood Concentrations:

Animal Concentrations:

Milk Concentrations:

Other Environmental Concentrations:

Probable Routes of Human Exposure: The most probable route of human exposure to 2-nitrotoluene are inhalation and dermal contact of workers involved in the production and use of this compound, dinitrotoluenes, and trinitrotoluene.

Average Daily Intake:

Occupational Exposures: A National Occupational Hazard Survey (1973-74) estimates that 285 workers are exposed to 2-nitrotoluene [14].

Body Burdens: Under controlled experimental conditions, 2-nitrotoluene was detected in human expired air, 387 samples from 54 people, 19.1% samples pos, 0.04 ng/L mean concn [10].

REFERENCES

1. Alexander M, Lustigman BK; J Agric Food Chem 14: 410-3 (1966)
2. Barltrop JA, Bunce NJ; J Chem Soc C: 1467-74 (1968)
3. Dunlap KL; Kirk-Othmer Encycl Chem Technol 3rd ed Vol 15 (1981)
4. GEMS; Graphical Exposure Modeling System. FAP. Fate of Atmos Pollut (1986)
5. Hansch C, Leo AJ; Medchem Project Issue No 26 Pomona College Claremont CA (1985)
6. Hine J, Mookerjee PK; J Org Chem 40: 292-7 (1975)
7. Howard PH et al; Investigation of Selected Potential Environ Contam, Nitroaromatics USEPA 560/2-76-010 (1976)
8. Kawasaki M; Ecotox Env Safety 4: 444-54 (1980)
9. Kool HJ et al; Criteria Rev Env Control 12: 307-57 (1982)
10. Krotoszynski BW et al; J Anal Tox 3: 225-30 (1979)
11. Lyman WJ et al; Handbook of Chemical Property Estimation Methods. Environmental Behavior of Organic Compounds. McGraw-Hill NY (1982)
12. McCormick NG et al; Appl Environ Microb 31: 949-58 (1976)
13. Meijers AP, Van Der Leer CR; Water Res 597-604 (1976)
14. NIOSH; National Occupational Hazard Survey (1974)
15. Nojima K, Kanno S; Chemosphere 6: 371-6 (1977)
16. Pellizari ED; Quantification of Chlorinated Hydrocarbons in Previously Collected Air Samples USEPA 450/3-78-112 (1978)
17. Piet GJ, Morra CF; pp 31-42 in Artificial Groundwater Recharge, Huisman L, Olsthorn Tn eds (1983)

2-Nitrotoluene

18. Pitter P; Water Res 10: 231-5 (1976)
19. Sadtler Research Lab; Sadtler Standard UV Specta no 1292 (1961)
20. Sasaki S; pp 283-98 in Aquatic Pollut, Transformation and Bio Effects, Hutzinger O, Van Leytoeld LH, Zoeteman BCJ eds (1978)
21. Spanggord RJ et al; Environ Sci Tech 16: 229-32 (1982)
22. Swann RL et al; Res Rev 85: 17-28 (1983)
23. Thom NS, Agg AR; Proc R Soc Lond B 189: 347-57 (1975)
24. Verschueren K; Handbook of Environmental Data on Organic Chemicals. 2nd ed Van Nostrand Reinhold NY (1983)
25. Zepp RG et al; Fresenius Z Anal Chem 319: 119-25 (1999)
26. Zoeteman et al; Chemosphere 9: 231-49 (1980)

4-Nitrotoluene

SUBSTANCE IDENTIFICATION

Synonyms: p-Nitrotoluene; 1-Methyl-4-nitrotoluene

Structure:

$$H_3C \text{—} \bigcirc \text{—} NO_2$$

CAS Registry Number: 99-99-0

Molecular Formula: $C_7H_7NO_2$

Wiswesser Line Notation: WNR D1

CHEMICAL AND PHYSICAL PROPERTIES

Boiling Point: 238.3 °C at 760 mm Hg

Melting Point: 53-54 °C

Molecular Weight: 137.14

Dissociation Constants:

Log Octanol/Water Partition Coefficient: 2.37 [5]

Water Solubility: 442 mg/L at 30 °C [21]

Vapor Pressure: 0.1 mm Hg at 20 °C [21]; 0.22 mm Hg at 30 °C [21]

Henry's Law Constant: 5×10^{-5} atm-m³/mole (calculated by structural contributions) [6]

461

ENVIRONMENTAL FATE/EXPOSURE POTENTIAL

Summary: The major sources of release of 4-nitrotoluene to the environment appear to be production and use facilities and plants which produce this compound as a by-product. This would include manufacturers of dinitrotoluene, trinitrotoluene, and azo and sulfur dye intermediates such as 4-toluidine, 4-nitrobenzaldehyde, and 4-nitro-2-chlorotoluene. 4-Nitrotoluene may also enter the environment from the disposal of waste products containing 4-nitrotoluene. If released to soil, 4-nitrotoluene should be resistant to oxidation and chemical hydrolysis. This compound is reported to biodegrade under anaerobic conditions to form toluidine, and one study on 4-nitrotoluene under aerobic conditions in mineral salts and a mixed culture of soil microorganisms resulted in persistence of >64 days. 4-Nitrotoluene is expected to be moderately to highly mobile in soil and to volatilize slowly from dry soil surfaces. If released to water, 4-nitrotoluene would be susceptible to photolysis, volatilization (estimated half-life 25 hours in a model river 1 m deep flowing 1 m/sec with a wind speed of 3 m/sec), and possibly aerobic biodegradation provided suitable acclimation has taken place. Oxidation, chemical hydrolysis, adsorption to suspended solids and sediments, and bioaccumulation in aquatic organisms are not expected to be significant aquatic fate processes. Toluidine has been identified as an anaerobic biodegradation product of 4-nitrotoluene; however, insufficient data are available to indicate the significance of anaerobic biodegradation as a possible removal mechanism. Based on monitoring data, the half-life of 4-nitrotoluene in a river 4 to 5 m deep has been estimated to be 2.7 days. If released to the atmosphere, 4-nitrotoluene is expected to exist entirely in the vapor phase. The dominant removal mechanisms would be reaction with photochemically generated hydroxyl radicals (estimated half-life 8 hours) and direct photolysis. 4-Methyl-2-nitrophenol is a photoproduct of 4-nitrotoluene. The most probable routes of human exposure to 4-nitrotoluene are inhalation and dermal contact of workers involved in the production and use of this compound, dinitrotoluene, and trinitrotoluene.

Natural Sources:

4-Nitrotoluene

Artificial Sources: In general, the major sources of release of nitroaromatic compounds to the environment appear to be production and use facilities and plants which produce these compounds as by-products [7]. 4-Nitrotoluene may be released to the environment by manufacturers of dinitrotoluene, trinitrotoluene, and various azo and sulfur dye intermediates such as 4-toluidine, 4-nitrobenzaldehyde, and 4-nitro-2-chlorotoluene [2]. 4-Nitrotoluene may also enter the environment from the disposal of waste products in which it is contained.

Terrestrial Fate: If released to soil, 4-nitrotoluene should be resistant to oxidation and chemical hydrolysis. This compound is reported to biodegrade under anaerobic conditions to form toluidine, and one study on 4-nitrotoluene under aerobic conditions in mineral salts and mixed culture of soil microorganisms resulted in persistence >64 days. 4-Nitrotoluene is expected to be moderately to highly mobile in soil and volatilize slowly from dry soil surfaces. 0.5 ug/L 4-nitrotoluene was detected in Rhine River water before infiltration and less than 0.01 ug/L was detected after bank filtration (filtration time 1-12 months) and after dune filtration (filtration time 2-3 months) [22].

Aquatic Fate: If released to water, 4-nitrotoluene would be susceptible to photolysis, volatilization (estimated half-life 25 hours in a model river 1 m deep flowing 1 m/sec with a wind speed of 3 m/sec), and possibly aerobic biodegradation provided suitable acclimation has taken place. Oxidation, chemical hydrolysis, adsorption to suspended solids and sediments, and bioaccumulation in aquatic organisms are not expected to be significant fate processes. Toluidine has been identified as an anaerobic biodegradation product of 4-nitrotoluene; however, insufficient data are available to indicate the significance of anaerobic biodegradation as a possible removal mechanism. Based on monitoring data, the half-life of 4-nitrotoluene in a river 4 to 5 m deep has been estimated to be 2.7 days [22].

Atmospheric Fate: If released to the atmosphere, 4-nitrotoluene is expected to exist mostly in the vapor phase. The dominant removal mechanisms would be reaction with photochemically generated hydroxyl radicals (estimated half-life 8 hours) and direct photolysis. 4-Methyl-2-nitrophenol is a photoproduct of 4-nitrotoluene.

4-Nitrotoluene

Biodegradation: 100 ppm 4-nitrotoluene inoculated with 30 ppm activated sludge under aerobic conditions at 25 °C was <30% degraded after 2 week [8,17]. 4-Nitrotoluene (200 mg/L COD) inoculated with adapted activated sludge under aerobic conditions at 20 °C underwent 98% degradation in 5 days as measured by COD removal [15]. 4-Nitrotoluene should be degraded by biological sewage treatment provided suitable acclimation can be achieved [20]. 10 ug/L 4-nitrotoluene inoculated in mineral salts and a mixed culture of soil microorganisms under aerobic conditions persisted >64 days as measured by UV absorbancy [1]. In general, anaerobic biodegradation of nitroaromatic compounds results in the reduction of the nitro group to an amino group [10]. Toluidine has been identified as an anaerobic biodegradation product of 4-nitrotoluene [4].

Abiotic Degradation: Chemical hydrolysis and oxidation of 4-nitrotoluene are not expected to be important removal processes since this compound contains no functional groups which are susceptible to these types of reactions [9]. Absorption of UV light in the environmentally significant range (>290 nm) by 4-nitrotoluene in cyclohexane [16] indicates that the potential exists for photolysis in water and air. Irradiation (at >300 nm) of 4-nitrotoluene vapor in air for 5 hours resulted in 38% loss of 4-nitrotoluene initially present and formation of 4-methyl-2-nitrophenol (6.1% yield) [12]. The half-life for 4-nitrotoluene vapor reacting with photochemically generated hydroxyl radicals in the atmosphere has been estimated to be 8.0 hours based on a reaction rate constant of 3.0×10^{-11} cm^3/molecules-sec at 25 °C and an ambient hydroxyl radical concn of 8.0×10^{-5} molecules/cm^3 [3].

Bioconcentration: The bioconcentration factor for 4-nitrotoluene has been experimently measured to be less than 100 in carp (Carprinus carpio) [8,17]. BCF values of 37 and 20 [9] have been calculated based on a log octanol/water partition coefficient of 2.37 [5] and a measured water solubility of 442 mg/L at 30 °C [21]. These BCF values suggest that 4-nitrotoluene will not bioaccumulate significantly in aquatic organisms.

4-Nitrotoluene

Soil Adsorption/Mobility: The soil adsorption coefficient for 4-nitrotoluene has been calculated to be 464 and 153 [9] based on the log octanol/water partition coefficient and the measured water solubility. These Koc values suggest that 4-nitrotoluene would be moderately to highly mobile in soil and would adsorb slightly to suspended solids and sediments in water [19].

Volatilization from Water/Soil: Based on the Henry's Law constant, the volatilization half-life of 4-nitrotoluene from a model river 1 m deep, flowing 1 m/sec with a wind speed of 3 m/sec has been calculated to be 25 hours [9]. A vapor pressure of 0.1 mm Hg at 20 °C [21] for 4-nitrotoluene suggests that this compound will volatilize slowly from dry soil surfaces.

Water Concentrations: During 1974, 2- and 4-nitrotoluene were detected in the river Waal (Netherlands), avg concn 4.5 ug/L, max concn 18.1 ug/L, and in the river Maas (Netherlands), max concn 0.3 ug/L [11]. 4-Nitrotoluene has been detected in Rhine River water at a concn of 10 ug/L [14].

Effluent Concentrations: 4-Nitrotoluene has been found in the effluent from a plant manufacturing trinitrotoluene (TNT) in Radford, VA, 0.12 to 9.2 mg/L detected [7]. 4-Nitrotoluene has been detected in the wastewater resulting from the production and purification of 2,4,6-trinitrotoluene, 0.01 to 0.17 mg/L detected, 43% of samples pos [18]. 4-Nitrotoluene has also been detected in the raw effluent from a plant manufacturing dinitrotoluene, 8.8 mg/L detected, and in a waste treatment lagoon of a chemical company, 0.04 mg/L [7].

Sediment/Soil Concentrations:

Atmospheric Concentrations: 4-Nitrotoluene has been detected in the ambient air at the Du Pont plant in Deepwater, NJ at a concn ranging from 59 to 89 ng/m^3 [13].

Food Survey Values:

Plant Concentrations:

Fish/Seafood Concentrations:

465

Animal Concentrations:

Milk Concentrations:

Other Environmental Concentrations:

Probable Routes of Human Exposure: The most probable routes of human exposure to 4-nitrotoluene are inhalation and dermal contact of workers involved in the production and use of this compound, dinitrotoluene, and trinitrotoluene.

Average Daily Intake:

Occupational Exposures:

Body Burdens:

REFERENCES

1. Alexander M, Lustigman BK; J Agric Food Chem 14: 410-3 (1966)
2. Dunlap KL; Kirk-Othmer Encycl of Chem Technol 3rd ed Vol.15 p 930 (1981)
3. GEMS; Graphical Exposure Modeling System. FAP. Fate of Atmos Pollut (1986)
4. Hallas LE, Alexander M; Appl Environ Microb 45: 1234-41 (1983)
5. Hansch C, Leo AJ; Medchem Project Issue No 26 Pomona College Claremont CA (1985)
6. Hine J, Mookerjee PK; J Org Chem 40: 292-7 (1975)
7. Howard PH et al; Investigation of Selected Potential Environ Contam, Nitroaromatics USEPA 560/2-76-010 (1976)
8. Kawasaki M; Ecotox Env Safety 4: 444-54 (1980)
9. Lyman WJ et al; Handbook of Chemical Property Estimation Methods. Environmental Behavior of Organic Compounds. McGraw-Hill NY (1982)
10. McCormick NG et al; Appl Environ Microb 31: 949-58 (1976)
11. Meijers AP, Van Der Leer CR; Water Res 597-604 (1976)
12. Nojima K, Kanno S; Chemosphere 6: 371-6 (1977)
13. Pellizzari ED; Quantification of Chlorinated Hydrocarbons in Previously Collected Air Samples pp 46 USEPA 450/3-78-112 (1978)
14. Piet GJ, Morra CF; pp 31-42 in Artificial Groundwater Recharge, Huisman L, Olsthorn TN eds (1983)
15. Pitter P; Water Res 10: 231-5 (1976)
16. Sadtler Res Lab; Sadtler Standard UV Spectra No.1293 (1961)

4-Nitrotoluene

17. Sasaki S; pp 283-98 in Aquatic Pollut, Transformation and Bio Effects, Hutzinger O, Van Leytoeld LH, Zoeteman BCJ eds (1978)
18. Spanggord RJ et al; Environ Sci Tech 16: 229-32 (1982)
19. Swann RL et al; Res Rev 85: 17-28 (1983)
20. Thom NS, Agg AR; Water Res 10: 231-5 (1976)
21. Verschueren K; Handbook of Environmental Data on Organic Chemicals. 2nd ed Van Nostrand Reinhold NY (1983)
22. Zoeteman et al; Chemosphere 9: 231-49 (1980)

Phenol

SUBSTANCE IDENTIFICATION

Synonyms: Hydroxybenzene

Structure:

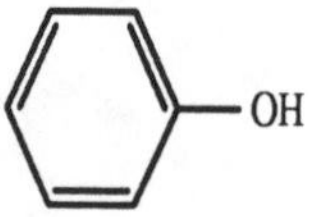

CAS Registry Number: 108-95-2

Molecular Formula: C_6H_6O

Wiswesser Line Notation: QR

CHEMICAL AND PHYSICAL PROPERTIES

Boiling Point: 181.75 °C at 760 mm Hg

Melting Point: 43 °C

Molecular Weight: 94.11

Dissociation Constants: pKa = 9.994 [51]

Log Octanol/Water Partition Coefficient: 1.46 [24]

Water Solubility: 87,000 mg/L at 25 °C [50]

Vapor Pressure: 0.524 mm Hg at 25 °C [10]

Henry's Law Constant: 3.97 x 10^{-7} atm-m^3/mole [28]

ENVIRONMENTAL FATE/EXPOSURE POTENTIAL

Summary: Phenol is a common and important industrial chemical and enters the environment in wastewater and spills connected with its use in resins, plastics, adhesives, or other uses. It is frequently found in wastewater from other commercial processes. If it is released to the environment, its primary removal mechanism is biodegradation which is generally rapid (days). Since phenol is a benchmark chemical for biodegradability studies, there is a large body of information on its degradation which concludes that phenol rapidly degrades in sewage, soil, freshwater, and sea water. Acclimation of resident populations of microorganisms is rapid. Under anaerobic conditions, degradation is slower and microbial adaptation periods longer. If phenol is released to soil, it will biodegrade in the soil and this degradation will be rapid (2-5 days) and will also occur in subsurface soils. Despite its high solubility and poor adsorption on soil, biodegradation is sufficiently rapid that most ground water is generally free of this pollutant. The exception would be in the cases of spills where high concentrations of phenol destroy degrading microbial populations. One would expect some of the phenol spilled on land to evaporate into the atmosphere. It will not be expected to significantly hydrolyze in water or soil under normal environmental conditions. If phenol is released to water, the primary removal process of it will be biodegradation, which will generally take on the order of hours to days in freshwater systems and up to a few weeks in estuarine waters. Direct photolysis should occur, but appears to be of minor importance compared with biodegradation under most circumstances. Phenols react relatively rapidly in sunlit natural water via reaction with photochemically produced hydroxyl and peroxyl radicals; typical half-lives for hydroxyl and peroxyl radical reactions are on the order of 100 and 19.2 hr of sunlight, respectively. Phenol will not be expected to significantly evaporate, hydrolyze, adsorb to sediment, or bioconcentrate in aquatic organisms. Based on the pKa, phenol will exist in a partially dissociated state in water and moist soils and, therefore, its transport and reactivity may be affected by pH. If phenol is released to the atmosphere, it will exist predominantly in the vapor phase. Phenol absorbs light in the region 290-330 nm and therefore might directly photodegrade. Phenol's estimated half-life by reaction with hydroxyl radicals in air is 0.61 days. Reaction of phenol with nitrate radicals during nighttime may be a significant removal process based on a rate constant of 3.8×10^{-12} cm^3/molecule-sec, which corresponds to a half-life of 15 minutes at an atmospheric concentration of $2 \times 10^{+8}$ nitrate radicals per cm^3. It may be subject to removal in rain based on its detection in rainwater. General exposure will occur through the use of disinfectants and cleaners containing phenol.

Natural Sources: Animal wastes; decomposition of organic wastes [62].

Artificial Sources: Wastewater from its manufacturers [62]; resins, plastics, fibers, adhesives, iron and steel, aluminum, leather, and rubber industries [62]; spills connected with its transport and use [8]. Phenol is also found in cigarette smoke and auto exhaust as well as disinfectants and medicinal products [22].

Terrestrial Fate: If phenol is released to soil, it will biodegrade in the soil and this degradation will be rapid (2-5 days) and will also occur in subsurface soils [3,16]. Degradation will be much slower under anaerobic conditions than under aerobic conditions [3]. Despite its high solubility and poor adsorption on soil, biodegradation is sufficiently rapid that most ground water is generally free of this pollutant [13]. The exception would be in the cases of spills where high concentrations of phenol destroy degrading microbial populations [13,16]. One would expect some of the phenol spilled on land to evaporate into the atmosphere. It will not be expected to significantly hydrolyze under normal environmental conditions.

Aquatic Fate: If phenol is released to water, the primary removal process will be biodegradation which will generally be rapid. Data suggest that degradation will take on the order of hours to days in freshwater systems and up to a few weeks in estuarine waters. Degradation rates are slower under anaerobic conditions [26]. Acclimation of resident microorganisms is rapid in aerobic waters and may take a few weeks under anaerobic conditions [26]. Evaporation is not a significant removal process for phenol. Direct photolysis should occur, but appears to be of minor importance compared with biodegradation under most circumstances. Phenols react relatively rapidly in sunlit natural water via reaction with photochemically produced hydroxyl radicals and peroxy radicals [39]; typical half-lives for hydroxyl and peroxyl radical reactions are on the order of 100 and 19.2 hr of sunlight, respectively [39]. In water, phenol will not be expected to significantly hydrolyze, adsorb to sediment or bioconcentrate in aquatic organisms. Based on the pKa, phenol will exist in a partially dissociated state in water and moist soils and, therefore, its transport and reactivity may be affected by pH.

Atmospheric Fate: If phenol is released to the atmosphere, it will exist predominantly in the vapor phase [17]. Phenol absorbs light in the region 290-330 nm [45] and therefore might directly photodegrade. Phenol's estimated half-life by reaction with hydroxyl radicals in air is 0.61 days [27]. Reaction of phenol with nitrate radicals during nighttime may be a significant removal process based on a rate constant of 3.8 x 10^{-12} cm^3/molecule-sec [2], which corresponds to a half-life of 15 minutes at an atmospheric concentration of 2 x 10^{+8} nitrate radicals per cm^3. It may be subject to removal in rain based on its detection in rainwater [35].

Biodegradation: Removal in aerobic activated sludge reactors is frequently better than 90% with retention of 8 hours [56]. Partial inhibition has been noted at concentrations as low as 50 ppm in aerobic reactors using industrial wastewater seed and activated sludge seed [12]. Utilization is also very high in anaerobic reactors although acclimation periods are longer and degradation slower (about 2 weeks) [4,26]. 95% degradation in 1 to 2 days using cultures obtained from garden soil, compost, river mud, and/or sediment from a petroleum refinery waste lagoon; the microorganisms present were common to each of the cultures utilized [59]. Complete degradation in <1 day in water from 3 lakes; rates increase with increase concn phenol and increase trophic levels of water; rate affected by concn of organic and inorganic nutrients [44]. Complete removal in river water after 2 days at 20 °C and after 4 days at 4 °C [36]. Degradation is somewhat slower in salt water and a half-life of 9 days has been reported in an estuarine river [34]. Degradation in soil is completed in 2-5 days even in subsurface soils [3]. Percent mineralization in an alkaline, para-brown soil under aerobic conditions was 45.5%, 48%, and 65% after 3, 7, and 70 days, respectively [23]. Half-lives for degradation of low concn of phenol in 2 silt loam soils were 2.70 and 3.51 hr [46].

Abiotic Degradation: Phenol absorbs light in the region 290-330 nm [45] and therefore might directly photodegrade. Natural sunlight

471

causes degradation in water [7]. Phenol's estimated half-life by reaction with hydroxyl radicals in air is 0.61 days [27]. It has also been shown to inhibit oxidant formation both in air [20] and in water [14]. It degrades on sand, possibly by a surface catalyzed reaction [3]. The estimated half-life for reaction of phenol with photochemically produced singlet oxygen in surface waters contaminated by humic substances is 83 days (assuming Switzerland summer sunlight and singlet oxygen concn 4 x 10^{-14} M) [49]. As a class, phenols react relatively rapidly in sunlit natural water via reaction with photochemically produced hydroxyl radicals and peroxy radicals [38]; typical half-lives for hydroxyl and peroxyl radical reactions are on the order of 100 and 19.2 hr of sunlight, respectively [38]. Reaction of phenol with nitrate radicals during nighttime may be a significant removal process based on a rate constant of 3.8 x 10^{-12} cm^3/molecule-sec, which corresponds to a half-life of 15 minutes at an atmospheric concentration of 2 x 10^{+8} nitrate radicals per cm^3 [2]. Based on the pKa, phenol will exist in a partially dissociated state in water and moist soils and, therefore, its transport and reactivity may be affected by pH.

Bioconcentration: BCF: Goldfish, (Carassius auratus), 1.9 [32]; water flea (Daphnia magna), 277 [11]; gold orfe, 20; algae (Chlorella fusca), 200 [19]; freshwater phytoplankter (Scenedesmus quadricauda), 3.5 [25]. Using the log Kow, an estimated BCF of 7.6 was calculated [37]. Based on the reported and estimated BCF's, phenol will not be expected to significantly bioconcentrate in aquatic organisms.

Soil Adsorption/Mobility: Low adsorptivity to clay soils and silt loam is reported [1]. Koc values for two silt loams are 39 and 91 [47]. No adsorption to aquifer material was observed [16]. Using the log Kow, an estimated Koc of 148 was calculated [37]. Based on the reported and estimated Koc, phenol will be expected to exhibit high to very high mobility in soil [58], and therefore may leach to ground water.

Volatilization from Water/Soil: Despite its moderate vapor pressure, phenol has a low Henry's Law constant and a low rate of evaporation from water [5,54]. The estimated half-life for evaporation from water is 3.2 months [5]. Using the Henry's Law constant, a half-life of 88 days was calculated for evaporation from

472

Phenol

a model river 1 m deep with a current of 3 m/sec and with a wind velocity of 3 m/sec [37]. Volatilization from near-surface soil should be relatively rapid due to its moderate vapor pressure.

Water Concentrations: DRINKING WATER: Not detected in finished drinking water in United States [61]. U.S. avg drinking water, 1 ppb [40]. Great Britain, March-Dec 1976, drinking water, from ground water sources, 4 sites, 50% pos, from surface water, 30% pos [18]. GROUND WATER: Maximum of 1130 ppm in 9 wells in Wisconsin after a spill [13]. 6.5-10,000 ppb in 2 aquifers 15 months after completion of coal gasification project [57]. SURFACE WATER: 0-5 ppb, industrial rivers in United States [52,53]. 3-24 ppb in Lake Huron [33]. Detected, not quantified, in 2 of 110 raw water supplies in 1977, U.S. EPA National Organics Monitoring Survey [61]. USEPA STORET database, 1754 data points, 13.0% pos, 5.0 ppb median concn [55]. RAIN/SNOW: 7 rain events in Portland, OR, Feb-Apr 1984, concn (ppt), in rain, 1200 to >75, avg >280; gas-phase concn (ng/m^3), 220-410, avg 320 [35].

Effluent Concentrations: 0.01-0.30 ppm, chemical specialties manufacturing (mfg) plant [31]; 7.0 ppm, 24-hr composite sample, plant on Delaware River [52]. Industries with mean concentrations >0.1 ppm in treated wastewater (max/avg concn, ppm): iron and steel mfg, 53 ppm max/5.7 ppm avg; leather tanning and finishing, 1.4/0.63; aluminum forming, 9.7/1.3; electrical/electronic components, 3.5/<0.74; foundries, 34/2.2; pharmaceuticals, 4.6/0.75; organics/plastics mfg, not reported/0.110; paint and ink formulation, 1.2/0.14; and rubber processing, 12/3.0 [62]. USEPA STORET database, 2068 data points, 42.1% pos, 10.0 ppb median concn [55]. U.S. Nationwide Urban Runoff Program, as of July 1982, 15 cities, 13.3% pos 86 samples, 4% pos, 3-10 ppb [9]. Ground water at 178 CERCLA hazardous waste sites, 13.6% pos [43].

Sediment/Soil Concentrations: SEDIMENT: Lake Huron, 13 ppm [33]. Not detected in unspecified industrial river in United States [31]. USEPA STORET database, 318 data points, 9% pos, <1000 ppb median concn (dry wt) [55]. Sediment collected 6 km NW of the Los Angeles county wastewater treatment plant discharge zone at Palos Verdes, CA, 10 ppb dry wt [21].

Phenol

Atmospheric Concentrations: Urban/suburban areas in United States (7 samples), 30 ppt avg [6]. 2 urban areas, 0.05-0.35 ug/m³, <20-289 ug/m³ (with 50% of all observations <30 ug/m³ [48]. Source dominated areas in United States (83 samples), 27,000 ppt avg [6]. Niagara River, Sept 1982, identified, not quantified [30].

Food Survey Values: 7 ppm - smoked summer sausage; 28.6 ppm - smoked pork belly; component of medicinal preparations such as throat lozenges [61]. Identified, not quantified, in: Mountain cheese [15]; fried bacon [29]; fried chicken [60].

Plant Concentrations:

Fish/Seafood Concentrations: Bottomfish, Commencement Bay, Tacoma, WA, June-Dec 1981, 5 sites, highest avg concn, 0.14 ppm, max concn, 0.22 ppm [40].

Animal Concentrations: Natural component of animal matter; rabbit (muscle) 0-1.6 ppm [61].

Milk Concentrations:

Other Environmental Concentrations:

Probable Routes of Human Exposure: Human exposure will be primarily from exposure to waste products and effluents associated with its manufacture and use in resins, plastics, fibers, adhesives, etc., and occurrence in other industries, especially iron and steel, aluminum, foundries, and rubber processing. It is also a normal constituent of animal wastes and the decomposition of organic matter. People will also be exposed to phenol through the use of disinfectants and the ingestion of medicinal products such as throat lozenges. Phenol is also found in cigarette smoke and auto exhaust but there are no data to ascertain the significance of exposure through inhalation of air contaminated by these or other sources [22].

Average Daily Intake: WATER INTAKE: Insufficient data; AIR INTAKE: (assume typical concn 30 ppt [6]) 2.7 ug ; FOOD INTAKE: Insufficient data.

474

Phenol

Occupational Exposures: NIOSH (NOES Survey 1981-1983) has statistically estimated that 192,739 workers are exposed to phenol in the United States [42]. Inhalation is largely restricted to occupational environments (e.g., 12.5 mg/m^3 in Bakelite factories) [61]. The number of workers exposed in commercial production, formulation of product, or distribution of concentrated products has been estimated by NIOSH to be 10,000 [41].

Body Burdens:

REFERENCES

1. Artiola-Fortuny J, Fuller WH; Soil Sci 133:18-26 (1982)
2. Atkinson R et al; Environ Sci Technol 21: 1123-6 (1987)
3. Baker MD, Mayfield CI; Water Air Soil Pollut 13:411 (1980)
4. Boyd SA et al; Appl Environ Microbiol 46:50-4 (1983)
5. Branson DR; Predicting the Fate of Chemicals in the Aquatic Environment from Laboratory Data, p 55-70 ASTM STP 657 (1978)
6. Brodzinsky R, Singh HB; Volatile Organic Compounds in the Atmosphere: An Assessment of Available Data. SRI International. EPA Contract 68-02-3452 198 pp (1982)
7. Callahan MA et al; Water Related Environmental Fate of Priority Pollutants p 83-3 EPA-440/4-79-029b (1979)
8. Chem Eng New p 9 Dec 19 (1983)
9. Cole RH et al; J Water Poll Control Fed 56: 898-908 (1984)
10. Daubert TE, Danner RP; Data Compilation Tables of Properties of Pure Compounds. American Institute of Chemical Engineers. pp 450 (1985)
11. Dauble DD et al; Bull Environ Contam Toxicol 37: 125-32 (1986)
12. Davis EM et al; Water Res 15:1125-7 (1981)
13. Delfino JJ, Dube DJ; J Environ Sci Health A11:345-55 (1976)
14. Draper WM, Crosby DG; Arch Environ Contam Toxicol 12:121-6 (1983)
15. Dumont JP, Adda J; J Agric Food Chem 26: 364-7 (1978)
16. Ehrlich GG et al; Groundwater 20:703-10 (1982)
17. Eisenreich SJ et al; Environ Sci Technol 15: 30-8 (1981)
18. Fielding M et al; Organic Micropollutants in Drinking Water. TR-159. Eng Water Res Center: Mendmenham (1981)
19. Freitag D et al; p 119 in: QSAR Environ Toxicol Proc Workshop Quant Struct-Act Relat. Kaiser KLE ed Reidel: Netherlands (1984)
20. Gitchell A et al; J Air Pollut Control Assoc 24:357-61 (1974)
21. Gossett RW et al; Marine Pollut Bull 14: 387-92 (1983)
22. Graedel TE; Chemical Compounds in the Atmosphere p 256 (1978)
23. Haider K et al; Arch Microbiol 96:183-200 (1974)
24. Hansch C, Leo AJ; Medchem Project Issue No 26. Claremont CA: Pomona College (1985)
25. Hardy JT et al; Environ Toxicol Chem 4: 29-35 (1985)
26. Healy JB, Jr, Young LY; Food Microbiol Toxicol 35:216-8 (1978)

27. Hendry DG, Kenley RA Atmospheric Reaction Products of Organic Compounds p 20,46 EPA-560/12-79-001 (1979)
28. Hine J, Mookerjee PK; J Org Chem 40: 292-8 (1975)
29. Ho CT et al; J Agric Food Chem 31: 336-42 (1983)
30. Hoff RM, Chan KW; Environ Sci Technol 21: 556-61 (1987)
31. Jungclaus GA et al; Environ Sci Technol 12:88-96 (1978)
32. Kobayashi K et al; Bull Jap Soc Sci Fish 45:173-5 (1979)
33. Konasewich D et al; Status Report on Organic and Heavy Metal Contaminants in Lakes Erie, Michigan, Huron and Superior Basin. Great Lakes Water Quality Board (1978)
34. Lee RF, Ryan C; Microbial Degradation of Pollutants in Marine Environments p 443-50 EPA-600/9-79-012 (1979)
35. Leuenberger C et al; Environ Sci Technol 19: 1053-8 (1985)
36. Ludzack FJ, Ettinger MB; J Water Pollut Control Fed 32:1173-200 (1960)
37. Lyman WJ et al; Handbook of Chem Property Estimation Methods. McGraw-Hill NY (1982)
38. Mill T, Mabey W; p. 208-11 in Environmental Exposure from Chemicals Vol I, Neely WR, Blau GE eds Boca Raton, FL: CRC Press (1985)
39. Mill T et al; Science 207:886-7 (1982)
40. Nicola RM; J Environ Health 49: 342-7 (1987)
41. NIOSH; NIOSH Criteria for recommended standard. Occupational Exposure to Phenol HEW 76-196 (1976)
42. NIOSH; The National Occupational Exposure Survey (NOES) (1983)
43. Plumb RHJr; Ground Water Monit Rev 7: 94-100 (1987)
44. Rubin HE, Alexander M; Environ Sci Tech 17:104-7 (1983)
45. Sadtler 258 UV (1988)
46. Scott HD et al; Environ Qual 11:107-11 (1983)
47. Scott HD et al; J Environ Qual 12:91-5 (1983)
48. Scow K et al; Exposure and Risk Assessment for Phenol (Revised) p 76 USEPA/440/4-85/013 (1981)
49. Scull FEJr, Hoigne J; Chemosphere 16: 681-94 (1987)
50. Seidell A; Solubities of Organic Compounds. New York,NY: D Van Norstrand Co Inc (1941)
51. Serjeant EP, Dempsey B; Ionisation Constants of Organic Acids in Aqueous Solution. IUPAC Chemical Data Series, New York,NY: Pergamon Press 989pp (1979)
52. Sheldon LS, Hites RA; Environ Sci Tech 13:574-9 (1979)
53. Sheldon LS, Hites RA; Environ Sci Tech 12:1188-94 (1978)
54. Shen TT; J Air Pollut Control Assoc 32:79-82 (1982)
55. Staples CA et al; Environ Toxicol Chem 4: 131-42 (1985)
56. Stover EL, Kincannon DF; J Water Pollut Control Fed 55:97-1 09 (1983)
57. Stuermer DH et al; Environ Sci Tech 16:582-7 (1982)
58. Swann RL et al; Res Rev 85: 17-28 (1983)
59. Tabak HH et al; J Bacteriol 87:910-9 (1964)
60. Tang JT et al; J Agric Food Chem 31: 1287-92 (1983)
61. USEPA; Ambient Water Quality Criteria Phenol, page C-3 EPA 440/5-80-066 (PB 81-17772) (1980)
62. USEPA; Treatability Manual pp I.8.1-1 to 1-5. EPA-600/2-82-001A (1981)

Phthalic Anhydride

SUBSTANCE IDENTIFICATION

Synonyms: 1,2-Benzenedicarboxylic acid anhydride

Structure:

CAS Registry Number: 85-44-9

Molecular Formula: $C_8H_4O_3$

Wiswesser Line Notation: T56 BVOVJ

CHEMICAL AND PHYSICAL PROPERTIES

Boiling Point: 295 °C

Melting Point: 130.8 °C

Molecular Weight: 148.11

Dissociation Constants:

Log Octanol/Water Partition Coefficient: Hydrolyzes rapidly in water [27]

Water Solubility: 6200 ppm at 25 °C [27]

Vapor Pressure: 2×10^{-4} mm Hg at 20 °C [28]

Henry's Law Constant: 6.2×10^{-9} atm-m^3/mol (calculated) [16]

ENVIRONMENTAL FATE/EXPOSURE POTENTIAL

Summary: Phthalic anhydride's release to the atmosphere will result from its manufacture and use in many products and its use in the manufacture of many materials including plasticizers, polyester and alkyd resins, phthaleins, phthalates, benzoic acid, synthetic indigo, artificial resins (glyptal), synthetic fibers, dyes, pigments, pharmaceuticals, insecticides, and chlorinated products. It is released from industrial plants which produce phthalic anhydride (PA) by oxidation of xylenes and naphthalene, from the incineration of industrial refuse and water sludges and slurries from plastic products and other manufacturing processes, and in leachate from municipal and industrial wastes containing plastics. If PA is released to soil, it will not be expected to sorb to the soil. It is expected to hydrolyze in moist soils, which will prevent its leaching to the ground water. If it is released to water, it will not be expected to bioconcentrate in aquatic organisms, sorb to sediments, or to evaporate. Hydrolysis will be a major fate process based on an estimated half-life of approx 1.5 minutes. If it is released to the atmosphere, it may be susceptible to direct photolysis. The estimated vapor phase half-life in the atmosphere is 1.00 day as a result of ring addition of photochemically produced hydroxyl radicals. Exposure to PA will result mainly from occupational exposure involving the inhalation of contaminated air. General exposure may occur as a result of the inhalation of contaminated air.

Natural Sources: There are no known natural sources of PA [1].

Artificial Sources: Identified industrial point sources include industrial plants which produce PA by oxidation of xylenes and naphthalene, incineration of industrial refuse and water sludges and slurries from plastic products and other manufacturing processes, and leachate from municipal and industrial wastes containing plastics [1]. PA leaches from tubings, dishes, paper, containers, etc. [1]. Major uses of PA include the manufacture of plasticizers (51%), polyester resins (25%), and alkyd resins (19%) [3]. Other reported uses of PA include the manufacture of phthaleins, phthalates, benzoic acid, synthetic indigo, artificial resins (glyptal) [18], synthetic fibers, dyes, pigments, pharmaceuticals, insecticides and chlorinated products [28].

Terrestrial Fate: If PA is released to soil, it will not be expected to sorb to the soil. PA is expected to hydrolyze in moist soils, which will prevent its leaching to the ground water.

Aquatic Fate: If PA is released to water, it will not be expected to bioconcentrate in aquatic organisms, sorb to sediments, or evaporate. Hydrolysis will be a major fate process based on an estimated half-life of approx 1.5 minutes calculated using a reported observed rate constant of 7.9 x 10^{-3} sec^{-1} for hydrolysis in aqueous solution at 25 °C [7]. PA absorbs light >290 nm and therefore may be susceptible to direct photolysis.

Atmospheric Fate: If PA is released to the atmosphere, it may be susceptible to direct photolysis since it absorbs light >290 nm. The estimated vapor phase half-life in the atmosphere is 1.00 days, as a result of addition of photochemically produced hydroxyl radicals.

Biodegradation: Percent theoretical BOD was reported to be 44-78 as a result of incubation of 1-4 ppm with sewage inoculum [8]. Reported degradation of an initial concn of 2 ppm PA was approx 21% after incubation with sewage (standard dilution method) and 18% (seawater dilution method) for 5 days [26]. Mineralization of 33% of an initial concn of 9 ppm PA incubated with activated sludge for 24 hr was reported based on COD [17]. PA was reported to be significantly degraded in Japanese MITI tests using activated sludge as inoculum [23]. All of the above tests may have tested phthalic acid rather than PA due to the rapid hydrolysis rate of PA. Percent theoretical BOD was reported to be 73.46% in 5 days using dilution water seeded with domestic sewage [25].

Abiotic Degradation: PA has been reported to hydrolyze in water [27] and an estimated half-life of approx 1.5 minutes was calculated using a reported observed rate constant of 7.9 x 10^{-3} sec^{-1} for hydrolysis in aqueous solution at 25 °C [7]. PA absorbs light >290 nm [22] and may therefore be susceptible to direct sunlight photolysis. The estimated vapor phase half-life in the atmosphere is 1.00 days, as a result of addition of photochemically produced hydroxyl radicals [6].

479

Bioconcentration: Reported BCF: <u>Oedogonium</u> (alga), 4053 [14]. PA did not bioconcentrate in <u>Daphnia</u>, <u>Physa</u> (snail), or <u>Gambusia</u> (fish) [14]. A BCF of 5.0 has been estimated from water solubility [10]. Based on this estimated BCF, PA will not be expected to bioconcentrate in aquatic organisms.

Soil Adsorption/Mobility: A Koc of 36 has been estimated [10]. Based on this estimated Koc, PA will not adsorb to soils or sediments.

Volatilization from Water/Soil: Based on the estimated Henry constant, evaporation from water should not be an important process [16].

Water Concentrations: DRINKING WATER: Identified, not quantified, in 1 of 14 treated waters, including treated water from rivers, reservoirs, and ground water (positive sample was a treated ground water after distribution; PA was not reported present in the source treated ground water) [5]. Identified, not quantified, in drinking water in the United States [12] and in unidentified drinking water [11]. Identified, not quantified, in drinking water concentrates [15]. U.S., 10 sites, 10% pos, identified, not quantified (only pos, Cincinnati, OH) [1].

Effluent Concentrations: Identified, not quantified, in advanced waste treatment concentrates [15]. Identified industrial point sources include industrial plants which oxidize xylenes and naphthalene for PA production; gross estimate of discharge 5000 tons/yr; waste treatment and disposal sources: incineration of industrial refuse and water sludges and slurries from plastic products and other manufacturing processes, leaches from municipal and industrial wastes containing plastics [1]. Identified, not quantified, in effluents from two unspecified chemical plants [24]. Concn (g/ton pulp) in spent chlorination liquor from bleaching of sulphite pulp: pulp with high lignin content, 0.2; normal lignin content after oxygen treatment and before and after alkali treatment, 0.4 [2]. Identified, not quantified, in effluents from efficient combustion of coal at power plants [9].

Sediment/Soil Concentrations:

Atmospheric Concentrations: URBAN: Identified, not quantified, in urban air particles collected over 18 months from St. Louis, MO [21]. Tsukuba, Japan, Apr 1985, identified, not quantified, in methanol-extraction fraction of atmospheric aerosols [29].

Food Survey Values: Identified, not quantified, in isolated volatile components of Idaho Russet Burbank potatoes [4]. Microcomponent of foods [1].

Plant Concentrations:

Fish/Seafood Concentrations:

Animal Concentrations:

Milk Concentrations:

Other Environmental Concentrations:

Probable Routes of Human Exposure: Exposure to PA will result mainly from occupational exposure involving the inhalation of contaminated air. General exposure may occur as a result of the ingestion of contaminated water and foods and the inhalation of contaminated air.

Average Daily Intake:

Occupational Exposures: NIOSH (NOES Survey 1981-1983) has statistically estimated that 40,518 workers are exposed to PA in the United States [20]. NIOSH (NOHS Survey 1972-1974) has statistically estimated that 142,657 workers are exposed to PA in the United States [19]. Occupational exposure, ug/m^3: PA production, 6 samples, 67% pos, 4-187, avg of pos, 53; Di-(2-ethylhexyl)phthalate (DEHP) production, 8 samples, 63% pos, 6-102, avg of pos, 38; PA and DEHP maintenance, 5 samples, 80% pos, 11-26, avg of pos, 24; batch ester plant production, 8 samples, 88% pos, 5-21, avg of pos, 11; batch ester plant maintenance, 3 samples, 67% pos, 21-44, avg of pos, 33; tank farm, 6 samples, 100% pos, 17-203, avg of pos 79 [13].

Body Burdens:

481

REFERENCES

1. Abrams EF et al; Identification of Organic Compounds in Effluents from Industrial Sources USEPA-560/3-75-002 (1975)
2. Carlberg GE et al; Sci Total Environ 48: 157-67 (1986)
3. Chemical Marketing Reporter July 7, p. 50 (1986)
4. Coleman EC et al; J Agric Food Chem 29: 42-8 (1981)
5. Fielding M et al; Organic Micropollutants in Drinking Water Medmenham, Eng Water Res Cent TR-159 (1981)
6. GEMS; Graphical Exposure Modeling System. Fate of atmospheric pollutants (FAP) data base. Office of Toxic Substances. USEPA (1986)
7. Hawkins MD; J Chem Soc Perkin Trans 2 75: 282-4 (1975)
8. Heukelekian H, Rand MC; J Water Pollut Contr Assoc 29: 1040-53 (1955)
9. Junk GA et al; ACS Symp Ser 319: 109-23 (1986)
10. Kenaga EE; Ecotox Environ Safety 4: 26-38 (1980)
11. Kool HJ et al; Crit Environ Control 12: 307-57 (1982)
12. Kopfler FC et al; Adv Environ Sci Technol 8: 419-33 (1977)
13. Liss GM et al; Scand J Work Environ Health 11: 381-7 (1985)
14. Lu PY, Metcalf RL; Environ Health Persp 10: 269-84 (1975)
15. Lucas SV; GC/MS (Gas Chromatography-Mass Spectrometry) Analysis of Organics in Drinking Water Concentrates and Advanced Waste Treatment Concentrates Vol.2 Battelle Columbus Labs, OH (1984)
16. Lyman WJ et al; Handbook of Chemical Property Estimation Methods Environ Behavior of Organic Compounds, McGraw-Hill NY (1982)
17. Matsui S et al; Prog Water Technol 7: 645-59 (1975)
18. Merck Index; An Encyclopedia of Chemicals, Drugs and Biologicals 10th ed p 1063 (1983)
19. NIOSH; The National Occupational Hazard Survey (NOHS) (1985)
20. NIOSH; The National Occupational Exposure Survey (NOES) (1985)
21. Ramdahl T et al; Environ Sci Technol 16: 861-5 (1982)
22. Sadtler UV No.18 (1966)
23. Sasaki S; pp 283-98 in Aquatic Pollutants: Transformation and Bio Effects Hutzinger O et al, eds Pergamon Press Oxford (1978)
24. Shackelford WM, Keith LH; Frequency of Organic Compounds Identified in Water ERL, ORD USEPA-600/4-76-062 (1976)
25. Swope HG, Kenna M; Sewage Ind Waste 21: 467-8 (1950)
26. Takemoto S et al; Suishitsu Odaku Kenkyu 4: 80-90 (1981)
27. Towle PH et al; Kirk-Othmer Encycl Chem Technol 2nd ed 15: 444 (1968)
28. Verschueren K; Handbook of Environ Data on Organic Chemicals 2nd ed Von Nostrand Reinhold NY (1983)
29. Yokouchi Y, Ambe Y; Atmos Environ 20: 1727-34 (1986)

1,2-Propylene Oxide

SUBSTANCE IDENTIFICATION

Synonyms: 1,2-Epoxypropane; Methyloxirane

Structure:

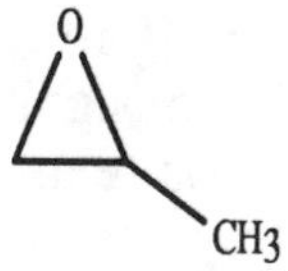

CAS Registry Number: 75-56-9

Molecular Formula: C_3H_6O

Wiswesser Line Notation: T3OTJ B1

CHEMICAL AND PHYSICAL PROPERTIES

Boiling Point: 34.23 °C

Melting Point: -112.13 °C

Molecular Weight: 58.08

Dissociation Constants:

Log Octanol/Water Partition Coefficient: 0.03 [7]

Water Solubility: 476,000 mg/kg at 25 °C [16]

Vapor Pressure: 532.1 mm Hg at 25 °C [4]

Henry's Law Constant: 8.54×10^{-5} atm-m³/mole (calculated from water solubility and vapor pressure)

ENVIRONMENTAL FATE/EXPOSURE POTENTIAL

Summary: Atmospheric emissions of propylene oxide from its commercial manufacturing and use processes were estimated to

483

total 1.34 million lb during 1978. It may also be emitted to the atmosphere in automobile exhaust and combustion exhausts of stationary sources that burn hydrocarbons. If released to the atmosphere, propylene oxide will react in the gas phase with photochemically produced hydroxyl radicals with an estimated half-life of approx 19.3 days. It is not expected to react significantly with ozone in the atmosphere. Localized atmospheric removal by rainfall may occur. If released to soil, propylene oxide is expected to be susceptible to significant leaching and chemical hydrolysis in moist soils. It is expected to evaporate relatively rapidly from dry soil surfaces; evaporation from wet soils may also occur, but at a rate diminished by leaching. If released to water, propylene oxide will hydrolyze at estimated half-life rates of 11.6 days (at pH's 7-9) and 6.6 days (at pH 5) at 25 °C. The presence of chloride ion accelerates the degradation in water and the chemical degradation half-lives in seawater are estimated to be 4.1 days (at pH's 7-9) and 1.5 days (at pH 5) at 25 °C. Reaction of propylene oxide with Cl ion in water yields approx 90% 1-chloro-2-propanol and 10% 2-chloro-1-propanol as products under neutral pH conditions. Volatilization of propylene oxide from the aquatic environment may be an important transport mechanism, as the calculated half-lives from a representative river and oligotrophic lake are 3 and 18 days, respectively. Adsorption to sediment, bioconcentration in aquatic organisms, and reaction with photochemically produced hydroxyl radicals in water are not expected to be environmentally significant processes. Occupational workers involved in the production and use of propylene oxide appear to have the greatest exposure to the chemical, particularly by the inhalation route.

Natural Sources: Propylene oxide is not known to occur as a natural product [8].

Artificial Sources: Atmospheric emissions of propylene oxide from propylene oxide manufacturing processes during 1978 were estimated to be about 1.16 million lb [8]. Atmospheric emissions resulting from the use of propylene oxide in the production of urethane polyols, propylene glycol, surfactant polyols, di- and tripropylene glycols, glycol ethers, and miscellaneous applications during 1978 were estimated to be about 147.7, 13.9, 15.8, 2.9, 1.2, and 3.8 thousand lb, respectively [8]. It has been suggested that propylene oxide may be emitted to the atmosphere from automobile exhaust and from combustion exhausts of stationary sources that burn hydrocarbons [1].

Terrestrial Fate: The aqueous hydrolysis of propylene oxide occurs at an environmentally significant rate; therefore, hydrolysis in moist soil conditions is likely to be important. The estimated Koc values of 3.6 and 30 indicate that propylene oxide is expected to be very mobile in soil. The relatively high vapor pressure of propylene oxide suggests that it should evaporate rapidly from dry soil surfaces. Propylene oxide is predicted to be moderately volatile from water; therefore, evaporation from wet soils may also be possible.

Aquatic Fate: In fresh water, propylene oxide will hydrolyze with estimated half-lives of 11.6 days (pH's 7-9) and 6.6 days (pH 5) at 25 °C. The presence of chloride ion accelerates the degradation, as the chemical degradation half-lives in seawater are estimated to be 4.1 days (pH's 7-9) and 1.5 days (pH 5) at 25 °C. Reaction of propylene oxide with Cl ion in water yields approx 90% 1-chloro-2-propanol and 10% 2-chloro-1-propanol as products under neutral pH conditions [1]. Reaction with photochemically produced hydroxyl radicals in natural water has no environmental significance, as the estimated half-life at room temperature is 9.15 years. The estimated Koc values of 3.6 and 30 suggest that partitioning of propylene oxide from the water column to sediments and particulate matter will not be important. Volatilization of propylene oxide from the aquatic environment may be an important transport mechanism, as the calculated half-life from a representative river and oligotrophic lake are 3 and 18 days, respectively. Calculated log BCF's of -0.20 and -0.40 suggest that bioconcentration in aquatic organisms will not be environmentally significant.

Atmospheric Fate: When released to the atmosphere, propylene oxide will react with photochemically produced hydroxyl radicals with a calculated half-life of 19.3 days. It is not expected to react significantly with ozone in the atmosphere. Physical removal of propylene oxide from the ambient atmosphere is not expected to be generally important [3], although localized wash out due to rainfall may occur.

1,2-Propylene Oxide

Biodegradation: Using the standard dilution method, a 5-day BODT of 8% was measured for propylene oxide using a filtered effluent seed from a biological sanitary waste treatment plant, while a 5-day BODT of 9% was measured using an adapted seed [2].

Abiotic Degradation: The hydrolysis half-life of propylene oxide in fresh water at 25 °C has been estimated to be 11.6 days at pH 7-9 and 6.6 days at pH 5 [1]. The half-life of propylene oxide seawater (3% NaCl concn) at 25 °C has been estimated to be 4.1 days at pH 7-9 and 1.5 days at pH 5 [1] with the formation of chloropropanols. The rate constant for the reaction of propylene oxide with photochemically produced hydroxyl radicals in water at room temperature has been experimentally determined to be 2.4 x 10^{+8} m-sec^{-1} [6]; assuming an average hydroxyl radical concentration of 1 x 10^{-17} M in natural water [10], a half-life of 9.15 years can be calculated indicating no environmental significance. The rate constant for the gas-phase reaction of propylene oxide with photochemically produced hydroxyl radicals in the atmosphere has been experimentally determined to be 5.2 x 10^{-13} cm^3/molecule-sec at room temperature [6]; assuming an average atmospheric hydroxyl radical concn of 8 x 10^{+5} molecules/cm^3 [5], a half-life of 19.3 days can be calculated for this reaction. The anticipated products of the atmospheric reaction with hydroxyl radicals have been cited as $CH_3C(O)OCHO$, $CH_3C(O)CHO$, H_2CO and $HC(O)OCHO$ [3]. Ozone is not expected to react with propylene oxide in the atmosphere [5].

Bioconcentration: Calculated log BCF's of -0.20 and -0.40 suggest that bioconcentration in aquatic organisms will not be environmentally significant.

Soil Adsorption/Mobility: Based on the water solubility, the Koc value for propylene oxide is estimated to be 4.2 from a regression equation [9]. Estimation of Koc from molecular topology and quantitative structure-activity relationships yields a Koc of 30 for propylene oxide [13]. Koc values of 4.2 and 30 indicate that propylene oxide is expected to be very mobile in soil [14].

Volatilization from Water/Soil: The calculated half-lives for the volatilization of propylene oxide from a representative or natural river and oligotrophic lake are 3 and 18 days, respectively [17].

Water Concentrations:

Effluent Concentrations: A propylene oxide concentration of 0.047 mg/L was detected in a water effluent from Olin Corporation's Brandenberg, KY chemical production facility on Feb 2, 1974 [15].

Sediment/Soil Concentrations:

Atmospheric Concentrations: Propylene oxide was tentatively identified in unspecified atmospheric air samples in the United States [8].

Food Survey Values:

Plant Concentrations:

Fish/Seafood Concentrations:

Animal Concentrations:

Milk Concentrations:

Other Environmental Concentrations: Propylene oxide has been found as a trace level impurity in poly(propylene oxide) [8].

Probable Routes of Human Exposure: Inhalation of contaminated air, especially near areas of commercial production or use, may provide a significant route of exposure to propylene oxide. The USFDA has approved the use of propylene oxide as a package fumigant for certain fruit products and as a fumigant for bulk quantities of several food products, provided residues of propylene oxide or propylene glycol do not exceed specific limits [8]; therefore, consumption of food contaminated by propylene oxide fumigation is another possible route of exposure. Occupational exposure by inhalation and dermal routes related to the production, storage, transport, and use of propylene oxide may be significant.

Average Daily Intake:

Occupational Exposures: A National Occupational Hazard Survey (NOHS) estimates that 268,433 workers in the United States are exposed to propylene oxide [12]. A National Occupational Exposure Survey (NOES) estimates that 194,342 U.S. workers are exposed to propylene oxide [11]. In a 1979 study by one U.S. manufacturer, the typical average daily exposure of workers to propylene oxide were 0.5-5 mg/m^3, with worst-case peak exposures of 59-9000 mg/m^3 (highest exposure being that of maintenance workers cleaning pumps) [8]. Levels of worker exposure were reported to be 0-5-5.9 mg/m^3 in a polymer polyol unit, not detectable to 1.2 mg/m^3 in an oxide adducts unit, and not detectable in a flexible polyol unit of a large chemical manufacturing facility producing many chemical products including propylene oxide derivatives [8]. A propylene oxide concn of 3.6 mg/m^3 was found near an operator at a flexible polyol unit in another large chemical manufacturing facility [8].

Body Burdens:

REFERENCES

1. Bogyo Da et al; Investigation of Selected Potential Environmental Contaminants: Epoxides pp 201 USEPA-560/11-80-005 (1980)
2. Bridie AL et al; Water Res 13: 627 (1979)
3. Cupitt L; Fate of Toxic Hazardous Materials in the Environment pp 7 USEPA-600/S3-80-084 (1980)
4. Daubert TE, Danner RP; Data Compilation Tables of Properties of Pure Compounds. Amer Institute of Chemical Engineers. pp 450 (1985)
5. GEMS; Graphical Exposure Modeling System. Fate of Atmospheric Pollutants (FAP) Data Base. Office of Toxic Substances. USEPA (1986)
6. Guesten H et al; Atmos Environ 15: 1763 (1981)
7. Hansch C, Leo AJ; Medchem Project Issue No 26. Claremont CA: Pomona College (1985)
8. IARC; Allyl Compounds, Aldehydes Epoxides and Peroxides 36: 227 (1985)
9. Lyman WJ et al; Handbook of Chemical Property Estimation Methods. Environmental Behavior of Organic Compounds. McGraw-Hill NY (1982)
10. Mill T et al; Science 207: 886 (1980)
11. NIOSH; National Occupational Exposure Survey (NOES) (1984)
12. NIOSH; National Occupational Hazard Survey (NOHS) (1973)

1,2-Propylene Oxide

13. Sabjlic A; J Agric Food Chem 32: 243 (1984)
14. Swann RL et al; Residue Reviews 85: 17 (1983)
15. USEPA; STORET (1985)
16. USEPA; Treatability Manual I. Treatability Data EPA-600/2-82-001a Washington,DC: USEPA (1981)
17. USEPA; EXAMS (1986)

Styrene

SUBSTANCE IDENTIFICATION

Synonyms: Vinylbenzene

Structure:

CAS Registry Number: 100-42-5

Molecular Formula: C_8H_8

Wiswesser Line Notation: 1U1R

CHEMICAL AND PHYSICAL PROPERTIES

Boiling Point: 145.2 °C

Melting Point: -30.63 °C

Molecular Weight: 104.16

Dissociation Constants:

Log Octanol/Water Partition Coefficient: 2.95 [22]

Water Solubility: 310 mg/L at 25 °C [70]

Vapor Pressure: 6.6 mm Hg at 25 °C (extrapolated) [8]

Henry's Law Constant: 2.81 x 10^{-3} atm m³/mol (estimated) [26]

Styrene

ENVIRONMENTAL FATE/EXPOSURE POTENTIAL

Summary: Significant amounts of styrene may be released to the environment from emissions generated by its production and use and from automobile exhaust. If released to the atmosphere, styrene will react rapidly with both hydroxyl radicals and ozone with a combined, calculated half-life of about 2.5 hours. If released to environmental bodies of water, styrene will volatilize relatively rapidly and may be subject to biodegradation, but is not expected to hydrolyze. If released to soil it will biodegrade and leach with a low to moderate soil mobility. While styrene has been detected in various U.S. drinking waters, it was not detected in a ground water supply survey of 945 U.S. finished water supplies which use ground water sources. Styrene has been detected in various U.S. chemical, textile, latex, oil refinery, and industrial wastewater effluents. Styrene has been frequently detected in the ambient air of source dominated locations and urban areas, has been detected in the air of a national forest in Alabama, and has been detected in the vicinity of oil fires. Food packaged in polystyrene containers has been found to contain small amounts of styrene.

Natural Sources: Styrene is not known to occur as a natural product [28]. Its presence in various food products is due to monomer leaching from polystyrene containers [28].

Artificial Sources: Styrene is released into the environment by emissions and effluents from its production and its use in polymer manufacture [54]. It has been found in exhausts from spark-ignition engines, oxy-acetylene flames, cigarette smoke, and gases emitted by pyrolysis of brake linings [54]. Stack emissions from waste incineration have been found to contain styrene [30]. Styrene is emitted in automobile exhaust [21]. Combustion of styrene polymer products may be an emission source [64]. Likely consumer exposure sources are use of products containing styrene, such as floor waxes and polishes, paints, adhesives, putty, metal cleaners, autobody fillers, fiberglass boats, and varnishes [43,64].

Terrestrial Fate: Styrene released to soils is subject to biodegradation. Degradation of 87-95% has been observed in sandy loam and landfill soil over a 16-week incubation and degradation of 2.3-12% per week has been observed with two subsurface

491

aquifers. Styrene may exhibit low to moderate soil mobility depending on soil conditions. It has been demonstrated that styrene buried in soil can leach into underlying ground water. Styrene which leaked into surrounding soil from buried drums persisted in the soil for up to two years [54].

Aquatic Fate: Volatilization and biodegradation may be dominant transport and transformation processes, respectively, for styrene in water. The volatilization half-life of styrene from a river 1 m deep with a current speed of 1 m/sec and wind velocity of 3 m/sec is about 3 hours. Although biodegradation studies utilizing only ambient waters are not available, various BOD and other studies have shown styrene to be biodegradable. Sufficient quantitative kinetic data are not available to predict the relative significance of aquatic photolysis. Hydrolysis is not expected to be important. Adsorption to particulate matter and sediment may have some significance (Koc of 270-550).

Atmospheric Fate: Styrene vapor in the atmosphere will react rapidly with hydroxyl radicals and with ozone. The reaction half-lives of styrene with hydroxyl radicals and ozone are calculated to be 3.5 and 9 hours, respectively. Atmospheric washout of styrene is not expected to be an important process, due to the rapid reaction of styrene with hydroxyl radicals and ozone and the high Henry's Law constant.

Biodegradation: 95% styrene degradation from a landfill soil and 87% degradation from a sandy loam soil in 16 weeks as measured by CO_2 evolution [58]. Degradation of 2.3-4.3% per week and 3.8-12.0% per week in subsurface soil was shown with samples taken directly above and below aquifers from Pickett, OK and Fort Polk, LA, respectively; No significant degradation in autoclave samples was observed [69]. Removal greater than 99% in an aerobic biofilm column with 20-minute detention time and 8% removal in a methanogenic biofilm column with a 2-day detention time was reported [9]. Five-day aqueous theoretical BOD (TBOD) of 80% in acclimated sewage seed and 42% TBOD in an unacclimated seed have been observed [10]. Five-day aqueous TBOD was 65% and 20-day TBOD was 87% with a filtered sewage seed, but TBOD's dropped to 8% (5-day) and 80% (20-day) in salt water [50]. In 17 days of incubation using a sewage seed, CO_2 production reached about 20% of theoretical [46]. Styrene was degraded in mixed propane-utilizing bacteria isolated from soil and lakes, with styrene oxide formed as a product [27].

Abiotic Degradation: Based on a rate constant of 5.3 x 10^{-11} cm^3/molecule-sec at 25 °C for the atmospheric reaction between styrene and hydroxyl radicals [6], the half-life in an average atmosphere is estimated to be 3.5 hours. Based on a rate constant of 2.16 x 10^{-17} cm^3/molecule-sec at 23 °C for the atmospheric reaction between styrene and ozone [3], the half-life in an average atmosphere is estimated to be 9 hours. The combined half-life of styrene in air due to reaction with both hydroxyl radical and ozone is estimated to be 2.5 hours. Alkenes and aromatic hydrocarbons are generally resistant to hydrolysis [38]. Styrene does not absorb solar radiation; therefore, it should not be directly photolyzed in the troposphere or surface water. However, it is a very active generator of photochemical smog due to indirect photochemical reactions [13].

Bioconcentration: Bioconcentration factor (BCF) of 13.5 for goldfish [45]. Log Kow of 2.95 [22] indicates moderate potential for bioaccumulation. Not likely to bioconcentrate in biological organisms due to relatively high water solubilities [54].

Soil Adsorption/Mobility: Relatively strong adsorption observed in a sand aquifer as breakthrough time took about 80 times longer than a nonadsorbing tracer [51]. Styrene buried in drums in soil was found to have leached into ground water [54], demonstrating that it is leachable in certain soils. Estimation of Koc from a water solubility of 160 ppm [4] yields a Koc value of about 270 [37]. Estimation of Koc from log Kow 2.95 [23] yields a Koc value of about 550 [37]. Koc values of 270-550 indicate moderate to low soil mobility [62].

Volatilization from Water/Soil: The calculated Henry's Law constant at 25 °C of 5.2 x 10^{-3} atm-m^3/mol [59] indicates rapid volatilization from environmental waters [36]. The volatilization half-life of styrene from a model river one m in depth with water current speed of one m/sec and wind velocity of 3 m/sec is calculated to be about 3 hours [36].

Styrene

Water Concentrations: DRINKING WATER: Water supply, Cincinnati, OH - 0.024 ppb [12]. Detected, but not quantified, in Evansville, IN and Cleveland, OH [33,53]. Not detected in 945 finished water supplies throughout U.S. which use ground water sources [68]. Detected in New Orleans drinking water that was filtered through commercial charcoal filter units [14]. Contamination may have been caused by the filter unit. GROUND WATER: Detected in Iowa well water at 1.0 ppb [32]. Detected but not quantified in a private well in Wisconsin and in ground water in England [15,34]. Maximum concentrations of 10 ppb found in Netherlands [71]. Well water adjacent to landfill containing buried styrene in drums at Gales Ferry, CT had concentrations of 100-200 ppb in 1962 [54]. SURFACE WATER: Water sample from lower TN River - 4.2 ppb [17]. Detected, but not quantified, in Delaware River, Waal River (Netherlands), England surface waters, and Great Lakes [15,19,40,57]. Concentrations of 1 ppb found in Kanawha River, WV and Scheldt River, Netherlands [41].

Effluent Concentrations: Detected, but not quantified, in various chemical, textile, and latex effluents in Louisville, KY, Calvert City, KY, Colliersville, TN, Memphis, TN, and other U.S. locations [56]. Wastewater effluent from LA oil refinery contained 31 ppb [31]. Unspecified industrial wastewater in U.S. - less than 10 ppb [49]. Air in vicinity of oil fire contained 0.5 ppm styrene [48].

Sediment/Soil Concentrations: Sediment sample from lower TN River contained 4.2 ppb [17]. Detected, but not quantified, in sediments from Lake Tobin, Saskatchewan, Canada [52].

Atmospheric Concentrations: SOURCE DOMINATED: Median concn of 2.1 ppb from 135 samples in U.S. [11]. Urban: 0.07-0.25 ppb mean concn in NJ cities [24,25,66]. 0.5-3 ppb concn in Los Angeles, CA and 0.09 ppb concn in Contra Costa, CA [20,66]. 1-15 ppb conc in four California cities [42]. Median concn in indoor air of New Jersey cities 3 times higher than outdoor air [67]. Avg concn less than or equal to 0.1 ppb (0.7 ppb max) in Delft, Netherlands [7]. Detected, but not quantified in 6 USSR cities [29]. Concn of 0.1-0.4 ppb in Dutch cities [60]. Concn of

494

0.2 ppb in Nagoya, Japan [28]. Concn of 26-71 ppb in indoor air of high-rise apartment [63]. RURAL: Detected in air of National Forest in Alabama [28]. OTHER: Detected in highway tunnel in PA [21]. Maximum concn of 15.5 with mean range 0.12-1.53 ppb detected in 7 sanitary and hazardous landfills in NJ [23]. Slightly higher level (1.85 ppb before occupancy vs 2.08 ppb after occupancy) in rooms of an office building after occupancy probably because of emissions from glued carpet [65]. Certain solvent-based adhesives used to finish interior of office buildings emits styrene [16].

Food Survey Values: Detected, but not quantified, in roasted filbert nuts and in 4 of 7 whiskey samples [54]. Tentatively identified in a commercial hickory-wood smoke flavor and in food and water stored in a refrigerator with a plastic interior [28]. Found in yogurt packaged in polystyrene containers at concn of 2.5-34.6 ug/kg with styrene content in yogurt increasing with time [28]. Concn of 59.2 ug/kg found in butterfat cream after 24 days, 9.3 ug/kg in cottage cheese after 27 days and 22.7 ug/kg in honey after 120 days [28].

Plant Concentrations:

Fish/Seafood Concentrations:

Animal Concentrations:

Milk Concentrations: Styrene concn of 17.2 ug/kg in homogenized milk after 19 days storage in polystyrene packaging [28].

Other Environmental Concentrations: Styrene has been detected at concn of 18.0 ug/cigarette in the smoke of American domestic, filter blend cigarettes [28].

Probable Routes of Human Exposure: Exposure of styrene to the general population is possible by oral ingestion of contaminated food which has been packaged in polystyrene, by oral ingestion of contaminated finished drinking water, by inhalation of air contaminated by industrial sources, auto exhaust, or incineration emission, and by inhalation of smoke from cigarettes.

Average Daily Intake: AIR INTAKE: (Assume atmos concn of 0.04-1.2 ug/m^3 [18] and daily inhalation rate of 20 m^3) = 0.8-24 ug; WATER INTAKE: Insufficient data; FOOD INTAKE: Insufficient data.

Occupational Exposures: A National Institute of Occupational Safety and Health (NIOSH) Survey (NOES Survey 1972-1974) has statistically estimated that at least 30,000 full-time workers are exposed to styrene in the U.S. [55]. NIOSH Survey (NOES Survey 1981-1983) has statistically estimated that 108,056 workers are exposed to styrene in the United States [44]. The full-shift time-weighted average (TWA) styrene exposures associated with styrene monomer and copolymer production are generally less than 10 ppm [55]. Average styrene exposures in reinforced plastics/composites plants can range from 40 to 100 ppm, with individual TWA and short-term exposures as high as 150-300 ppm and 1000-1500 ppm, respectively [55]. Workers manufacturing boats and yachts, truck parts, tubs and showers, and tanks and pipes that use reinforced plastics may be substantially exposed to styrene [1,35]. Workers using certain polyester resins may be exposed to styrene [5,39] and the measured TWA ranged from 3.93 to 45.96 ppb [5].

Body Burdens: Detected, but not quantified in 8 of 8 human breast milk samples from U.S. women in 4 cities [47]. Six of 250 patients with suspected environmental exposure-related symptoms showed significantly elevated level of styrene in the blood. Concn ranged from none detected - 1.9 ppb, with a mean value of 0.6 ppb [2]. A National Human Adipose Tissue Survey (NHATS) by EPA during fiscal year 1982 detected styrene in wet adipose tissue with a frequency of 100% at conc range 8-350 ppb [61].

REFERENCES

1. Anderson KE; Diss. Abstr. Int. B 47: 979 (1986)
2. Antoine SR et al; Bull. Environ. Contam. Toxicol. 36: 364-371 (1986)
3. Atkinson R, Carter WPL; Chem Rev 84: 437-70 (1984)
4. Banerjee S et al; Environ Sci Technol 11: 1227-9 (1980)
5. Bartolucci GB et al; Appl. Ind. Hyg. 1: 125-131 (1986)
6. Bignozzi CA et al; Int J Chem Kinet 13: 1235-42 (1981)

7. Bos R et al; Sci Total Environ 7: 269-81 (1977)
8. Boublik T et al; The Vapor Pressures of Pure Substances Vol 17 Amsterdam, Netherlands: Elsevier Science Publ (1984)
9. Bouwer EJ, McCarty PL; Ground Water 22: 433-40 (1984)
10. Bridie AL et al; Water Res 13: 627-30 (1979)
11. Brodzinsky R, Singh HB; Volatile Organic Chemicals in the Atmosphere: An Assessment of Available Data. Menlo Park, CA: Atmospheric Science Center. SRI Intl Contract 68-02-3452 (1982)
12. Coleman WE et al; Arch Environ Contam Toxicol 13: 171-8 (1984)
13. Darnell KR et al; Environ. Sci. Technol. 10: 692-696 (1976)
14. Dowty BJ et al; Environ. Sci. Technol. 9: 762-765 (1985)
15. Fielding M et al; Organic Micropollutants in Drinking Water Medmenham, Eng Water Res Cent p 49 TR-159 (1981)
16. Girman JR et al; Environ International 12: 317-321 (1986)
17. Goodley PC, Gordon M; Kentucky Academy of Science 37: 11-15 (1976)
18. Graedel TE; Chemical Compounds in the Atmosphere p440, Academic Press, NY (1978)
19. Great Lakes Water Quality Board; An Inventory of Chemical Substances Identified in the Great Lakes Ecosystem Vol.1 - Summary p 195 Report to the Great Lakes Water Quality Board, Windsor, Ontario, Canada (1983)
20. Grosjean D, Fung K; J Air Pollut Control Assoc 34: 537-43 (1984)
21. Hampton CV et al; Environ Sci Technol 16: 287-98 (1982)
22. Hansch C, Leo AJ; Medchem Project Issue No 26. Claremont CA: Pomona College (1985)
23. Harkov R et al J Environ Sci Health A20: 491-501 (1985)
24. Harkov R et al; Sci Total Environ 38: 259-74 (1984)
25. Harkov R et al; J Air Pollut Control Assoc 33: 1177-83 (1983)
26. Hine J, Mookerjee PK; J Org Chem 40: 292-8 (1975)
27. Hou CT et al; Appl Environ Microbiol 46: 171-7 (1983)
28. IARC; Some Monomers, Plastics and Synthetic Elastomers and Acrolein 19: 231 (1979)
29. Ioffe BV et al; Environ Sci Technol 13: 864-8 (1979)
30. Junk GA, Ford CS; Chemosphere 9: 187-230 (1980)
31. Keith LH; Sci Total Environ 3: 87-102 (1974)
32. Kelley RD; Synthetic Organic Compounds Sampling Survey of Public Water Supplies p 38 NTIS PB 85-214427 (1985)
33. Kleopfer RD, Fairless BJ; Environ Sci Technol 6: 1036-7 (1972)
34. Krill RM and Sonzogni WC J Amer Water Assoc 78: 70-75 (1986)
35. LeMasters GE et al; Am. Ind. Hyg. Assoc. J 46: 434-441 (1985)
36. Lyman WJ et al; Handbook of Chemical Property Estimation Methods p 15-1 to 15-33 McGraw-Hill NY (1982)
37. Lyman WJ et al; Handbook of Chemical Property Estimation Methods p 4-9 McGraw-Hill NY (1982)
38. Lyman WJ et al; Handbook of Chemical Property Estimation Methods. Environmental Behavior of Organic Compounds. McGraw-Hill NY (1982)
39. Malek RF et al; Am. Ind. Hyg. Assoc. J 47: 524-529 (1986)
40. Meijers AP, Vanderlee RC; Water Res 10: 597-604 (1976)
41. National Academy of Sciences; The Alkyl Benzenes. Washington, D.C., National Academy Press (1980)
42. Neligan RE et al; ACS Natl Mtg pp 118-21 (1965)

Styrene

43. NIOSH; Criteria Document: Styrene, DHEW Pub. NIOSH 83-119, (1983)
44. NIOSH; National Occupational Exposure Survey (preliminary as of 9-20-85) (1985)
45. Ogata M et al; Bull Environ Contam Toxicol 33: 561-7 (1984)
46. Pahren HR, Bloodgood DE; Water Pollut Contr Fed J 33: 233-8 (1961)
47. Pellizzari ED et al; Bull Environ Contam Toxicol 28: 322-8 (1982)
48. Perry R; Mass Spectrometry in the Detection and Identification of Air Pollutants. Int Symp Ident Mass Environ Pollut pp 130-7 (1971)
49. Perry DL et al; Identification of Organic Compounds in Industrial Effluent Discharges USEPA-600/4-79-016 (1979)
50. Price KS et al; J Water Pollut Control Fed 46: 63-77 (1974)
51. Roberts PV et al; J Water Pollut Control Fed 52: 161-71 (1980)
52. Samolloff MR et al; Environ Sci Technol 17: 329-334 (1983)
53. Sanjivamurthy VA; Water Res 12: 31-3 (1978)
54. Santodonato J et al; Investigation of Selected Potential Environmental Contaminants: Styrene, Ethylbenzene and Related Compounds p 261 USEPA-560/11-80-018 (1980)
55. Santodonato J et al; Monograph on Human Exposure to Chemicals in the Workplace: Styrene p 3-1 to 3-12 Bethesda, MD: National Cancer Institute NO1-CP-26002-03 (1985)
56. Shackelford WM, Keith LH; Frequency of Organic Compounds Identified in Water p 213-4 USEPA-600/4-76-062 (1976)
57. Sheldon LS, Hites RA; Environ Sci Technol 12: 1188-94 (1978)
58. Sielicki M et al; Appl Environ Microbiol 35: 124-8 (1978)
59. Singh HB et al; Reactivity/volatility classification of selected organic chemicals: existing data. p 190 USEPA-600/3-84-082 (1984)
60. Smeyers-Verbeke J et al; Atmos Environ 18: 2471-8 (1984)
61. Stanley JS; Broad Scan Analysis of Human Adipose Tissue Vol. 1. Executive Summary p22 Washington, D.C., EPA-560/5-86-035 (1986)
62. Swann RL et al; Res Reviews 85: 17-28 (1983)
63. Tanaka T; Kogai 19: 121-128 (1984)
64. USEPA; Health Assessment Document: Styrene (DRAFT), (1985)
65. Wallace LA et al; Atmos Environ 21: 385-393 (1987)
66. Wallace LA; Toxicol Environ Chem 12: 215-236 (1986)
67. Wallace LA et al; J Occup Med 28: 603-607 (1986)
68. Westrick JJ et al; J Amer Water Works Assoc 76: 52-9 (1984)
69. Wilson JT et al; Devel Indust Microbiol 24: 225-33 (1983)
70. Yalkowsky S.; Arizona Data Base of Water Solubility (1987)
71. Zoetemann BCJ; Sci Total Environ 21: 187-202 (1981)

Toluene-2,4-diamine

SUBSTANCE IDENTIFICATION

Synonyms: 4-Methyl-1,3-benzenediamine; 1,3-Diamino-4-methylbenzene; 2,4-Diaminotoluene

Structure:

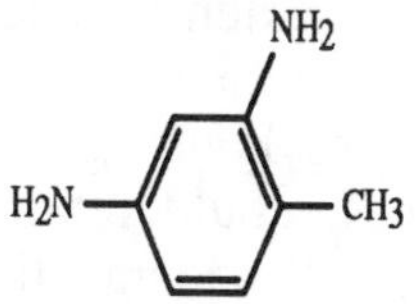

CAS Registry Number: 95-80-7

Molecular Formula: $C_7H_{10}N_2$

Wiswesser Line Notation: ZR CZ D1

CHEMICAL AND PHYSICAL PROPERTIES

Boiling Point: 292 °C

Melting Point: 99 °C

Molecular Weight: 122.17

Dissociation Constants:

Log Octanol/Water Partition Coefficient: 0.337 (calculated) [13]

Water Solubility: 7,470 mg/L (calculated from the octanol/water solubility) [6]

Vapor Pressure: 5.2 x 10^{-5} mm Hg at 20 °C [6]

Henry's Law Constant: 1,2 x 10^{-9} atm-m³/mole (calculated from water solubility and vapor pressure)

ENVIRONMENTAL FATE/EXPOSURE POTENTIAL

Summary: The major release of toluene-2,4-diamine to the environment is expected to occur during the production of the compound and during its major use, the production of diisocyanates. Minor releases may occur from the other uses to which the compound is put. Release of toluene-2,4-diamine to the soil will likely result in extensive leaching due to its high estimated water solubility and low estimated soil sorption partition coefficient, although covalent bonding to soil, which is known to occur with aromatic amines, could retard leaching. Biodegradation in the soil may also occur. Biodegradation and photooxidation are expected to be the major means of degradation in water and soil, although photolysis may also occur to some extent. Volatilization from the water and soil are not expected to be significant. Bioconcentration will not be important. In the atmosphere, toluene-2,4-diamine may photolyze and will react with hydroxyl radicals with an estimated half-life of 7.99 hr.

Natural Sources:

Artificial Sources: Toluene-2,4-diamine is produced in significant quantities (1000 - 100,000,000 lb) [15]. An estimated 16.5 million lb of toluene-2,4-diamine were released during production in 1977 [14]. Releases of toluene-2,4-diamine associated with production are, therefore, expected to constitute the bulk of the toluene-2,4-diamine released to the environment [14]. Releases may result from such uses of toluene-2,4-diamine as the preparation of toluene isocyanates (the major use), dyes, impact resins, polyimides with superior wire coating properties, benzimidazolethiols (antioxidants), hydraulic fluids, urethane foams, and fungicide stabilizers [4]. Very small amounts of toluene-2,4-diamine may be released from boil-in-bags and retortable pouches upon prolonged boiling [12].

Terrestrial Fate: No data on the biodegradation of toluene-2,4-diamine in the soil were available. However, one activated ' sludge screening test does suggest that toluene-2,4-diamine is fairly biodegradable [8] and information on structural analogs would support this belief [3,5,6]. It is expected, therefore, that biodegradation will be an important fate process in soil. Due to the very low estimated vapor pressure and the high estimated water solubility, toluene-2,4-diamine is not expected to volatilize from the soil. With an absorption maximum of 294 nm, toluene-2,4-diamine may directly photolyze under environmental conditions, but this would only occur on the top surface and, therefore, is not expected to be a significant process in soil. Based on the estimated log soil sorption coefficient of 1.56, toluene-2,4-diamine is expected to leach extensively, and may reach ground water, although covalent bonding to soil, which is known to occur with aromatic amines, could retard leaching. Biodegradation in the soil may also occur.

Aquatic Fate: Toluene-2,4-diamine is not expected to hydrolyze significantly [6]. Toluene-2,4-diamine biodegrades in the presence of activated sludge [8]. Toluene-2,4-diamine may, therefore, biodegrade in water under environmental conditions. Singlet oxygen, alkyl peroxyl radicals, alkoxy radicals, and hydroxyl radicals, which are generated by sunlight, are capable of oxidizing toluene-2,4-diamine. The reaction of aniline with alkyl peroxyl radicals is rapid, with a half-life of <several days [7]. Toluene-2,4-diamine may, therefore, be subject to photooxidations of this type in water. With an absorption maximum of 294 nm [11], toluene-2,4-diamine may directly photolyze under environmental conditions. Volatilization from water is expected to be insignificant since the value of the Henry's Law constant indicates that toluene-2,4-diamine will be less volatile than water [6].

Atmospheric Fate: An atmospheric half-life of 7.99 hr has been estimated for the reaction of toluene-2,4-diamine with hydroxyl radicals [2]. It is possible that toluene-2,4-diamine directly photolyzes in the atmosphere since its absorption maximum is at 294 nm.

Biodegradation: A decline of 45% in the theoretical organic carbon of toluene-2,4-diamine was observed after 4-hr exposure to activated sludge in a fill and draw apparatus [8]. Although the experimental method used did not preclude losses due to volatilization and hydrolysis, toluene-2,4-diamine is not expected to either volatilize [6] or hydrolyze [6] significantly, leaving biodegradation as the most likely mechanism responsible for this result. The presence of amine groups confers some degree of biodegradability on aromatic rings [5]. Aniline has been classified as a relatively biodegradable compound [6] and 2-methylaniline and 4-methylaniline have been confirmed to be biodegradable [3]. It is expected, therefore, that toluene-2,4-diamine will be biodegraded in the environment.

Abiotic Degradation: Toluene-2,4-diamine is not expected to hydrolyze significantly [6]. Singlet oxygen, alkyl peroxyl radicals, alkoxy radicals, and hydroxyl radicals, which are generated by sunlight, are capable of oxidizing toluene-2,4-diamine. The reaction of aniline with alkyl peroxy radicals is rapid, with a half-life of <several days [7]. Toluene-2,4-diamine may, therefore, be subject to photooxidations of this type in the water. With an absorption maximum of 294 nm (E = 2891 L/M cm) in methanol [11], toluene-2,4-diamine may also be subject to direct photolysis under environmental conditions. An atmospheric half-life of 7.99 hr has been estimated for the reaction of toluene-2,4-diamine with hydroxyl radicals [2].

Bioconcentration: A bioconcentration factor (BCF) of 91 was determined for an unspecified isomer of diaminotoluene using fathead minnows exposed for 32 days [16]. Using the log octanol/water partition coefficient, a BCF of 1.06 was estimated [6]. Based on these BCF values, toluene-2,4-diamine is not expected to bioconcentrate significantly in aquatic organisms.

Soil Adsorption/Mobility: The estimated log octanol/water partition coefficient was used to estimate a log soil sorption coefficient of 1.56 [6]. Based on this Koc value and the high estimated water solubility of toluene-2,4-diamine, extensive leaching and very little adsorption to soil or sediment is expected. However, aromatic amines are known to form covalent bonds to humic materials (quinone-like compounds), followed by slow oxidation [9]. Toluene-2,4-diamine may also react in this way, although no data were available indicating that it does so. Any toluene-2,4-diamine thus chemically bound will most likely remain in the soil much longer than the purely physical interactions suggest.

Volatilization from Water/Soil: Based on the Henry's Law constant value, toluene-2,4-diamine should be less volatile than water and volatilization is not expected to be significant [6].

Water Concentrations:

Effluent Concentrations:

Sediment/Soil Concentrations:

Atmospheric Concentrations: Samples of air collected around wire coating machines were found to contain ND(<0.0005)-8.13 mg/m^3 with an average concentration of 0.99 mg/m^3 [10]. Samples of working atmosphere collected at Olin Chemical Company, a producer of toluene-2,4-diamine, contained toluene-2,4-diamine at 0.008-0.39 mg/m^3 in one study and ND-0.038 mg/m^3 in a subsequent study [1].

Food Survey Values:

Plant Concentrations:

Fish/Seafood Concentrations:

Animal Concentrations:

Milk Concentrations:

Other Environmental Concentrations:

Probable Routes of Human Exposure:

Average Daily Intake:

Occupational Exposures: Samples collected around a wire coating machine and in the working atmosphere of a producer of toluene-2,4-diamine have shown detectable quantities of the chemical. From this data and the other fate information, it is expected that the major human exposure will be in occupational settings where toluene-2,4-diamine is produced or used in large quantities.

Body Burdens:

REFERENCES

1. Ahrenholz SH; Gov Rep Announce Index 81: 2765-89 (1980)
2. GEMS; Graphical Exposure Modeling System. Fate of atmospheric pollutants (FAP) data base. Office of Toxic Substances USEPA (1985)
3. Hutzinger O et al; Aquatic Pollutants: Transformation and Biological Effects, Pergammon Press pp.283-98 (1978)
4. Kirk-Othmer Encycl Chem Tech; 3rd ed John Wiley and Sons New York 2: 321-29 (1978)
5. Leisinger T et al; Microbial Degradation of Xenobiotics and Recalcitrant Compounds; Academic Press London p.41 (1981)
6. Lyman WJ et al; Handbook of Chemical Property Estimation Methods. Environmental behavior of organic compounds, McGraw-Hill New York (1982)
7. Mabey WR et al; Aquatic Fate Process Data for Organic Priority Pollutants USEPA-440/4-81-014 (1981)
8. Matsui S et al; Prog Water Technol 7: 645-59 (1975)
9. Parris GE; Environ Sci Technol 14: 1099-1106 (1980)
10. Rosenberg C; Analyst (London) 109: 859-66 (1984)
11. Sadtler 2177 UV
12. Snyder RC, Breder CV; J Chromatogr 236: 429-40 (1982)
13. USEPA; Graphical Exposure Modeling System; Medchem software release 3.32, CLOGP3 (1985)
14. USEPA; Materials balance for toluene-2,4-diamine. Office of Toxic Substances (1980)
15. USEPA; Production statistics for chemicals in the nonconfidential initial TSCA inventory. Office of Pesticides and Toxic Substances (1985)
16. Veith GD et al; J Fish Res Board Can 36: 1040-48 (1979)

Toluene-2,4-diisocyanate

SUBSTANCE IDENTIFICATION

Synonyms: 2,4-Diisocyanato-1-methylbenzene; 2,4-Diisocyanatoluene

Structure:

NCO

H₃C—⟨benzene ring⟩—NCO

CAS Registry Number: 584-84-9

Molecular Formula: $C_9H_6N_2O_2$

Wiswesser Line Notation: OCNR B1 ENCO

CHEMICAL AND PHYSICAL PROPERTIES

Boiling Point: 251 °C at 760 mm Hg

Melting Point: 19.5-21.5 °C

Molecular Weight: 174.17

Dissociation Constants:

Log Octanol/Water Partition Coefficient:

Water Solubility:

Vapor Pressure: 0.01 mm Hg at 20 °C [5]

Henry's Law Constant:

Toluene-2,4-diisocyanate

ENVIRONMENTAL FATE/EXPOSURE POTENTIAL

Summary: 2,4-Toluene diisocyanate constitutes roughly 80% of commercial toluene diisocyanate (TDI), the other 20% being the 2,6-isomer. It may be released to the environment as fugitive emissions and from stack exhaust during the production, transport, and use of TDI in the manufacture of polyurethane foam products and coatings as well as from spills. Much of the environmental fate and monitoring data does not distinguish between the isomers of toluene diisocyanate; however, since the types of reactions are similar, some of the data presented below is for the commercial TDI mixture. The reactivity of the 2,4-toluene diisocyanate with compounds containing active hydrogen, such as water, alcohols, or acids is greater than with the 2,6-isomer; therefore, the percentage of 2,4-toluene diisocyanate in monitored TDI is generally much less than 80% in the commercial mixture. If spilled on wet land, TDI is rapidly degraded. In one experiment simulating a spill, 5.5% of the original material remained after 24 hours, and in a field situation the concentration of TDI had declined to the ppm level in 12 weeks. If released into water, a crust forms around the liquid TDI and < 0.5% of the original material remains after 35 days. Low concentrations of TDI hydrolyze in the aqueous environment in approx a day. In the atmosphere, TDI reacts with photochemically produced hydroxyl radicals (half-life 3.3 hr) and is also removed by dry deposition. Human exposure is primarily occupational via inhalation and dermal contact.

Natural Sources: 2,4-Toluene diisocyanate is not known to occur as a natural product [5].

Artificial Sources: 2,4-Toluene diisocyanate may be released to the environment, primarily as stack and fugitive emissions into air during its commercial production, handling, and processing prior to polyurethane foam production [5]. It may also be released during the manufacture of polyurethane foam products and coatings, from sprays, insulation materials, and polyurethane foam coated fabrics [5]. About 0.005% of the toluene diisocyanate (TDI) used in flexible foam slabstock plants is released in vent stack exhaust and this contains <50% of the 2,4-isomer [1]. Large quantities of TDI are transported by rail, road, and sea and there is a possibility of spills during loading or unloading [1].

Toluene-2,4-diisocyanate

Terrestrial Fate: Ten days after a spill of 13 tons of toluene diisocyanate (TDI) onto swampy, wet forest soil, TDI and toluenediamines were found in the soil [1,5]. The TDI solidified and the area was covered with sand. The soil concn of TDI and toluenediamine combined declined from the parts per thousand to parts per million range between 10 days and 12 weeks after the spill [1]. No TDI was detected in a connecting brook 10 days after the spill [1]. After 6 years, only TDI-derived polyureas were found at the site [5]. In a simulated spill, 5 kg of TDI in a container was covered with 50 kg of sand and 5 kg of water at ambient temperatures and samples taken from the top and bottom of the sand pile. After 24 hr, 5.5% of the TDI remained unreacted and after 8 days 3.5% remained [1]. The reaction product was largely polyureas [1].

Aquatic Fate: When low concns of TDI are released into model river or seaway systems, it is hydrolyzed within a day [2]. In order to simulate a spill of TDI in stagnant water, 0.5 L of TDI was poured into 20 liters of water at pH 5, 7, or 9 at 20 °C. A hard compact crust formed at the water TDI interface [1]. This crust thickened over a period of 30 days until no liquid TDI remained. The solid crust contained < 0.5% of the original TDI after 35 days [1]. Another experiment was performed to simulate a spill into running water by pouring 0.5 L of TDI into 20 L of slowly overflowing water. Barely detectable amounts of toluenediamine was present in the overflow samples. As in the stagnant water simulation, a crust formed that contained < 0.5% of the original TDI after 35 days [1].

Atmospheric Fate: Removal of TDI from the atmosphere is due to reaction with photochemically produced hydroxyl radicals (half-life 3.3 hr) as well as through dry deposition. Earlier ideas that reaction with water vapor was of major importance have not been substantiated in recent experiments.

Biodegradation:

Abiotic Degradation: 2,4-Toluene diisocyanate reacts readily with compounds containing active hydrogens, such as water, acids, and alcohols [5]. Contact with bases may cause uncontrollable

507

polymerization with rapid evolution of heat [5]. Hydrolysis of one of the isocyanate groups to the amine will be followed by rapid reaction of the amine with an isocyanate group on another molecule leading to dimers, oligomers, and polymers unless the solution is very dilute [5]. In one experiment the hydrolysis product yield was 20% diamine and 80% polyurea [12]. When 50 ppm of toluene diisocyanate was added to a model river and seawater systems, the concn was 0.1 ppm or less at the end of 1 day [2]. Gas phase loss of TDI was originally thought to be due to reaction of TDI with water vapor [4]. One early work showed that the percent disappearance of TDI in air depended almost solely on the water vapor concentration, increasing 3.2% per unit increase in absolute humidity (g water/kg dry air), so that at 15 g water/kg dry air, a 50% reduction in TDI was obtained [4]. These reductions were obtained after 8 sec and did not differ appreciably after 75 sec [4]. The reaction product of 2,4-toluene diisocyanate vapor with limited quantities of moisture was reported to be 3,3'-diisocyanto-4,4'-dimethylcarbanilide [6]. Other work contradicted these earlier findings and a program was instituted to study the gas-phase reactions between TDI and water using a room sized environmental chamber [2]. They found that over a relative humidity range of 7-70%, the loss rate of TDI was independent of humidity and no toluenediamine or TDI-urea products could be detected [2]. In addition deposition onto the chamber walls was an important removal process [2]. Subsequent experiments were performed in the environmental chamber to assess the importance of photolysis, reaction with free radicals, and adsorption onto particulate matter as atmospheric removal processes [3]. The loss rate of TDI in irradiated clean air was first order with a half-life of 3.3 hr [3]. By using a free radical scavenger, it was shown that free-radicals and not photolysis was responsible for the removal. The half-life is consistant with the reaction with photochemically generated hydroxyl radicals [3]. The addition of a urban surrogate hydrocarbon mixture (polluted urban air) or ammonium sulfate particulate matter, a major component of atmospheric aerosols, did not significantly alter removal rates [3]. Because the concentration of hydroxyl radicals generally increases in polluted air, shorter half-lives are expected.

Bioconcentration:

Toluene-2,4-diisocyanate

Soil Adsorption/Mobility:

Volatilization from Water/Soil: The vapor pressure would suggest that spilled TDI will not readily volatilize from soil and surfaces.

Water Concentrations:

Effluent Concentrations: Toluene diisocyanate (TDI), no isomer specified, has been reported in waste water from furniture manufacturing in the concn range of 0.1-4.1ppm [5]. Stack exhaust from a polyurethane foam production plant was reported to contain 100-17,700 ug/m^3 of TDI (no isomer specified) [5].

Sediment/Soil Concentrations:

Atmospheric Concentrations:

Food Survey Values:

Plant Concentrations:

Fish/Seafood Concentrations:

Animal Concentrations:

Milk Concentrations:

Other Environmental Concentrations: Commercial toluene diisocyanate typically contains 80% 2,4-toluene diisocyanate [5]. TDI monomer has been found in a urethane foam fabric coating in concn of <200 ppm [5].

Probable Routes of Human Exposure: Exposure to 2,4-toluene diisocyanate is primarily occupational via inhalation contact with aerosol or vapor. Dermal exposure occurs through skin contact with freshly cured foam or as a result of spills or splashes.

Average Daily Intake:

Occupational Exposures: NIOSH has estimated that 29,284 workers are potentially exposed to 2,4-toluene diisocyanate based on statistical estimates derived from a survey conducted in 1972-1974 in the United States [8]. This survey sampled 5000 businesses in 67 metropolitan areas throughout the United States for the manufacture and use of chemicals, trade name products known to contain the compound and generic products suspected of containing the compound. NIOSH has estimated that 2103 workers are potentially exposed to 2,4-toluene diisocyanate based on statistical estimates derived from a survey conducted in 1981-1983 in the United States [7]. Airborne toluene diisocyanate in two plants producing flexible polyurethane foam for use in mattresses, carpet pads, and air filters by a continuous slab process were determined [10]. In this process, the starting materials are mixed internally in a foaming nozzle and are poured onto moving kraft paper, at which time the material foams and heats up (mixing end). The foam slab is then carried into a ventilated curing tunnel, during which time the unreacted TDI and other material is driven off due to the high internal temperature (finishing end). The paper liner is subsequently stripped off and the slab cut into pieces. Concentration medians for 2,4-toluene diisocyanate were 0.70 and 0.33 ppb for the mixing and finishing ends, respectively [10]. Only an insignificant number of samples exceeded the 20 ppb ceiling recommended by the ACGIH [10]. In a similar Finish study, TDI vapor concentrations ranged from 25 to 80 ug/m^3 for pouring, 14 to 44 ug/m^3 for paper stripping, and 1-8 ug/m^3 for cutting [11]. Personal monitors measured 9-34 ug/m^3 and 5-30 ug/m^3 of the 2,4-isomer for pouring and paper stripping, respectively [11]. While 2,4-toluene diisocyanate constitutes 80% of TDI, and 2,6-toluene diisocyanate 20%, the 2,4- isomer is more reactive so that the concentration of the 2,6-isomer is higher in workplace air [11]. The percentage of 2,4-toluene diisocyanate to total toluene diisocyanate measured with the personal monitors ranged from 29 to 69% [11]. Mean concn ranges of TDI reported in ambient workplace air were 0.7-710 ug/m^3 in TDI production; ND-1490 ug/m^3 during polyurethane foam production; 70-140 mg/m^3 during elastomers production; 13-1050 ug/m^3 during polyurethane foam use; 10-710 ug/m^3 during polyurethane spray paint use; and <1-740 ug/m^3 during the production of polyurethane-coated wire [5]. Mean concn in personal samples ranged from ND to 540 ug/m^3 during polyurethane foam production and 2 to 1220 ug/m^3 during polyurethane spray foam use [5]. Additionally, mean concn of TDI released from insulation in a ship's hold was reported to be

Toluene-2,4-diisocyanate

120-150 ug/m^3 and that released from coated fabric in a workplace 2-10 ug/m^3 [5]. During the outdoor application of TDI foam to a 40-ft diameter storage tanks, the mean atmospheric TDI concentration downwind from the spray gun was 0.3 ppm at 8 ft, 0.02 ppm at 40 ft, and 0.002 ppm at 150 ft [9]. TDI was not detected more than 8 ft upwind of the spray gun [9]. Air samples taken within 2 ft of the foam surface immediately after spraying had ceased were found to have 0.03 ppm [9]. Mean TDI concentrations near the pumping equipment during material transfer was 0.02 ppm [9].

Body Burdens:

REFERENCES

1. Brochhagen FK, Grieveson BM; Cellular Polymers 3: 11-17 (1984)
2. Duff PB; pp. 408-12 In: Polyurethane - New Paths to Progress Marketing Technology (1983)
3. Duff PB; Proc of the SPI Tech/Mark Conf 29: 9-14 (1985)
4. Dyson WL, Hermann ER; Amer Ind Hyg Assoc J 32: 741-4 (1971)
5. IARC Monograph on the Evaluation of the Carcinogenic Risk of Chemicals to Humans. Some Monomers, Plastics and Synthetic Elastomers, and Acrolein 19: 303-11 (1979)
6. Marcali K; Anal Chem 29: 552 (1957)
7. NIOSH; National Occupational Exposure Survey (1985)
8. NIOSH; National Occupational Health Survey (1975)
9. Peterson JE et al; Amer Ind Hyg Assoc J 23: 345-52 (1962)
10. Rando RJ et al; Am Ind Hyg Assoc J 45: 199-203 (1984)
11. Rosenberg C, Savolainen H; J Chromatography 367: 385-92 (1986)
12. Sopach ED, Boltromeyuk LP; Gig Sanit 7: 10-13 (1974)

Toluene-2,6-diisocyanate

SUBSTANCE IDENTIFICATION

Synonyms: 1,3-Diisocyanato-2-methylbenzene

Structure:

CAS Registry Number: 91-08-7

Molecular Formula: $C_9H_6N_2O_2$

Wiswesser Line Notation: OCNR B1 CNCO

CHEMICAL AND PHYSICAL PROPERTIES

Boiling Point: 129-133 °C at 18 mm Hg

Melting Point:

Molecular Weight: 174.17

Dissociation Constants:

Log Octanol/Water Partition Coefficient:

Water Solubility:

Vapor Pressure: 0.02 mm Hg at 20 °C (commercial toluene diisocyanate - 80 % 2,4-isomer,20% 2,6-isomer) [6]

Henry's Law Constant:

ENVIRONMENTAL FATE/EXPOSURE POTENTIAL

Summary: 2,6-Toluene diisocyanate constitutes roughly 20% of commercial toluene diisocyanate (TDI), the other 80% being the

2,4-isomer. TDI may be released to the environment through stack and fugitive emissions to the air during its commercial production and handling and its use in producing polyurethanes. It has been released to the environment through accidental spillage. Much of the fate and monitoring data do not distinguish between the isomers of TDI and reflect results of the commercial mixture of the 2,4- and 2,6-isomers. If spilled on wet land, TDI is rapidly degraded. In one experiment simulating a spill, 5.5% of the original material remained after 24 hours and in a field situation, the concn of TDI had declined to the ppm level in 12 weeks. If released into water in spill situation, a crust forms around the liquid TDI and <0.5% of the original material remains after 35 days. Low concentrations of TDI hydrolyze in the aqueous environment in approx a day. If released to the atmosphere, TDI reacts rapidly with photochemically produced hydroxyl radicals (half-life of 3.3 hr). Human exposure is primarily occupational via inhalation and dermal contact.

Natural Sources: It is not known whether 2,6-toluene diisocyanate occurs as a natural product [6].

Artificial Sources: Toluene diisocyanate (both the 2,4- and 2,6-isomers) may be released to the environment through stack and fugitive emissions to the air during its commercial production and handling and its use in producing polyurethanes [3,6]. About 0.005% of the TDI used in flexible foam slabstock plants is released in vent stack exhaust; this exhaust contains >50% of the 2,6-isomer [2]. Release of TDI to the environment may also occur from accidental spillage to land or surface waters during transit [3].

Terrestrial Fate: Ten days after a spill of 13 tons of toluene diisocyanate (TDI) (2,4- plus 2,6-isomers) onto a swampy, wet forest soil, TDI and toluenediamines were found in the soil [2,6]. The TDI solidified and the area was covered with sand. The soil concn of TDI and toluenediamine combined declined from the parts per thousand to parts per million range between 10 days and 12 weeks after the spill [2]. One year later, neither TDI nor toluenediamine could be found in soil samples [3]. After 6 years, only TDI-derived polyureas were found at the site in soil samples at depths to 100 cm [3]. In a simulated spill, 5 kg of TDI in a container was covered with 50 kg of sand and 5 kg of water at ambient temperatures and samples were taken from the top and bottom of the sand pile [2]; after 24 hours, 5.5% of the TDI remained unreacted and after 8 days, 3.5% remained [2]; the reaction product was largely polyureas [2].

Aquatic Fate: When low concn of TDI (2,4- plus 2,6-isomers) are released into model river or seaway systems, it is hydrolyzed within a day. In order to simulate a spill of TDI in stagnant water, 0.5 liters of TDI was poured into 20 liters of water at pH 5, 7, or 9 at 20 °C. A hard compact crust formed at the water-TDI interface [2]. This crust thickened over a period of 30 days until no liquid TDI remained. The solid crust contained <0.5% of the original TDI after 35 days [2]. Another experiment was performed to simulate a spill into running water by pouring 0.5 liters of TDI into 20 liters of slowly overflowing water. Barely detectable amounts of toluenediamine were present in the overflow samples. As in the stagnant water simulation, a crust formed that contained <0.5% of the original TDI after 35 days [2].

Atmospheric Fate: Removal of TDI from the atmosphere is due to reaction with photochemically produced free radicals (hydroxyl radicals), which proceeds at a half-life rate of 3.3 hr in typical clean air. Reaction of TDI with water vapor in the ambient atmosphere is not an important removal process as shown by experimental studies.

Biodegradation:

Abiotic Degradation: In concentrated solutions, hydrolysis of a single isocyanate function to the amine is followed by rapid reaction of the amine, with an isocyanate function in a second molecule resulting in the formation of dimers, oligomers, and polymers [6]. When 50 ppm of toluene diisocyanate (2,4- and 2,6-isomers) was added to a model river and seawater systems, the concn fell to 0.1 ppm or less at the end of 1 day [3]. Experimental studies considering the gas-phase reaction of water vapor and toluene diisocyanate (2,4- and 2,6-isomers) in the atmosphere have determined that this reaction does not proceed rapidly (and possibly not at all) in air [3,5]; reaction between water vapor and TDI does not control the lifetime and fate of TDI in air [4]. Experiments were performed in an environmental chamber to assess the importance of photolysis, reaction with free radicals, and adsorption onto particulate matter as atmospheric removal processes [4]; the loss rate of TDI (2,4- plus 2,6-isomers) in irradiated clean air was first order with a half-life of 3.3 hr [4]; it was shown that free radicals, and not photolysis, was responsible for the removal [4]; the half-life was consistent with estimated reaction rates via hydroxyl radicals [4]; adsorption to particulate matter was not significant [4].

Bioconcentration:

Soil Adsorption/Mobility:

Volatilization from Water/Soil: The vapor pressure of commercial toluene diisocyanate (80% 2,4- and 20% 2,6- isomer) is 0.02 mm Hg at 20 °C [6], suggesting that spilled TDI will not readily volatilize from soil and surfaces. TDI reacts with water; therefore, volatilization from water is not expected to be important.

Water Concentrations:

Effluent Concentrations: Toluene diisocyanate (TDI), no isomer specified, has been reported in wastewater from furniture manufacturing in the concn range 0.1-4.1 ppm [6]. Stack exhaust from a polyurethane foam production plant was reported to contain 100-17,700 ug/m^3 of TDI (no isomer specified).

Sediment/Soil Concentrations:

Atmospheric Concentrations:

Food Survey Values:

Plant Concentrations:

Fish/Seafood Concentrations:

Animal Concentrations:

Milk Concentrations:

Toluene-2,6-diisocyanate

Other Environmental Concentrations: Commercial toluene diisocyanate typically contains 20% 2,6-toluene diisocyanate [6]. TDI monomer has been found in a urethane foam fabric coating at concentrations below 200 ppm [6].

Probable Routes of Human Exposure: Exposure to toluene diisocyanate is primarily occupational via inhalation contact with aerosol or vapor [6]. Dermal exposure can occur through skin contact with freshly cured foam or as a result of spills or splashes.

Average Daily Intake:

Occupational Exposures: NIOSH has estimated that 18,820 workers are potentially exposed to 2,6-toluene diisocyanate based on statistical estimates derived from a survey conducted between 1972-1974 in the United States [8]. NIOSH has estimated that 549 workers are potentially exposed to the 2,6-toluene diisocyanate isomer mixtures based on statistical estimates derived from a survey conducted between 1981-1983 in the United States [7]. Median 2,6-TDI values of 6.4 and 7.8 ug/m^3 were detected at the initial mixing and finishing ends, respectively, of a plant producing polyurethane foam, with a max concn above 450 ug/m^3 at the finishing end [9]. Workplace air of a factory manufacturing polyurethane-coated wire contained TDI (2,4- plus 2,6-isomers) levels ranging from <1-740 ug/m^3 [10]. TDI levels of 1.3-2.8 ppb were detected in the air inside a factory manufacturing polyurethanes [1]. Mean TDI levels (2,4- plus 2,6-isomers) of 0.7-180 ug/m^3 were reported for workplace air in U.S. factories (1973-8) manufacturing TDI [6]. Mean TDI levels (2,4- plus 2,6-isomers) ranging from below detection limits to 540 ug/m^3 for personal and workplace air in U.S. factories (1972-81) producing polyurethane foam [6].

Body Burdens:

REFERENCES

1. Audunsson G et al; Intern J Environ Anal Chem 20: 85-100 (1985)
2. Brochhagen FK, Grieveson BM; Cellular Polymers 3: 11-17 (1984)
3. Duff PB; p.408-12 in Polyurethane-New Paths to Progress-Marketing-Technology. Proc of the SPI 6th Inter Tech/Market Conf (1983)

Toluene-2,6-diisocyanate

4. Duff PB; Proc SPI Annu Tech/Mark Conf 29: 9-14 (1985)
5. Holdren MW et al; Am Ind Hyg Assoc J 45: 626-33 (1984)
6. IARC Monograph on the Evaluation of the Carcinogenic Risk of Chemicals to Humans. Some Chemicals Used in Plastics and Elastomers, 39: 287-323 (1985)
7. NIOSH; National Occupational Exposure Survey (NOES) (1983)
8. NIOSH; National Occupational Hazard Survey (NOHS) (1974)
9. Rando RJ et al; Am Ind Hyg Assoc J 45: 199-302 (1984)
10. Rosenberg C; Analyst 109: 859-66 (1984)

1,2,4-Trichlorobenzene

Synonyms:

Structure:

CAS Registry Number: 120-82-1

Molecular Formula: $C_6H_3Cl_3$

Wiswesser Line Notation: GR BG DG

CHEMICAL AND PHYSICAL PROPERTIES

Boiling Point: 213.5 °C at 760 mm Hg; 84.8 °C at 10 mm Hg

Melting Point: 17 °C.

Molecular Weight: 181.46

Dissociation Constants:

Log Octanol/Water Partition Coefficient: 4.02 [13]

Water Solubility: 48.8 mg/L water at 20 °C [6]

Vapor Pressure: 0.29 mm Hg at 25 °C; 1.0 mm Hg at 38.4 °C [52]

Henry's Law Constant: 1.42×10^{-3} atm-m^3/mol (calculated from water solubility and vapor pressure [22]

1,2,4-Trichlorobenzene

ENVIRONMENTAL FATE/EXPOSURE POTENTIAL

Summary: 1,2,4-Trichlorobenzene's (1,2,4-TCB) release to the environment will occur through its manufacture and use as a dye carrier (major use), intermediate in the manufacture of herbicides and higher chlorinated benzenes, dielectric fluid, solvent, heat-transfer medium, and its use in degreasing agents, septic tank and drain cleaners, wood preservatives, and abrasive formulations. If it is released to soil, it will probably adsorb to soil and therefore will not leach appreciably to ground water. However, 1,2,4-TCB has been detected in some ground water samples, which indicates that it can be transported there by some process. 1,2,4-TCB will not hydrolyze or biodegrade in ground water, but it may biodegrade slowly in the soil based upon the data from one experiment. If 1,2,4-TCB is released to water it will adsorb to the sediments and may bioconcentrate in aquatic organisms. It will not hydrolyze in surface waters but it may be subject to significant biodegradation. It is expected to significantly evaporate from water with half-lives of 11-22 days for evaporation from a study of a mixed, 5.4 m deep seawater microcosm, and a half-life of 4.2 hr predicted for evaporation from a model river. Adsorption to sediments or absorption by microorganisms may retard the rate of evaporation. It will not appreciably directly photolyze in surface waters based on a reported half-life for sunlight photolysis in surface water at 40 deg latitude in summer of 450 yrs. If 1,2,4-TCB is released to the atmosphere it will react with photochemically produced hydroxyl radicals with a resulting estimated vapor phase half-life in the atmosphere of 18.5 days. Exposure to 1,2,4-TCB will result mainly from occupational exposure during its manufacture and use, while general population exposure will result from the ingestion of contaminated drinking water and food, especially contaminated fish.

Natural Sources: 1,2,4-TCB is a synthetic organic chemical and is not known to occur as a natural product.

Artificial Sources: 1,2,4-TCB's release to the environment will occur through its manufacture and use as a dye carrier (major use), intermediate in the manufacture of herbicides and higher chlorinated benzenes, dielectric fluid, solvent, heat-transfer medium

[50], and its use in degreasing agents, septic tank and drain cleaners, wood preservatives, and abrasive formulations [27].

Terrestrial Fate: If 1,2,4-TCB is released to soil, it will be expected to adsorb to the organic matter in soil and therefore should not leach appreciably to the ground water. However, 1,2,4-TCB has been detected in some ground water samples, which indicates that it can be transported there by some process. It will not hydrolyze but it may biodegrade slowly in the soil based upon the data from one experiment [25]; it will not be expected to biodegrade in the ground water [40].

Aquatic Fate: If 1,2,4-TCB is released to water, it will be expected to adsorb to the sediment. Bioconcentration in aquatic organisms has been measured and values ranging from 51 to 2800 have been reported. It will not hydrolyze. Although no data were found concerning biodegradation in natural waters, 1,2,4-TCB may be subject to biodegradation in such waters based on limited laboratory tests using various wastewaters as sources for microbes. 1,2,4-TCB may be subject to significant evaporation from water with half-lives of 11-22 days for evaporation from a study of a mixed, 5.4 m deep seawater microcosm [58]. Using the Henry's Law constant, a half-life of 4.2 hr was predicted for evaporation from a model river. Adsorption to sediments or absorption by microorganisms may retard the evaporation process. It will not be expected to appreciably directly photolyze in surface waters based on a reported half-life for sunlight photolysis in surface water at 40 deg latitude in summer of 450 yr [8]. Half-lives of 2.1, 1.5, and 28 days were estimated in rivers in The Netherlands based upon monitoring data [61].

Atmospheric Fate: If 1,2,4-TCB is released into the atmosphere, it will be subject to reaction with photochemically produced hydroxyl radicals with an estimated vapor phase half-life in the atmosphere of 18.5 days. It will not be expected to be subject to appreciable direct photolysis.

Biodegradation: Percent biodegradation of 1,2,4-TCB at concn 5 ppm (10 ppm) by microbes in settled domestic wastewater in original culture, 1st, 2nd, and 3rd subculture, respectively (7 days between each measurement and subculture): 54(43), 70(54), 59(14), and 24(0). Significant biodegradation, gradual adaptation followed by a reduction of the biodegradation rate and an accumulation of 1,2,4-TCB in the culture media [48]. BOD_{20} (type of treatment plant source of microbes): 0% (municipal wastewater); 78%, 100%, and 55% (industrial wastewater); percent disappearance of last BOD_{20} value by GC detection, 99% after 10 days [42]. Percent remaining after 135 hr incubation with adapted (unadapted) domestic wastewater: 44% (100%) [11]. Very slow biodegradation rate for 1,2,4-TCB at concn of 50 ug/g Nixon sandy loam soil (mineralization rate of 0.181 ug/day/20 g soil) [25]. No degradation reported in study of the transport and degradation of 1,2,4-TCB injected into ground water and monitoring wells at different distances from the injection well [40]. Oxidation by benzene-acclimated activated sludge led to 1.0% theoretical COD in 192 hr [23]. Very little mineralization (<.2% for concn of 1,2,4-TCB of 78 or 434 ppm/g of saturated subsurface soil), with no adaptation period observed in aerobically incubated ground water from Lulu, OK [1].

Abiotic Degradation: 1,2,4-TCB will not hydrolyze under environmental conditions [41]. Exposure of 20 g 1,2,4-TCB in borosilicate glass containers to sunlight for 56 days resulted in formation of 9770 ppm polychlorinated biphenyl [55]. The half-life for sunlight photolysis in surface water at 40 deg latitude in summer was 450 yr [8]. Recovery of 1,2,4-TCB from isopropanol solution in Pyrex glass tubing (cutoff 285 nm) irradiated with 300 and 310 nm fluorescent lamps for 30 min was 89.4% (under N_2) and 8.1% (aerated); products of photodegradation were 1,3- and 1,4-trichlorobenzene [2]. Using a reaction rate of 0.532 x 10^{-12} cm^3/molecule-sec [3], an estimated vapor phase half-life in the atmosphere of 18.5 days was calculated as a result of reaction with photochemically produced hydroxyl radicals (concn 8 x 10^{+5} molecules/cm^3).

Bioconcentration: BCF: Fathead minnow, 2800 [57]; rainbow trout (fingerling muscle): 8 hr static conditions, 51; 35 day flowing conditions, 89 [45]; rainbow trout, 1200 (Lake Ontario), 1300 (lab data) [29]; 8 species of fish: 124-1300 (data points), avg 863, median 810 [12]; bluegill sunfish, 812 [4]; golden ide (Leuciscus idus melanotus), 490 [9].

1,2,4-Trichlorobenzene

Soil Adsorption/Mobility: Percent desorbed from 3 sediments after 8 days at 20 °C: 19%, 19%, and 25% [34]. Reported Koc: 1441 [10]; 2042 [50]; 5000 [6]; and 1000 [59]. Percent eluted through column of sandy soil over 21 days: 39-46% (54-61% degraded or not accounted for) [59]. 1,2,4-TCB reportedly migrated quickly through the soil at a major spill site where 1500 gallons of mixed solvents were spilled [27]; The undetermined effects on soil mobility of the size of the spill and the presence of many other organic solvents diminishes the relevance of this data. The Koc has been estimated to be 670 [18].

Volatilization from Water/Soil: Complete evaporation from distilled water solutions: non-aerated, within 72 hr; aerated, within 4 hr; Aerated in the presence of mixed cultures of microorganisms, 2% remained after 80 hr [50]. Half-lives for evaporation from mesocosms with seawater, 5.4 m deep and mixed four times a day to simulate waves: spring, 22 days (8-16 °C); summer, 11 days (20-22 °C); winter, 12 days (3-7 °C) [58]. Using the Henry Law constant, a half-life of 4.2 hr was predicted for evaporation from a model river 1 m deep, flowing at 1 m/sec with a wind velocity of 3 m/sec [22].

Water Concentrations: DRINKING WATER: U.S., 1976-77, 11 cities, 91% pos, 110 samples, 0.01-0.53 ppb, avg 0.09 ppb [27]. U.S. drinking water from treatment plants which use surface water as source, 11.5% pos [50]. Ontario, Canada, 1980, 3 cities, 1-4 ppt, avg 2 ppt [31]. Bank filtered Rhine River tapwater, >5 ppb [38]. Unidentified isomers; detected (not quantified) in drinking water [19]. GROUND WATER: The Netherlands, 1976, 250 ground water pumping stations, max concn, 1.2 ppb [61]. Unidentified isomers; New Jersey, detected at <1-100 ppb in 3% of 396 samples in ground water survey [50,51]. SURFACE WATER: Niagara Falls and Buffalo, NY, avg 0.5 ppt [37]. In 1980: Lake Ontario, 5 stations, 0.3-1 ppt, avg 0.6 ppt; Lake Huron, 0.1-0.4 ppt, avg 0.2 ppt; Grand River, Ontario, not detected-8 ppt, avg 2 ppt; Niagara River, 4 sites, 0.1-107 ppt (max just below chemical manufacturer's effluent discharge), avg of lowest 3 sites 6.4 ppt [31]. Niagara River, Niagara-on-the-Lake: 1981-83, 104 samples, 100% pos, 5.8-120 ppt, avg 16 ppt, median 12 ppt [30]; 1981, 4.9-52 ppt, avg 11 ppt, median 9.1 ppt [32]. Rhine River: 1977-82, detected (<0.1 ppb) [24]; 0.05 ppb [44]. Unspecified isomers of trichlorobenzene: USEPA STORET database, ambient water, 882 samples, 0.3% pos, median <10 ppb [46]; Merrimack River, MA, 100-1000 ppt; Delaware River, not detected (ND)-1000 ppt; Niagara, NY, drainage streams, 100-8000 ppt; all isomers: Great Lakes, 0.1-1.6 ppt, avg 0.5 ppt; Grand River, Canada, ND-8.7 ppt, avg 2.1 ppt [50]. Unidentified isomers of trichlorobenzene: Niagara River, 1981, total water samples, 60% pos, detected-1.0 ppt, avg 1.9 ppt [20]. RAINWATER/SNOW: Portland, OR, 1982, 4 rain events, 25% pos, 0.10 ppt [35]; 1984, 7 sampling periods, 43% pos, 17 total days, 0.13-0.45 ppt, avg pos 25 ppt [21]. Southeast Portland, 1982, 5 rain events, 40% pos, 0.086 ppt and 0.11 ppt [35]. SEAWATER: Southern North Sea (Rhine/Meuse Estuary, Aug 1983-Jul 1984) <0.3-24 ppt, 1.0 ppt median concn [56].

Effluent Concentrations: Concentration in advanced wastewater treatment plant - influent 0.11-0.46 ppt, effluent not detectable - 0.01 ppt, 97.8 - >99% removed [39]. U.S., several industrial wastewater surveys 1978-79, 3, 266 samples, 0.92% pos, 12-607 ppb, avg 161 ppb [54]. Major Southern California municipal wastewaters, 1976, 6 sites: 2 test periods each, 5 sites: 0.018-100 ppb, avg 1.2 ppb, last site, 43 and 100 ppb, avg mass emission rates to Southern California Bight, 1975-76, 10-1580 kg/yr [60]. Hazardous waste incinerator effluents, concn in extracts of effluent gases, 220,000 ppm [14]. Refuse incinerator flue gas: Municipal refuse-fired steam boiler, Virginia, 19 ug/m^3; refuse-derived fuel fired plant, Ohio, 56 ng/m^3 [49]. U.S. National Urban Runoff Program, 15 cities, 86 samples, not detected [7]. USEPA STORET database, 1256 samples, 2.1% pos, median <10 ppb [46]. Unspecified isomers: Georgia, 4 municipal treatment plants (including treated wastewater from synthetic carpet mills), influent, 1-60 ppb; effluent, 0.13 ppb [60]. Major Southern California municipal wastewaters, 1978, 8 stations, summer, 0.007-2.2 ppb, winter, <0.01-0.03 ppb [54]. Detected at 1.2 ng/m^3 in stack effluent from coal-fired power plants in Ames, IA [17]. Detected in U.S. industrial wastewaters (number of samples/% pos, max/avg concn (ppb)): foundries, raw water, 2/100% pos, 1000 max/500 avg, treated water, 2/100% pos, 570/290. Textile mills, raw water, 50/16% pos, 2700/410; treated water, 50/32% pos, 1400/14; electrical/electronic components, 2/100%, 27,000/16,600 (4,500 min concn) [53].

1,2,4-Trichlorobenzene

Sediment/Soil Concentrations: SEDIMENTS: Lake Ontario, Niagara River area, 6 sites, 20 m sediment traps, 1982, 22-55 ppb; 3 stations, sediment traps at 4 depths 20-68 m deep, avgs 34-41 ppb [20]. Niagara River, Niagara-on-the-Lake: 1980, 10 samples, 3 sites, 33% pos, 10% pos, 3.4 ppb dry wt [26]. West Germany, 7 lakes, 29% pos, detected, not quantified [5]. USEPA STORET database, 353 samples, 0.6% pos, median <500 ppm [46]. Concn (ppm dry wt) in sediments: Southern Lake Huron, 9 samples, 1.5-6.6, 4.3 avg; Lake St. Clair, 2 samples, 1.7-2.5, 2.1 avg; Western Lake Erie, 9 samples, 1.3-14, 5.3 avg; Central Lake Erie, 22 samples, 1.2-3.4, 2.3 avg; and Eastern Lake Erie, 15 samples, 1.3-3.5, 2.4 avg [33].

Atmospheric Concentrations: Portland, OR, concn in air during 7 rain events, 3.4-4.7 mg/m^3, avg 3.8 mg/m^3 [21]. U.S., 3 cities, 1979: Los Angeles, CA, 19.7-339.4 ppb, avg 69.3; Phoenix, AZ, 8.7-101.5 ppb, avg 30.8 ppb; and Oakland, CA, 9.7-151-2 ppb, avg 29.5 ppb [43]. Unspecified isomers, avg (ng/m^3): U.S., 35 localities, 136, rural/remote, not detected, urban/ suburban, 128, areas of production, 181; U.S., 12 sites, 17% pos, not detected-4346 ng/m^3 [50].

Food Survey Values: Concn (ppm) in crude oils: Yugoslavia, corn, 0.010, sunflower, 0.003, rape seed, walnut and poppy, trace; United States, soybean, trace; China, peanut, 0.005, sesame, trace; Turkey, hazelnut, trace [36].

Plant Concentrations:

Fish/Seafood Concentrations: Great Lakes trout, 1980, 5 sites, 100% pos, 0.5-5 ppb [16]. Livers of flatfish collected off Southern California municipal wastewater outfalls, 4 sites, ppb wet wt, avg <3.4-26, medians <0.7-29; Control site, not detected [60]. Lake Ontario rainbow trout, 1983, 10 samples, 0.6 ppb [29]. Slovenia, Yugoslavia, 4 rivers, 5 species fish, 1-15 ppb, avg 5 ppb [16].

Animal Concentrations: Herring gull eggs, 1979, Detroit River, 10 pos samples, avg 0.01 ppm; Niagara River, 10 samples, 100% pos, avg 0.02 ppm [47].

Milk Concentrations:

Other Environmental Concentrations:

Probable Routes of Human Exposure: Occupational exposure to 1,2,4-TCB will result mainly from inhalation during its manufacture and use. General population exposure will result from ingestion of contaminated drinking water and food, especially contaminated fish [27].

Average Daily Intake: WATER INTAKE: Insufficient data. AIR INTAKE: Estimated inhalation dosage (ug/day, based on intake of 23 m³/day at 25 °C, 1 atm pressure), 1979: Los Angeles, CA, 11.8; Phoenix, AZ, 4.8; and Oakland, CA, 6.3 [43]. FOOD INTAKE: Insufficient data.

Occupational Exposures: A National Occupational Hazard Survey (NOHS) estimated that 80,648 workers are exposed to 1,2,4-TCB [28]. A National Occupational Exposure Survey (NOES) by NIOSH estimated that 4032 workers were exposed to 1,2,4-TCB [28].

Body Burdens: Slovenia, Yugoslavia: Human milk, (Sep-Oct 1981), 12 samples, ND-4 ppb ("as is basis"), avg 1 ppb; avg 25 ppb (fat basis); human adipose tissue (1979-80), 15 samples, 2-15 ppb (fat basis), avg 9 ppb [15]. Unspecified isomers: Love Canal, Niagara Falls, NY: 9 residents, breath, 22% pos, trace-90 ng/m³, urine, 0% pos [50].

REFERENCES

1. Aelion CM et al; Appl Environ Microbiol 53: 2212-7 (1987)
2. Akermark B et al; Acta Chem Scand B 30: 49-52 (1976)
3. Atkinson R et al; Environ Sci Tech 19: 87-9 (1985)
4. Barrows ME et al; pp 379-92 in Dyn Exposure Hazard Assess Toxic Chem Ann Arbor, MI Ann Arbor Sci (1980)
5. Buchert H et al; Chemosphere 10: 945-56 (1981)
6. Chiou PE et al; Environ Sci Tech 17: 227-31 (1983)
7. Cole RH et al; J Water Pollut Control Fed 56: 898-908 (1984)
8. Dulin D et al; Environ Sci Technol 20: 72-7 (1986)
9. Freitag D et al; Chemosphere 14: 1589-1616 (1985)
10. Friesel P et al; Fresenius Z Anal Chem 319: 160-4 (1984)
11. Gaffney PE; J Water Pollut Control Fed 48: 2731-7 (1976)
12. Geyer H et al; Chemosphere 14: 545-55 (1985)

1,2,4-Trichlorobenzene

13. Hansch C, Leo AJ; Medchem Project Issue No 26. Claremont CA: Pomona College (1985)
14. James RH et al; Proc APCA 77th Ann Meet 1: 84-18.5 (1984)
15. Jan J; Bull Environ Contam Toxicol 30: 595-9 (1983)
16. Jan J, Malnersic S; Bull Environ Contam Toxicol 24: 824-7 (1980)
17. Junk GA et al; ACS Symp Ser 319: 109-23 (1986)
18. Kenaga EE; Ecotox Env Safety 4: 26-38 (1980)
19. Kool HJ et al; Crit Environ Control 12: 307-57 (1982)
20. Kuntz KW; Toxic Contaminants in the Niagara River, 1975-82; Burlington, Ontario Bulletin No. 134 Burlington, Ontario (1984)
21. Ligocki MP et al; Atmos Environ 19: 1609-17 (1985)
22. Lyman WJ et al; Handbook of Chem Property Estimation Methods Environ Behavior of Org Compounds McGraw-Hill NY pp 15-1 to 34 (1982)
23. Malaney GW, McKinney RE; Water Sew Works 113: 302-9 (1966)
24. Malle KG; Z Wasser Abwasser Forsch 17: 75-81 (1984)
25. Marinucci AC, Bartha R; Appl Environ Microbiol 38: 811-7 (1979)
26. McFall JA et al; Chemosphere 14: 1561-9 (1985)
27. McNamara PW et al; Exposure and Risk Assessment for 1,2,4-trichlorobenzene USEPA-440/4-85-017 (1981)
28. NIOSH; National Occupational Hazard Survey (NOHS) (1972-1974), National Occupational Exposure Survey (NOES) (1981-1983)
29. Oliver BG, Niimi AJ; Environ Sci Technol 19: 842-8 (1985)
30. Oliver BG, Nicol KD; Sci Tot Environ 39: 57-70 (1984)
31. Oliver BG, Nicol KD; Environ Sci Technol 16: 532-6 (1982)
32. Oliver BG; Symp Amer Chem Soc, Div Environ Chem 186th Natl Mtg 23: 421-2 (1983)
33. Oliver BG, Bourbonniere RA; J Great Lakes Res 11: 366-72 (1985)
34. Oliver BG; Chemosphere 14: 1087-106 (1985)
35. Pankow JF et al; Environ Sci Technol 18: 310-8 (1984)
36. Peattie ME et al; Sci Total Environ 34: 73-86 (1984)
37. Pellizzari ED et al; Formulation of Preliminary Assess of Halogenated Org Compounds in Man and Environ Media USEPA-560/13-79-006 (1979)
38. Piet GJ et al; pp 69-80 in Hydrocarbon Halo Hydrocarbon Aquatic Environ Afghan BK, MacKay D eds (1980)
39. Reinhard M, McCarty PL; Chem Water Reuse 2: 33-53 (1981)
40. Roberts RV et al; J Water Pollut Control Fed 52: 161-72 (1980)
41. Schmidt-Bleek F et al; Chemosphere 11: 383-415 (1982)
42. Simmons P et al; Text Chem Color 9: 211-3 (1977)
43. Singh HB et al; Atmos Measurements of Selected Toxic Organic Chem PB80-198989 (1980)
44. Sontheimer H et al; Sci Tot Environ 47: 27-44 (1985)
45. Spehar RL et al; J Water Pollut Control Fed 53: 1028-76 (1981)
46. Staples CA et al; Environ Toxicol Chem 4: 131-42 (1985)
47. Struger J et al; J Great Lakes Res 11: 223-30 (1985)
48. Tabak HH et al; J Water Pollut Control Fed 53: 1503-18 (1980)
49. Tiernan TO et al; Environ Health Persp 59: 145-58 (1985)
50. USEPA: pp. 4-5 to 4-39 in Health Assessment Document for Chlorinated Benzenes Part 1 USEPA-600/8-84-015A (1984)
51. USEPA; pp. 36-7 in Assessment of Testing Needs: Chlorinated Benzenes USEPA-660/11-80-014 (1980)

52. USEPA; Ambient Water Quality Criteria Doc: Chlorinated Benzenes p.C-34 EPA 440/5-80-028 (1980)
53. USEPA; Treatability Manual - Vol 1 USEPA-600/8-80-042 (1980)
54. USEPA; TSCA Chem Assess Series Assess of Testing Needs Chlorinated Benzenes USEPA-560/11-80-014 (1980)
55. Uyeta M et al; Nature 264: 583-4 (1976)
56. VandeMeent D et al; Wat Sci Technol 18: 73-81 (1986)
57. Veith GD et al; J Fish Res Board Can 36: 1040-8 (1979)
58. Wakeham SG et al; Environ Sci Technol 17: 611-17 (1983)
59. Wilson JT et al; J Environ Qual 10: 501-6 (1981)
60. Young DR et al; in Water Chlorination Environ Impact and Health Effects 3: 471-86 (1980)
61. Zoeteman BCJ et al; Chemosphere 9: 231-49 (1980)

1,3,5-Trichlorobenzene

SUBSTANCE IDENTIFICATION

Synonyms:

Structure:

CAS Registry Number: 108-70-3

Molecular Formula: $C_6H_3Cl_3$

Wiswesser Line Notation: GR CG EG

CHEMICAL AND PHYSICAL PROPERTIES

Boiling Point: 208 °C at 763 mm Hg

Melting Point: 63-4 °C

Molecular Weight: 181.45

Dissociation Constants:

Log Octanol/Water Partition Coefficient: 4.49 [10]

Water Solubility: 6.01 mg/L at 25 °C [2]

Vapor Pressure: 0.578 mm Hg at 25 °C [18]

Henry's Law Constant: 1.9 x 10^{-3} atm-m³/mol [19]

ENVIRONMENTAL FATE/EXPOSURE POTENTIAL

Summary: 1,3,5-Trichlorobenzene's (1,3,5-T) release to the environment will occur from its manufacture and use as an industrial chemical, chemical intermediate, solvent, and emulsifier.

1,3,5-Trichlorobenzene

If released to the soil, it will adsorb to the soil organic matter and will not be expected to appreciably leach to the ground water; however, its presence in some ground water samples illustrates that it can be transported. It will not hydrolyze and will not be expected to significantly biodegrade. It may be subject to evaporation from the surface of the soil; however, adsorption to the soil may retard this process. If released to the water, it will adsorb to sediment and bioconcentrate in aquatic organisms. It will not hydrolyze and will not be expected to undergo significant biodegradation nor directly photolyze. Evaporation from water may be significant with a half-life of 4.5 hr predicted for evaporation from a model river 1 m deep, flowing at 1 m/sec with a wind velocity of 3 m/sec. A half-life of 18 days was predicted for 1,3,5-T in a model river based upon monitoring data. If 1,3,5-T is released to the atmosphere, it will not be expected to directly photolyze. The estimated vapor phase half-life in the atmosphere is 6.17 months as a result of reaction with photochemically produced hydroxyl radicals. Exposure to 1,3,5-T will be mainly through occupational exposure. Minor exposure may occur through the ingestion of contaminated water and fish, and inhalation of contaminated air near industrial areas where it is manufactured and used.

Natural Sources: 1,3,5-T is a synthetic organic chemical and is not known to occur as a natural product.

Artificial Sources: Release of 1,3,5-T to the environment mainly occurs through its manufacture and use as an industrial chemical, chemical intermediate, solvent, and emulsifier.

Terrestrial Fate: If 1,3,5-T is released to soil, it will adsorb to the soil and will not appreciably leach to the ground water; however, its presence in some ground water samples illustrates that it can be transported there by some process. It will not hydrolyze nor will it be expected to biodegrade. It may be subject to evaporation from the surface of the soil; however, adsorption to the soil may retard this process.

Aquatic Fate: If 1,3,5-T is released to water, it will be expected to adsorb to sediment and to bioconcentrate in aquatic organisms. It will not hydrolyze and will not be expected to undergo

529

significant biodegradation or direct photolysis. Evaporation from water may be significant, with a half-life of 4.5 hr predicted for evaporation from a model river 1 m deep, flowing at 1 m/sec with a wind velocity of 3 m/sec. An overall half-life of 18 days was predicted for 1,3,5-T in a river in The Netherlands based upon monitoring data; no experimental details or possible mechanisms of dissipation were reported [34].

Atmospheric Fate: If 1,3,5-T is released to the atmosphere, it may be susceptible to direct photolysis. The estimated vapor phase half-life in the atmosphere is 6.17 months as a result of reaction with photochemically produced hydroxyl radicals [8].

Biodegradation: 1,3,5-T was resistant to biodegradation when exposed to mixed cultures of microorganisms adapted to phenol for an unspecified period of time [27]. Percent 1,3,5-T remaining after incubation with normal domestic wastewater (adapted wastewater): 24 hr, 100%(80%); 135 hr, 100%(53%) [7]. Oxygen uptake by phenol-adapted bacteria with(without) 1,3,5-T at concentration of 100 ppm: 146 uL(125 uL) [4]. Product of biodegradation by Pseudomonas sp. was 2,4,6-trichlorophenol [1].

Abiotic Degradation: 1,3,5-T will not hydrolyze under normal environmental conditions based on the resistance to hydrolysis exhibited by other chlorinated benzenes, such as the dichlorobenzenes and hexachlorobenzene [3]. When 20 g of pure 1,3,5-T was irradiated by sunlight in a borosilicate glass-stoppered Erlenmeyer flask (wavelengths <300 nm were excluded) for 56 days, 160 ppm of polychlorinated biphenyl was detected in the product mixture [31]. 1,3,5-T will probably not be susceptible to direct photolysis based on the behavior of 1,2,4-trichlorobenzene, which exhibited a half-life of 450 yr for sunlight photolysis in surface water at 40 deg latitude in summer [5]. The estimated vapor phase half-life in the atmosphere is 6.17 months as a result of reaction with photochemically produced hydroxyl radicals [8].

Bioconcentration: BCF: Rainbow trout, 1800-4100 [24]; guppy, 760 [13]; guppy (female), 13,000 [26]. Using a reported log octanol/water partition coefficient of 4.49 [10], an estimated BCF of 1520 was calculated [17]. Based on the reported and estimated BCF, 1,3,5-T will be expected to bioconcentrate in aquatic organisms.

Soil Adsorption/Mobility: Koc: 126,000 [20]; 1800 [6]. Using a reported log octanol/water partition coefficient of 4.49 [10], an estimated Koc of 6600 was calculated [17]. Based on this estimated Koc, 1,3,5-T will be expected to adsorb to soils and sediments of high organic content.

Volatilization from Water/Soil: Using the Henry's Law constant, a half-life of 4.5 hr was estimated for evaporation from a river 1 m deep, flowing at 1 m/sec with a wind velocity of 3 m/sec [17].

Water Concentrations: DRINKING WATER: Unidentified isomers: detected (not quantified) in drinking water [14]. GROUND WATER: Unidentified isomers of trichlorobenzene: New Jersey, detected at <1-100 ppb in 3% of 396 samples in ground water survey [29,30]. SEAWATER: Southern North Sea (Rhine/Meuse Estuary, Aug 1983- Jul 1984) <0.2-4 ppt, 0.6 ppt median concn [32].

Effluent Concentrations: Unspecified isomers of trichlorobenzene: Georgia, 4 municipal treatment plants (including treated wastewater from synthetic carpet mills), influent, 1-60 ppb, effluent, 0.13 ppb [29]; refuse incinerator flue gas, municipal refuse-fired steam boiler, Virginia, 19 ug/m^3, refuse-derived fuel fired power plant, Ohio, 56 ng/m^3 [28]. Major Southern California municipal wastewaters, 1978, 8 stations, summer, 0.007-2.2 ppb; winter, <0.01-0.03 ppb [33]. Los Angeles County effluent (11/80 to 8/81) 0.035 ppb avg [9].

Sediment/Soil Concentrations: SEDIMENTS: Niagara River, suspended sediments (seds): Niagara-on-the-Lake, 1980, 28 samples, 93% pos, avg 20 ppb dry wt [16]; 1981, settling particles, 5.3 ppb [22]. 1981, 5 stations, range of avgs, 3-53 ppb [15]. Surficial seds, Lake Superior, 13 sites, 46% pos, not detected (ND)-0.4 ppb, avg of pos 0.2 ppb; Lake Huron, 42 sites, 90% pos, ND-4 ppb, avg pos 0.7 ppb, Lake Erie, 5 sites, 100% pos, 0.1-5 ppb, avg pos 1 ppb; Lake Ontario, 11 sites, 100% pos, 7-250 ppb, avg pos 60 ppb [23]. Lake Ontario, Niagara River vicinity, 1982:

1,3,5-Trichlorobenzene

20 m sediment traps, 6 stations, 83% pos, 2.4-5.2 ppb; 3 sites, avgs, ppb - depths in m, 5.5 - 18, 7.0 - 40, 8.6 - 60, 9.5 - 68 [20]. Concn (ppm dry wt) in seds: Southern Lake Huron, 9 samples, 0.1-0.5, 0.3 avg; Lake St. Clair, 2 samples, 3.6-10, 6.8 avg; Western Lake Erie, 9 samples, 0.8-12, 3.1 avg; Central Lake Erie, 22 samples, ND-1.0, 0.6 avg; Eastern Lake Erie, 15 samples, ND-1.7, 0.9 avg [21].

Atmospheric Concentrations: U.S., rural/remote, not detected (ND) [29]. Unspecified isomers of trichlorobenzene: urban, 12 sites, 17% pos, ND-4346 ng/m³ (max Niagara Falls, NY [29]; U.S., urban-suburban areas, avg concn 128 ng/m³ [29]. Unspecified isomers of trichlorobenzene: U.S., areas of production, 181 ng/m³ avg [29].

Food Survey Values: Concn (ppm) in crude oils: Yugoslavia, corn, 0.7, sunflower, 0.02, walnut, 0.01, rape seed and poppy, trace; United States, soybean, trace; China, peanut, 0.01, sesame, 0.005; Turkey, hazelnut, 0.01 [25].

Plant Concentrations:

Fish/Seafood Concentrations: Slovenia, Yugoslavia, 1978, 5 river sites, 3 species of fish, 0.001-0.005 ppm (fat basis), avg 0.003 ppm; Gulf of Trieste, 2 species, 0.021 ppm and 0.015 ppm (fat basis) [12].

Animal Concentrations:

Milk Concentrations:

Other Environmental Concentrations:

Probable Routes of Human Exposure: General population exposure will mainly occur through the inhalation of contaminated air in areas near plants that manufacture and use 1,3,5-T. Exposure may also occur from the consumption of contaminated water and fish.

Average Daily Intake:

1,3,5-Trichlorobenzene

Occupational Exposures:

Body Burdens: Love Canal, Niagara Falls, NY, 9 residents, breath, 22% pos, trace-90 ng/m^3, urine, not detected [29]. Slovenia, Yugoslavia, human milk (Sept-Oct 1981) 12 samples, ND-3 ppb ("as us" basis), avg 1 ppb; avg (fat basis), 25 ppb; adipose tissue, 15 samples (1979-80), 8-20 ppb (fat basis), avg 16 ppb [11].

REFERENCES

1. Ballschmiter K, Scholz C; Chemosphere 9: 457-67 (1980)
2. Banerjee S; Environ Sci Technol 18: 587-91 (1984)
3. Callahan MA et al; p 98-8 in Water-Related Environ Fate of 129 Priority Pollutants Vol 2 USEPA-440/4-79-029b (1979)
4. Chambers CW et al; J Water Pollut Control Fed 35: 1517-28 (1963)
5. Dulin D et al; Environ Sci Technol 120: 72-7 (1986)
6. Fiesel P et al; Fresenius Z Anal Chem 319: 160-4 (1984)
7. Gaffney PE; J Water Pollut Control 48: 2731-7 (1976)
8. GEMS; Graphical Exposure Modeling System. Fate of atmospheric pollutants. Office of Toxic Substances. USEPA (1986)
9. Gossett RW et al; Marine Pollut Bull 14: 387-92 (1983)
10. Hansch C, Leo AJ; Medchem Project Issue No 26. Claremont CA: Pomona College (1985)
11. Jan J; Bull Environ Contam Toxicol 30: 595-9 (1983)
12. Jan J, Malnersic S; Bull Environ Contam Toxicol 24: 824-7 (1980)
13. Konemann H, Van Leeuwen K; Chemosphere 12: 1159-67 (1980)
14. Kool HJ et al; Crit Environ Control 12: 307-57 (1982)
15. Kuntz KW; Toxic Contaminants in the Niagara River, 1975-82 Burlington, Ontario Tech Bull No. 134 (1984)
16. Kuntz KW, Wary ND; J Great Lakes Res 9: 241-8 (1983)
17. Lyman WJ et al; Handbook of Chemical Property Estimation Methods. Environmental Behavior of Organic Compounds. McGraw-Hill NY pp 15-1 to 34 (1982)
18. Mackay D, Shiu WY; J Phys Chem Ref Data 10: 1175-99 (1980)
19. Oliver BG; Chemosphere 14: 1087-106 (1985)
20. Oliver BG, Charlton MN; Environ Sci Technol 18: 903-8 (1986)
21. Oliver BG, Bourbonniere RA; J Great Lakes Res 11: 366-72 (1985)
22. Oliver BG; Symp Amer Chem Soc Div Environ Chem 186th Natl Mtg 23: 421-2 (1983)
23. Oliver BG, Nicol KD; Environ Sci Technol 16: 532-6 (1982)
24. Oliver BG, Niimi AJ; Environ Sci Technol 17: 287-91 (1983)
25. Peattie ME et al. Sci Total Environ 34: 73-86 (1984)
26. Spehar RL et al; J Water Pollut Control Fed 53: 1028-76 (1981)
27. Tabak HH et al; J Bacteriol 87: 910-9 (1964)
28. Tiernan TO et al; Environ Health Persp 59: 145-58 (1985)
29. USEPA; pp 4-18 to 4-35 in Health Assessment Document for Chlorinated Benzenes Part 1 USEPA-600/8-84-015A (1984)

1,3,5-Trichlorobenzene

30. USEPA; pp 36-7 in Assessment of Testing Needs: Chlorinated Benzenes USEPA-660/11-80-014 (1980)
31. Uyeta M et al; Nature 264: 583-4 (1976)
32. VendeMeent D et al; Wat Sci Technol 18: 73-81 (1986)
33. Young DR et al; Water Chlorination: Environ Impact and Health Effects 3: 471-86 (1980)
34. Zoeteman BCJ et al; Chemosphere 9: 231-49 (1980)

2,4,6-Trichlorophenol

SUBSTANCE IDENTIFICATION

Synonyms:

Structure:

CAS Registry Number: 88-06-2

Molecular Formula: $C_6H_3Cl_3O$

Wiswesser Line Notation: QR BG DG FG

CHEMICAL AND PHYSICAL PROPERTIES

Boiling Point: 246 °C

Melting Point: 69 °C

Molecular Weight: 197.45

Dissociation Constants: $Ka = 3.8 \times 10^{-8}$ at 25 °C [24]

Log Octanol/Water Partition Coefficient: 3.69 [17]

Water Solubility: 0.9 g/L at 25 °C [26]

Vapor Pressure: 0.0084 mm Hg at 24 °C [37]

Henry's Law Constant: 6.14×10^{-8} atm-m³/mole (calculated from water solubility and vapor pressure [30]

ENVIRONMENTAL FATE/EXPOSURE POTENTIAL

Summary: 2,4,6-Trichlorophenol (2,4,6-T) will enter the environment as emissions from the combustion of fossil fuels and incineration of municipal wastes, as well as emissions from its manufacture and use as a wood and leather preservative and biocide. Significant amounts may result from the chlorination of phenol-containing wastewaters. Release to soil may decrease in concentration due to biodegradation, depending upon the temperature, availability of oxygen, and the presence of appropriate organisms. Adsorption to soil will be significant in soils with high organic content; leaching to ground water may be significant in sandy soils and in soils where biodegradation is not rapid. Volatilization and photomineralization may contribute to losses at the surface of the soil. Releases to water will biodegrade, photolyze, and adsorb to sediments. Volatilization from water may also contribute to losses. Although extensive bioaccumulation is not expected, it is important in some species of fish and invertebrates. Release to the atmosphere can decrease due to photolysis and reaction with hydroxyl radicals as well as wet and dry deposition mechanisms. Human exposure will result from consumption of contaminated drinking water and occupational or other contact associated with its use. Inhalation of air contaminated by emissions from combustion of wastes and fossil fuels as well as occupational contact with chlorinated phenol containing wastewater may also be significant sources of human exposure.

Natural Sources: 2,4,6-T is not known to occur as a natural product [20]; however, it may be formed during forest fires.

Artificial Sources: 2,4,6-T may be released through its production and use as a bactericide, wood and glue preservative, insecticide ingredient, and anti-mildew treatment agent for textiles, as well as its use in the leather tanning and finishing industry [20]. It may be released through use as a fungicide, herbicide, and defoliant [18] as well as through the burning of municipal wastes and of fossil fuels including incinerators, power generators, and automobiles. The chlorination of wastewaters containing phenol and other phenolic wastes may also be a significant source of 2,4,6-T [50].

2,4,6-Trichlorophenol

Terrestrial Fate: If released on land, 2,4,6-T would be subject to biodegradation, the rate of which will depend upon conditions, such as the temperature, availability of oxygen, and the presence of appropriate organisms. Total degradation in soils in as little as 3 days has been reported. No biodegradation was observed in soil under anaerobic conditions or in sterile soil, although significant decreases were observed by an unidentified mechanism. Moderate to extensive adsorption to soil would be expected, especially in soils with high organic content. Therefore, substantial leaching to ground water is not expected, except in sandy soils where mobility may be significant and in soils where rapid biodegradation does not occur. The presence of 2,4,6-T in some samples of ground water and relatively deep in soils near pulp mills indicates that leaching through soil can be significant. Its fate in ground water is unknown. Volatilization and photomineralization on the soil surface may also be significant mechanisms of loss whereas hydrolysis should not be a factor.

Aquatic Fate: If released in water, 2,4,6-T will be subject to appreciable biodegradation; the average half-life for river water was 6.3 days, with the rate affected by the amount of sediment present. It will not hydrolyze but will be subject to appreciable adsorption to sediments. It would not be expected to appreciably bioaccumulate in most aquatic organisms, although significant bioaccumulation in certain species has been reported. Significant photodegradation near the surface is expected, with a half-life of 2.1 hr reported [28]. Volatilization may also be significant, with a half-life of 2 days reported from an experimental pool [46].

Atmospheric Fate: If released into the atmosphere, 2,4,6-T will be subject to photodegradation, with a half-life of less than 17 hr reported for the compound coated on silica gel and irradiated with >290 nm light. Reaction with photochemically produced hydroxyl radicals will also contribute losses (estimated half-life of 2.7 days). Significant deposition in snow is expected, and rainout as well as dry deposition may also be significant. Detection of 2,4,6-T in relatively pristine areas suggests that significant transport is possible or that 2,4,6-T may be formed in forest fires.

Biodegradation: 2,4,6-T is readily biodegradable in South Trenton Delaware River water with suspended sediment, avg half-life 6.3

days; half-life range of 2.75 days (10 g suspended sediment/L) to 70 days (1 g suspended sediment/L) [5]. Rate of biodegradation in water from 11 sites on the Delaware River affected by concn; 100% decrease in filtered water, 10-14 days (8-day lag time) and 8-10 days in water with sediment (2-day lag time) [4]. Total degradation in soil in 5 days (Dunkirk silt loam) to 13 days (Mardin silt loam) [1]. Only 67% decrease reported over 60 days in Labisch soil [57]. Aerobic degradation in clay loam led to 95% degradation in 3 days (non-sterile soil) and 27% degradation in 80 days (sterile soil); anaerobic degradation led to 80 day decreases of 28% (non-sterile) and 25% (sterile); microbial degradation, volatilization, and photodecomposition were ruled out in the sterile soils and non-sterile soil under anaerobic conditions, indicating that other mechanisms contribute to degradation in soils [2]. Anaerobic fluidized bed reactor - 62% decrease in 12 hr [38]. Activated sludge made from soil resulted in 100% removal in 3 days [21]; sewage seed resulted in complete removal in 7 days [47]; 39% removal by activated sludge in 14 days has been reported [38]. 2,4,6-T was readily degraded in an anaerobic microcosm using subsurface soils from both unsaturated and saturated zones [43], and in an aerobic microcosm using soils from the saturated zone [45].

Abiotic Degradation: Halogenated aromatics and phenols are known to be resistant to hydrolysis [30]. 2,4,6-T is a weak acid in aqueous solution (pKa 5.99) and it will exist appreciably in the ionized state where it exhibits a UV maximum at 311 nm, making it susceptible to photodegradation [10]. Photomineralization of 2,4,6-T coated on silica gel was 65.8% (based on CO_2 production; 17 hr irradiation) [27]. A half-life of 2.1 hr was reported for an aqueous solution irradiated at >290 nm [28]. Photochemically produced hydroxyl radicals will add to the aromatic ring in the vapor phase (estimated half-life 21.7 days) [15]. A half-life of 62 hr has been estimated for reaction with singlet oxygen in eutrophic waters based on experimental rate constants, steady state singlet oxygen concn of 4 x 10^{-14} M, and noontime summer sunlight (Switzerland) [40].

Bioconcentration: Bioconcentration factors (BCF) for the golden orfe (fish) ranged from 250-310 [14,26] and for the mussel Mytilus edulis, 35-60 [16]. The BCF for the algae Chlorella fusca was 51

[14]. Using the recommended value for the log octanol/water partition coefficient for 2,4,6-T of 3.69 [17], the estimated BCF is 273 [30]. Based on these BCF data, a low to moderate potential for bioconcentration in aquatic systems would be predicted. However, significant bioaccumulation in Lymnea (snails) and Poecilia (fish) has been reported [52].

Soil Adsorption/Mobility: Measured Koc values - 1070 average (3 values: lake sediment - 830; river sediment - 1310; aquifer material - 1070) [39]. These values indicate that moderate to considerable adsorption to soil and sediment is expected. Extensive leaching to ground water is not expected, although the low level presence of 2,4,6-T in some samples of ground water indicates that leaching through soil is possible.

Volatilization from Water/Soil: The half-life for evaporation from water of 2,4,6-T at 20.7 °C in an indoor experimental pond 10.2 cm deep with a wind velocity of 1.7 m/sec and an initial concentration of 177 ppb was 48 hr [46]. The volatilization rate from water was 1.41% in first hour and 1.52% in second hour (per mL H_2O evaporated) [23]. The volatilization rates from three types of soil in %/mL water evaporated over the first two hours - 1.05% and 1.13% (sandy soil), 0.73% and 0.62% (loamy soil), and 0.15% and 0.13% (humus soil) [23]. Outdoor volatilization rate from loamy soil was 36%, applied over one vegetation period [23]. Evaporation accounted for 17% weight loss over 60 days during lysimeter biodegradation tests [57].

Water Concentrations: DRINKING WATER: Tap water: Janakka, Finland - 30 ppt 2,4,6-T; Jyvaskyla, Finland - 14 ppt [36]. Identified in finished U.S. drinking water (not quantified) [8,9,25]. Not detected in Dade County, FL drinking water (12 samples; limit detection 0.2 ppb) [32]. Not detected in midwestern U.S. city [7]. Unspecified trichlorophenol isomers 2 and 4 ppt [20]. Detected at 16-60 ppt (31 ppt avg) in the finished drinking water of 3 of 6 Canadian cities in Feb 1985 (at 12 ppt in the raw water of 1 of 6 cities) [41]. Frequency of detection/maximum level (ppt) in samples from 40 potable water treatment plants in Canada: treated water, 11/719 (Oct-Dec 1984); 11/61 (Feb-Mar 1985); 8/148 (May-Jul 1985); raw water, 3/23 (Oct/Dec 1984); nd (Feb-Mar May-Jul 1985) [42]. GROUND WATER: Finnish sawmill - 0.03 ppb (average 7 samples positive out of 12, 12 ppb maximum found 360-600 m from dripping pit [51]. Conroe, TX creosote waste site - well #2 - 91.3 ppb (avg 3 of 5 samples pos; 50 ft from nearest waste pit); well #1 - not detected in 5 samples (300 ft from nearest waste pit) [3]. Western suburbs of Melbourne, Australia - identified, not quantified, 110 and 400 m from contaminated quarry [44]. Near Mt St Helens (1981-1982) - 1.1 ppb wells dug for testing (1 out of 20 samples positive [19]. SURFACE WATER: Netherlands 1976 and 1977: Rhine river - 0.19 and 0.18 ppb median (94% and 87% of 52 samples positive, 2.5 ppb max); 5 other rivers, Netherlands - 0.03-0.18 ppb range of medians (27-92% range of positive samples; 0.37 ppb max level found) [56]. Ijssel River, Netherlands - 0.13 ppb median (all 13 samples pos, 0.74 ppb max) [55]. Weser River, Germany 5.8 ppt (avg of 6 sites; 10.4 ppt max, 30 - 100 km from Bremen); German Bight - 0.01 ppt (avg for 2 sites) [54]. Lake Vatia, Finland - 30 (high water) to 110 (normal to low water) ppt 5 km downstream from pulp mill; 40 km downstream - not detected; and 65 km downstream - 7 ppt [35]. Lake Saimaa, Finland near kraft pulping mill, range of 1.2 ppb (surface) to 1.6 ppb (bottom) [38]. St. John River estuary, New Brunswick, Canada - not detected above and below pulp and paper mill outfalls [33]. Near Mt St Helens (1981-1982) - lake - 9.3 ppb (surface) to 2.2 ppb (bottom) (2 of 4 pos) plant effluent (1 out of 7 pos); lake - 9.3 ppb (surface) to 2.2 ppb; not detected in 9 river samples or 10 spring samples [19]. SNOW: South Finland - 0.509 ppb avg (melted snow, 19 samples, all pos), 0.931 ppb max (possible origin - heavy traffic); Jyvaskyla, central Finland - 0.108 ppb (1 of 3 samples pos) [36].

Effluent Concentrations: 2,4,6-T in a Swiss pulp mill wastewater treatment plant: influent - 19.8 ppb (avg of 8 samples; 14.1-26.0 ppb range); effluent - 14.2 ppb (avg of 8 samples; 7.6-20.8 ppb range) [29]. Finnish pulp mill drainage ditch to creek - 800 ppm [38]. Finnish pulp mill waste liquors - 22 ppb (average 4 sites); combustion ash samples: 280 and 78 ppb (municipal waste incinerator); 19 ppb (wood burning); 12 and <0.5 ppb (oil power plants); and 18 ppb (coal power plant) [36]. UK treatment works final effluent - 0.144 ppb; detected, not quantified, in sludge and leachate of treatment works [6]. Identified, not quantified, in effluent from municipal waste incinerator in Ontario, Canada [48]. Detected in industrial wastewaters (number of samples/% pos, max/avg in samples (ppb)): leather tanning and finishing, raw water, 18/61% pos, 5900 max/2200 avg, treated water, 6/67%, 4300/<1100; foundries, raw water, 14/100% pos, 1400/240, treated water, 14/100%, 600/110; aluminum forming, treated water, 20/45% pos, 1800/<260 [49]. Midwest U.S. source raw sewage - 12 ppb (1 sample of 48 positive); not detected in influent and effluent of sewage treatment facility [8]. Detected, not quantified in 2 of 10 samples of U.S. wastewater treatment effluent [12] and effluent from Oak Ridge diffusion plant [31]. Near Mt St Helens (1981-1982) - 1.4 ppb, treatment plant effluent (1 out of 7 samples pos); not detected in treatment plant influent [19].

Sediment/Soil Concentrations: SEDIMENT: Lake Saimaa, Finland, 3 km from pulp mill - 190 ppb (0-5 cm), 70 ppb (5-10 cm), and 50 ppb (10-15 cm) [38]. Baltic Sea, 2 km from Swedish sulphate mill - 0.4 ppb; 10 km from mill 0.1 ppb [58]. Lake Ketelmeer, Netherlands - 1.9 ppb median (16 of 17 samples pos; 3.7 ppb maximum); Haringvliet River, Netherlands - 0.9 ppb median [55]. 2,4,6-Trichlorophenol was detected in three lakes in Middle Finland near pulp and saw mills - 14.4 ppb (0-2 cm avg of 15 samples; 27.7 ppb maximum avg of 5 samples from Lake Vatia) [36]. Weser estuary, Germany - moist sediment - 86.8 ppt (avg 6 of 10 samples pos); suspended matter - 105 ppt (avg 4 of 4 samples positive) [11]. SOIL: Finnish sawmill, soil within mill area Site A near dripping pit - 1780 ppb (0-5 cm); 1810 ppb (120-230 cm, deepest sample); 14,000 ppb (80-100 cm; maximum level found); site B - 281 ppb (0-5 cm); 1162 ppb (80-100 cm, deepest sample), 1763 ppb (40-60 cm, max level found) [51].

Atmospheric Concentrations:

Food Survey Values:

Plant Concentrations: 2,4,6-T was detected in plankton from Finnish lake-river system: 2.45 ppb (avg of 4 samples pos out of 23 total) [36].

Fish/Seafood Concentrations: New Brunswick, Canada: marine and brackish water aquatic organisms from waters receiving pulp mill effluent - 0.51 ppm 2,4,6-T avg (16 of 18 samples positive; 3.9 ppm max found in flounder viscera) [33], Finnish lake-river system near pulp mills and saw mills: muscles of aquatic organisms 13.6 ppb (avg of 68 samples pos out of 87 total samples; 55.9 ppb max avg for 10 samples of roach muscle) [36]. Baltic salmon - muscle tissue: 2.7 ppb (avg of 74 samples; 5.9 ppb max) [35]. Levels (ppb)/number of samples in the muscles, liver and eggs, respectively, of Baltic fish in the Kemi River, Finland: salmon, 1.9/13, 5.9/15, and 5.4/15; trout, 5.9/11, 15.0/11, and 10.7/11 [53].

Animal Concentrations:

Milk Concentrations:

Other Environmental Concentrations:

Probable Routes of Human Exposure: Exposure to 2,4,6-trichlorophenol will be mainly occupational; for example, in hospitals where it is used as a bactericide and in the leather industry [20], as well as in the pulp and wood industry. The general public may be exposed through ingestion of contaminated drinking water and possibly through inhalation of particulate matter near municipal waste incinerators, areas of high automobile traffic and near wood and coal burning facilities.

Average Daily Intake: AIR INTAKE: Insufficient data; WATER INTAKE: (assume 14-30 ppt) 28-60 ng; FOOD INTAKE: Insufficient data.

Occupational Exposures: A National Occupational Hazard Survey (NOHS) (1973-1974) estimates that 112 workers may be exposed to 2,4,6-trichlorophenol [34]. Worker exposures in ten Finnish saw mills - range varies with work phase - 0.25 ppb to 15 ppb range (38 of 67 samples pos; 2 of 19 work phases negative); 3.5 ppb median, 5.0 ppb avg of 30 samples [22]. Very high levels with exposure of short duration inside kilns during drying - 1.0 ppb to 1360 ppb, median 76 ppb, avg 235 ppb (12 samples) [22].

Body Burdens: Detected in 8 of 10 human urine samples at 25-200 ppb, avg of pos 88 ppb [13].

2,4,6-Trichlorophenol

REFERENCES

1. Alexander M, Aleem MIH; J Agric Food Chem 9: 44-7 (1961)
2. Baker MD, Mayfield CI; Water Air Soil Pollut 13: 411 (1980)
3. Bedient PB; Groundwater 22: 318-29 (1984)
4. Blades-Filmore LA; The Biodegradation of 2,4,6-Trichlorophenol in the Delaware River Water and Sediments and in Urban Runoff Master's Thesis Rutgers State Univ, NJ p.136 (1980)
5. Blades-Filmore LA et al; J Env Sci Health A17: 797-818 (1982)
6. Buisson RSK; J Chromat Sci 22: 339-42 (1984)
7. Callahan MA et al; pp 55-61 in Proc Natl Conf Munic Sludge Manag 8th (1979)
8. Callahan MA et al; Water-related Environmental Fate of 129 Priority Pollutants - Vol II USEPA-440/4-79-029b p 86-1 (1979)
9. Crathorne B et al; Environ Sci Technol 18: 797-802 (1984)
10. Drahonovsky J, Vacek Z; Coll Czech Chem Comm 36: 3431-40 (1971)
11. Eder G, Weber K; Chemosphere 9: 111-8 (1980)
12. Ellis DD et al; Arch Environ Contam Toxicol 11: 373-82 (1982)
13. Fatiadi AJ; Environ Inter 10: 175-205 (1984)
14. Freitag D et al; Ecotox Environ Safety 6: 60-81 (1982)
15. GEMS; Graphical Exposure Modeling System Fate of Atmospheric Pollutants (FAP) Data Base Office of Toxic Substances USEPA (1985)
16. Geyer H et al; Chemosphere 11: 1121-34 (1982)
17. Hansch C, Leo AJ; Medchem Project Issue No 26. Claremont CA: Pomona College (1985)
18. Hawley GG; The Condensed Chemical Dictionary 10th ed Van Nostrand Reinhold New York p.1043 (1981)
19. Hindin E; Occurrence of organic compounds in water and sediments due to Mt St Helens eruptions OWRT-C-10108-V(1458)(1) p 101 (1983)
20. IARC: Some Halogenated Hydrocarbons 20: 349-67 (1979)
21. Ingols RS et al; J Water Pollut Control Fed 38: 629-35 (1966)
22. Kauppinen T, Lindroos L; Am Ind Hyg Assoc J 46: 34-38 (1985)
23. Kilzer L et al; Chemosphere 8: 751-61 (1979)
24. Kirk-Othmer. Encyc Chem Tech 3rd ed 5: 865 (1978)
25. Kool HJ et al; Rev Env Control 12: 307-57 (1982)
26. Korte F et al; Chemosphere 1: 79-102 (1978)
27. Korte F, Klein W; Ecotox Environ Saf 6: 311-27 (1982)
28. Kotzias D et al; Naturwiss 69: 444-5 (1982)
29. Leuenberger C et al; Water Res 19: 885-94 (1985)
30. Lyman WJ et al; Handbook of Chemical Property Estimation Methods. Environmental Behavior of Organic Compounds McGraw-Hill New York p.5.1-5.30; 7-4 (1982)
31. McMahon LW; Organic priority pollutants in wastewater NTIS DE83010817 (1983)
32. Morgade C et al; Bull Environ Contam Toxicol 24: 257-64 (1980)
33. New Brunswick Res Prod Council; Bioaccumulation of toxic compounds in pulp mill effluents by aquatic organisms in receiving waters CPAR Project Report 675-1 (1978)
34. NIOSH; National Occupational Hazard Survey (NOHS) (1975)
35. Paasivirta J et al; Chemosphere 14: 469-91 (1985)

2,4,6-Trichlorophenol

36. Paasivirta J et al; Chemosphere 9: 441-56 (1980)
37. Politzki GR et al; Chemosphere 11: 1217-29 (1982)
38. Salkinoja-Salonen MS et al; pp. 668-72 in Current Perspectives in Microbiology Klug MJ, Reddy CA eds (1984)
39. Schellenberg K et al; Environ Sci Technol 18: 652-7 (1984)
40. Scully FE, Hoigne J; Chemosphere 16: 681-94 (1987)
41. Sithole BB, Williams DT; J Assoc Off Anal Chem 69: 807-10 (1986)
42. Sithole BB et al; J Assoc Off Anal Chem 69: 466-73 (1986)
43. Smith JA, Novak JT; Water Air Soil Pollut 33: 29-42 (1987)
44. Stepan S; Australian Water Resources Council Conf Ser 1: 415-24 (1981)
45. Suflita JM, Miller GD; Environ Toxicol 4: 751-8 (1985)
46. Sugiura K et al; Arch Environ Contam Toxicol 13: 745-58 (1984)
47. Tabak HH et al; J Water Pollut Control Fed 53: 1503-18 (1981)
48. Tong HY et al; J Chrom 285: 423-41 (1984)
49. USEPA Treatability Manual - Vol 1 USEPA 600/8-80-042 (1980)
50. USEPA; Ambient Water Quality Document: Chlorinated Phenols p.C-47 USEPA-440/5-80-032 (1980)
51. Valo R et al; Chemosphere 13: 835-44 (1984)
52. Virtanen MT, Hattula ML; Chemosphere 11: 641-9 (1982)
53. Vourinen PJ et al; Chemosphere 14: 1729-40 (1985)
54. Weber K, Ernst W; Chemosphere 11: 873-79 (1978)
55. Wegman RCC, Vandenbroek HH; Water Res 17: 227-30 (1983)
56. Wegman RCC, Hofstee AWM; Water Res 13: 651-57 (1979)
57. Wilkinson RR, Hopkins FC; Pest Chem Hum Welfare Environ Proc Int Cong Pest Chem 4: 391-6 (1982)
58. Xie TM; Chemosphere 12: 1183-91 (1983)

Vinyl Acetate

SUBSTANCE IDENTIFICATION

Synonyms: 1-Acetoxyethylene

Structure:

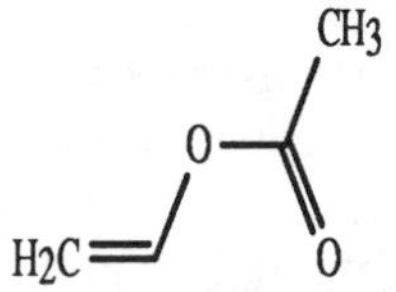

CAS Registry Number: 108-05-4

Molecular Formula: $C_4H_6O_2$

Wiswesser Line Notation: 1VO1U1

CHEMICAL AND PHYSICAL PROPERTIES

Boiling Point: 72-73 °C

Melting Point: -93.2 °C

Molecular Weight: 86.09

Dissociation Constants:

Log Octanol/Water Partition Coefficient: 0.73 [9]

Water Solubility: 20,000 mg/L at 20 °C [16]

Vapor Pressure: 85 mm Hg at 20 °C [25]

Henry's Law Constant: 4.81×10^{-4} atm m^3/mole at 20 °C (calculated from the water solubility and vapor pressure)

ENVIRONMENTAL FATE/EXPOSURE POTENTIAL

Summary: Vinyl acetate is primarily released to the environment from industrial emissions. It is degraded relatively rapidly in the

545

environment by chemical processes, and appears to be susceptible to biodegradation as well. If released to the atmosphere, vinyl acetate is degraded rapidly by reaction with photochemically produced hydroxyl radicals (estimated half-life of 12 hours in an average atmosphere). If released to water, degradation by hydrolysis (half-life of 7.3 days at 25 °C and pH 7) and by photochemically produced oxidants will occur. Volatilization from water may be significant (half-lives of 4.4 hr and 2.2 days have been estimated for a model river, 1 m deep, and an environmental pond, respectively). If released to soil, hydrolysis will occur in the presence of moisture. Although significant leaching is possible, concurrent hydrolysis will decrease its importance. Evaporation from dry surfaces will occur. If released in a spill situation, significant polymerization may occur. Primary human exposure to vinyl acetate most likely results through inhalation at occupational settings.

Natural Sources: With the exception of an isolated report of an occurrence in trace quantities in plants (watercress), it is not known whether vinyl acetate occurs as a natural product [10].

Artificial Sources: Vinyl acetate can be released to the environment from industrial sources and biomass combustion [7]. Waste gases from scrubbers (generated during the industrial manufacture of vinyl acetate) may contain trace levels of vinyl acetate [11].

Terrestrial Fate: Based on an aqueous hydrolysis half-life of 7.3 days (25 °C and pH 7), hydrolysis should be a significant process in moist soils. Hydrolysis rates will increase as the soil becomes more alkaline. Various screening studies (not utilizing soil media) have indicated that vinyl acetate is readily biodegradable; therefore, significant biodegradation may occur in soil. Vinyl acetate's vapor pressure indicates that evaporation from dry surfaces is likely to occur. Estimated Koc value of 19-59 indicate that significant leaching is possible; however, concurrent hydrolysis should decrease the environmental importance of leaching. Vinyl acetate readily polymerizes; therefore, if vinyl acetate is released to the environment in a spill situation, significant polymerization may occur.

Vinyl Acetate

Aquatic Fate: Vinyl acetate is chemically degraded in natural water by hydrolysis and reaction with photochemically produced oxidants. The estimated hydrolysis half-life at 25 °C and pH 7 is 7.3 days; the hydrolysis rate increases as the pH increases. Reaction with hydroxyl radicals and singlet oxygen in natural water may proceed at half-life rates of approx 13 and 8 days, respectively. Removal from water via volatilization may be very significant. Volatilization half-lives of 4.4 hr and 2.2 days have been estimated for a model river (1 m deep) and an environmental pond, respectively. Various screening studies have indicated that vinyl acetate is readily biodegradable. Aquatic adsorption to sediment, bioconcentration, and direct photolysis are not important.

Atmospheric Fate: Based on the vapor pressure, vinyl acetate is expected to exist almost entirely in the vapor phase in the ambient atmosphere [4]. Vapor-phase vinyl acetate is degraded rapidly in the atmosphere by reaction with photochemically produced hydroxyl radicals (estimated half-life of 12 hours in an average atmosphere).

Biodegradation: A 5-day 42% BODT in marine water and a 5-day 51.3% BODT using a sewage inocula [23]. A 62% BODT in 5 days and a 72% BODT in 20 days using an acclimated sewage inoculum [21]; 51 and 69% BODTs in 5 and 15 days, respectively, in marine water containing a synthetic sewage seed [21]. CO_2 evolutions of 27 and 49% over 19 and 38 days incubation, respectively, using non-acclimated sewage inocula [19]; a 58% CO_2 evolution in 22 days using an acclimated sewage inocula [19]. CO_2 evolution of 42% in 10 days using an acclimated sewage inocula [13]. A 100% degradation after a 3-day lag period using the Hungate Serum Bottle technique (anaerobic conditions) and enriched methane cultures [2].

Abiotic Degradation: The rate constant for the vapor-phase reaction of vinyl acetate with photochemically produced hydroxyl radicals has been estimated to be 26.32×10^{-12} cm^3/molecule-sec at 25 °C [1]; assuming an average atmospheric hydroxyl radical concn of $5 \times 10^{+5}$ molecules/cm^3, the half-life for this reaction is estimated to be 14.6 hours [1]. The aqueous hydrolysis half-life of vinyl acetate at 25 °C and pH 7 has been reported to 7.3 days [15]; the hydrolysis rate will increase as the pH increases [15]. The hydrolysis rate at pH 4.4 has been reported to be minimal [3]. The half-lives for olefinic structures in sunlit natural waters are about 13 and 8 days with respect to reaction via hydroxyl radicals and singlet oxygen [17]. Vinyl acetate does not absorb UV light significantly above 250 nm in ethanol solvent [3] and, therefore, it should not be susceptible to direct sunlight photolysis. Vinyl acetate readily polymerizes [3,16].

Bioconcentration: Based on the water solubility and the log Kow, the log BCF of vinyl acetate can be estimated to range from 0.32 to 0.37 from the recommended regression-derived equation [14] which indiates that vinyl acetate should not bioconcentration in aquatic organisms.

Soil Adsorption/Mobility: Based on the water solubility and the log Kow, the Koc of vinyl acetate can be estimated to range from 19 to 59 from regression-derived equation [14]. These estimated Koc values are indicative of very high to high soil mobility [22].

Volatilization from Water/Soil: The Henry's Law constant suggests that volatilization from environmental waters can be significant [14]. The volatilization half-life from a model river (1 m deep flowing 1 m/sec with a wind speed of 3 m/sec) can be estimated to be 4.4 hr [14]. The volatilization half-life from an environmental pond can be estimated to be 2.1 days [24].

Water Concentrations: SURFACE WATER: Vinyl acetate has reportedly been detected in river water from Great Britain [5].

Effluent Concentrations: Vinyl acetate was qualitatively detected in wastewater effluents collected from the advanced waste treatment facility in Lake Tahoe, CA in Oct 1974 [12]. A concn of 50 ppm was detected in a wastewater effluent from a polyvinyl acetate plant [10].

Sediment/Soil Concentrations:

Atmospheric Concentrations: Ambient vinyl acetate air levels of 0.07-0.57 ppm were detected in Houston, TX air during June and July 1974 [6]. Samples of ambient air collected in the vicinity of the Kin-Buc waste disposal site (Edison, NJ) between Jun 29 and Jul 1, 1976 contained 0.5 ug/m^3 vinyl acetate [20].

Food Survey Values:

Plant Concentrations:

Fish/Seafood Concentrations:

Animal Concentrations:

Milk Concentrations:

Other Environmental Concentrations: Drums of chemical waste collected from the North Sea were found to contain vinyl acetate [8]. Residual levels of vinyl acetate monomer as high as 5 g/kg can remain in the polymer [10].

Probable Routes of Human Exposure: The general population may be exposed to vinyl acetate through inhalation of contaminated air in the vicinity of its commercial production and use, and in the vicinity of waste disposal sites containing the compound. Dermal contact from residual monomer remaining in polyvinyl acetate products may be possible. Although never demonstrated experimentally, residual monomer (in polyvinyl acetate containers used to store food products) may leach into food products and result in oral consumption. Primary human exposure to vinyl acetate, however, most likely results through inhalation at occupational settings.

Average Daily Intake:

Occupational Exposures: NIOSH has estimated that 102,233 workers are potentially exposed to vinyl acetate based upon statistical estimates derived from a survey conducted between 1972-74 in the United States [18]. NIOSH has estimated that 13,230 workers are exposed to vinyl acetate based upon statistical estimates derived from a survey conducted between 1981-1983 in the United States [18]. Air concns of 0-173 mg/m^3 (mean concn 30 mg/m^3) were detected in personal air samples from a vinyl acetate production plant [10]; short-term concns as high as 1150 mg/m^3

were measured during hopper cleaning [10]. Personal and area air sample concns of <0.4-126 mg/m^3 have been reported for various production and polymerization sites [10].

Body Burdens:

REFERENCES

1. Atkinson R; Inter J Chem Kinet 19: 799-828 (1987)
2. Chou WL et al; Biotechnol Bioeng Symp 8: 391-414 (1979)
3. Daniels W; Kirk-Othmer Encycl Chem Tech 3rd ed NY: Wiley 23: 817-20 (1983)
4. Eisenreich SJ et al; Environ Sci Technol 15: 30-8 (1981)
5. Fielding M et al; Organic Micropollutants in Drinking Water. TR-159. Medmenham, England: Water Research Center (1981)
6. Gordon SJ, Meeks SA; AICHE Symp Ser 73: 84-94 (1977)
7. Graedel TE et al; Atmospheric Chemical Compounds. Sources, Occurrence, and Bioassay. Orlando, FL: Academic Press p. 324 (1986)
8. Greve PA; Science 173: 1021-2 (1971)
9. Hansch C, Leo AJ; Medchem Project Issue No. 26 Claremont, CA: Pomona College (1985)
10. IARC; Monographs on the Evaluation of the Carcinogenic Risk of Chemicals to Humans. 39: 113-31 (1986)
11. Liepins R et al; Industrial Process Profiles for Environmental Use. USEPA-600/2-77-023f Chpt 6 p.6-261 (1977)
12. Lucas SV; GC/MS Analysis of Organics in Drinking Water Concentrates and Advanced Waste Treatment Concentrates: Vol 1. USEPA-600/1-84-020a p.46,175 (1984)
13. Ludzack FJ, Ettinger MB; J Water Pollut Control Fed 32: 1173-1200 (1960)
14. Lyman WJ et al; Handbook of Chemical Property Estimation Methods NY:McGraw-Hill p.4-9 (1982)
15. Mabey W, Mill T; J Phys Chem Ref Data 7: 383-415 (1978)
16. Merck Index; An Encyclopedia of Chemicals, Drugs and Biologicals 10th ed p.1429 (1983)
17. Mill T, Mabey W; p.208-10 in Environ Exposure from Chemicals Vol I; Neely WB, Blau GE eds Boca Raton, FL: CRC Press (1985)
18. NIOSH; National Occupational Hazard Survey (NOHS) (1972-1974), National Occupational Exposure Survey (NOES) (1981-1983)
19. Pahren HR, Bloodgood DE; Water Pollut Control Fed J 33: 233-8 (1961)
20. Pellizzari ED; Environ Sci Technol 16: 781-5 (1982)
21. Price KS et al; J. Water Pollut Control Fed 46: 63-77 (1974)
22. Swann RL et al; Res Rev 85: 16-28 (1983)
23. Takemoto s et al; Suishitsu Odaku Kenkyu 4: 80-90 (1981)
24. USEPA; EXAMS II Computer Simulation (1987)
25. Weber RC et al; Vapor Pressure Distribution of Selected Organic Chemicals. USEPA-600/2-81-021 p.31 (1981)

Vinyl Chloride

SUBSTANCE IDENTIFICATION

Synonyms: Chloroethene

Structure:

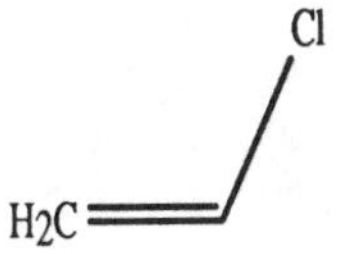

CAS Registry Number: 75-01-4

Molecular Formula: C_2H_3Cl

Wiswesser Line Notation: G1U1

CHEMICAL AND PHYSICAL PROPERTIES

Boiling Point: -13.37 °C

Melting Point: -153.8 °C

Molecular Weight: 62.50

Dissociation Constants:

Log Octanol/Water Partition Coefficient: 1.38 (calculated) [21]

Water Solubility: 2763 mg/L at 25 °C [27]

Vapor Pressure: 2660 mm Hg at 25 °C [48]

Henry's Law Constant: 1.07×10^{-2} atm-m³/mole [26]

ENVIRONMENTAL FATE/EXPOSURE POTENTIAL

Summary: Although vinyl chloride is produced in large quantities, almost all of it is used captively for the production of polyvinyl chloride (PVC) and other polymers. Therefore, its major release to the environment will be as emissions and wastewater at these production and manufacturing facilities. If vinyl chloride is released to soil, it will be subject to rapid volatilization with reported half-lives of 0.2 and 0.5 days for evaporation from soil at 1 and 10 cm incorporation, respectively. Any vinyl chloride which does not evaporate will be expected to be highly to very highly mobile in soil and it may leach to the ground water. It may be subject to biodegradation under aerobic conditions such as exists in flooded soil and ground water. If vinyl chloride is released to water, it will not be expected to hydrolyze, to bioconcentrate in aquatic organisms, or to adsorb to sediments. It will be subject to rapid volatilization, with an estimated half-life of 0.805 hr for evaporation from a river 1 m deep with a current of 3 m/sec and a wind velocity of 3 m/sec. In waters containing photosensitizers such as humic acid, photodegradation will occur fairly rapidly. Limited existing data indicates that vinyl chloride is resistant to biodegradation in aerobic systems and, therefore, it may not be subject to biodegradation in aerobic soils and natural waters. It will not be expected to hydrolyze in soils or natural waters under normal environmental conditions. If vinyl chloride is released to the atmosphere, it can be expected to exist mainly in the vapor phase in the ambient atmosphere and to degrade rapidly in air by gas-phase reaction with photochemically produced hydroxyl radicals with an estimated half-life of 1.5 days. Products of reaction in the atmosphere include chloroacetaldehyde, HCl, chloroethylene epoxide, formaldehyde, formyl chloride, formic acid, and carbon monoxide. In the presence of nitrogen oxides, e.g., photochemical smog situations, the half-life would be reduced to approx a few hours. Since vinyl chloride is primarily used in limited number of locations, it is unlikely that contamination will be widespread. Major human exposure will be from inhalation of occupational atmospheres and from ingestion of contaminated food and drinking water which has come into contact with PVC packaging material, or pipe which has not been treated adequately to remove residual monomer.

Natural Sources: Vinyl chloride monomer is not known to occur in nature [29].

Vinyl Chloride

Artificial Sources: Air emission from vinyl chloride production and use as a feedstock in the plastics industry (principally for PVC production and wastewater from these industries) [28]. Spills [7].

Terrestrial Fate: If vinyl chloride is released to soil, it will be subject to rapid volatilization based on the reported vapor pressure; half-lives of 0.2 and 0.5 days were reported for volatilization from soil at 1 and 10 cm incorporation, respectively [30]. Any vinyl chloride which does not evaporate will be expected to be highly to very highly mobile in soil and it may leach to the ground water. It may be subject to biodegradation under aerobic conditions such as exists in flooded soil and ground water; however, limited existing data indicates that vinyl chloride is resistant to biodegradation in aerobic systems and, therefore, it may not be subject to biodegradation in natural waters. It will not be expected to hydrolyze in soils under normal environmental conditions.

Aquatic Fate: If vinyl chloride is released to water, it will not be expected to hydrolyze, to bioconcentrate in aquatic organisms, or to adsorb to sediments. It will be subject to rapid volatilization with an estimated half-life of 0.805 hr for evaporation from a river 1 m deep with a current of 3 m/sec and a wind velocity of 3 m/sec [36]. In waters containing photosensitizers such as humic acid, photodegradation will occur fairly rapidly. Limited existing data indicates that vinyl chloride is resistant to biodegradation in aerobic systems and, therefore, it may not be subject to biodegradation in natural waters.

Atmospheric Fate: If vinyl chloride is released to the atmosphere, it can be expected to exist mainly in the vapor phase in the ambient atmosphere [14] based on the reported vapor pressure. Gas phase vinyl chloride is expected to degrade rapidly in air by reaction with photochemically produced hydroxyl radicals with an estimated half-life of 1.5 days [46]. Products of reaction in the atmosphere include chloroacetaldehyde, HCl, chloroethylene epoxide, formaldehyde, formyl chloride, formic acid, and carbon monoxide [39]. In the presence of nitrogen oxides, e.g., photochemical smog situations, the half-life would be reduced to approx a few hours.

Vinyl Chloride

Biodegradation: Limited existing data indicates that vinyl chloride is resistant to biodegradation in aerobic systems [5,25]. Vinyl chloride was approx 50% (20%) and 100% (55%) degraded in 4 and 11 weeks, respectively, in the presence (absence) of sand by methanogenic microorganisms under anaerobic conditions in laboratory scale experiments [3].

Abiotic Degradation: The rate constant for the vapor phase reaction of vinyl chloride with photochemically produced hydroxyl radicals has been determined in laboratory experiments to be 6.60 x 10^{-12} cm^3/molecule-sec at 26 °C [46], which corresponds to an atmospheric half-life of 1.5 days at an atmospheric concentration of 8 x 10^{+5} hydroxyl radicals per cm^3. Disappearance of approx 50% of vinyl chloride exposed to sunlight outdoors in air occurred in 0.5 and 2 days in Sept and Dec, respectively [6]. The products of reaction include chloroacetaldehyde, HCl, chloroethylene epoxide, formaldehyde, formyl chloride, formic acid, and carbon monoxide [31,39]. In the presence of nitrogen oxides, its reactivity is higher with a half-life of 3-7 hr [6,17,54]. In water, no photodegradation was observed in 90 hours; however, degradation is rapid in the presence of sensitizers, such as might be the case in humic waters, etc., or free radicals as might be found in PVC manufacturing effluent streams [5]. Hydrolysis will not be a significant loss process [37].

Bioconcentration: Based on the reported water solubility, a BCF of 7 was estimated [36]. Based on the estimated BCF, vinyl chloride will not be expected to significantly bioconcentrate in aquatic organisms. While not reporting actual bioconcentration factors, a lack of appreciable bioconcentration in extractable fractions of fish and aquatic invertebrates was reported in an ecosystem study [35].

Soil Adsorption/Mobility: A Koc of 0.40 was reported in "standard soil" [30]. Based on the reported water solubility, a Koc of 56 was estimated [36]. Based on the reported and estimated Koc values, vinyl chloride will be expected to be highly to very highly mobile in soil [51] and therefore it may leach to the ground water.

Volatilization from Water/Soil: Using the reported Henry's Law constant, a half-life of 0.805 hr was calculated for evaporation from a river 1 m deep with a current of 3 m/sec and with a wind velocity of 3 m/sec [36]. Due to its high Henry's Law constant and high vapor pressure, volatilization from soil would be rapid; half-lives of 0.2 and 0.5 days were reported for volatilization from soil at 1 and 10 cm incorporation, respectively [30].

Water Concentrations: DRINKING WATER: In the National Organic Monitoring Survey (1976-7), 2 samples out of 113 contained detectable levels (>0.1 ppb) and these averaged 0.14 ppb [12]. Highest value found in U.S. drinking water is 10 ppb [15,32]. 23% of 133 U.S. cities using finished surface water were pos, 0.1 to 9.8 ppb, 0.4 ppb median of pos samples [8]. A finished ground water survey in 25 U.S. cities resulted in 4.0% pos, 9.4 ppb mean [8,11]. One contaminated drinking water well contained 50 ppb [4]. Drinking water from PVC pipes contained 1.4 ppb in a recent installation, while a 9-yr-old system had 0.03 to 0.06 ppb [52]. DRINKING WATER: U.S: National Screening Program, 1977-1981, 142 water supplies, 4.9% pos, trace to 76 ppb [10]; state sampling data, 1033 supplies sampled, 7.1% pos, trace to 380 ppb [10]. GROUND WATER: 4 of 1060 wells in New Jersey were positive [42] and vinyl chloride was present in the 10 most polluted wells from 408 New Jersey samples; however, it was not quantified [19]. 15.4% of 13 U.S. cities sampled were pos - 2.2 to 9.4 ppb, 5.8 ppb median [8,11]. In a 9-state survey, 7% of the wells tested were positive, with a maximum value of 380 ppb reported [13]. After train derailment in Manitoba on Mar 10, 1980, in which large amounts of VC was spilled in the snow, 10 ppm max occurred in ground water, which decreased to below 0.02 ppm by 10 weeks after the spill [7]. U.S. 1982 National Ground Water Supply Survey, 466 samples, 1.1% pos, 1.1 ppb median, 1.1 ppb max (1 ppb quantification limit) [9]. SURFACE WATER: 9.8 ppb max value found in a 1981, 9-state survey [4,13]. It was not detected in winter or summer samples from the Delaware River [49]. Vinyl chloride has been detected in 21 out of 606 samples from New Jersey [42] and other U.S. samples [15]. 7.6% of 105 U.S. cities were positive, with pos samples ranging from 0.2 to 5.1 ppb, 3.25 ppb median [8].

Effluent Concentrations: The only industry with appreciable waste water effluents of vinyl chloride is the organic chemicals mfg/plastic industry, where mean levels are 750 ppb [53].

Vinyl Chloride

Wastewater from 12 PVC plants in 7 U.S. areas ranged from 0.05 to 20 ppm with typical levels being 2 to 3 ppm [52]. Vinyl chloride has been detected in effluents from chemical and latex plants in Long Beach, CA [15]. It was not detected in effluents from major municipal waste water discharges in Southern California [55]. Ground water from hazardous waste sites, CERCLA Database, 178 sites, 8.7% pos [47].

Sediment/Soil Concentrations: After March 10, 1980 train derailment in Canada in which large quantities of vinyl chloride were spilled in the snow, soil samples reached levels as high as 500 ppm between 1 and 2 meters below the soil surface [7].

Atmospheric Concentrations: RURAL: <5 ppt detected in rural northwest United States [20,24]; <10 ppb at the Whiteface Mtn in New York [33]. No vinyl chloride was detected in 7 samples taken in the Grand Canyon, AZ, detection limit = 2.8 ppb [43] and <10 ppb at the Whiteface Mtn in New York [33]. URBAN/SUBURBAN: NJ area, 36 samples, 42% pos, trace-3132 ng/m^3 [45]. Baton Rouge, LA, area, 16 samples, 56% pos, trace-1334 ng/m^3 [45]. URBAN/SUBURBAN: Vinyl chloride has a low frequency of occurrence in studies done in New Jersey [1,2,22,23], with a concn range of a trace to <10 ppb avg [1,2,22,23,33,34]. 2 of 8 sites (1 - Staten Island and 7 - New Jersey) were pos - 0.15 and 46 ppb [44]; Baltimore, MD - 38.5 ppb avg [33]. SOURCE DOMINATED AREAS: Houston, TX - Gulf Coast area (18 samples) 3.1-1250 ppb, max in Texas City [18]; Niagara Falls, NY, upwind from plant - 0 ppb, downwind - 28 ppb, and 40 ppb measured in a nearby residential area [16]. Delaware City, DE highway intersection - 790 ppb avg [33]. Houston, TX area which is the site of 40% of U.S. production capacity of PVC, had a range of 3.1-1250 ppb [15,38], with 33.0 ppm being detected 0.5 km from the center of a VC plant [15]. Eight highly industrialized U.S. areas 0 to 0.513 ppb [43]. Not detected in 23 samples from the Kin Buc disposal site in New Jersey [43]. SOURCE DOMINATED: Ambient air near 2 vinyl chloride plants in Long Beach, CA - 1.3-3.4 ppm [28]. New auto interiors, <0.05% pos, 0.4 to 1.2 ppm [15]; however, <10 ppb was measured in another study involving 16 new or used cars and 4 new or old mobile homes [28]. After the Mar 10, 1980 train derailment in Canada in which quantities of VC were spilled during a blizzard, levels in excess of 200 ppm were found at ground levels near some freight cars but levels outside of the spill were <0.02 ppm, the detection limit [7]. Ambient air near waste site, 2-7.3 ppb, 1 background site, 2-3 ppb [50]; air in homes in neighborhood surrounding the landfill, 108 samples from 19 homes, 4 ppb avg, 7 ppb max; 420 samples from 50 different homes, 4 ppb avg, 9.3 ppb max [50].

Food Survey Values: 20 mg/kg were detected in alcoholic beverages which were packaged in products containing vinyl chloride [15,29]. Alcoholic beverages - 0.025 to 1.60 ppm, 0.44 ppm avg; edible oils - 0.3 to 3.29 ppm, 2.16 ppm avg; vinegars 0 (red wine) to 8.40 ppm (apple cider); detected but not quantified in butter and margarine when these products were packaged in PVC containers [15,52].

Plant Concentrations:

Fish/Seafood Concentrations:

Animal Concentrations:

Milk Concentrations:

Other Environmental Concentrations: Vinyl chloride monomer has been found in polyvinyl chloride resins, but these levels can be reduced by new processing techniques in food grade resins [28]. For example, PVC delivered to a fabricator contained 250 ppm vinyl chloride monomer which was reduced to 0.5-20 ppm after fabrication [28]. Residual vinyl chloride monomer found in food packing material ranged from 0.043 to 71 ppb for film and up to 7.9 ppm for plastic bottles [28]. It has been found in domestic and foreign cigarettes and little cigars in concentrations of 5.6-27 mg/cigarette [28].

Probable Routes of Human Exposure: Inhalation is the major route of exposure for nearby residents and workers [52]. Exposure is also possible by ingestion of contaminate foods, drinking water, and absorption through skin from cosmetics [52].

Average Daily Intake: WATER INTAKE: Insufficient data. AIR INTAKE: Insufficient data. FOOD INTAKE: Insufficient data.

Occupational Exposures: Potential risk groups include workers in VC production or use facilities and nearby residents, people coming in contact with recently manufactured PVC in enclosed quarters (e.g., new cars), consumers of food products packaged in PVC, and drinking water from PVC pipes [15]. The total worldwide work force in VC and PVC industries exceeds 70,000 [15]. NIOSH (NOES Survey 1981-1983) has statistically estimated that 18,368 workers are exposed to vinyl chloride in the United States [40]. NIOSH (NOHS Survey 1972-1974) has statistically estimated that 239,375 workers are exposed to vinyl chloride in the United States [41]. Average exposure of a random person living within a 5-mi radius of a plant - 17 ppb [53]. The estimated number of persons at risk is 4.6 million [53].

Body Burdens:

REFERENCES

1. Bozzelli JN et al; Analysis of selected toxic and carcinogenic substances in ambient air in New Jersey. New Jersey Dept Environ Protect (1980)
2. Bozzelli JW, Kebbehus BB; Analysis of selected volatile substances in ambient air, final report Apr-Nov 1978. Newark, NJ: New Jersey Institute Tech 80 p. (1979)
3. Brauch HJ et al; Vom Wasser 68: 23-32 (1987)
4. Burmaster DE; Environ 24: 6-13, 33-6 (1982)
5. Callahan MA et al; Water-related environmental fate of 129 priority pollutants Vol 2 p.49-1 to 49-10 USEPA-440/4-79-029b (1979)
6. Carassiti V et al; Ann Chim 67: 499-512 (1978)
7. Charleton J et al; p. 245-67 in Hazard Assessment of Chemicals Vol 2; Saxena J ed (1983)
8. Coniglio WA et al; The occurrence of volatile organics in drinking water. EPA exposure assessment project (1980)
9. Cotruvo JA; Sci Total Environ 47: 7-26 (1985)
10. Cotruvo JA et al; pp.511-30 in Organic Carcinogens in Drinking Water (1986)
11. Council Environmental Quality; Contamination on ground water by toxic organic chemicals. Washington, DC 84 (1981)
12. Drury JS, Hammons AS; Investigations of Selected Environmental Pollutants 1,2-dichloroethane. p.63 EPA-560/2-78-006 (1979)
13. Dyksen JE, Herr AF III; J Amer Water Work Assoc 1982, 394-403 (1982)
14. Eisenreich SJ et al; Environ Sci Technol 15: 30-8 (1981)

Vinyl Chloride

15. Fishbein L; Sci Total Environ 11: 111-61 (1979)
16. Gay BW Jr, Noonan RC; Ambient air measurements of vinyl chloride in the Niagara Falls area; environmental monitoring series. 19 p. USEPA-650/4-75-020 (1975)
17. Gay BW Jr et al; Environ Sci Technol 10: 58-66 (1976)
18. Gordon SJ, Meeks SA; AICHE Symp Ser 73: 84-94 (1977)
19. Greenberg M et al; Environ Sci Technol 16:14-9 (1982)
20. Grimsrud EP, Rasmussen RA; Atmos Environ 9: 1014-7 (1975)
21. Hansch C, Leo AJ; Medchem Project Issue No 26. Claremont CA: Pomona College (1985)
22. Harkov R et al; Toxic and carcinogenic air pollutants in New Jersey - volatile organic substances. Trenton, NJ: Office of Cancer and Toxic Substances Research (1981)
23. Harkov R et al; J Air Pollut Control Assoc 33: 1177-83 (1983)
24. Harsch DE et al; J Air Pollut Control Assoc 29: 975-6 (1979)
25. Helfgott TB et al; An index of refractory organics, p.21 USEPA-600/2-77-174 (1977)
26. Hine J, Mookerjee PK; J Org Chem 40: 292-8 (1975)
27. Horvath AL; Halogenated Hydrocarbons: Solubility-Miscibility with Water. New York,NY: Marcel Dekker, Inc. pp 889 (1982)
28. IARC; Monograph Some Monomers, Plastics and Synthetic Elastomers and Acrolein 19: 377-83 (1979)
29. IARC; Monograph Some anti-thyroid and related substances, nitrofurans and industrial chemicals 7: 291-318 (1974)
30. Jury WA et al; J Environ Qual 13: 573-9 (1984)
31. Kagiya T et al; Japan Chem Soc Spring Term Mtg, 32nd Tokyo Japan paper 1035 (1975)
32. Kraybill HF; NY Acad Sci Annals 298:80-9 (1977)
33. Lillian D et al; Environ Sci Technol 9: 1042-8 (1975)
34. Lioy PY et al; J Water Pollut Control Fed 33: 649-57 (1983)
35. Lu PY et al; Arch Environ Contam Toxicol 9: 1042-8 (1977)
36. Lyman WJ et al; Handbook of Chem Property Estimation Methods Environ Behavior of Org Compounds McGraw-Hill NY p. 4-9 (1982)
37. Mabey WR et al; Aquatic Fate Process Data for Organic Priority Pollutants p.156 USEPA-440/4-81-014 (1981)
38. McMurray JR, Tarr J; IES 24 Ann Mtg Fort Worth, Tx 18-20 Apr 78 p 149-53 (1978)
39. Muller JPH, Korte F; Chemosphere 6: 341-6 (1977)
40. NIOSH; The National Occupational Exposure Survey (NOES) (1983)
41. NIOSH; The National Occupational Hazard Survey (NOHS) (1974)
42. Page GW; Environ Sci Technol 15:1475-81 (1981)
43. Pellizzari ED; Quantification of chlorinated hydrocarbons in previously collected air samples USEPA-450/3-78-112 (1978)
44. Pellizzari ED; The measurement of carcinogenic vapors in ambient atmospheres. p.288 USEPA-600/7-77-055 (1977)
45. Pellizzari ED et al; Formulation of preliminary assessment of halogenated organic compounds in man and environmental media. USEPA-560/13-79-006 (1979)
46. Perry RA et al; J Chem Phys 67: 458-62 (1977)
47. Plumb RH Jr; Ground Water Monit Rev 7: 94-100 (1987)

48. Riddick JA et al; Organic Solvents: Physical Properties and Methods of Purification. Techniques of Chemistry. 4th ed. Wiley-Interscience pp.1325 (1986)
49. Sheldon LS, Hites RA; Environ Sci Technol 12:1188-94 (1978)
50. Stephens RD et al; pp. 265-87 in: Pollutants in a Multimedia Environment. Cohen Y Ed. Plenum Press: New York (1986)
51. Swann RL et al; Res Rev 85: 17-28 (1983)
52. USEPA; Ambient Water Quality Criteria for Vinyl Chloride. USEPA-440/5-80-078 (1980)
53. USEPA; Treatability Manual. p.I.12.12-1 to I.12.12-4 USEPA-600/2-82-001a (1981)
54. Woldbaek T, Klaboe P; Spectrochim Acta A 34: 481-7 (1978)
55. Young DR; Annual Rep Southern California Coastal Water Res Proj p.103-12 (1978)

Vinylidene Chloride

SUBSTANCE IDENTIFICATION

Synonyms: 1,1-Dichloroethylene; 1,1-Dichloroethene

Structure:

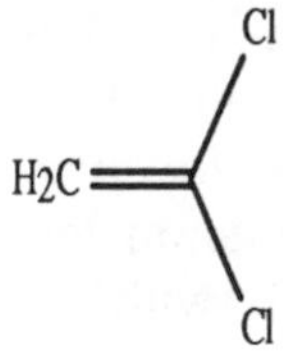

$$H_2C = C\begin{cases} Cl \\ Cl \end{cases}$$

CAS Registry Number: 75-35-4

Molecular Formula: $C_2H_2Cl_2$

Wiswesser Line Notation: GYGU1

CHEMICAL AND PHYSICAL PROPERTIES

Boiling Point: 31.7 °C

Melting Point: -122.5 °C

Molecular Weight: 96.95

Dissociation Constants:

Log Octanol/Water Partition Coefficient: 2.13 [18]

Water Solubility: 0.25 g/100 at 25 °C [22]

Vapor Pressure: 591 mm Hg at 25 °C [22]

Henry's Law Constant: 0.0301 atm-m³/mole (calculated)

ENVIRONMENTAL FATE/EXPOSURE POTENTIAL

Summary: Vinylidene chloride enters the atmosphere from its production and use in the manufacture of plastics such as Saran wrap. It is released in wastewater from plastics manufacturing and

561

metal finishing. Releases to water will primarily be lost to the atmosphere through evaporation. Once in the atmosphere, it will degrade rapidly by photooxidation, with a half-life of 11 hours in relatively clean air or under 2 hours in polluted air. If spilled on land, part of the vinylidene chloride will evaporate and part will leach into the ground water, where its fate is unknown. Vinylidene chloride would not be expected to bioconcentrate into fish. Major human exposure is from occupational atmospheres. The general population may be exposed to low levels of vinylidene chloride in ambient air, indoor air, contaminated drinking water, and food which has come in contact with plastic wrap which contains residual monomer.

Natural Sources: Vinylidene chloride is not known to occur as a natural product [22].

Artificial Sources: Vinylidene chloride may be released into the environment as emissions or in wastewater during its production and use in the manufacture of plastic wrap, adhesives, and synthetic fiber [20]. Vinylidene chloride is formed by a minor pathway during the anaerobic biodegradation of trichloroethylene and also by the hydrolysis of 1,1,1-trichloroethane [7]. Therefore there is a potential for it to form in ground water that has been contaminated by chlorinated solvents. Vinylidene chloride is also produced by the thermal decomposition of 1,1,1-trichloroethane, a reaction that is catalyzed by copper [15]. 1,1,1-Trichloroethane is used as a degreasing agent in welding shops, so there is a potential for vinylidene chloride to be formed in these shops as well as in other industrial environments where 1,1,1-trichloroethane is used near sources of heat [15].

Terrestrial Fate: When spilled on land, vinylidene chloride will be lost partially by evaporation and partially by percolation into the ground water. In the ground water, very slow hydrolysis and biodegradation should occur.

Aquatic Fate: When released into water, vinylidene chloride will primarily be lost by evaporation into the atmosphere with a half-life of 1-6 days. Little of the chemical would be lost by adsorption onto the sediment.

Vinylidene Chloride

Atmospheric Fate: Vinylidene chloride is a photochemically reactive compound, and when released to the atmosphere, it will degrade by reaction with hydroxyl radicals with a half-life of 11 hours. Under photochemical smog conditions, its half-life is much shorter (<2 hr).

Biodegradation: Few studies on the biodegradation of vinylidene could be found. In one study, 45-78% of the chemical was lost in 7 days when incubated with a wastewater inoculum; however, a sizeable fraction of the loss was due to volatilization [1]. 97% of vinylidene chloride was reported to be removed in a municipal wastewater plant, but again the fraction lost by evaporation is unknown [40]. Under anaerobic conditions in microcosms designed to simulate the anaerobic conditions in ground water [1] and landfills [17], vinylidene chloride undergoes reductive dechlorination to vinyl chloride. In the microcosms designed to simulate a ground water environment, 50% of the vinylidene chloride disappeared in 5-6 mo [1]. Under the simulated landfill conditions, degradation occurred in 1-3 weeks [17].

Abiotic Degradation: Vinylidene chloride reacts with photochemically produced hydroxyl radicals with an atmospheric half-life of 11 hr [12]. Under photochemical smog situations, when nitrogen dioxide is present, vinylidene chloride decomposes more rapidly (half-life < 2 hr) [14]. Products which are formed in the photooxidation of vinylidene chloride in the presence of nitrogen oxides include chloracetyl chloride, phosgene, formaldehyde, formic acid, hydrochloric acid, carbon monoxide, and nitric acid [12,14]. When adsorbed on silica gel, vinylidene chloride undergoes photolysis; approx 72% of it degrading on exposure to 170 hr of sunlight [32]. In water, the photooxidation of vinylidene chloride is insignificant [5,26]. A hydrolysis half-life of 6-9 months has been observed with no significant difference in hydrolysis rate between pH 4.5 and 8.5 [7]. This value differs markedly from the estimated hydrolytic half-life of 2 yr at pH 7 [36].

Bioconcentration: No experimental data could be found on the bioconcentration of vinylidene chloride in fish or aquatic invertebrates. Based on its low octanol/water partition coefficient, one would not expect any significant bioconcentration.

Soil Adsorption/Mobility: No experimental data is available on the adsorption of vinylidene chloride. A low Koc of 150 is calculated from a regression equation based on its octanol/water partition coefficient [23].

Volatilization from Water/Soil: The mass transfer coefficient between water and the atmosphere of vinylidene chloride relative to oxygen has been measured to be 0.62 [27]. Using data for the oxygen reaeration rate of typical bodies of water [28], one can calculate the half-life for evaporation of vinylidene chloride to be 5.9, 1.2, and 4.7 days from a pond, river, and lake, respectively.

Water Concentrations: DRINKING WATER: In a nationwide survey, vinylidene chloride was detected in 7.1% of finished supplies from ground water sources [11]. In 1979, the highest reported concn was 0.1 ppb [13,21]. Of 103 U.S. cities sampled, 1.9% pos, 0.36 mean ppb mean, 0.2-0.51 ppb range in finished surface water [9]. 13 U.S. cities sampled, 7.7% pos, 0.2 ppb mean and max, in finished ground water [9,10]. In a screening of 1174 community wells and 617 private wells in Wisconsin, 1 community and 3 private wells had detectable levels of vinylidene chloride [24]. U.S. Groundwater Supply Survey (945 supplies derived from ground water chosen both randomly and on the basis that they may contain VOCs) - 24 samples positive for vinylidene chloride, max 6.3 ppb [45]. Mean and max conc of vinylidene chloride in 2 New Jersey supplies serving roughly 100,000 persons each ranged from 0.1-0.2 and 0.9-2.5 ppb, respectively [44]. GROUND WATER: Contaminated drinking water wells in New Jersey, Massachusetts, and Maine had max vinylidene chloride concn of 280, 118, and 70 ppb, respectively [4]. A 13-U.S. city survey of raw ground water supplies resulted in 15.4% pos, and 0.5 ppb avg and max [9]. Miami, FL had 0.1 ppb vinylidene chloride in their raw drinking water supply [8]. As reported by Aerojet-General Corp, vinylidene chloride was detected in several domestic and industrial well water samples in Sacramento, CA [42]. SURFACE WATER: 3 tributaries and 7 of 8 sites on the Ohio River pos (4972 samples, 343 pos); 304 samples 0.1-1.0 ppb, 36 samples; 1.0-10 ppb; and 3 samples >10 ppb [31]. 2 of 4 cities with surface water contaminated with industrial, municipal, agricultural, and natural waste as a source of drinking water supply contained vinylidene chloride in the raw water; of the pos supplies, one contained <0.1

ppb and one was not quantified [8]. In a survey of 105 U.S. cities using surface water supplies, no vinylidene chloride was detected in the raw water [9].

Effluent Concentrations: Detected, not quantified, in effluent from U.S. latex and chemical manufacturing plants [13,21]. 32 ppb - discharged from a chemical manufacturing plant, The Netherlands [13,21]. Samples from the 4 largest publicly owned treatment plants in Southern CA were as follows: primary effluent, 3 of 4 pos, <10 to 20 ppb; secondary effluent, 2 of 3 pos, <10 ppb; 7 mile sludge and concentrate, 2 of 3 pos, <10 ppb [46]. Detected in 1 of 2 municipal treatment plants [6]. Industries with mean raw wastewater concn >100 ppb - metal finishing (760 ppb), nonferrous metal mfg (200 ppb), and organic chemicals mfg/plastics (200 ppb) [41]. 17% of 48 samples of influent to a sewage treatment plant in United States pos, 5.0 ppb avg when found above detection limit [6]. In a comprehensive survey of wastewater from 4000 industrial and publicly owned treatment works (POTWs) sponsored by the Effluent Guidelines Div of the U.S. EPA, vinylidene chloride was identified in discharges of the following industrial category (frequency of occurrence, median concn in ppb): timber products (2, 10.8); steam electric (2, 38.8); petroleum refining (1, 8.0); nonferrous metals (3, 2.9); paint and ink (1, 4.6); printing and publishing (1, 152.6); organics and plastics (31, 35.7); inorganic chemicals (2, 20.7); pulp and paper (4, 9.3); rubber processing (1, 137.7); auto and other laundries (6, 32.8); pesticides manufacture (2, 246.8); organic chemicals (2, 675.8); transportation equipment (1, 238.0); publicly owned treatment works (40, 23.0) [37]. The highest effluent concn was 3636 ppb in the auto and other laundries industry [37].

Sediment/Soil Concentrations:

Atmospheric Concentrations: RURAL/REMOTE: Not detected in 2 U.S. samples [3]. <5 ppt rural northwest (Pullman, WA) [16]. Grand Canyon, AR - 0.065 ppb [34]. URBAN/SUBURBAN: 325 U.S. samples, 8.0 ppt mean, 5 ppt median, 25% of samples >7.5 ppt max 2400 ppt [3]. Detected but not quantified in representative cities in NJ [19,25]. 3 western U.S. urban areas - 4.9-28.8 ppt mean concn [39]. 5 U.S. cities 0-31 ppt avg, 34-224 ppt max [38]. SOURCE DOMINATED: 14 samples in U.S. mean concn

3800 ppt, median - 3600 ppt, max 6700 ppt [3]. Kin-Buc Disposal site, Edison, NJ 114-148 ppt, and was not detected downwind of several petroleum facilities in Baton Rouge, LA [34]. 5 Dow plant sites, one at Freeport, TX, four at Plaquemine, LA, showed a detection range of 9 to 249 ppt [2]. Not detected at 7 other Dow sites at Plaquemine, LA [2]. INDOOR AIR: Vinylidene chloride was found in 4 of 15 samples of indoor air taken during the summer, 1.8-93.6 ppb; and 4 of 16 samples taken during the winter, 5.0-9.6 ppb [35]. None of the chemical was found in parallel samples taken out-of-doors [35]. No correlation was found between the presence of vinylidene chloride in indoor air and structural characteristics of the dwelling or activity [35].

Food Survey Values: Although no monitoring data could be found, vinylidene chloride is a known contaminant in plastic wrap made from this monomer; the maximum amount possible that could be adsorbed by food from such food wraps has been estimated to be less than or equal to the detection limit (<10 ppb) [13].

Plant Concentrations:

Fish/Seafood Concentrations:

Animal Concentrations:

Milk Concentrations:

Other Environmental Concentrations:

Probable Routes of Human Exposure: Humans are exposed to vinylidene chloride from ambient air, particularly near industrial sources and contaminated drinking water. Indoor air sometimes contains vinylidene chloride, although its source is unknown. Exposure can also occur from ingestion of food wrapped in plastic with residue vinylidene chloride monomer.

Average Daily Intake: AIR INTAKE (assume 0.005 ppb) 0.40 ug; WATER INTAKE (assume 0 ppb) 0 ug; from contaminated sources (assume 0.2-0.36 ppb) 0.4-0.7 ug; FOOD INTAKE - insufficient data but probably minimal.

Vinylidene Chloride

Occupational Exposures: Number of exposed workers not available; however, the estimated population residing near plants manufacturing monomers and polymers of vinylidene chloride in 1976 is 3,573,385 [2,13]. Levels of 8 mg/m³ (2 ppm) were reported to occur as contaminants in atmospheres of submarines, and levels of 0-2 ppm have been found in spacecraft [21]. NIOSH has estimated that 56,887 workers are potentially exposed to vinylidene chloride based on statistical estimates derived from the National Occupational Hazard Survey (NOHS) conducted in 1972-74 in the United States [30]. NIOSH has estimated that 2,675 workers are potentially exposed to vinylidene chloride according to statistical estimates derived from the National Occupational Exposure Survey (NOES) conducted in 1981-83 in the U.S. [29].

Body Burdens: 12 percent of approx 300 breath samples from Elizabeth and Bayonne, NJ contained quantifiable levels (0.2-2 ug/m³) of vinylidene chloride [43].

REFERENCES

1. Barrio-Lage G et al; Environ Sci Technol 20: 96-9 (1986)
2. Basu D et al; Health assessment document for vinylidene chloride USEPA External Review Draft p 7-1 to 7-21 (1983)
3. Brodzinsky R, Singh HB; Volatile organic chemicals in the atmosphere: An assessment of available data 198 p SRI Inter Contract 68-02-3452 (1982)
4. Burmaster DE; Environ 24: 6-13,33-6 (1982)
5. Callahan MA et al; Water-related environmental fate of 129 priority pollutants p. 50-1 to 50-10 USEPA 440/4-79-029a (1979)
6. Callahan MA et al; Proc Natl Conf Munic Sludge Manag 8th p 55-61 (1979)
7. Cline PV, Delfino JJ; Am Chem Soc Div Environ Chem Preprint New Orleans LA 27: 577-9 (1987)
8. Coleman WE et al; p.305-27 in Analysis and Identification of organic substances in water Keith L ed Ann Arbor, MI: Ann Arbor Sci (1976)
9. Coniglio WA et al; The occurrence of volatile organics in drinking water Exposure Assessment project. Criteria and Standards Division, Science and Technology Branch (1980)
10. Council on Environmental Quality; Contamination of Groundwater by Toxic Organic Chemicals (1981)
11. Dyksen JE, Hess AF III; J Amer Water Works Assoc 74:394-403 (1982)
12. Edney E et al; Atmospheric chemistry of several toxic compounds USEPA-600/53-82-092 (1983)
13. Fishbein L; Sci Total Environ 11: 111-61 (1979)
14. Gay BW et al; Environ Sci Technol 10: 58-67 (1976)
15. Glisson BT; Am Ind Hyg Assoc J 47: 427-35 (1986)

Vinylidene Chloride

16. Grimsrud EP, Rasmussen RA; Atmos Environ 9: 1014-7 (1975)
17. Hallen RT et al; Am Chem Soc Div Environ Chem 26th Natl Mtg 26: 344-6 (1986)
18. Hansch C, Leo AJ; Medchem Project Issue No 26. Claremont CA: Pomona College (1985)
19. Harkov R et al; Toxic and carcinogenic air pollutants in New Jersey - volatile organic substances Unpublished work Trenton, NJ: Office of Toxic Substances Res (1981)
20. Hawley GG; Condensed Chem Dict 10th ed Von Nostrand Reinhold NY (1981)
21. IARC; Monograph Some monomers, plastics and synthetic elastomers, and acrolein 19: 439-59 (1979)
22. IARC; Monograph Some chemicals used in plastics and elastomers 39: 195-226 (1986)
23. Kenaga EE, Goring CAI; Aquatic Toxicology 3rd Annual Symp on Aquatic Toxicology Philadelphia, PA ASTM (1980)
24. Krill RM, Sonzogni WC; J Am Water Works Assoc 78: 70-5 (1986)
25. Lioy PJ et al; J Water Pollut Control Fed 33: 649-57 (1983)
26. Mabey WB et al; Aquatic Fate Process Data for Organic Priority Pollutants p 157 USEPA 440/4-81-014 (1981)
27. Matter-Mueller C et al; Water Res 15: 1271-9 (1981)
28. Mill T et al; Aquatic Fate Process Data for Organic Priority Pollutants p 255 USEPA-440/4-80-014 (1982)
29. NIOSH; National Occupational Exposure Survey (1985)
30. NIOSH; National Occupational Health Survey (1975)
31. Ohio River Valley Water Sanit Comm; Assessment of water quality conditions, Ohio River Mainstream 1980-81 Cincinnati, OH (1982)
32. Parlar H; Fresenius Z Anal Chem 319: 114-8 (1984)
33. Patterson JW, Kodukala PS; Chem Eng Prog 77: 48-55 (1981)
34. Pellizzari ED; Quantification of chlorinated hydrocarbons in previously collected air samples USEPA-450/3-78-112 (1978)
35. Pleil JD; Volatile organic compounds in indoor air: A survey of various structures USEPA-600/D-85-100 (1985)
36. Schmidt-Bleek F et al; Chemosphere 11: 383-415 (1982)
37. Shackelford WM et al; Analyt Chim Acta 146: 157 (1983)
38. Singh HB et al; Environ Sci Technol 16: 872-80 (1982)
39. Singh HB et al; Atmos Environ 15: 601-12 (1981)
40. Tabak HH et al; J Water Pollut Control Fed 53: 1503-18 (1981)
41. USEPA; Treatability Manual page I.12.24-1 to I.12.24-5 USEPA 600/2-82-001a (1981)
42. USEPA; Subst Risk Not, 8(e) 35 USEPA 560/11-80-020 (1980)
43. Wallace L et al; J Occu Med 28: 603-7 (1986)
44. Wallace LA et al; Environ Res 43: 290-307 (1987)
45. Westrick JJ et al; J Am Water Works Assoc 76: 52-9 (1984)
46. Young DR; Ann Rep Southern Calif Coastal Water Res Proj p 103-12 (1978)

Index of Synonyms

Index by CAS Registry Number

Index by Chemical Formula